PETERSON FIELD GUIDE TO

REPTILES AND AMPHIBIANS

of Eastern and Central North America

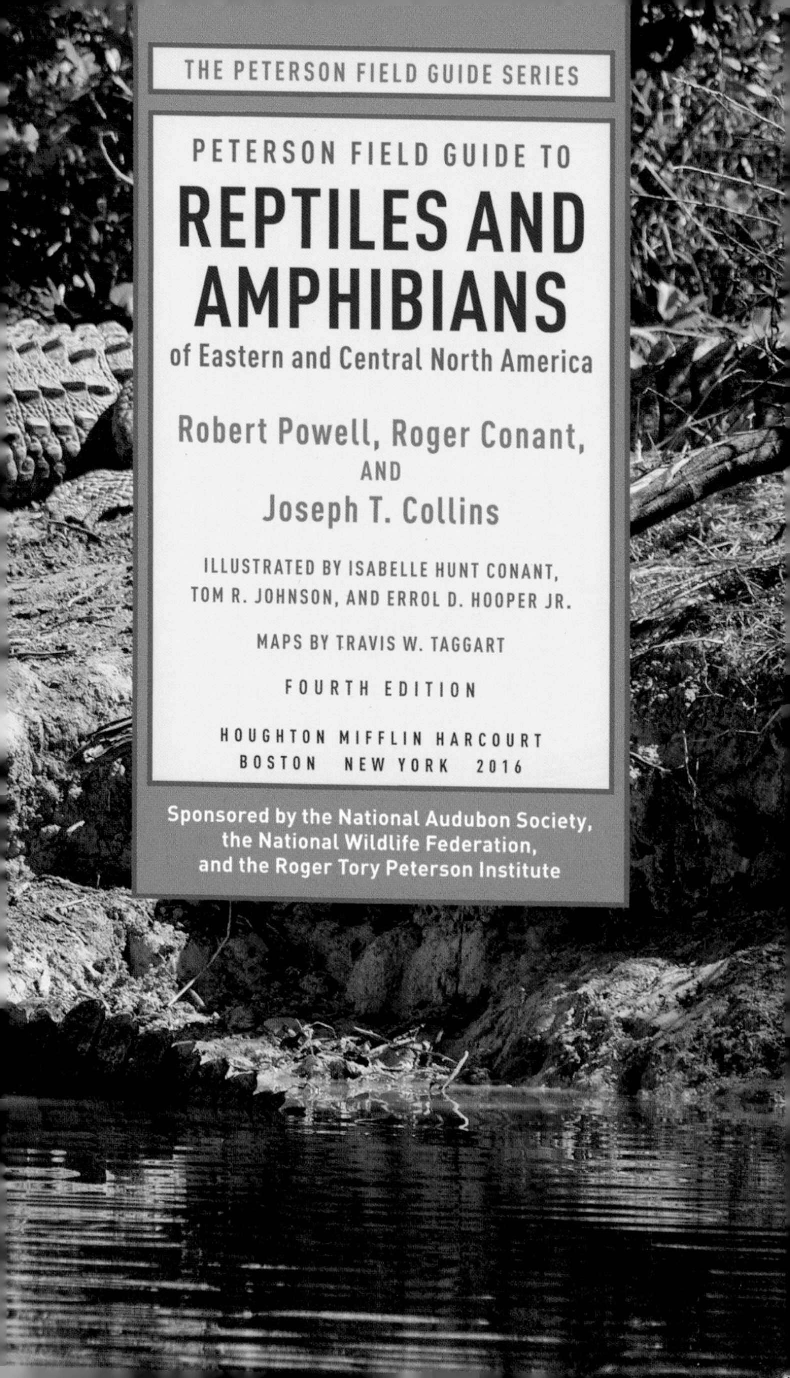
THE PETERSON FIELD GUIDE SERIES
PETERSON FIELD GUIDE TO
REPTILES AND AMPHIBIANS
of Eastern and Central North America
Robert Powell, Roger Conant,
AND
Joseph T. Collins
ILLUSTRATED BY ISABELLE HUNT CONANT,
TOM R. JOHNSON, AND ERROL D. HOOPER JR.
MAPS BY TRAVIS W. TAGGART
FOURTH EDITION
HOUGHTON MIFFLIN HARCOURT
BOSTON NEW YORK 2016
Sponsored by the National Audubon Society,
the National Wildlife Federation,
and the Roger Tory Peterson Institute

The fourth edition includes newly created text and illustrations as well as text and artwork from each of the earlier editions.

www.hmhco.com

Library of Congress Cataloging-in-Publication Data is available.
ISBN 978-0-544-12997-9

Book design by Eugenie S. Delaney

Printed in China

SCP 10 9 8 7 6 5 4 3
4500666091

DEDICATION

An interest in herpetology continues to proliferate among scientists who employ reptiles and amphibians for studies on many subjects, including evolutionary relationships, life histories, behavior, and the roles of these animals in natural and increasingly altered ecosystems—as well as among individuals who derive pleasure and knowledge from observing these animals in nature or by keeping them as pets. In all cases, accurate identification of species is essential. We have written this book for those people—and for all naturalists.

This book is dedicated to
Roger Conant
(1909–2003),
who inspired generations of herpetologists,
and to
Joseph T. Collins
(1939–2012),
who recruited countless young people to herpetology by treating each person, regardless of age or experience, with respect.

The legacy of America's greatest naturalist and creator of the field guide series, Roger Tory Peterson, is kept alive through the dedicated work of the Roger Tory Peterson Institute of Natural History (RTPI). Established in 1985, RTPI is located in Peterson's hometown of Jamestown, New York, near the Chautauqua Institution in the southwestern part of the state.

Today RTPI is a national center for nature education that maintains, shares, and interprets Peterson's extraordinary archive of writings, art, and photography. The institute, housed in a landmark building by world-class architect Robert A. M. Stern, continues to transmit Peterson's zest for teaching about the natural world through leadership programs in teacher development as well as outstanding exhibits of contemporary nature art, natural history, and the Peterson Collection.

Your participation as a steward of the Peterson Collection and supporter of the Peterson legacy is needed. Please consider joining RTPI at an introductory rate of 50 percent of the regular membership fee for the first year. Simply call RTPI's membership department at (800) 758-6841 ext. 226, or e-mail membership@rtpi.org to take advantage of this special membership offered to purchasers of this book. For more information, please visit the Peterson Institute in person or virtually at www.rtpi.org.

PREFACE

Since the third edition of this field guide was published in 1991 (an expanded version with colored maps and photographs was published in 1998), our understanding of our area's amphibians and reptiles has changed dramatically. The number of currently recognized species in our area has increased from 379 to 501, including 60 established non-native species. Those numbers continue to grow, with several species known to compose species complexes remaining to be defined, and newly established non-native species discovered regularly. The increase in the number of native taxa reflects the dynamic nature of modern systematics and the use of new (especially molecular) techniques to elucidate relationships and redefine species boundaries. The increase in non-native species reflects an inability to control animal imports.

The area covered by this guide is illustrated on page 9. It includes all of the United States and Canada from the Atlantic Seaboard to the western limits of Texas, Oklahoma, Kansas, Nebraska, the two Dakotas, Manitoba, and Nunavut. This artificial boundary is designed to meet the eastern edge of the territory embraced by Robert C. Stebbins's *Peterson Field Guide to Western Reptiles and Amphibians.*

CONTENTS

INTRODUCTION

Animals covered in this guide represent two groups of vertebrates (animals with backbones). Amphibians have moist, glandular skin, and their toes lack claws. The young of most species pass through an aquatic larval stage ("tadpoles" in frogs) before metamorphosing into the adult form. The word *amphibious* is derived from Greek and means "living a double life." Amphibians include salamanders, frogs, and limbless burrowing or aquatic caecilians (the last restricted to tropical regions and not included in this guide).

Reptiles have skin with scales, shields, or plates, and their toes bear claws (the clawless Leatherback Sea Turtle is an exception, and soft-shelled turtles have only a few scales on their limbs). Young are miniature adults in general appearance but may vary in color and pattern. The word *reptile* is derived from Latin and means "to creep or crawl." When extinct forms and their various descendants are considered, reptiles should include birds, which descended from dinosaurs. Living non-avian reptiles include crocodilians, turtles, and lepidosaurs (scaly reptiles). The last contains squamates (lizards and snakes) and lizardlike rhynchocephalians, the only surviving member of which, the Tuatara (*Sphenodon punctatus*), was confined to small islets off New Zealand until it recently was reintroduced onto the North Island.

Although many amphibians and reptiles are thermal conformers (meaning that their body temperatures fluctuate with those of their surroundings), a large number are ectotherms, animals that behaviorally regulate their body temperatures by using environmental sources of heat. Although often called "cold-blooded," some reptiles thrive at temperatures well above those maintained by birds and mammals (so-called warm-blooded animals). Many reptiles manipulate their body temperatures with great precision by exploiting cooler or warmer microhabitats or merely altering their posture or orientation to sources of heat.

Conservation

Populations of many reptiles and amphibians are declining, and extinctions are increasingly common—and most are attributable to humans. The greatest threat is habitat loss and alteration. As human populations grow, they exert insatiable demands for infrastructure, food, and energy—and neither parking lots nor cornfields provide suitable conditions for reptiles and amphibians. Dams flood lowlands and alter stream dynamics, roads bisect shrinking patches of natural habitats, and urban sprawl replaces woodlands and prairies with manicured lawns and shopping centers. Frequently interacting with and exacerbating habitat loss are pollution and invasive species. Amphibians in particular are susceptible to agricultural and industrial pollutants and those that drain from suburban lawns. Non-native species, many of which prey on or compete with native forms, are appearing in North America in ever-increasing numbers. Most non-native species are byproducts of the live animal trade ("pets" that have escaped or been set "free"). However, intentional releases or inadvertent introductions with shipments of building materials or ornamental plants account for some additions to our herpetofauna.

Less evident threats are diseases that spread rapidly and are affecting more species than ever. Many declines in amphibian populations around the world are attributable to chytridiomycosis (fungal diseases affecting the skin) and ranavirus (a viral disease that causes internal hemorrhaging and which occurs in some reptiles as well). Complicating matters is that some resistant species transmit pathogens. For example, the American Bullfrog, widely introduced outside its native range as a table delicacy, is resistant to both diseases but readily passes them to other species. One explanation for recent epidemics is the increased movement of animals by humans; however, as Earth's climate on average becomes warmer and drier, disease symptoms are exacerbated, and increasingly stressed animals become even more vulnerable.

Furthermore, many people, often by overstating the danger of venomous species, fear all reptiles—and use that fear to justify the persecution of not only dangerous snakes but all reptiles and even some amphibians. Folklore is rife with tales of evil serpents and toads causing warts. Instead of accepting these animals as vital components of natural (and altered) ecosystems, we allow hearsay and dread to prevail.

Contributing further to the tarnished images of amphibians and reptiles is that fewer and fewer people interact with nature. Human populations are increasingly urban, and a growing reliance on technology for entertainment means that many North Americans rely solely on television and the Internet for information. Quality programs and accurate, fact-based websites exist. Unfortunately, too many programs and sites use a superficially documentary style to hide overt exploitation. Instead of nurturing respect for and an appreciation of nature, they emphasize danger, cater to our worst instincts and biases, and feature thrill-seeking, ignorance, and irresponsibility.

What Can You Do?

Although many problems are global and seem beyond the reach of individuals, you can do your part by being "green." "Think globally, act locally" is good advice. Minimize your impact on habitats, climate, and the use and spread of pollutants by being a responsible citizen of planet Earth.

Also, treat nature with respect. Learn about natural systems and their inhabitants. Visit zoos, aquariums, and museums. Read and view selectively, distinguishing between the factual and the irresponsible. In general, publications and websites of educational institutions (such as zoos, aquariums, and natural history museums), governmental agencies, and professional and conservation-oriented nonprofit entities are reliable.

Never release animals that have been kept in captivity, even for short periods. This is especially true for non-native species. Not only are such animals likely to suffer and die when introduced into alien habitats, but should they survive, their impact on local wildlife cannot be anticipated. For example, the Burmese Python, now well established in southern Florida, has assumed top-predator status, eating native birds, mammals, and other reptiles, including many endangered species. The person who released just one snake after it outgrew its cage probably never considered the chaos unleashed on a vulnerable ecosystem.

Learn More

In the species accounts, we list the conservation status of species that the International Union for Conservation of Nature has listed on its Red List (www.iucnredlist.org/) as being threatened with extinction (Vulnerable, Endangered, or Critically Endangered). Following that, in parentheses, is the species' status (Threatened or Endangered) as listed by the U.S. Fish and Wildlife Service. However, some populations may be extirpated and declines in some portions of the range might be substantive even for species not currently considered threatened.

Many organizations are dedicated to conserving amphibians and reptiles. Some websites are listed below. Most also provide links to other relevant sites.

AMPHIBIAN ARK (www.amphibianark.org)
AMPHIBIAWEB (www.amphibiaweb.org)
CANADIAN AMPHIBIAN AND REPTILE CONSERVATION NETWORK (www.carcnet.ca)
CENTER FOR NORTH AMERICAN HERPETOLOGY (www.cnah.org)
FLORIDA NATURAL AREAS INVENTORY (www.fnai.org)
IUCN SSC AMPHIBIAN SPECIALIST GROUP (ASG) AND AMPHIBIAN SURVIVAL ALLIANCE (ASA) (www.amphibians.org)
IUCN-SSC CROCODILE SPECIALIST GROUP (CSG) (www.iucncsg.org)
IUCN SSC MARINE TURTLE SPECIALIST GROUP (MTSG) (www.iucn-mtsg.org)
IUCN/SSC TORTOISE AND FRESHWATER TURTLE SPECIALIST GROUP (TFTSG) (www.iucn-tftsg.org)
PARTNERS IN AMPHIBIAN AND REPTILE CONSERVATION (PARC) (www.parcplace.org)
SAVANNAH RIVER ECOLOGY LABORATORY HERPETOLOGY PROGRAM (www.srelherp.uga.edu/index.htm)
TURTLE SURVIVAL ALLIANCE (TSA) (www.turtlesurvival.org)
UNITED STATES FISH AND WILDLIFE SERVICE (www.fws.gov)

Captive Reptiles and Amphibians

Fascination with a pet lizard or snake was the beginning of a lifelong passion for many professional herpetologists and serious hobbyists. Previous editions of this guide included the chapters "Making and Transporting the Catch" and "Care in Captivity" and relevant advice

in many species accounts. We omitted such information from this edition to accommodate accounts of newly discovered or described species, but also because resources now widely available on the Internet did not exist when the third edition was published in 1991 (the 1998 expanded version did not include updated text). Searches will quickly reveal links to literature and other resources, ranging from short "how to keep" guides to entire books devoted to certain species or herpetological husbandry in general.

If you plan to keep harmless amphibians or reptiles as pets, in addition to finding reliable resources on acquisition and care, you might consult a local, regional, or state herpetological society, zoo, nature center, wildlife rehabilitation center, or natural history museum about sources of healthy animals and advice on how best to maintain them. You also should check with local, state, or provincial agencies to make sure you are aware of all pertinent laws and regulations. Some cities prohibit possession of "dangerous" species (often including large constrictors in addition to venomous snakes). ***Under no circumstances should venomous animals be kept as pets.***

If you are considering a wild-caught animal, note that fish and game, nongame, or wildlife departments of state and provincial conservation agencies can provide lists of species protected by state, provincial, or federal laws and other restrictions that might apply. In many states and provinces, you need a permit or license to catch a wild animal. Numbers and kinds of animals might be restricted, and closed seasons might apply to some species. Transporting wild-caught animals across state, provincial, or national borders often is illegal, and buying and selling them might be prohibited.

Also, be kind to habitats. Some species occur only in restricted areas. By removing or destroying rocks, logs, and other shelters, you can do irreparable damage or even exterminate a population. Always leave the countryside in the same or better condition than you found it.

For Serious Students

If you have a deep interest in herpetology, you may wish to read further or even join one of the organizations devoted to the subject. We provide a short list of important references on page 456. Three major North American societies are devoted to herpetology: the Society for the Study of Amphibians and Reptiles (www.ssarherps.org), the Herpetologists' League (www.herpetologistsleague.org), and the

American Society of Ichthyologists and Herpetologists (www.asih.org). Additional sources of information are Amphibian Species of the World (www.research.amnh.org/vz/herpetology/amphibia/) and the Reptile Database (www.reptile-database.org), both of which provide constantly updated information and relevant references to species around the world. The Center for North American Herpetology (www.cnah.org) provides a regularly updated checklist of North American species with ranges north of Mexico, announcements of new publications, and relevant research news.

Venomous Snakebite

Previous editions of this field guide included a short chapter titled "In Case of Snakebite." We omitted that from this edition because treatment of venomous snakebites is too complex to cover adequately in a few words. Instead, we suggest resources available on the Internet. The Arizona Poison and Drug Information Center provides information on "Venomous Creatures" (www.azpoison.com/venom). Although the Arizona Coralsnake and many rattlesnakes found in Arizona do not occur in the area covered by this guide, the general information at the website remains applicable. Another reliable source of information about snakebite prevention, symptoms, and first aid (especially what *not* to do) is MedlinePlus (www.nlm.nih.gov/medlineplus/ency/article/000031.htm).

The best treatment is to prevent snakebites in the first place—and the truth is that a large majority of venomous snakebite victims were engaged in irresponsible behavior when bitten. Most bites can be prevented by employing common sense and taking a few simple precautions. Do not keep venomous snakes as pets. Leave wild animals alone, especially if you are unsure if they are dangerous. When in the range of venomous snakes, always be aware of where you put your hands and feet, and avoid places where snakes are likely to hide.

If you are bitten, the best first-aid kit includes your car keys and a cell phone. Get help quickly. Remain calm, remove tight clothing and jewelry, and elevate and immobilize the affected limb, but do not do anything that will delay your trip to the nearest medical facility. Call 911. If the staff of the medical facility is unfamiliar with snakebite treatment, call the American Association of Poison Control Centers (1-800-222-1222) for assistance.

Accurate identification of most reptiles and amphibians often requires careful, hands-on examination. Only a few snakes are venomous, and all of those known from our area are illustrated on Plates 44–47 (pp. 350–357). Learn to recognize them on sight, and leave them alone!

Begin with pictures. Plates are grouped by type of animal (salamanders, frogs, crocodilians, turtles, lizards, snakes) and are supplemented by figures and photographs. If the animal you see or have in hand does not match an image, select the illustration it most closely resembles, paying particular attention to key characters indicated on plates, and consult the text for details pertaining to that species. Read the species account and examine the relevant map(s) to confirm your identification or eliminate similar species. Unfamiliar terms are listed in the Glossary (p. 452).

Some animals will defy all efforts to identify them. Individuals vary considerably. Such variation is normal in nature, and some species show a bewildering array of colors and patterns. Also, some frogs and lizards change colors. Many variations are mentioned and some are illustrated, but aberrant individuals or hybrids between species can confound even experts. If you have difficulty identifying an animal, take a photograph to the nearest natural history museum, zoo, or university and ask for help.

Area and Maps

Geography is important. Knowing where an animal was found usually is critical in making a correct identification. Although increasingly moved about by humans, most reptiles and amphibians rarely wander far from natural habitats. So, note the origin of your subject. If you take a photograph to a herpetologist and ask for help, his or her first question will probably be "Where did you get it?"

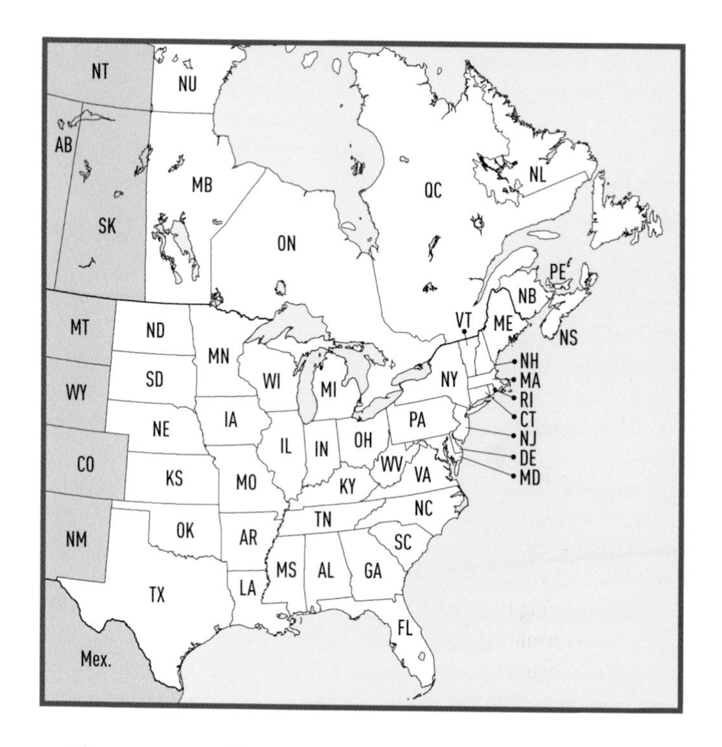

The area covered by this book is shown above. Ranges are summarized in accounts and shown in greater detail on maps. Note that the maps are largely conservative estimates based on documented records, although some indicate known historical distributions and thus may include areas from which a species has been extirpated. Also, reptiles and amphibians are not evenly distributed throughout areas indicated. North American Watersnakes (*Nerodia*), for example, seldom wander far from water, and in areas where rivers are far apart, many miles can separate localities where such snakes occur.

Illustrations

Color illustrations of animals on any one plate (or portion of a plate) are presented in the same scale. Plates are supplemented by photographs and line drawings showing key characters. The late Isabelle Hunt Conant and Tom R. Johnson drew the illustrations in the color plates. Many of the line drawings are by Isabelle Hunt Conant. Some additions or alterations by Ann Musser first appeared in the third edition. Images new to this edition are by Errol D. Hooper Jr. Many drawings are originals, but others have been adapted from figures in scientific publications. Photo credits begin on page 458, and captions for photos that introduce sections or are not in the species accounts begin on page 461.

Names

Most of the scientific and common names used in this guide are those in the seventh edition of the list published by the Society for the Study of Amphibians and Reptiles (SSAR) in 2012 (www.ssarherps.org/wp-content/uploads/2014/07/HC_39_7thEd.pdf) and the SSAR's regularly updated online North American checklist (www.ssarherps.org/publications/north-american-checklist/north-american-checklist-of-scientific-and-common-names/). These lists are recognized by all major North American herpetological organizations.

Exceptions in this guide include a few interpretations that differ from those lists and names for some non-native species for which we use common English names applicable to the region from which they originated. For example, for some taxa native to the West Indies, we use common names from the website Caribherp: Amphibians and reptiles of Caribbean Islands (www.caribherp.org).

Scientific names reflect evolutionary relationships. Species consist of individuals that freely interbreed and share a relatively recent (in geological terms) common ancestor. Related species are grouped into genera (singular: genus), and related genera are grouped into families. A species' scientific name (the binomen) consists of two Latinized words that are italicized (e.g., *Trachemys scripta*). The first word is the name of the genus, which is always capitalized. The second is a descriptive term identifying the species and is not capitalized. If subspecies are recognized (see facing page), another descriptor (e.g., *Trachemys scripta elegans*) is appended. A species epithet repeated as the subspecific name (e.g., *Kinosternon subrubrum subrubrum*) indicates that this is the "nominate" subspecies (i.e., the first formally described member of this species came from within the range of this subspecies). When used repeatedly in the same paragraph, generic and species names are sometimes abbreviated (e.g., *K. s. subrubrum*). Family names end in "-idae" (e.g., Emydidae) and are capitalized but not italicized.

Science is dynamic and, although scientific names reflect our best efforts to portray relationships that exist in nature, names change as new evidence comes to light. We note some recent nomenclatural changes or alternative interpretations under Remark(s).

Subspecies

The subspecies concept has been variously applied to everything from "incipient species" (itself an undefined entity) to localized pattern variants and geographic isolates of uncertain status. Because of this ambiguity, the entire concept has fallen into disfavor in recent years, few new subspecies of amphibians and reptiles have been formally described in recent decades, and we suspect that such taxa will not be recognized in the future. That said, we include most currently recognized subspecies that occur in our area because such variation within a species might be an indicator of a species complex (a group of closely related species).

Species Accounts

Within the major groups of reptiles and amphibians, the species accounts generally are arranged alphabetically using the scientific names of the family, genus, and species. Exceptions occur when morphologically indistinguishable, similar, or related taxa are grouped. For example, lizard families are assembled into related groups so that all gecko and skink families are listed sequentially.

The familial and generic accounts provide general information. Species accounts begin with the common and scientific names. Relevant illustrations are referenced, and established ***non-native*** species are clearly marked. We include only species known to reproduce and maintain populations in our area. Many additional introductions have been reported, and at least some are likely to establish breeding populations.

Species accounts usually consist of four sections. A list of characteristics necessary for identification is followed by Similar Species, Habitat, and Range. The Similar Species section is omitted for species easily identified and unlikely to be confused with others. The Range section is augmented by a separate description of the Natural Range for non-native species. Other sections that sometimes are included are Voice; Subspecies (currently recognized subspecies are listed as appropriate and their ranges marked on the map); Conservation (see "Learn More," p. 4); and Remark(s).

The section on characteristics begins with size, followed by other descriptive features. Reptiles and amphibians often continue to grow as long as they live, rapidly at first but more slowly after maturity. Unless otherwise noted, all sizes (including those for hatchlings and juveniles) refer to total length. For species with a tail, we sometimes provide a "head-body" length along with total length. The head-body length is the distance from the tip of the snout to the vent (the cloacal aperture from which wastes and reproductive cells are expelled). In salamanders and crocodilians, which have a longitudinal vent, measurements are to the posterior end of the vent. When available, we also provide the greatest total length ("record") we believe to be authentic. For turtles, sizes are for the length of the carapace (upper part of the shell). We sometimes give separate measurements for species in which sexes differ in size. Proper methods for taking measurements of animals in each major group are illustrated in Fig. 1.

The Similar Species section provides comparisons with species similar in appearance and found in the same general area. The information under Habitat may be precise or general, depending on the species, and may include information on breeding seasons for amphibians. Range provides a brief description of each species' geographic distribution and is supplemented by a map.

Some amphibians and reptiles in our area vocalize. Crocodilians grunt, roar, and bellow, and some geckos utter mouselike squeaks. Male frogs are quite vocal, and different species have distinctive calls; many can be identified by voice alone. Describing calls accurately is difficult, and recordings can be immensely helpful. You can find recordings at the Animal Diversity Web at the University of Michigan Museum of Zoology (http://animaldiversity.ummz.umich.edu/collections/frog_calls/), the Macaulay Library at the Cornell Lab of Ornithology (www.macaulaylibrary.org/browse/taxa/anura), and the U.S. Geological Survey's Patuxent Wildlife Research Center (www.pwrc.usgs.gov/Frogquiz/). An excellent book on the subject by Elliot, Gerhardt, and Davidson (full citation provided in References, p. 456) includes a compact disk with calls of most North American species.

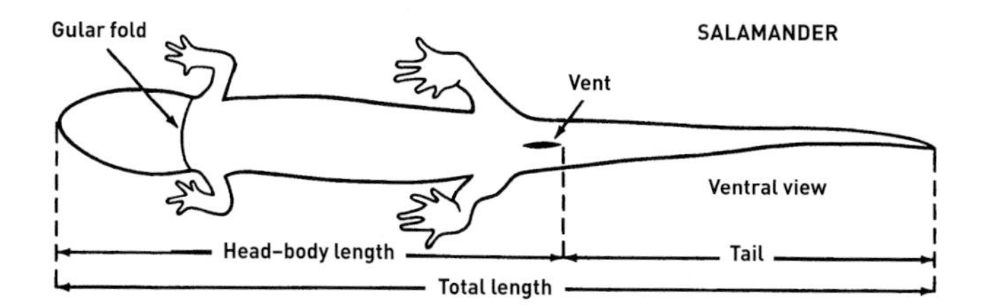

FIG. 1. *Taking measurements of salamanders, frogs, turtles, lizards, and snakes. Head-body length often is called snout-vent length. Crocodilians have a longitudinal vent (cloacal aperture) and are measured in the same fashion as salamanders.*

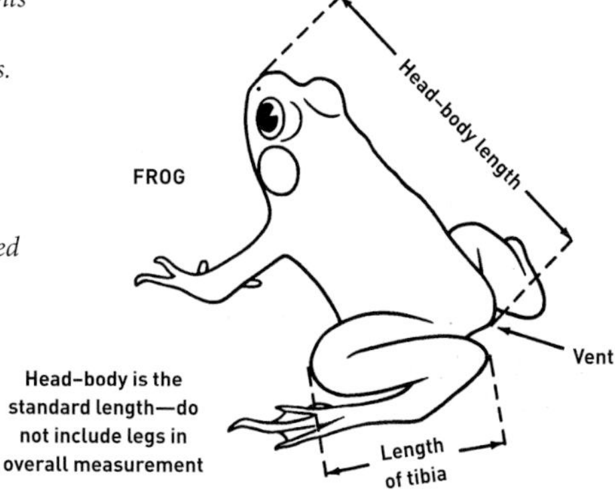

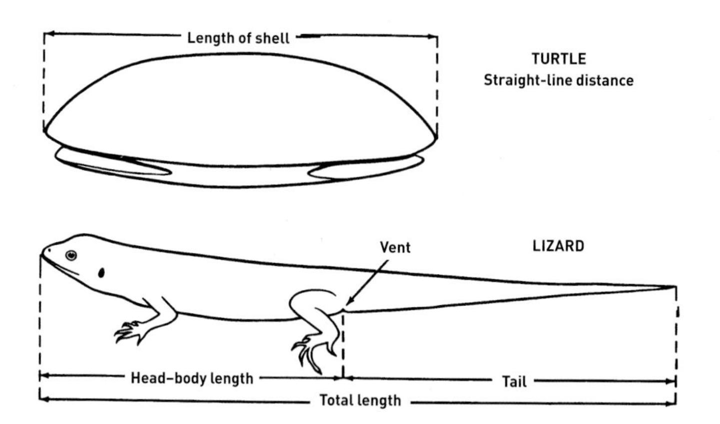

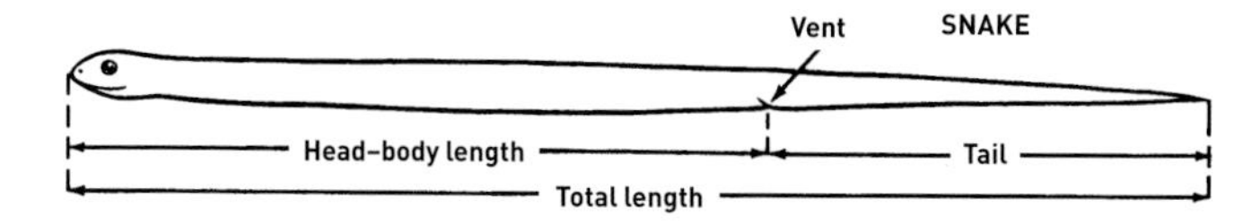

SALAMANDERS

PLATE 1

MOLE SALAMANDERS (*Ambystoma*)

RETICULATED FLATWOODS SALAMANDER *A. bishopi* **P. 31**
Reticulate pattern or series of narrow light rings, occasionally "frosted" (see Frosted Flatwoods Salamander).

FROSTED FLATWOODS SALAMANDER *A. cingulatum* **P. 32**
"Frosted" or lichenlike pattern, occasionally with a reticulate pattern (see Reticulated Flatwoods Salamander).

MOLE SALAMANDER *A. talpoideum* **P. 36**
Short, chunky body; large head.

MABEE'S SALAMANDER *A. mabeei* **P. 33**
Profuse light speckling on sides; small head.

MARBLED SALAMANDER *A. opacum* **P. 35**
White or silvery crossbars; black on lower sides and belly.

SMALL-MOUTHED SALAMANDER *A. texanum* **P. 36**
Very short snout; markings indistinct or with a profuse lichenlike pattern.

BLUE-SPOTTED SALAMANDER *A. laterale* **P. 32**
Long toes; numerous blue or bluish white spots and flecks.

JEFFERSON SALAMANDER *A. jeffersonianum* **P. 32**
Long toes; long snout; usually with small bluish flecks on sides.

RINGED SALAMANDER *A. annulatum* **P. 30**
Bold yellow rings; small head.

SPOTTED SALAMANDER *A. maculatum* **P. 33**
Round yellow spots in irregular dorsolateral rows. Spots on head often orange.

BARRED TIGER SALAMANDER *A. mavortium mavortium* **P. 34**
Large light bars or blotches characterize this subspecies of the Western Tiger Salamander.

EASTERN TIGER SALAMANDER *A. tigrinum* **P. 36**
Light markings small, not forming a definite pattern.

FIG. 2. *Characteristics of salamanders.*

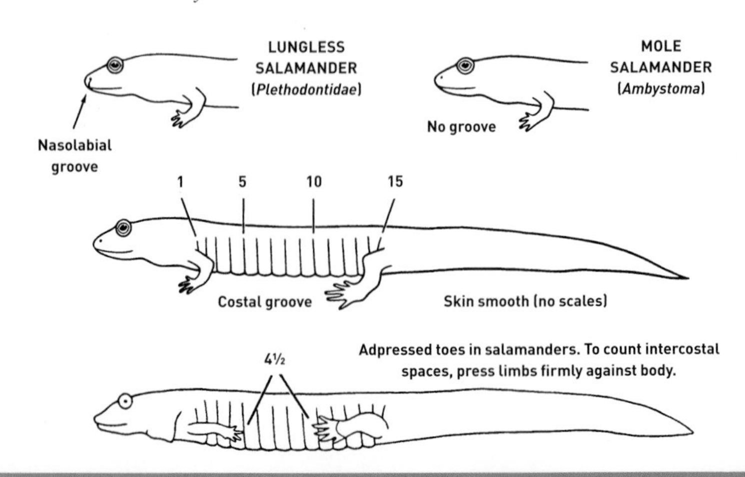

RETICULATED FLATWOODS
MOLE
FROSTED FLATWOODS
MABEE'S
MARBLED
SMALL-MOUTHED
pattern variants
BLUE-SPOTTED
JEFFERSON
RINGED
SPOTTED
BARRED TIGER
EASTERN TIGER

AMPHIUMAS (*Amphiuma*)

Tiny fore- and hindlimbs, no external gills. Three species virtually identical (distinguish by counting toes). **P. 37**

HELLBENDERS (*Cryptobranchus*)

Broad, flat head; skinfolds along sides.

EASTERN HELLBENDER *C. alleganiensis* **P. 38**
Dark markings few and small.

OZARK HELLBENDER *C. bishopi* **P. 38**
Dark markings large and blotchlike.

CLIMBING SALAMANDERS (*Aneides*)

GREEN SALAMANDER *A. aeneus* **P. 39**
Green, lichenlike markings.

DUSKY SALAMANDERS (*Desmognathus*)

Usually with diagonal light lines behind eyes.

ALLEGHENY MOUNTAIN DUSKY SALAMANDER *D. ochrophaeus* **P. 48**
Coloration and pattern highly variable; young brightly colored, old adults dark with faint markings. Middorsal light stripe with nearly straight edges. Tail round, not keeled.

IMITATOR SALAMANDER *D. imitator* **P. 46**
Like Allegheny Mountain Dusky Salamander, but with red, orange, or yellow cheek patches.

OCOEE SALAMANDER *D. ocoee* **P. 48**
Borders of middorsal stripe wavy, irregular, or interrupted. Tail round, not keeled or with a very low keel.

SPOTTED DUSKY SALAMANDER *D. conanti* **P. 44**
Six to eight pairs of golden or reddish spots; tail laterally compressed, knife-edged above.

OUACHITA DUSKY SALAMANDER *D. brimleyorum* **P. 43**
Usually a uniformly colored dorsum; tail keeled, compressed near tip.

ALLEGHENY MOUNTAIN
(*D. ochrophaeus*)

CAROLINA MOUNTAIN
(*D. carolinensis*)

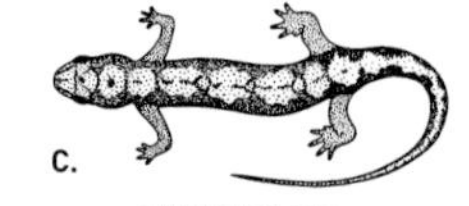

CUMBERLAND
(*D. abditus*)

FIG. 3. *Variations in dorsal patterns of some dusky salamanders.* **A.** *The Allegheny Mountain Dusky Salamander* (Desmognathus ochrophaeus) *often has chevronlike dorsal spots and a straight-edged light middorsal stripe.* **B.** *The Carolina Mountain Dusky Salamander* (D. carolinensis) *has dark dorsolateral stripes that usually are straight but sometimes wavy or strongly undulating.* **C.** *The Cumberland Dusky Salamander* (D. abditus) *has dorsal stripes that often are broken into pairs of spots.*

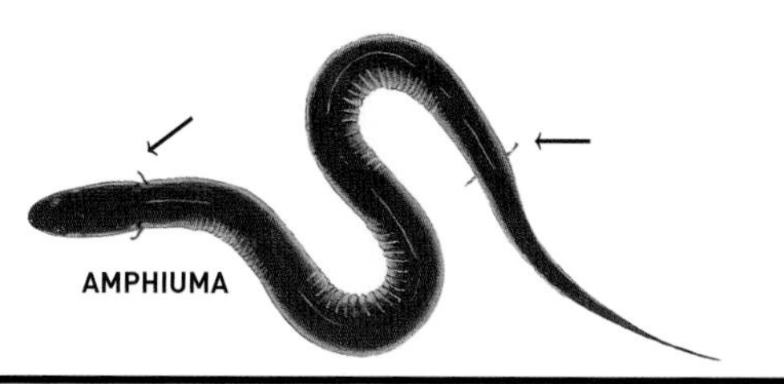
AMPHIUMA

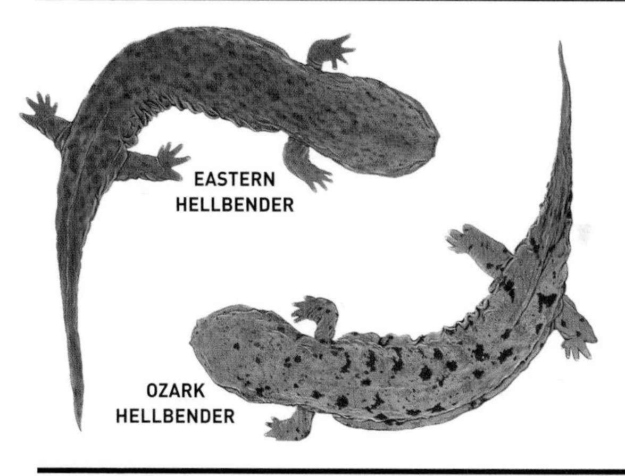
EASTERN
HELLBENDER
OZARK
HELLBENDER

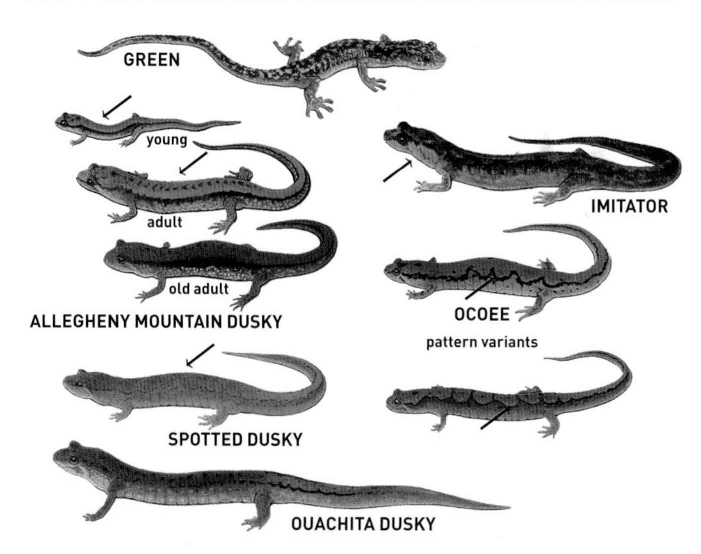
GREEN
young
adult
old adult
ALLEGHENY MOUNTAIN DUSKY
IMITATOR
OCOEE
pattern variants
SPOTTED DUSKY
OUACHITA DUSKY

DUSKY SALAMANDERS (*Desmognathus*) (cont.)

Usually with diagonal light lines behind eyes.

PYGMY SALAMANDER *D. wrighti* **P. 52**
Dorsal herringbone pattern.

SEAL SALAMANDER *D. monticola* **P. 47**
Heavy dark markings on dorsum; venter pale. Tail laterally compressed, knife-edged above.

NORTHERN DUSKY SALAMANDER *D. fuscus* **P. 45**
Markings variable; venter usually lightly pigmented.

BLACK-BELLIED SALAMANDER *D. quadramaculatus* **P. 50**
Two rows of light dots on sides most evident in young; venter black in adult.

SHOVEL-NOSED SALAMANDER *D. marmoratus* **P. 47**
Head flattened; dorsum variable, usually spotted or blotched. Check internal openings of nostrils if in doubt (Fig. 4). Tail short, laterally compressed, knife-edged above.

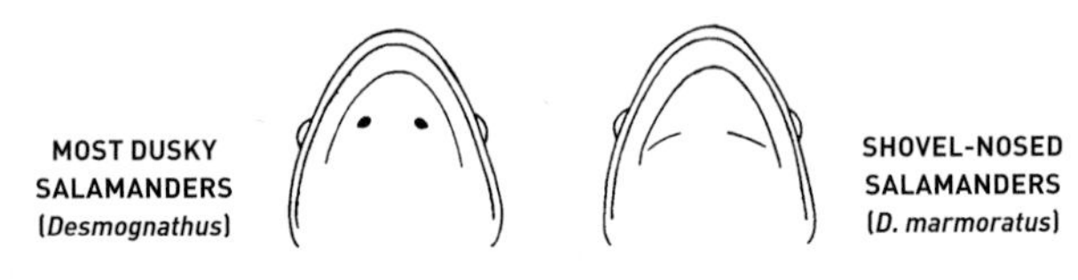

FIG. 4. *The internal openings of the nostrils are clearly visible in the roof of the mouth of most dusky salamanders* (Desmognathus). *However, the Shovel-nosed Salamander* (D. marmoratus) *has narrow slitlike openings that often are not visible.*

BROOK SALAMANDERS (*Eurycea*)

BLUE RIDGE TWO-LINED SALAMANDER *E. wilderae* **P. 60**
Bright coloration; dark lines clear-cut. Male with conspicuous cirri.

NORTHERN TWO-LINED SALAMANDER *E. bislineata* **P. 53**
Dark lines from eyes to tail; belly plain yellow.

BROWN-BACKED SALAMANDER *E. aquatica* **P. 53**
Brown back, dusky-black sides; relatively short tail.

DWARF SALAMANDER *E. quadridigitata* **P. 57**
Four toes on all limbs; unicolored or with dark dorsolateral stripes.

MANY-RIBBED SALAMANDER *E. multiplicata* **P. 57**
Light spots on dark sides; undersurfaces bright yellow.

OKLAHOMA SALAMANDER *E. tynerensis* **P. 59**
Sides often darker than back; light spots on dark sides; belly gray with dark specks or blotches. Some populations transform, others are neotenic.

GROTTO SALAMANDER *E. spelaea* **P. 58**
Adult very pale, no gills. Larva with high tail fin, longitudinal streaks on sides.

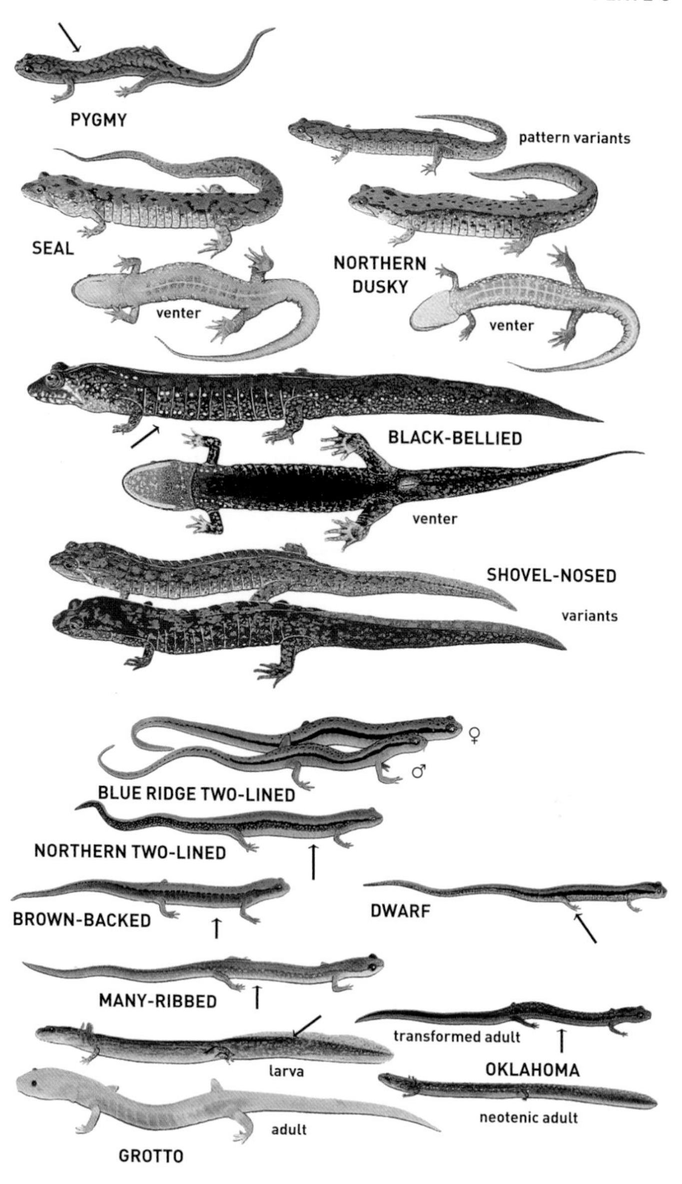
PYGMY
pattern variants
SEAL
NORTHERN
DUSKY
venter
venter
BLACK-BELLIED
venter
SHOVEL-NOSED
variants
♀
♂
BLUE RIDGE TWO-LINED
NORTHERN TWO-LINED
BROWN-BACKED
DWARF
MANY-RIBBED
transformed adult
larva
OKLAHOMA
neotenic adult
adult
GROTTO

BROOK SALAMANDERS (*Eurycea*) (cont.)

CAVE SALAMANDER *E. lucifuga* **P. 57**
Orange or reddish with scattered black spots; tail long.

LONG-TAILED SALAMANDER *E. longicauda* **P. 56**
EASTERN LONG-TAILED SALAMANDER (*E. l. longicauda*): "Dumbbells" or herring-bone pattern on sides of tail.

DARK-SIDED SALAMANDER (*E. l. melanopleura*): Light middorsal stripe, dark sides with many light flecks.

THREE-LINED SALAMANDER *E. guttolineata* **P. 55**
Middorsal and two dorsolateral dark stripes; tail long.

TEXAS SALAMANDER *E. neotenes* **P. 63**
Yellowish; dark bars from nostrils to eyes; rows of light spots on sides. See section on neotenic salamanders from Texas (p. 61).

SAN MARCOS SALAMANDER *E. nana* **P. 62**
Brown with rows of small light spots on sides.

TEXAS BLIND SALAMANDER *E. rathbuni* **P. 63**
Shovel-nosed snout; toothpick legs; pale coloration.

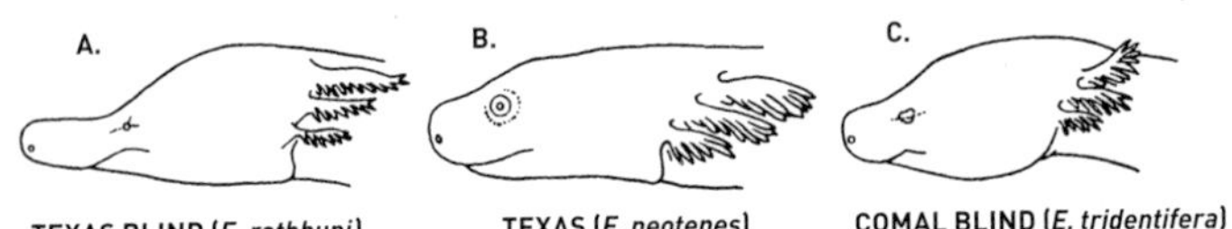

FIG. 5. *Neotenic brook salamanders* (Eurycea) *from Texas.* **A.** *The Texas Blind Salamander* (E. rathbuni) *has a strongly shovel-nosed snout, small eyespots covered by skin, and 12 costal grooves.* **B.** *The Texas Salamander* (E. neotenes) *has a flattened head (but lacks a shovel-nosed snout), relatively large eyes, and 15–17 costal grooves. Note that several other very similar species often are confused with the Texas Salamander (see text).* **C.** *The Comal Blind Salamander* (E. tridentifera) *has a shovel-nosed snout, eyespots usually covered by skin, and 11–12 costal grooves.*

RED, MUD, AND SPRING SALAMANDERS (*Pseudotriton* and *Gyrinophilus*)

RED SALAMANDER *P. ruber* **P. 72**
NORTHERN RED SALAMANDER (*P. r. ruber*): Black spots numerous, irregular, and often running together; irises yellowish. Old adults with dark ground color and large indistinct spots.

BLACK-CHINNED RED SALAMANDER (*P. r. schencki*): Black pigment on chin.

SOUTHERN RED SALAMANDER (*P. r. vioscai*): Small white flecks, especially on head.

MUD SALAMANDER *P. montanus* **P. 70**
EASTERN MUD SALAMANDER (*P. m. montanus*): Few black spots round.

RUSTY MUD SALAMANDER (*P. m. floridanus*): Markings obscure; often with dark streaks on side.

TENNESSEE CAVE SALAMANDER *G. palleucus* **P. 68**
Pinkish (sometimes darker; varies by subspecies); external gills; large tail fin.

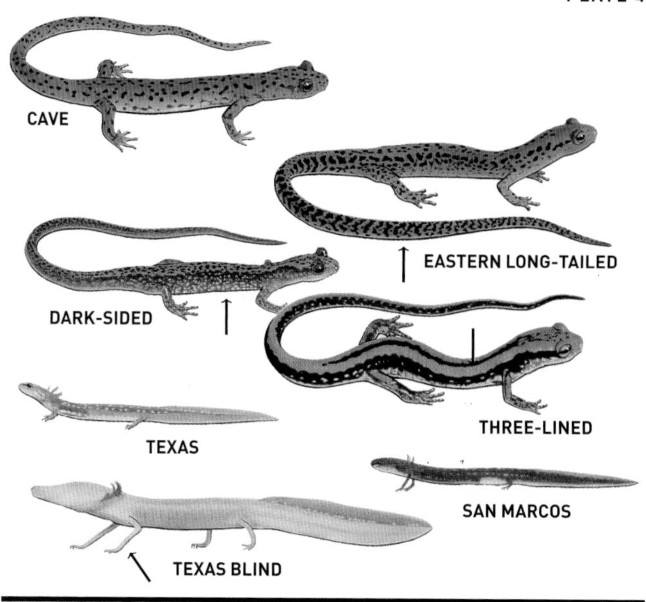
CAVE
EASTERN LONG-TAILED
DARK-SIDED
THREE-LINED
TEXAS
SAN MARCOS
TEXAS BLIND

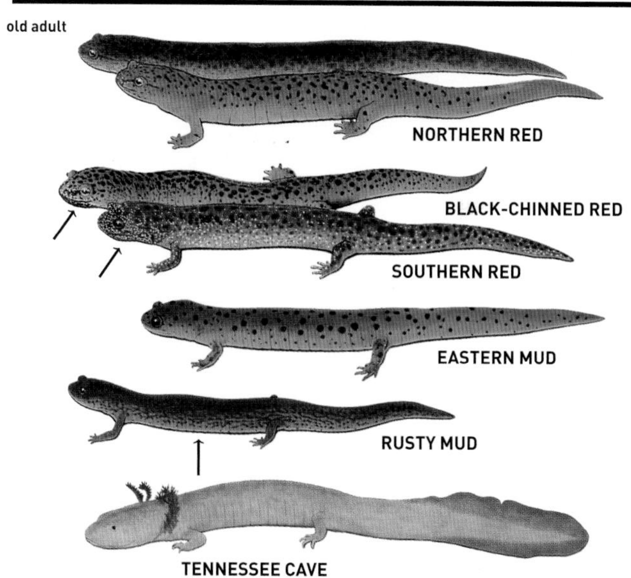
old adult
NORTHERN RED
BLACK-CHINNED RED
SOUTHERN RED
EASTERN MUD
RUSTY MUD
TENNESSEE CAVE

SPRING SALAMANDERS (*Gyrinophilus*) (cont.)

SPRING SALAMANDER *G. porphyriticus* **P. 69**

Light lines from eyes to nostrils bordered below by black or gray.

NORTHERN SPRING SALAMANDER (*G. p. porphyriticus*): Eye lines and pattern not clear-cut.

BLUE RIDGE SPRING SALAMANDER (*G. p. danielsi*): Eye lines distinct; dorsum with scattered black spots.

FOUR-TOED SALAMANDERS (*Hemidactylium*)

FOUR-TOED SALAMANDER *H. scutatum* **P. 73**

Four toes on all limbs; constriction at base of tail; belly white with black spots (Fig. 37, p. 80).

RED HILLS SALAMANDERS (*Phaeognathus*)

RED HILLS SALAMANDER *P. hubrichti* **P. 74**

Very long body, short legs; dorsum uniformly dark brown.

WOODLAND SALAMANDERS (*Plethodon*)

WEHRLE'S SALAMANDER *P. wehrlei* **P. 92**

White, bluish white, or yellow spots along sides. Dixie Caverns variant has a purplish brown back frosted with small white flecks and bronzy mottling.

WELLER'S SALAMANDER *P. welleri* **P. 92**

Gold or brassy blotches.

CHEAT MOUNTAIN SALAMANDER *P. nettingi* **P. 86**

Small brassy dorsal flecks; sometimes with larger white spots on flanks; belly dark gray to black.

EASTERN RED-BACKED SALAMANDER *P. cinereus* **P. 78**

Salt-and-pepper effect on belly (Fig. 37, p. 80); middorsal stripe (when present) straight-edged.

SOUTHERN RED-BACKED SALAMANDER *P. serratus* **P. 88**

Serrated middorsal stripe often narrow at base of tail (also a lead-backed variant; not shown).

NORTHERN ZIGZAG SALAMANDER *P. dorsalis* **P. 80**

Middorsal stripe generally narrow with straight edges for at least part of length, or wider with wavy or scalloped margins (also a lead-backed variant; not shown).

CADDO MOUNTAIN SALAMANDER *P. caddoensis* **P. 78**

Small whitish dorsal and lateral spots; brassy flecks on back; throat and chest light, belly dark.

FOURCHE MOUNTAIN SALAMANDER *P. fourchensis* **P. 82**

Two irregular dorsal rows of large whitish spots; variable number of smaller spots; chin light, chest and belly dark.

RICH MOUNTAIN SALAMANDER *P. ouachitae* **P. 87**

Back usually chestnut with scattered white specks; chin light, chest and belly dark.

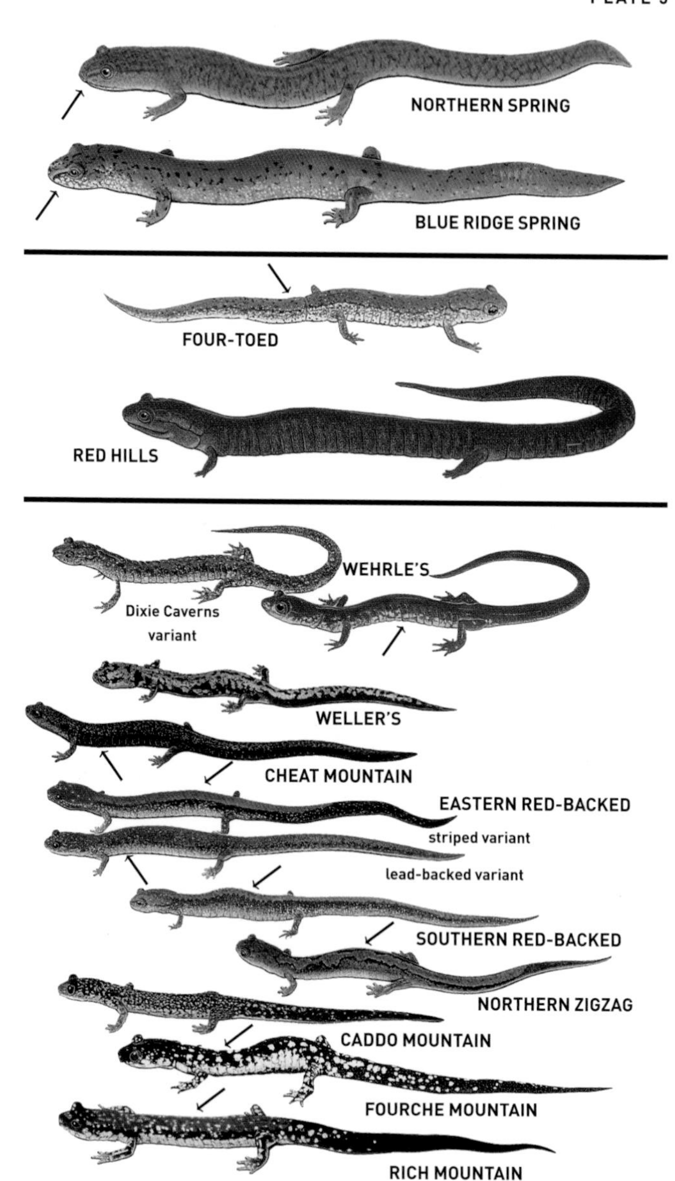
NORTHERN SPRING
BLUE RIDGE SPRING
FOUR-TOED
RED HILLS
WEHRLE'S
Dixie Caverns
variant
WELLER'S
CHEAT MOUNTAIN
EASTERN RED-BACKED
striped variant
lead-backed variant
SOUTHERN RED-BACKED
NORTHERN ZIGZAG
CADDO MOUNTAIN
FOURCHE MOUNTAIN
RICH MOUNTAIN

PLATE 6

WOODLAND SALAMANDERS (*Plethodon*) (cont.)

SOUTHERN GRAY-CHEEKED SALAMANDER *P. metcalfi* **P. 75**
Cheeks gray; belly light in northern portions of range; belly dark, sometimes with white spots on sides and brassy dorsal flecks, in southern portions of range.

RED-LEGGED SALAMANDER *P. shermani* **P. 90**
Limbs red or with red blotches; scattered light spots on back and sides.

RED-CHEEKED SALAMANDER *P. jordani* **P. 85**
Red on cheeks; scattered light spots on back and sides.

SOUTHERN RAVINE SALAMANDER *P. richmondi* **P. 81**
Long, slender body; no conspicuous markings; belly dark (Fig. 37, p. 80).

PIGEON MOUNTAIN SALAMANDER *P. petraeus* **P. 87**
Back reddish brown or brown; white or yellowish spots on sides; belly black.

NORTHERN SLIMY SALAMANDER *P. glutinosus* **P. 82**
Numerous white spots or brassy flecks or both; throat dark.

YONAHLOSSEE SALAMANDER *P. yonahlossee* **P. 92**
Chestnut dorsum bordered laterally by light sides. Bat Cave variant has reduced chestnut color; ventrolateral light sides dark-bordered above.

MANY-LINED SALAMANDERS (*Stereochilus*)

MANY-LINED SALAMANDER *S. marginatus* **P. 93**
Dark lines on lower sides sometimes reduced to streaks or spots; head small; tail short.

WATERDOGS AND MUDPUPPIES (*Necturus*)

All have external gills and four well-developed legs, each with four toes.

GULF COAST WATERDOG *N. beyeri* **P. 94**
Dark spots numerous and close together.

DWARF WATERDOG *N. punctatus* **P. 98**
Almost uniformly dark above; paler below.

COMMON MUDPUPPY *N. maculosus* **P. 97**
Dark spots few and well-separated.

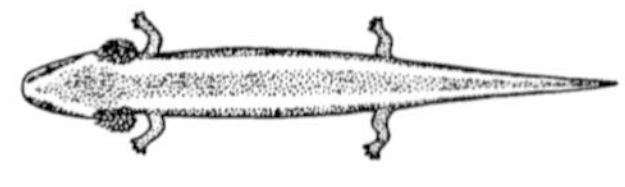

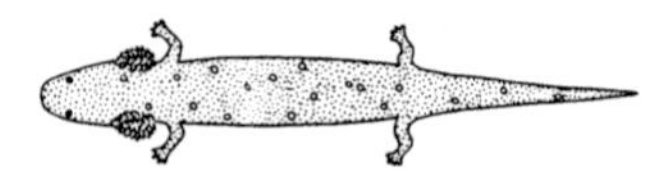

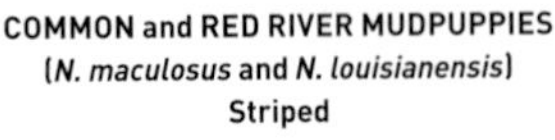

COMMON and RED RIVER MUDPUPPIES
(*N. maculosus* and *N. louisianensis*)
Striped

GULF COAST WATERDOG
(*N. beyeri*)
Spotted

FIG. 6. *Larvae of the Common Mudpuppy* (Necturus maculosus) *and Red River Mudpuppy* (N. louisianensis) *compared with that of the Gulf Goast Waterdog* (N. beyeri).

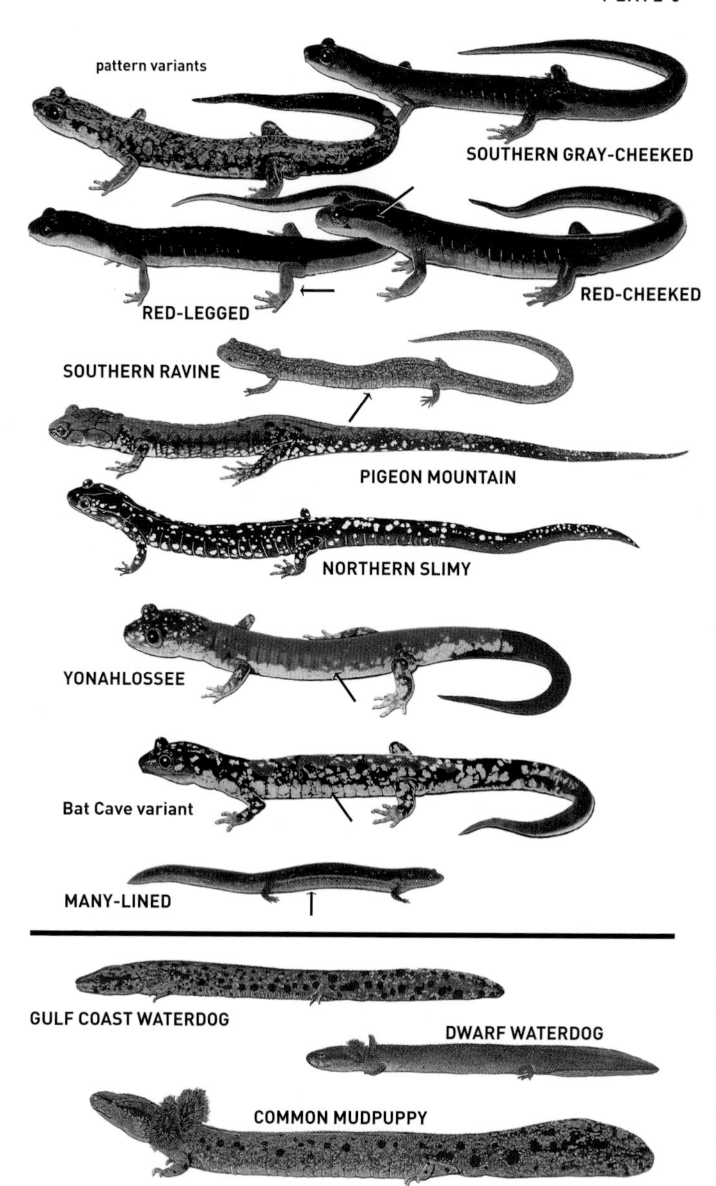
pattern variants
SOUTHERN GRAY-CHEEKED
RED-LEGGED
RED-CHEEKED
SOUTHERN RAVINE
PIGEON MOUNTAIN
NORTHERN SLIMY
YONAHLOSSEE
Bat Cave variant
MANY-LINED
GULF COAST WATERDOG
DWARF WATERDOG
COMMON MUDPUPPY

EASTERN NEWTS (*Notophthalmus*)

BLACK-SPOTTED NEWT *N. meridionalis* P. 98
Large scattered black spots; irregular yellowish stripes; orange belly.

STRIPED NEWT *N. perstriatus* P. 98
Red dorsolateral stripes.

EASTERN NEWT *N. viridescens* P. 99

RED-SPOTTED NEWT (*N. v. viridescens*): Red spots in all phases. Red eft (land form): Red or orange, transforming to dark olive (or other colors). Aquatic ♀: Olive; belly yellow, spotted with black. Aquatic ♂ (in breeding condition): High tail fin; black horny growths on hindlimbs and toes.

CENTRAL NEWT (*N. v. louisianensis*): Dorsal and ventral coloration in sharp contrast; normally no red markings.

BROKEN-STRIPED NEWT (*N. v. dorsalis*): Red dorsolateral stripes broken.

PENINSULA NEWT (*N. v. piaropicola*): Dorsum dark; belly peppered with black.

SIRENS and DWARF SIRENS (*Siren* and *Pseudobranchus*)

Eel-like salamanders with external gills, tiny forelimbs, and no hindlimbs.

LESSER SIREN *S. intermedia* P. 102
Virtually no markings or with scattered black dots.

GREATER SIREN *S. lacertina* P. 103
Back darker than sides, usually with some light markings.

SLENDER DWARF SIREN *P. striatus spheniscus* P. 101
Yellowish lateral stripes; darker stripes above.

NARROW-STRIPED DWARF SIREN *P. axanthus axanthus* P. 100
Grayish with faint stripes.

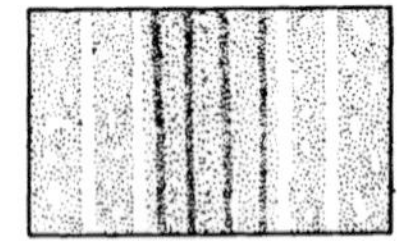

NARROW-STRIPED (*P. a. axanthus*)
Dark stripes narrow, subdued

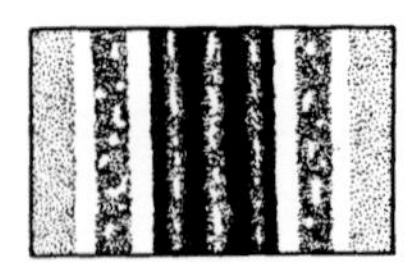

EVERGLADES (*P. a. belli*)
Lateral stripe buff; belly gray

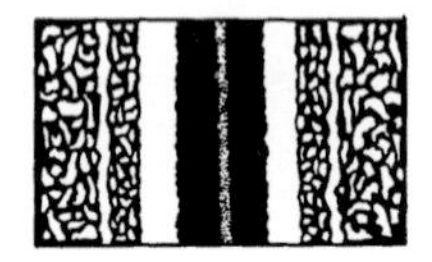

BROAD-STRIPED (*P. s. striatus*)
Dark middorsal stripe flanked by broad yellow stripes

GULF HAMMOCK (*P. s. lustricolus*)
Three narrow yellow stripes in dark middorsal region; belly blackish

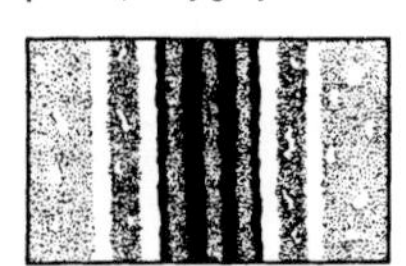

SLENDER (*P. s. spheniscus*)
Two tan or yellow stripes on each side

FIG. 7. *Diagrammatic patterns of subspecies of the Southern Dwarf Siren* (Pseudobranchus axanthus) *and the Northern Dwarf Siren* (P. striatus). *Each diagram shows a section of skin removed from an animal and flattened out—dorsum in the center, belly at the sides.*

BLACK-SPOTTED NEWT
STRIPED NEWT
RED-SPOTTED NEWT
red eft
transforming
♀
CENTRAL NEWT
BROKEN-STRIPED NEWT
PENINSULA NEWT
LESSER SIREN
GREATER
SIREN

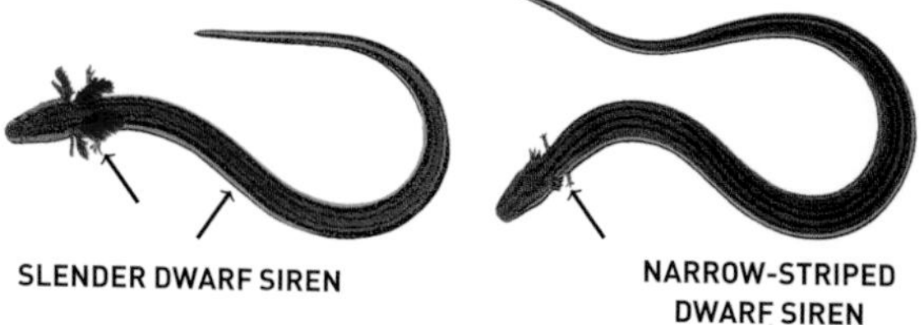
SLENDER DWARF SIREN
NARROW-STRIPED
DWARF SIREN

Nearly 700 species are currently recognized, and that number is increasing rapidly with the use of molecular tools capable of distinguishing cryptic species (genetically unique but similar in appearance). The Americas have more species of salamanders than the rest of the world combined. A few species remain aquatic throughout life, retaining larval features such as external gills while becoming sexually mature (this is known as neoteny).

Counting costal grooves is frequently necessary for identification, as is adpressing the limbs as an indicator of relative body or limb length (Fig. 2, p. 16). Identifying larval salamanders is often difficult, as many salamanders acquire the diagnostic features of adults only after metamorphosis. We rarely mention larvae in the following accounts, instead referring readers to the book on larval amphibians by Altig and McDiarmid (see References, p. 456). Additional information on larvae is available in the book by Petranka (also in References) and in several online guides.

MOLE SALAMANDERS: Ambystomatidae

Thirty-two currently recognized species, all North American, range from southern Labrador and southeastern Alaska to the southern edge of the Mexican Plateau. Mole salamanders spend most of their lives underground but congregate primarily in temporary pools and ponds after early-spring or fall and winter rains for courtship and deposition of eggs. Finding these salamanders before or after the breeding season is difficult except where water tables are near the surface.

RINGED SALAMANDER *Ambystoma annulatum* **PL. 1, FIG.8**

5½–7 in. (14–18 cm); record 9⅜ in. (23.8 cm). Dark brown to black, with narrow, widely spaced buff, yellow, or whitish *rings*; rings may be interrupted across the back or represented solely by vertical light bars or elongated spots. Often with light gray, rather irregular stripes along lower sides. Belly slate-colored, with small whitish spots. Head small. **SIMILAR SPECIES:** (1) Marbled Salamander is shorter and stouter, gray or white crossbands are wider, and the belly is black. (2)

FIG. 8. *The Ringed Salamander* (Ambystoma annulatum) *is a fall breeder endemic to the Central Highlands (the Ozark Plateau, Boston and Ouachita Mountains, and associated ranges). The narrow, widely spaced rings and the small head easily distinguish this species from other mole salamanders.*

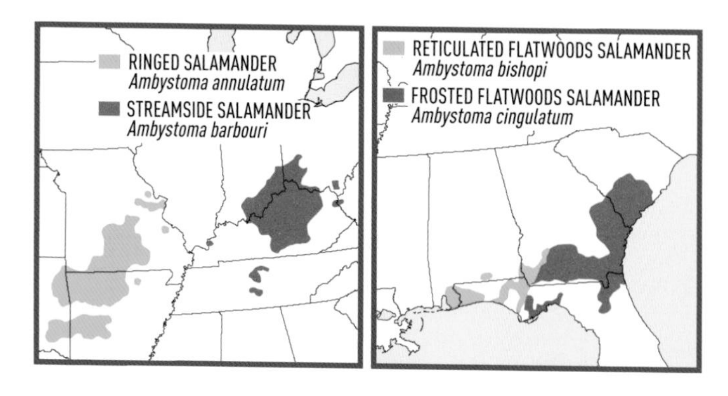

Eastern and Western Tiger Salamanders have a wide head, and light marks do not form narrow, widely spaced rings. **HABITAT:** Forested areas near breeding pools. Breeds in fall. **RANGE:** Central Highlands in MO, AR, and OK.

STREAMSIDE SALAMANDER *Ambystoma barbouri* **NOT ILLUS.**

4–5½ in. (10–14 cm); record 6⅝ in. (16.8 cm). Like the Small-mouthed Salamander in appearance but breeds primarily in streams and lays eggs singly. Head small. Dorsum dark gray to black or very dark brown, usually with grayish lichenlike markings varying in intensity. **SIMILAR SPECIES:** See Small-mouthed Salamander. **HABITAT:** Upland deciduous forests. Breeds Dec.–Mar. in headwater streams, ditches, vernal pools, and small ponds. **RANGE:** N.-cen. KY, sw. OH, se. IN; isolated populations in w. KY, cen. TN, and farther east along the Ohio R.

RETICULATED FLATWOODS SALAMANDER *Ambystoma bishopi*

PL. 1, FIG. 9

3–4 in. (7.6–10.2 cm). Black to chocolate brown with gray or brownish gray reticulations sometimes forming series of narrow light rings, but occasionally "frosted" like the Frosted Flatwoods Salamander.

FIG. 9. *The Reticulated Flatwoods Salamander* (Ambystoma bishopi) *varies considerably in appearance. Some individuals are crossbanded (as shown here), others have a reticulate pattern, and occasional individuals are "frosted" like the Frosted Flatwoods Salamander* (A. cingulatum), *with which they until recently were considered conspecific.*

Usually 14 or more costal grooves. Belly black with *scattered pearl gray spots*. **SIMILAR SPECIES:** See Frosted Flatwoods Salamander. **HABITAT:** Pine-wiregrass flatwoods. Scatters eggs terrestrially in fall or early winter near water. **RANGE:** Sw. GA and FL Panhandle west of the Appalachicola R. to Mobile Bay, AL. **CONSERVATION:** Vulnerable (USFWS: Endangered).

FROSTED FLATWOODS SALAMANDER *Ambystoma cingulatum* **PL. 1**

3½–5 1/16 in. (9–12.9 cm). Black to chocolate brown with small gray or brownish gray specks forming "frosted" or lichenlike patterns, occasionally with a reticulate pattern, rarely with series of narrow light rings. Jaw teeth in 2 or more rows. Usually 14 or fewer costal grooves. Belly with a salt-and-pepper appearance. **SIMILAR SPECIES:** (1) Reticulated Flatwoods Salamander is sometimes frosted and best distinguished by range, usually has 14 or more costal grooves and a belly with discrete spots. (2) Mabee's Salamander has light specks most conspicuous along sides (not scattered on body), jaw teeth in a single row. **HABITAT:** Like that of the Reticulated Flatwoods Salamander. **RANGE:** S. SC to n. FL west to the Apalachicola R. Populations in SC and most in FL and GA are extirpated. **CONSERVATION:** Vulnerable (USFWS: Threatened).

JEFFERSON SALAMANDER *Ambystoma jeffersonianum* **PL. 1**

4¼–7 in. (10.7–18 cm); record 8¼ in. (21 cm). Dark brown or gray, dorsum distinctly darker than belly, and usually with small bluish flecks, chiefly on limbs and lower sides (most conspicuous on smaller individuals). Belly *pale*, especially south of the hybrid zone with the Blue-spotted Salamander. Toes *very long*. **SIMILAR SPECIES:** (1) Blue-spotted Salamander is smaller, profusely spotted, with shorter toes and a black belly. (2) Small-mouthed and Streamside Salamanders have a short snout and short toes. **HABITAT:** Deciduous forests, also caves. Breeds in late winter and very early spring. **RANGE:** South of the area occupied largely by unisexual populations, "pure" populations occur in NY southwest to cen. KY and extreme e. IL. **REMARKS:** Jefferson and Blue-spotted Salamanders interbreed from NS west to n. WI and south to IN and OH. Some hybrids are unisexual and reproduce by parthenogenesis; occasionally interbreeds with Small-mouthed and Eastern Tiger Salamanders.

BLUE-SPOTTED SALAMANDER *Ambystoma laterale* **PL. 1**

4–5½ in. (10–14 cm); record (probably a hybrid) 6 5/16 in. (16 cm). Bluish black, usually profusely patterned with white and blue flecks and spots, especially on sides and tail. Belly *black*, especially north of the hybrid zone with the Jefferson Salamander; even in the hybrid zone, area around vent almost always *black*. Toes long. **SIMILAR SPECIES:** See Jefferson Salamander. **HABITAT:** Forested areas, lowland swamps, marshes, surrounding uplands, overgrown pastures. Breeds in spring. **RANGE:** North of the area occupied largely by unisexual populations,

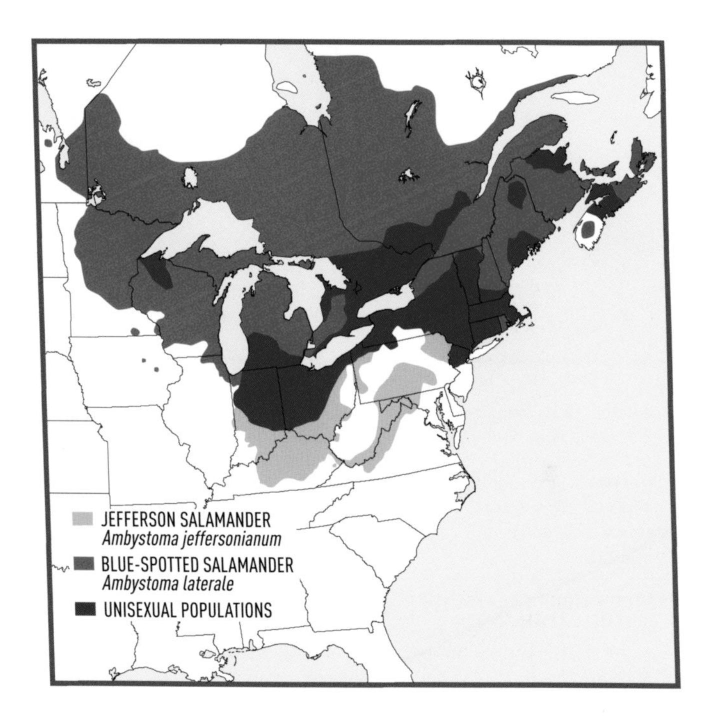

"pure" populations occur from the northern shore of the Gulf of St. Lawrence west to se. MB and south to e.-cen. IA. **REMARK:** See Jefferson Salamander.

MABEE'S SALAMANDER *Ambystoma mabeei* **PL. 1**

3–4 in. (7.5–10.2 cm); record 4½ in. (11.4 cm). Deep brown to black with light specks palest and most conspicuous along sides. Belly dark brown or gray. Head small, toes long, jaw teeth in a *single row*. **SIMILAR SPECIES:** (1) Mole Salamander has a larger head. (2) Frosted Flatwoods Salamander is patterned all over, not more conspicuously along sides, and has multiple rows of jaw teeth. **HABITAT:** Usually near ephemeral breeding pools and ponds in tupelo and cypress bottoms in pinewoods, open fields and savannas, lowland deciduous forests, and swamps. Breeds late winter through early spring. **RANGE:** Coastal Plain of se. VA and the Carolinas.

SPOTTED SALAMANDER *Ambystoma maculatum* **PL. 1**

4⅜–7¾ in. (11.2–19.7 cm); record 9¾ in. (24.8 cm). Black or slate, with round yellow or yellowish white spots in irregular dorsolateral *rows* from eyes to tailtip (rarely without spots). Spots on head often

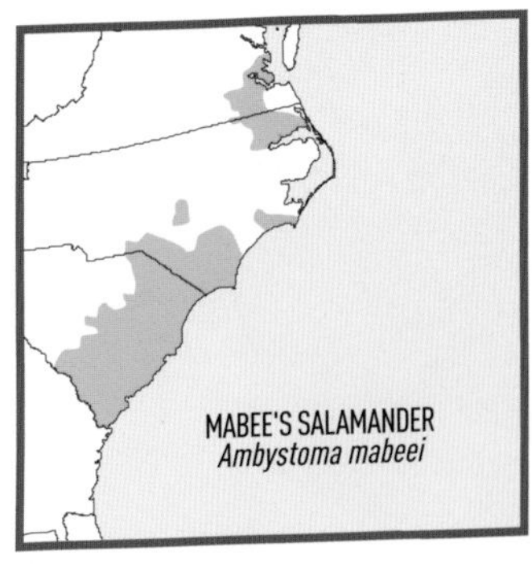

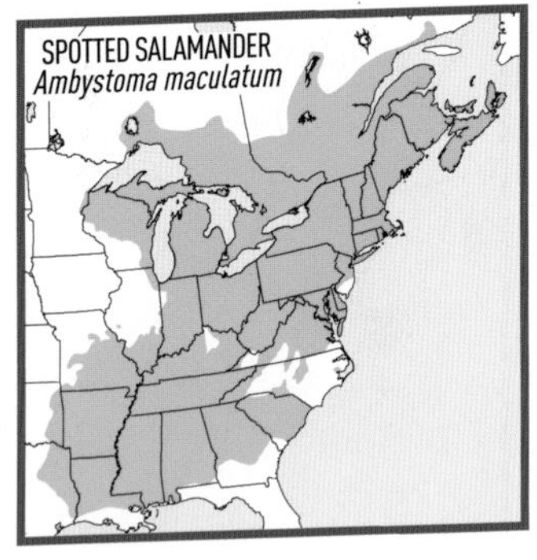

orange. Belly pale to slate gray. **SIMILAR SPECIES:** Eastern and Western Tiger Salamanders have an olive-yellow belly and irregular light dorsal spots or bars extending far down onto sides and not in rows. **HABITAT:** Hardwood and mixed forests, often near swamps and vernal pools. Breeds in early spring. **RANGE:** Atlantic Provinces to s. ON and south to GA and e. TX.

WESTERN TIGER SALAMANDER *Ambystoma mavortium* PL. 1

6–8½ in. (15.2–21.5 cm); record (neotenic individual; see below) 12⅞ in. (32.7 cm). Pattern variable, ranging from yellowish bars and lines, often brightest on sides, on dark ground color to dull yellow spots with vague borders surrounded by a network of narrow black or dark brown lines or scattered dark spots on light olive to dark brown ground color. Belly black and yellowish. Similar species: (1) Eastern Tiger Salamander has an olive-yellow belly marbled with darker pigment; dorsal pattern variable, but light dorsal markings tend to be smaller and greater in number. (2) Spotted Salamander has light spots in irregular rows, belly light to dark gray. (3) Ringed Salamander has a smaller head, and black dorsal color does not extend onto the belly. (4) Marbled Salamander has a solid black belly. **HABITAT:** Bottomland deciduous forests, coniferous forests and woodlands, open fields and bushy areas, alpine and subalpine meadows, grasslands, semideserts and deserts, rarely streams. Occasionally neotenic, especially in western portions of the range. **RANGE:** S. Can. to n. Mex.; many western populations introduced. **SUBSPECIES:** (1) **BARRED TIGER SALAMANDER** (*A. m. mavortium*); black or dark brown, yellowish markings brightest on sides, often diffused with dark pigment on back; either dark or light bars may extend upward from belly to the dorsal midline. (2) **GRAY TIGER SALAMANDER** (*A. m. diaboli*); olive to dark brown, scattered small dark brown to black spots on back and sides. (3) **BLOTCHED TIGER SALAMANDER** (*A. m. melanostictum*); brown to black ground color reduced to a network surrounding dull yellow areas with indefinite borders. **REMARK:** Until recently considered conspecific with the Eastern Tiger Salamander.

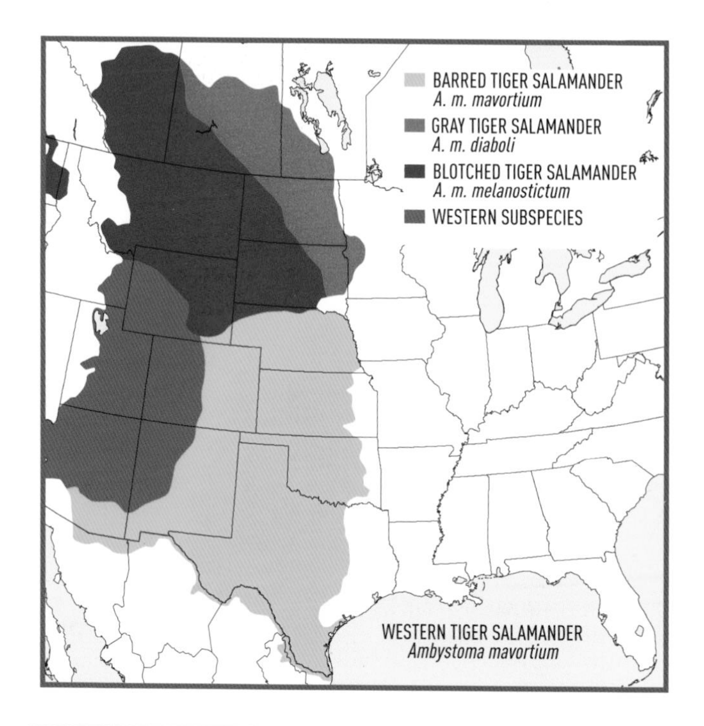

MARBLED SALAMANDER *Ambystoma opacum* **PL. 1**

3½–4¼ in. (9–10.7 cm); record 5 5/16 in. (13.6 cm). "Chunky," gray to black, with light gray (female) or white (male) crossbars, ranging from incomplete to running together or enclosing dark spots to form a ladderlike pattern; rarely with light stripes along or parallel to the middorsal line. Belly plain black. Juvenile with scattered light flecks. **SIMILAR SPECIES:** See Ringed Salamander. **HABITAT:** Woodlands and proximity of swamps; rocky bluffs and slopes and wooded sand dunes (tolerant of dry conditions). Breeds in fall. **RANGE:** S. New England to n. FL west to e. OK and TX.

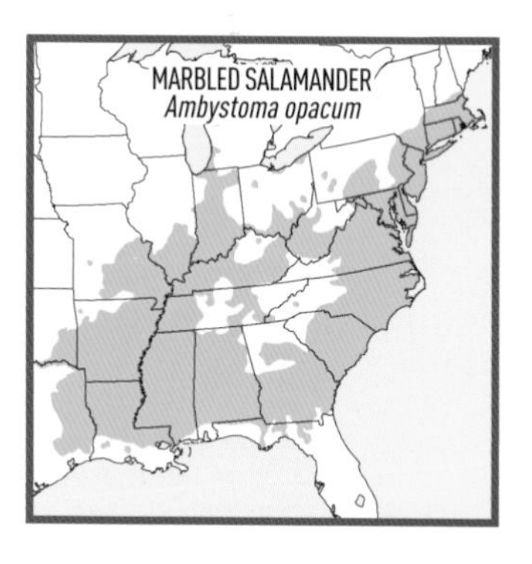

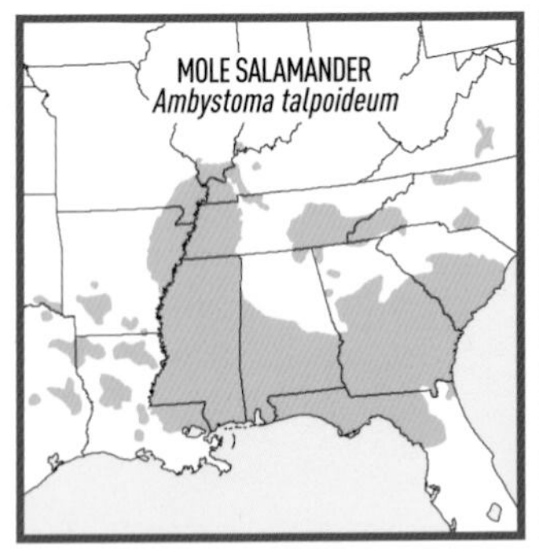

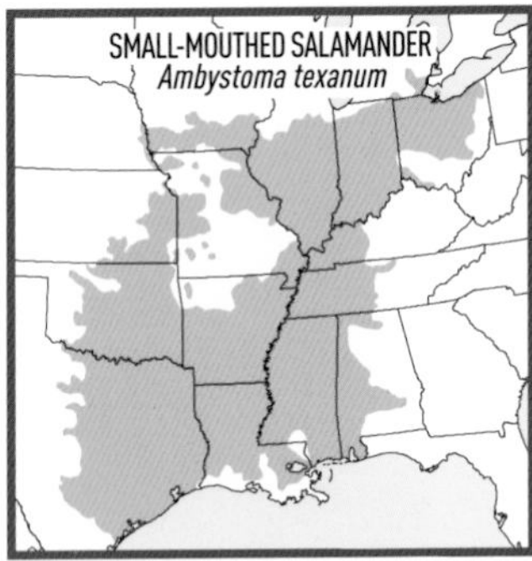

MOLE SALAMANDER *Ambystoma talpoideum* **PL. 1**

3–4 in. (7.5–10 cm); record 4 15/16 in. (12.5 cm). Black, brown, or gray with pale bluish white flecks. Short and stocky, 10–11 costal grooves, *head and feet disproportionately large*. **SIMILAR SPECIES:** (1) Small-mouthed Salamander is more slender and head is not enlarged. (2) Mabee's Salamander is more slender, has a smaller head, and light specks are palest and most conspicuous along sides. **HABITAT:** Forests, especially in lowlands and valleys. Breeds in late winter and early spring. **RANGE:** VA to n. FL west to e. TX and OK; north in the Mississippi River Valley to s. IN and IL.

SMALL-MOUTHED SALAMANDER *Ambystoma texanum* **PL. 1**

4–5½ in. (10–14 cm); record 7½ in. (19.1 cm). Head small. Dark gray to black or very dark brown, usually with grayish lichenlike markings varying in intensity. Markings sometimes concentrated on the back and upper sides, some individuals almost plain black (especially in northeastern parts of the range), others heavily speckled with light markings especially prominent on lower sides (especially in TX). **SIMILAR SPECIES:** (1) Streamside Salamander is essentially identical but often breeds in streams and lays eggs singly. (2) Jefferson and Blue-spotted Salamanders have a longer snout and head and longer toes. (3) Mole Salamander has a large head. **HABITAT:** Woodlands, tallgrass prairies, intensely farmed areas. Breeds in late winter and early spring. **RANGE:** OH and se. MI south to the Gulf and west to OK and TX.

EASTERN TIGER SALAMANDER *Ambystoma tigrinum* **PL. 1**

7–8¼ in. (18–21 cm); record 13 in. (33 cm). Dull black to deep brown, with irregular light olive or yellowish brown spots or blotches extending well down onto sides. Belly *olive-yellow* marbled with darker pigment. **SIMILAR SPECIES:** See Western Tiger Salamander. **HABITAT:** Hardwood and mixed forests, usually in lowlands. Breeds in winter

and early spring. **RANGE:** Long Island, NY, to n. FL west to the Great Plains, but absent from most of the Appalachian Highlands. **REMARK:** See Western Tiger Salamander.

AMPHIUMAS: Amphiumidae

Restricted to the southeastern United States, these slender aquatic salamanders resemble eels (and frequently are called congo eels or ditch eels), but they have four tiny legs with one to three digits. External gills disappear, but gill slits are retained through life.

TWO-TOED AMPHIUMA *Amphiuma means* **PL. 2**

14½–30 in. (36.8–76 cm); record 45¾ in. (116.2 cm). Dorsum dark brown or black, venter dark gray, no sharp transition between back and belly. *Two toes* on each limb. **SIMILAR SPECIES:** (1) One-toed and Three-toed Amphiumas have 1 and 3 toes, respectively, on each limb. (2) Sirens have external gills, no hindlimbs. (3) True eels (which are fish, not amphibians) have lateral fins behind the head and no legs. **HABITAT:** Swamps, bayous, margins of muddy sloughs, cypress heads, drainage ditches, sluggish streams, wet meadows, muddy lakes. **RANGE:** Coastal Plain from se. VA through FL and west to e. LA. Remark: Interbreeds with the Three-toed Amphiuma.

ONE-TOED AMPHIUMA *Amphiuma pholeter* **PL. 2**

8½–12½ in. (21.5–32 cm); record 13 in. (33 cm). Back and belly dark gray to black. Head and limbs proportionately shorter and eyes proportionately smaller than in other amphiumas. *One toe* on each limb. **SIMILAR SPECIES:** See Two-toed Amphiuma. **HABITAT:** Deep muck in alluvial swamps, spring runs, occasionally swampy streams in floodplains. **RANGE:** Gulf Coast from n. FL to se. MS.

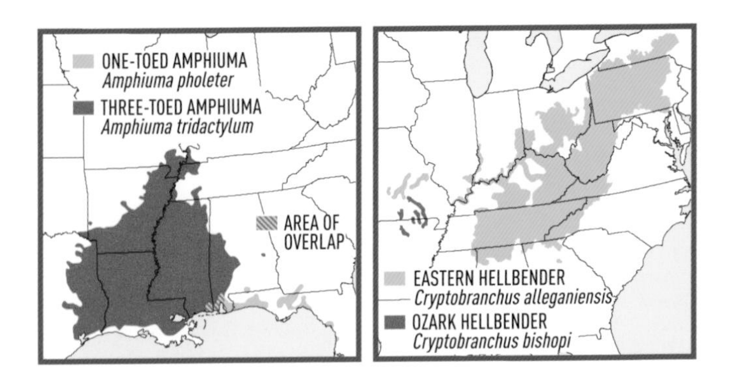

THREE-TOED AMPHIUMA *Amphiuma tridactylum* **PL. 2**

18–30 in. (45.7–76 cm); record 41¾ in. (106 cm). Dorsum dark brown, contrasting sharply with the light gray venter. Dark throat patch. *Three toes* on each limb. **SIMILAR SPECIES:** See Two-toed Amphiuma. **HABITAT:** Oxbows, lakes, ponds, wooded alluvial swamps, calcareous streams, marshes, swampy banks of bayous and cypress sloughs. **RANGE:** Lower Mississippi River Valley east to AL and west to e. TX. **REMARK:** Interbreeds with the Two-toed Amphiuma.

HELLBENDERS: Cryptobranchidae

Two genera of entirely aquatic salamanders occur in the eastern United States and in China and Japan. Adults lose the prominent larval gills but retain a small pair of gill slits. Asian species in the genus *Andrias* are the world's largest living salamanders; individuals to nearly 6 ft. (180 cm) in length have been reported.

EASTERN HELLBENDER *Cryptobranchus alleganiensis* **PL. 2, P. 446**
OZARK HELLBENDER *Cryptobranchus bishopi* **PL. 2**

11–22⅜ in. (28–56.8 cm); records male 27 in. (68.6 cm), female 29⅛ in. (74 cm). Head *flattened*; sides with *wrinkled, fleshy skinfolds*. Usually gray (sometimes yellowish brown to almost black). Eastern Hellbender has small vague dorsal spots and a lower lip spotted lightly or not at all. Ozark Hellbender has conspicuous dark dorsal blotches and a lower lip heavily spotted with black. **SIMILAR SPECIES:** Mudpuppies and waterdogs have external gills. **HABITAT:** Mostly clear, flowing rivers and streams with riffles. **RANGE:** Eastern Hellbender: s.-cen. NY southwest to extreme ne. MS; disjunct populations in e.-cen. MO. Ozark Hellbender: se. MO and adjacent AR. **REMARKS:** These two species sometimes are considered to be conspecific. Two lineages of the Ozark Hellbender are probably distinct species. **CONSERVATION:** Ozark Hellbender (USFWS: Endangered).

LUNGLESS SALAMANDERS: Plethodontidae

About 450 species range from southern Canada to Bolivia, a few species occur in Mediterranean Europe, and a single species occurs on the Korean Peninsula. Nasolabial grooves extend from the nostrils to the upper lip; in some species with cirri, projections of the upper lip below the nostrils, these grooves extend to near the tip of each cirrus.

CLIMBING SALAMANDERS: *Aneides*

One of eight currently recognized species occurs in our area. Six others range from British Columbia to northern Baja California, and one species occurs only in southern New Mexico.

GREEN SALAMANDER

Aneides aeneus **PL. 2**

3¼–5 in. (8.3–12.5 cm); record 5¹⁰⁄₁₆ in. (14.8 cm). Dark, with *green* lichenlike markings. Head and body flattened. Toes *square-tipped*. **HABITAT:** Crevices in rock outcrops and ledges, beneath loose bark, in or under fallen logs; occasionally arboreal. **RANGE:** Appalachia from sw. PA to n. AL and extreme ne. MS.

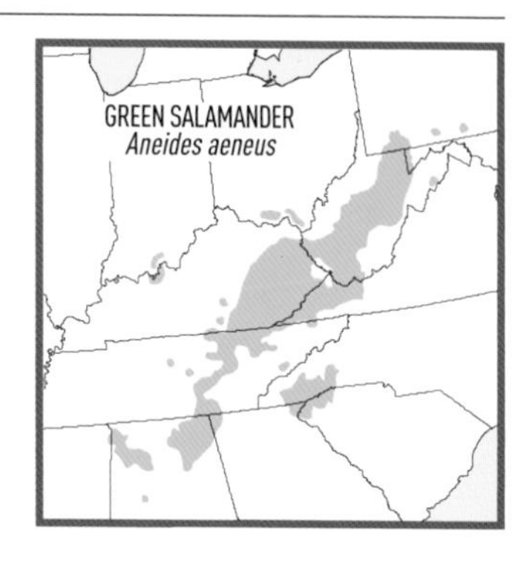

DUSKY SALAMANDERS: *Desmognathus*

All 21 currently recognized species occur in our area. Bodies are generally relatively short and stout, *necks distinctly wider than heads, hindlimbs larger and stouter than forelimbs*, and most species have *pale diagonal lines from each eye to the angle of the jaw*.

Species in closely related groups with contiguous or overlapping ranges might be impossible to distinguish without molecular data. Start by eliminating species not found at a given location, then look at tail shape and body size. Eliminating superficially similar salamanders in other families or genera will help: (1) Mole salamanders lack nasolabial grooves. (2) Woodland salamanders have long, slender bodies, lack light lines from the eyes to the angles of the jaw, and have fore- and hindlimbs about equal in size. (3) The Red Hills Salamander has 20–22 costal grooves (13–15 in dusky salamanders) and no light lines from the eyes to the angles of the jaw.

CUMBERLAND DUSKY SALAMANDER *Desmognathus abditus* **FIGS. 3, 10**

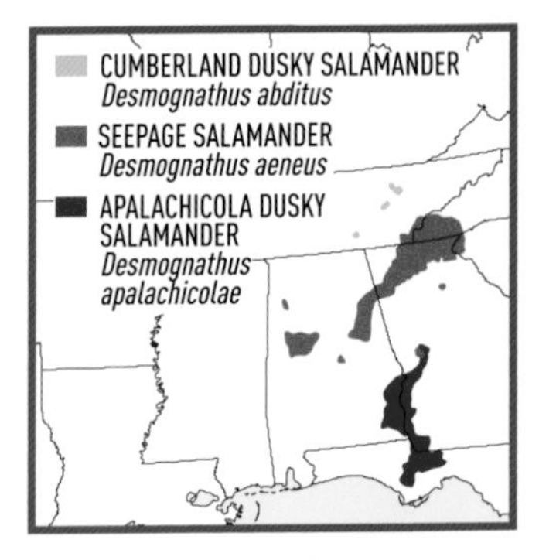

2³/₈–3⁷/₈ in. (6–8 cm). Tail *round, no trace of keel,* about one-half total length. Brown, usually with *8 pairs of reddish or yellowish spots* bordered by dark pigment forming *strongly undulating* dorsolateral stripes (Fig. 3, p. 18; Fig. 10, below). Mouthline of adult male sinuous (Fig. 11, facing page). Costal grooves 14. **SIMILAR SPECIES:** See Allegheny Mountain Dusky Salamander. **HABITAT:** Near streams. **RANGE:** Cumberland Plateau of TN. **REMARK:** Once considered conspecific with the Allegheny Mountain Dusky Salamander.

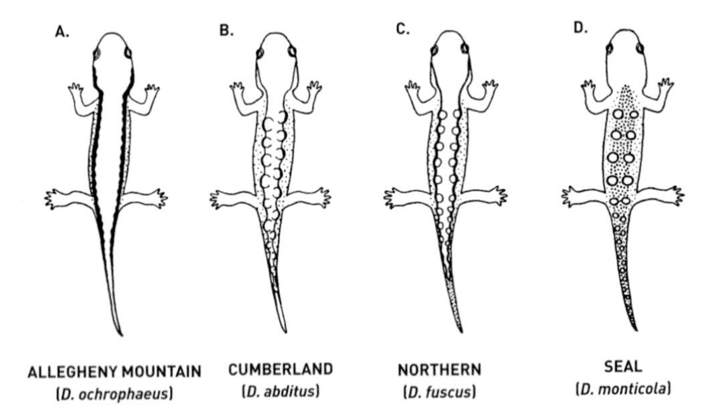

FIG. 10. *Dorsal patterns of young dusky salamanders* (Desmognathus) *are extremely variable.* **A.** *The Allegheny Mountain Dusky Salamander* (D. ochrophaeus) *frequently has a distinct middorsal stripe with nearly straight edges.* **B.** *The Cumberland Dusky Salamander* (D. abditus) *often has rounded spots set in a zigzag fashion (illustrated) or in pairs.* **C.** *The Northern Dusky Salamander* (D. fuscus) *typically has five to eight pairs of spots on the back and more on the tail.* **D.** *The Seal Salamander* (D. monticola) *usually has four pairs of spots on the body and more on the tail.*

FIG. 11. *Mouthlines of adult male dusky salamanders* (Desmognathus). *That of the Allegheny Mountain Dusky Salamander* (D. ochrophaeus; *top*) *and related species* (D. abditus, D. apalachicolae, D. carolinensis, D. imitator, D. ocoee, D. orestes) *is strongly sinuous, whereas that of the Northern Dusky Salamander* (D. fuscus; *bottom*) *and related species* (D. auriculatus, D. conanti, D. planiceps, D. santeetlah) *is less sinuous.*

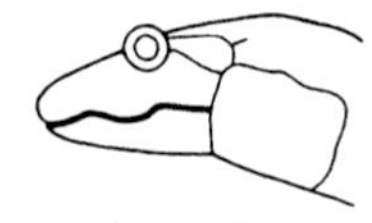

Strongly sinuous

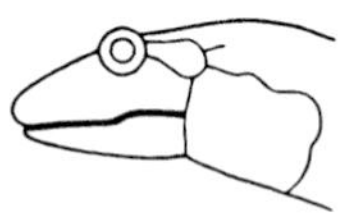

Less sinuous

SEEPAGE SALAMANDER *Desmognathus aeneus* **FIGS. 12, 13**

1¾–2¼ in. (4.4–5.7 cm); record 2½ in. (6.4 cm). Tail *round*, about one-half total length. *Top of head smooth.* Male with a kidney-shaped mental gland (Fig. 13, below). Wide, wavy to almost straight-sided pale dorsal stripe yellow or tan to reddish brown; stripe often flecked or smudged with darker pigment, sometimes suggesting a herringbone pattern; no continuous dark dorsolateral lines, but sometimes the dark middorsal line is continuous with a Y-shaped mark on the head (Fig. 12, below). Sides dark where they meet the dorsal stripe, paler toward the belly. Undersurfaces plain to mottled with brown and white. Costal grooves 13–14. **SIMILAR SPECIES:** Pygmy and Northern Pygmy Salamanders generally have a distinct middorsal herringbone pattern, top of head is rugose, mental gland of male is U-shaped. **HABITAT:** Mixed hardwood forests and shaded ravines near creeks, springs, seepage areas. **RANGE:** Sw. NC to AL.

FIG. 12. *Dorsal patterns in the Seepage Salamander* (Desmognathus aeneus) *compared with that of the Northern Pygmy Salamander* (D. organi) *and the Pygmy Salamander* (D. wrighti).

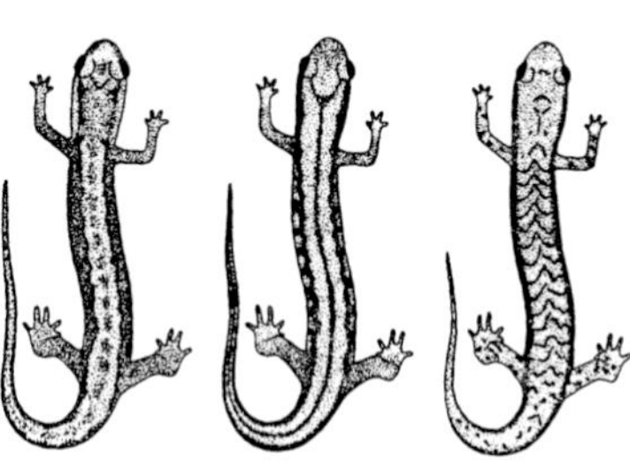

SEEPAGE (*D. aeneus*)
Two variations

PYGMY
(*D. organi* and *D. wrighti*)

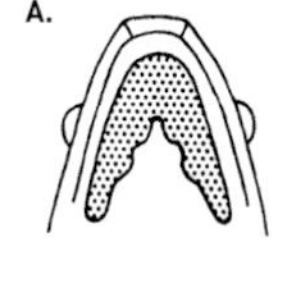

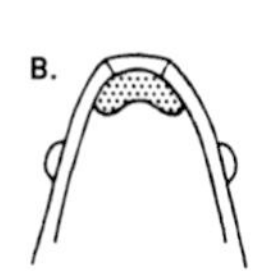

FIG. 13. A. *The Northern Pygmy Salamander* (Desmognathus organi) *and the Pygmy Salamander* (D. wrighti) *have a U-shaped mental gland on the chin.* **B.** *The Seepage Salamander* (D. aeneus) *has a small, kidney-shaped mental gland.*

FIG. 14. *The Apalachicola Dusky Salamander* (Desmognathus apalachicolae) *has a round tail that tapers substantially near the tip.*

APALACHICOLA DUSKY SALAMANDER *Desmognathus apalachicolae* **FIG. 14**

3¼–4 in. (8.3–10 cm). Tail *round*, over one-half total length, tapering to a thin, laterally compressed filament. Occasionally uniformly brown (especially large individuals), usually with 10–14 pairs of indistinct, light, round, and often coalescing dorsal blotches set off by black fringes; belly white or smudged with a thin veneer of dark pigment. Costal grooves 14. **SIMILAR SPECIES:** Southern Dusky Salamander has a black or very dark brown belly sprinkled with white. **HABITAT:** Seepage streams in forested areas, mucky floodplains, bottomland forests. **RANGE:** Apalachicola and Ochlockonee R. drainages north along the AL-GA border.

SOUTHERN DUSKY SALAMANDER *Desmognathus auriculatus* **FIG. 15**

3–5 in. (7.5–12.5 cm); record 6⅜ in. (16.3 cm). Tail stout at base, compressed posteriorly, *knife-edged above*. Dark brown or black, with rows of *widely spaced* whitish or reddish spots between fore- and hindlimbs and along sides of the tail; spots irregular in placement, arranged in double rows, or obscured in dark individuals (Fig. 15, facing page). Belly *black or very dark brown*, sprinkled with distinct white dots. Some individuals (especially in ravine or spring habitats of peninsular FL) have a reddish dorsum that looks somewhat like a light middorsal stripe. Costal grooves 14; 4½–5½ costal folds between adpressed limbs. **SIMILAR SPECIES:** Spotted Dusky Salamander has a lighter belly with dark flecks and is stockier (2–4 costal folds between adpressed limbs). **HABITAT:** Mucky

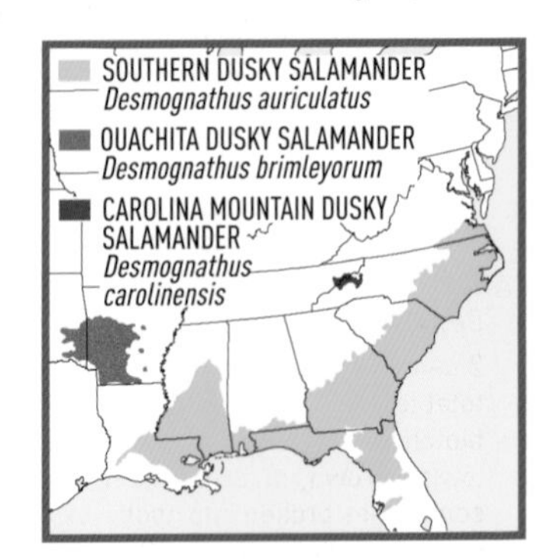

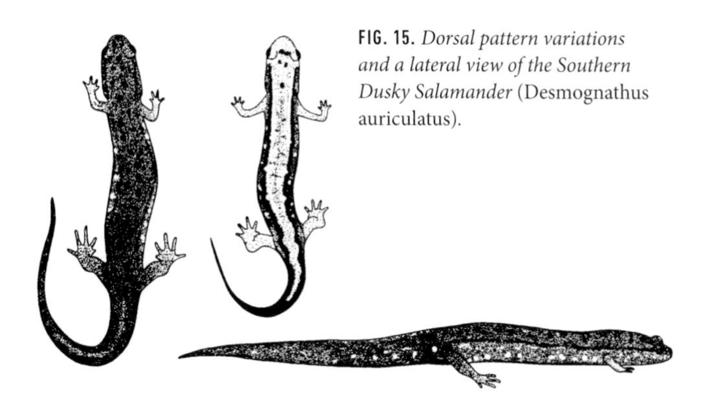

FIG. 15. *Dorsal pattern variations and a lateral view of the Southern Dusky Salamander* (Desmognathus auriculatus).

areas near springs, cypress ponds, or pools in floodplains and coastal swamps. **RANGE:** Coastal Plain from se. VA to cen. FL west to LA. **REMARKS:** Some populations might represent divergent lineages. May interbreed with the Spotted Dusky Salamander.

OUACHITA DUSKY SALAMANDER *Desmognathus brimleyorum* PL. 2, FIG. 16

3⅛–5½ in. (8–14 cm); record 7 in. (17.8 cm). Tail distinctly *keeled*, compressed near tip. Brown or gray, often uniformly dark brown (especially adult male). Belly pinkish white to yellowish, virtually unmarked or faintly stippled with pale brown. Juvenile usually with rows of faint pale spots along sides (Fig. 16, below) and in parallel rows extending from fore- to hindlimbs. **HABITAT:** Rocky, gravelly streams. **RANGE:** S. AR and se. OK.

FIG. 16. *Dorsal view of a young Ouachita Dusky Salamander* (Desmognathus brimleyorum).

CAROLINA MOUNTAIN DUSKY SALAMANDER
Desmognathus carolinensis FIG. 3

2¾–4$^{5}/_{16}$ in. (7–11 cm). Tail *round, without keel*, slightly over one-half total length. Mouthline of adult male sinuous (Fig. 11, p. 41). Dorsum blotched or striped in various shades of greenish gray, brown to yellowish brown, or bright to brownish red. Dark dorsolateral stripes sometimes broken into spots, usually straight but sometimes wavy or strongly undulating (Fig. 3, p. 18). Belly uniformly gray to dark gray.

Adult often melanistic. Costal grooves 14. **SIMILAR SPECIES:** See Ocoee and Blue Ridge Dusky Salamanders. Other dusky salamanders in range have a distinctly keeled tail. **HABITAT:** Under cover near springs, seepage areas, streams; more terrestrial than most dusky salamanders, sometimes wandering far into woodlands. **RANGE:** W.-cen. NC and adjacent ne. TN. **REMARK:** Until recently considered conspecific with the Allegheny Mountain Dusky Salamander.

SPOTTED DUSKY SALAMANDER *Desmognathus conanti* PL. 2

2½–5 in. (6.4–12.7 cm). Tail *compressed, knife-edged above*, higher than wide at posterior edge of vent, slightly less than one-half total length. Mouthline of adult male slightly sinuous (Fig. 11, p. 41). Very dark to light gray or brown; frequently with 6–8 pairs of golden to reddish dorsal spots normally separated, occasionally fused to form a light dorsal band with saw-toothed, wavy, or straight dark margins. Lines from eyes to angles of the jaw yellow or orange. Venter *light*, mottled with black and white (or gold) flecks. Costal grooves 14. **SIMILAR SPECIES:** See Southern and Northern Dusky Salamanders. **HABITAT:** Floodplains, sloughs, and mucky sites along upland streams. **RANGE:** W. SC west to e. TX. **REMARKS:** Once considered a subspecies of the Northern Dusky Salamander. May interbreed with the Southern Dusky Salamander.

DWARF BLACK-BELLIED SALAMANDER *Desmognathus folkertsi* FIG. 17

3⅜–5½ in. (8.5–14 cm); record 6 in. (15.3 cm). Tail stout at base, less than one-half total length, *knife-edged above*. Dorsum with brown and black vermiculate markings or irregular or alternating brown blotches. Adult *belly black*, dark but flecked with yellow in young. Juvenile usually with *double rows of light dots* along sides. **SIMILAR SPECIES:** See Black-bellied Salamander. **HABITAT:** Boulder-strewn streams or where cold water drips or flows. **RANGE:** N. GA and adjacent sw. NC.

FIG. 17. *The Dwarf Black-bellied Salamander* (Desmognathus folkertsi) *is characterized by dorsal vermiculations or dark dorsal blotches. This species was until recently considered conspecific with the Black-bellied Salamander* (D. quadramaculatus).

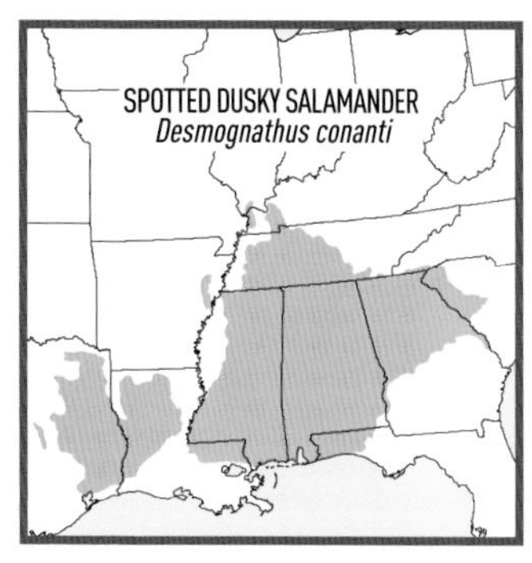

NORTHERN DUSKY SALAMANDER *Desmognathus fuscus* **PL. 3, FIG. 10**

2½–4½ in. (6.4–11.5 cm); record 5⁹⁄₁₆ in. (14.1 cm). Tail *keeled*, slightly less than one-half total length, *compressed, knife-edged above*, higher than wide at posterior edge of vent. Mouthline of adult male slightly sinuous (Fig. 11, p. 41). Gray or brown, with markings often barely darker than ground color. Very young with pairs of round yellowish dorsal spots bordered by dark *wavy* bands (Fig. 10, p. 40) continuing onto tail. Pattern broken in older animals, darker remnants appearing as spots or streaks. Base of tail usually lighter (olive, yellowish, or bright chestnut) than rest of dorsum and bordered by dark scallops. Venter usually lightly (sometimes heavily) mottled with gray or brown; dark pigment on sides may encroach onto edges of the belly. Costal grooves 14. **SIMILAR SPECIES:** (1) Spotted Dusky Salamander is more colorful (frequently with 6–8 pairs of golden or reddish dorsal spots, lines from eyes to angles of jaw often yellow or orange), but both species are highly variable and many individuals in the contact zone will be indistinguishable. (2) Santeetlah Dusky Salamander has a slightly less keeled tail, yellowish wash on undersides of limbs and tail, and generally occurs at higher elevations. (3) Flat-headed Salamander differs in tooth structure and is best distinguished by range. (4) Seal Salamander is larger, usually has very dark dorsal spots, a pale, uniformly colored or lightly mottled venter, dorsal and ventral colors sharply demarcated, and dark cornified toetips. (5) Black Mountain Salamander has pale, uniformly colored or lightly mottled venter and dark cornified toetips. (6) Black-bellied Salamander is larger, adult with a black belly, and young with 2 rows of conspicuous light dots on sides.

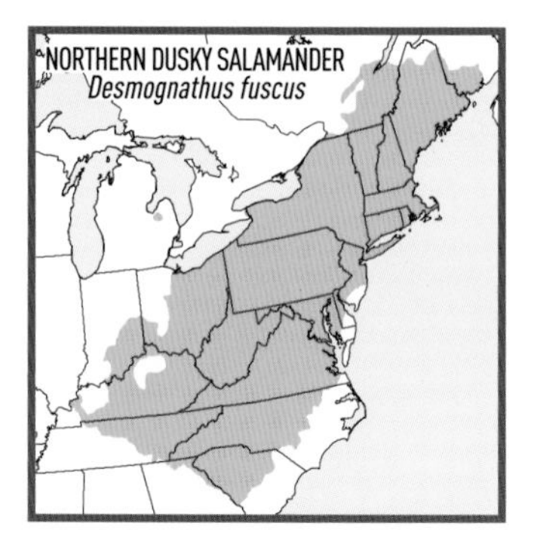

(7) Other dusky salamanders in range have a round tail and the mouthline of adult males is sinuous. **HABITAT:** Rock-strewn woodland streams, seepages, springs in northern portions of range; floodplains, sloughs, and mucky sites along upland streams farther south. **RANGE:** S. NB south to SC and west to w. KY; isolated population in MI. **REMARKS:** Until recently considered conspecific with Spotted Dusky, Flat-headed, and Santeetlah Dusky Salamanders; interbreeds with the Santeetlah Dusky Salamander.

IMITATOR SALAMANDER *Desmognathus imitator* **PL. 2, FIG. 18**

2½–4 in. (6.4–10 cm); record 4⅜ in. (11 cm). Yellow, orange, or red *cheek patches*. Usually with strongly undulating, often interrupted, dorsolateral dark stripes but no broad middorsal stripe. Old adults often black, patternless, and lacking cheek patches. Tail *round*; belly gray; suggestion of *dark* (occasionally light) lines from eyes to angles of the jaw. Costal grooves 14. **SIMILAR SPECIES:** (1) Other dusky salamanders with a round tail; no red, orange, or yellow cheek patches; lines from eyes to angles of the jaw are light; and, if patterned, they often have a middorsal stripe or a series of blotches merging to form a line. (2) Woodland salamanders, including the Red-cheeked Salamander, have a narrow neck (narrower or barely wider than head), lack enlarged hindlimbs, and do not have lines from eyes to angles of the jaw. **HABITAT:** Small streams, wet rock faces, seepages, springs. **RANGE:** W. NC and adjacent e. TN.

FIG. 18. *As implied by its name, the Imitator Salamander* (Desmognathus imitator) *"imitates" a Red-cheeked Salamander* (Plethodon jordani) *in having yellow, orange, or red cheek patches.*

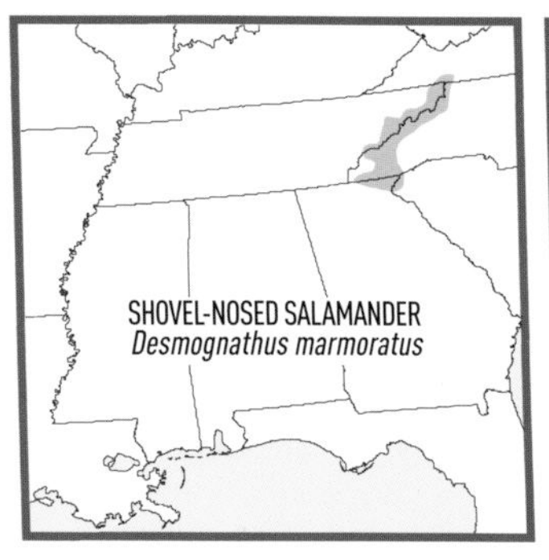

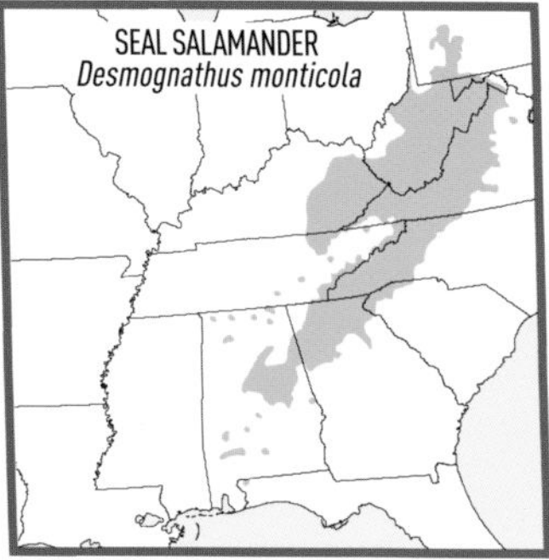

SHOVEL-NOSED SALAMANDER *Desmognathus marmoratus* **PL. 3**

3½–5 in. (9–12.5 cm); record 5¾ in. (14.6 cm). Tail *compressed, knife-edged* above, less than one-half total length. Black, brown, or gray, usually with 2 rows of irregular, often weakly indicated, pale gray, olive, yellowish, or whitish spots or blotches varying in size, intensity, and color. Lines from eyes to angles of the jaw often poorly defined. Belly dark gray (never intense black). Head flattened and wedge-shaped, slope starting downward from *well behind* the small eyes. Internal openings of nostrils (internal nares or choanae) slitlike and scarcely noticeable (Fig. 4, p. 20). **SIMILAR SPECIES:** Adult Black-bellied Salamander has an intensely black belly; it and other dusky salamanders have larger, more prominent eyes, more conspicuous light lines from eyes to angles of the jaw, and internal openings of nostrils are rounded and clearly evident. **HABITAT:** Highly aquatic in rocky mountain brooks. **RANGE:** Sw. VA to n. GA. **REMARKS:** Almost certainly a complex of species. Previously placed in the genus *Leurognathus*.

SEAL SALAMANDER *Desmognathus monticola* **PL. 3, FIG. 10**

3–5 in. (7.5–12.5 cm); record 5⅞ in. (14.9 cm). Stout-bodied. Tail *compressed*, approximately one-half total length, *knife-edged* above, especially near tip. Eyes relatively large. Buff, gray, or light brown, with prominent black or dark brown markings ranging from vermiculate or reticulate and roughly circular rings to scattered dark or light spots or streaks; sometimes a *single* lateral row of light dots between legs. Old adults often purplish brown, with few obscure dark markings. Patterns usually more evident in western parts of the range, least obvious farther south, where adults are virtually unicolored. Distinct separation between dorsal and ventral coloration. Juvenile with about 4 pairs of rounded chestnut or orange-brown dorsal spots (Fig. 10, p. 40). Juvenile belly white; light gray or brown, usually *uniformly* pigmented or lightly mottled in old adults. Toetips *dark and cornified* in all but very young. Costal grooves 14. **SIMILAR SPECIES:** Other dusky salamanders in range have a belly at least partly but distinctly mottled or entirely black, many have a round tail, and most lack cornified toetips. **HABITAT:** Mountain streams, seepages, often muddy areas. **RANGE:** Sw. PA southwest to AL and extreme w. FL.

ALLEGHENY MOUNTAIN DUSKY SALAMANDER

PL. 2, FIGS. 3, 10

Desmognathus ochrophaeus

2¾–4 in. (7–10 cm); record 4⅜ in. (11.1 cm). Tail *round*, about one-half total length, as wide or wider than height at posterior edge of vent. Mouthline of adult male sinuous (Fig. 11, p. 41). Typically with a middorsal row of dark, often *chevronlike* spots and a *straight-edged* light stripe down the back and tail (Fig. 3, p. 18; Fig. 10, p. 40); stripe yellow, orange, olive, gray, tan, brown, or reddish, flanked by very dark pigment that usually fades into mottled lower sides. Old individuals nearly plain dark brown and virtually patternless. Costal grooves 14. **SIMILAR SPECIES:** (1) Cumberland Dusky Salamander usually has distinct reddish or yellowish spots bordered by dark pigment that forms strongly undulating dorsolateral stripes. (2) Blue Ridge Dusky Salamander sometimes has wavy dorsolateral stripes, but individuals with straight-edged stripes are very difficult to distinguish without relying on geographic or molecular data. (3) Imitator Salamander usually has bright cheek patches and dark lines from eyes to angles of the jaw. (4) Other dusky salamanders in range have a tail that is knife-edged above. **HABITAT:** Springs, seepage areas, streams, wet rock faces; sometimes far from water. **RANGE:** Extreme s. QC south to TN.

OCOEE SALAMANDER *Desmognathus ocoee*

PL. 2

2¾–4 5/16 in. (7–11 cm). Tail *round*, about one-half total length, often with a low keel, especially toward the tip. Mouthline of adult male sinuous (Fig. 11, p. 41). Light to very dark brown, usually with 4–6 pale spots fusing to varying degrees to form a wavy to straight-edged brown to grayish green to yellow or red middorsal stripe. Occasionally with red legs. Belly uniformly light to dark gray. Juvenile usually with a distinct wavy red or brown middorsal stripe. Costal grooves 14. **SIMILAR SPECIES:** (1) Carolina Mountain Dusky Salamander has a longer tail (slightly more than one-half total length) but is difficult to distinguish without using geographic or molecular data. (2) Cumberland Dusky Salamander has a round tail with no trace of a keel (Ocoee Salamanders on the Cumberland Plateau almost always have a distinct keel near the tailtip) and generally have distinct reddish or yellowish spots bordered by dark pigment that forms strongly undulating dorsolateral stripes. (3) Imitator Salamander lacks a middorsal stripe and usually has bright cheek patches and dark lines from eyes to angles of the jaw. (4) Other dusky salamanders in range have a tail that is knife-edged above. **HABITAT:** Low gorges to high mountaintops; near springs, seepage areas, streams, also wet rock faces; sometimes far from water. **RANGE:** Disjunct; w. NC and TN south into

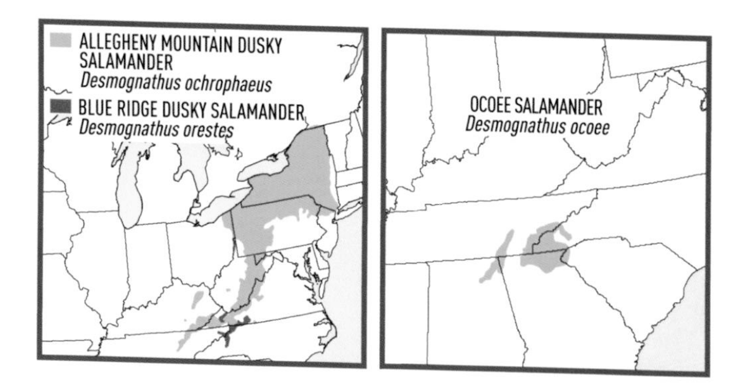

GA and nw. SC; also Cumberland and Appalachian Plateaus of TN and ne. AL. **REMARKS:** Until recently considered conspecific with the Allegheny Mountain Dusky Salamander. Might comprise at least two species.

BLUE RIDGE DUSKY SALAMANDER *Desmognathus orestes* NOT ILLUS.

2¾–4⁵⁄₁₆ in. (7–11 cm). Tail *round, without keel*, slightly over one-half total length. Mouthline of adult male sinuous (Fig. 11 p. 41). Brown, with a brown, yellow, or red straight-edged middorsal stripe or a wavy-edged series of blotches. Belly uniformly gray to blackish gray. Large adults often melanistic. Juvenile with a distinct pattern. Costal grooves 14. **SIMILAR SPECIES:** Allegheny Mountain Dusky Salamander has a shorter tail (about one-half total length), but that species and the Carolina Mountain Dusky Salamander are essentially impossible to distinguish without geographic or molecular data. **HABITAT:** See Allegheny Mountain Dusky Salamander. **RANGE:** Disjunct; sw. VA south into TN and NC. **REMARK:** Might comprise at least two species.

NORTHERN PYGMY SALAMANDER *Desmognathus organi* FIGS. 12, 13

1½–2 in. (3.8–5.1 cm). Tail round, less than one-half total length. Broad reddish brown to tan middorsal stripe usually with a dark median *herringbone* pattern (Fig. 12, p. 41). Venter pinkish and unmarked, although often with scattered gold pigment cells. Top of head (especially snout) *rugose*; mental gland of male U-shaped (Fig. 13, p. 41). **SIMILAR SPECIES:** See Pygmy Salamander. **HABITAT:** Spruce-fir and hardwood forests, usually near seepage areas. **RANGE:** Extreme sw. VA south to the French Broad River Valley in w. NC. **REMARK:** Until recently considered conspecific with the Pygmy Salamander.

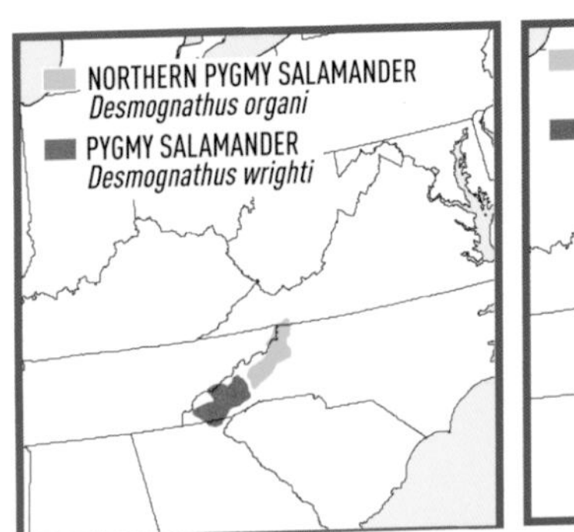

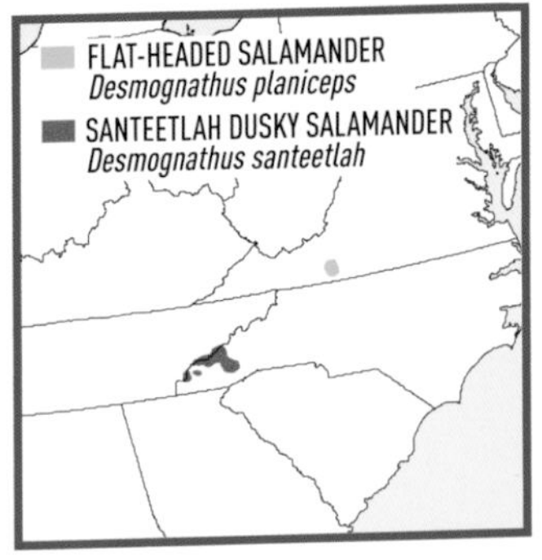

FLAT-HEADED SALAMANDER *Desmognathus planiceps* **FIG. 19**

2½–4½ in. (6.4–11.5 cm). Essentially identical to the Northern Dusky Salamander, except for tooth structure (evident only with dissection and magnification). Best distinguished by range. **SIMILAR SPECIES:** See Northern Dusky Salamander. **HABITAT:** Rock-strewn woodland streams, seepages, springs. **RANGE:** S. VA. **REMARK:** Until recently considered conspecific with the Northern Dusky Salamander.

BLACK-BELLIED SALAMANDER *Desmognathus quadramaculatus* **PL. 3**

4–6⅞ in. (10–17.5 cm); record 8¼ in. (21 cm). Tail very stout at base, less than one-half total length, *knife-edged* above. Brown or dark green with scattered black patches. Adult *belly black*, dark but flecked with yellow in young. Juvenile usually with a conspicuous *double row of light dots* along sides (less evident in adult), reddish stripe on top of tail, and light snout and feet, especially in southern parts of the range. Costal grooves 14. **SIMILAR SPECIES:** (1) Dwarf Black-bellied Salamander is much smaller, with a brown and black vermiculate pattern or irregular or alternating brown and black dorsal blotches. (2) Shovel-nosed Salamander has a long sloping snout, small eyes, slitlike internal nares, and a dark gray (never black) belly. (3) Other dusky salamanders tend to have a lighter belly, but old adults may be almost uniformly dark (belly included); all, however, are smaller and have a proportionately longer tail. **HABITAT:** Boulder-strewn streams or places where cold water drips or flows.

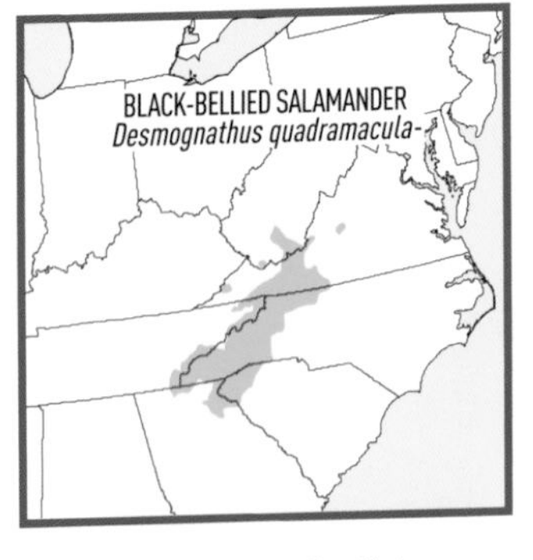

FIG. 19. *The Flat-headed Salamander* (Desmognathus planiceps) *is very similar and closely related to the Northern Dusky Salamander* (D. fuscus). *These two species and their close relatives* (D. auriculatus, D. conanti, D. santeetlah) *all have a laterally compressed tail.*

RANGE: S. WV and w. VA south to n. GA. **REMARKS:** Almost certainly a complex of species. Until recently considered conspecific with the Dwarf Black-bellied Salamander.

SANTEETLAH DUSKY SALAMANDER *Desmognathus santeetlah* **NOT ILLUS.**

2½–3¾ in. (6.4–9.5 cm). Tail *weakly keeled*, slightly less than one-half total length, *compressed and knife-edged above*, higher than wide at posterior edge of vent. Mouthline of adult male slightly sinuous (Fig. 11, p. 41). Light brown to yellowish or greenish brown, pattern subdued, often little more than faint mottling, occasionally with small dark-bordered red spots. Faint dorsolateral stripes sometimes present, although usually thin and often interrupted. Belly light gray, mottled, and *washed with yellow*, as are sides, which exhibit a salt-and-pepper effect and usually have a single ventrolateral row of light spots between fore- and hindlimbs. Lines from eyes to angles of the jaw yellow. Juvenile usually with ten dorsal spots between fore- and hindlimbs. Costal grooves 14. **SIMILAR SPECIES:** See Northern Dusky Salamander. **HABITAT:** Stream headwaters, seepage areas. **RANGE:** Higher elevations of w. NC and adjacent TN. **REMARKS:** Until recently considered conspecific with Northern Dusky Salamander; hybrids are known.

FIG. 20. *Most Black Mountain Salamanders* (Desmognathus welteri) *have irregular, sometimes faint, dark dorsal markings.*

BLACK MOUNTAIN SALAMANDER *Desmognathus welteri* **FIG. 20**

3–6 in. (7.5–15.2 cm); record 6$^{11}/_{16}$ in. (17 cm). "Chubby." Tail stout at base but posteriorly *compressed and knife-edged* above. Pale to medium brown, with small dark brown spots or streaks sometimes in irregular circles. *No sharp lateral separation* of dorsal and ventral coloration. Lateral lines of faint pale dots between fore- and hindlimbs. Belly whitish, usually finely to heavily stippled with gray or brown. Toetips *dark and cornified*, except in very young. Costal grooves 14. **SIMILAR SPECIES:** Seal Salamander usually has distinct lateral separation between dorsal and ventral coloration and a lightly pigmented venter. **HABITAT:** Mountain brooks, spring runs. **RANGE:** E. KY and s. WV south into TN.

PYGMY SALAMANDER *Desmognathus wrighti* **PL. 3, FIGS. 12, 13**

1½–2 in. (3.8–5.1 cm). Tail round, less than one-half total length. Broad reddish brown to tan middorsal stripe usually with a dark median *herringbone* pattern (Fig. 12, p. 41). Venter pinkish and unmarked except for gold pigment cells arranged in an *elliptical pattern* extending from between the forelimbs to just anterior to the vent. Top of head (especially snout) *rugose*; mental gland of male U-shaped (Fig. 13, p. 41). Costal grooves 13–14. **SIMILAR SPECIES:** (1) Northern Pygmy Salamander is very similar and best distinguished by range; however, gold pigment cells are scattered or absent. (2) Seepage Salamander generally lacks a middorsal herringbone pattern, top of head is smooth, and mental gland of male is kidney-shaped. **HABITAT:** Spruce-fir and hardwood forests, frequently near seepage areas.

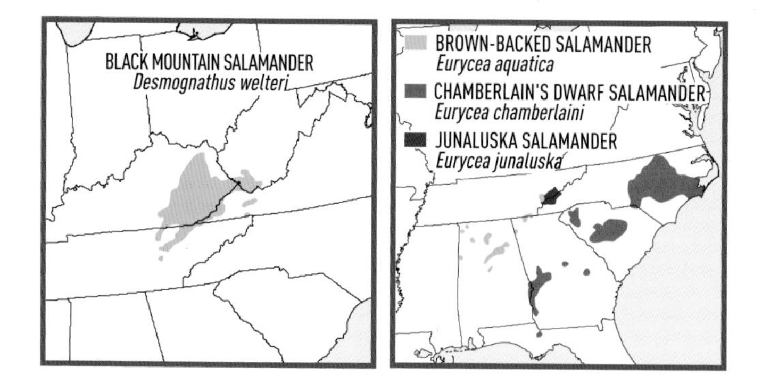

RANGE: South of the French Broad River Valley in w. NC and adjacent TN and GA. **REMARK:** Until recently considered conspecific with the Northern Pygmy Salamander.

BROOK SALAMANDERS: *Eurycea*

All 28 currently recognized species are endemic to our area. Some are neotenic, several frequently occur in caves, and a few cave specialists are blind (eyes rudimentary and covered by skin). Breeding males of many species have well-developed cirri (downward projections below the nostrils) bearing extensions of the nasolabial grooves. Many males also have an enlarged mental gland. These salamanders usually are near water, but some range far into surrounding woodlands during wet weather.

BROWN-BACKED SALAMANDER *Eurycea aquatica* PL. 3

2½–3⅝ in. (6.4–9.2 cm). Brown, somewhat stocky, short-tailed (tail about one-half total length). A *dark-bordered light brown middorsal band* extends from head to tailtip, often with scattered dark specks and faint median lines frequently broken into series of dots and dashes. Sides darker than back. Belly yellow, virtually unmarked. Costal grooves 13–14; 2–4 costal grooves between adpressed limbs. **HABITAT:** Near springs and small rocky streams. **RANGE:** Disjunct; cen. AL northeast to e. TN.

NORTHERN TWO-LINED SALAMANDER *Eurycea bislineata* PL. 3

2½–3¾ in. (6.4–9.5 cm); record 4¾ in. (12.1 cm). Yellowish (sometimes tinged brownish, greenish, bronzy, or bordering on orange), relatively long-tailed (55–60 percent total length). The *broad, light middorsal stripe bordered by 2 dark lines* generally unbroken, but sometimes reduced to dots or dashes near the tailtip. Middorsal stripes usually with small black spots, sometimes forming a narrow median line. Sides mottled and as dark as dorsolateral lines or barely

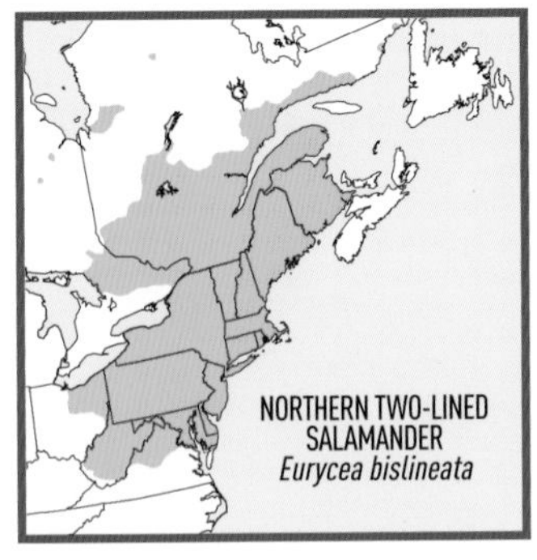

darker than the middorsal stripe; belly yellow. Costal grooves 15–16. **SIMILAR SPECIES:** (1) Southern Two-lined Salamander is best distinguished by range but usually has only 14 costal grooves. (2) Three-lined and Long-tailed Salamanders are larger and have 13–14 costal grooves. (3) Cave Salamander is larger, lacks dorsolateral lines, and generally has a longer tail (about 60 percent total length). **HABITAT:** Rocky brooks, springs, seepages, surrounding saturated areas. **RANGE:** QC south to WV and VA.

CHAMBERLAIN'S DWARF SALAMANDER *Eurycea chamberlaini* NOT ILLUS.

2–2¾ in. (5–7 cm). *Four toes on all limbs.* Yellowish to light brown, uniformly colored or with dark, occasionally broken dorsolateral stripes bordering a middorsal area that is unmarked or with scattered dark flecks, faint herringbone pattern, or a median line sometimes broken into a series of spots. Sides often with scattered light flecks or streaks. Belly yellow and *unmarked*, chin and throat plain or lightly pigmented, underside of tail often orange or orange-yellow. Costal grooves 15–16, usually 16; 3–6 (usually 4) costal folds between adpressed limbs. **SIMILAR SPECIES:** See Dwarf Salamander. **HABITAT:** Seepage areas near streams or ponds, damp woodlands, lowland swamps. **RANGE:** Disjunct in the Piedmont and Coastal Plain of the Carolinas, also GA and extreme e. AL. **REMARK:** Until recently considered a yellow variant of the Dwarf Salamander.

SOUTHERN TWO-LINED SALAMANDER *Eurycea cirrigera* FIG. 21

2½–3¾ in. (6.4–9.5 cm); record 4⁵⁄₁₆ in. (11 cm). Like the Northern Two-lined Salamander but with 14 costal grooves. **SIMILAR SPECIES:** (1) Blue Ridge Two-lined Salamander is best distinguished by range but is more yellowish or orange and dorsolateral lines generally break up about halfway down the tail. (2) Brown-backed and Junaluska Salamanders have a shorter tail (about one-half total length). See also Northern Two-lined Salamander. **HABITAT:** Rocky brooks, springs, seepages, river swamps, forested floodplains with stagnant

FIG. 21. *As in many other brook salamanders, the breeding male Southern Two-lined Salamander* (Eurycea cirrigera) *has nasolabial grooves that extend on cirri beyond the lip.*

pools. **RANGE:** VA to n. FL west to e. IL and LA; isolated populations in MI, ne. IL, and n. FL.

THREE-LINED SALAMANDER *Eurycea guttolineata* PL. 4

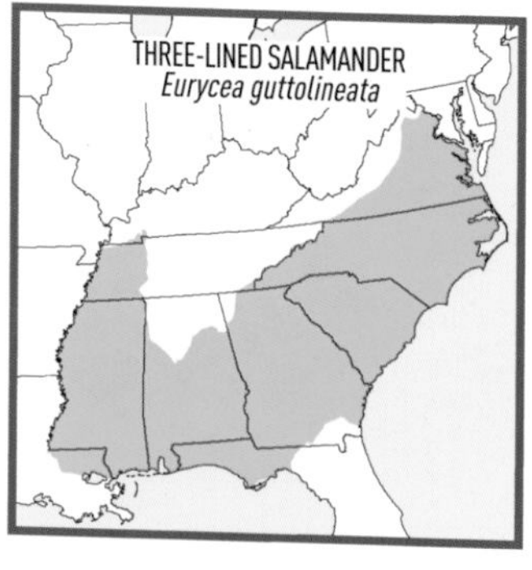

4–6¼ in. (10–15.9 cm); record 7⅞ in. (20 cm). Yellow, tan, brownish, or orangish. Long-tailed (often two-thirds total length). *Three dark longitudinal stripes*; middorsal stripe extending from behind head onto base of tail, often broken into a series of elongated dark spots. Belly cream to dull yellow, mottled with greenish gray. Costal grooves 13–14. **SIMILAR SPECIES:** See Long-tailed Salamander. **HABITAT:** Forested floodplains, ditches, damp streamsides, seepage springs. **RANGE:** VA to the FL Panhandle and west to the Mississippi R.

JUNALUSKA SALAMANDER *Eurycea junaluska* FIG. 22

3–4 in. (7.5–10 cm). Relatively short-tailed (about one-half total length). Yellowish brown, with small irregular dark brown spots on back and sides, those on the latter forming wavy, frequently broken

FIG. 22. *The Junaluska Salamander* (Eurycea junaluska), *like many amphibians, is sensitive to alterations of aquatic habitats. Already small populations are potentially threatened by siltation due to logging, urban development, and other activities negatively affecting water quality.*

dorsolateral lines. Costal grooves 14. Limbs long, *no more than one costal groove* between adpressed limbs. **SIMILAR SPECIES:** See Northern and Southern Two-lined Salamanders. **HABITAT:** In and along streams. **RANGE:** W. NC and e. TN. **CONSERVATION:** Vulnerable.

LONG-TAILED SALAMANDER *Eurycea longicauda* **PL. 4**

3⅝–6¼ in. (9.2–15.9 cm); record 7½ in. (19.7 cm). Long-tailed (often two-thirds total length). Yellow (sometimes tinged with orange or brown), with numerous black spots on back and sides; *vertical black markings* on sides of tail; tail markings vary in intensity but usually conspicuous, frequently forming a herringbone pattern or even coalescing into almost solid dark bands. Belly yellowish, unmarked. Young yellow, with a relatively short tail. Costal grooves 13–14. **SIMILAR SPECIES:** Three-lined Salamander is sometimes similar, but ranges generally do not overlap. **HABITAT:** Streamsides, spring runs, cave mouths, abandoned mines, even ponds in n. NJ. **RANGE:** S. NY southwest to ne. OK. **SUBSPECIES:** (1) **Eastern Long-tailed Salamander** (*E. l. longicauda*); spots on sides often form dark, broken dorsolateral lines, but those on the back do not. (2) **Dark-sided Salamander** (*E. l. melanopleura*); spots on sides of body and tail coalesce to form dark lateral stripes that border the light middorsal stripe that is marked with scattered spots that sometimes form 1–2 irregular median or paramedian lines; dark pigment on sides grayish in juvenile, deep reddish brown in adult; middorsal light areas bright yellow in juvenile, dull brownish or even greenish yellow in adult; often associated with caves in the Central Highlands of MO, AR, and OK.

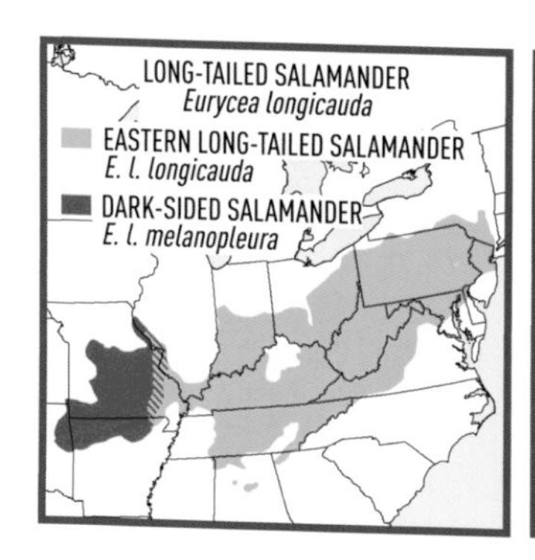

CAVE SALAMANDER *Eurycea lucifuga* **PL. 4**

4–6 in. (10–15.2 cm); record 7⅛ in. (18.1 cm). Orange to reddish orange. Head wide and flat, tail long (about 60 percent total length). *Profusion of black spots* usually scattered but sometimes forming 2–3 irregular longitudinal rows. Belly white to yellowish, normally unmarked. Juvenile yellowish with a proportionately shorter tail. Costal grooves 14–15. **HABITAT:** Caves, especially twilight zones, rocky streams and springs, wooded areas and fields, usually near caves or limestone outcrops. **RANGE:** W. VA to e. OK.

MANY-RIBBED SALAMANDER *Eurycea multiplicata* **PL. 3**

2½–3¼ in. (6.4–8.3 cm); record 3 9/16 in. (9 cm). Yellowish to brownish; sides somewhat darker than middorsal area and often with a row of faint light spots. Rows of darker spots form broken, sometimes faint dorsolateral lines; dark spots or V-shaped marks may form a broken, often faint middorsal line. *Undersurfaces bright yellow.* Costal grooves 19–20. **SIMILAR SPECIES:** See Oklahoma Salamander. **HABITAT:** Cold, clear streams, often cave springs and runs. **RANGE:** Sw. AR and se. OK, mostly south of the Arkansas R. **REMARK:** Two or three lineages might represent distinct species.

DWARF SALAMANDER *Eurycea quadridigitata* **PL. 3**

2⅛–3 in. (5.4–7.5 cm); record 3 9/16 in. (9 cm). *Four toes on all limbs.* Brown to light brown or yellowish buff, uniformly colored or with dark, occasionally broken dorsolateral stripes bordering a middorsal area that is unmarked or bears scattered dark flecks, a faint herringbone pattern, or a median line sometimes broken into a series of spots. Sides often with scattered light flecks or streaks. Belly gray or silvery with some dark pigmentation; chin and throat moderately to heavily pigmented. Costal grooves 16–18. **SIMILAR SPECIES:** (1) Chamberlain's Dwarf Salamander has an unmarked yellow belly and unmarked or lightly pigmented chin and throat; also, at least in areas where both

species occur, usually with 16 or fewer costal grooves and 4 (rarely 5) costal folds between adpressed limbs. (2) Four-toed Salamander has a white belly with black spots. All other salamanders in range have 5 toes on hindlimbs. **HABITAT:** Low swampy areas, margins of ponds, bottomland forest. **RANGE:** E. NC through FL and west to e. TX. **REMARK:** Light variants in AL and LA, similar in appearance to Chamberlain's Dwarf Salamander, probably represent another species.

GROTTO SALAMANDER *Eurycea spelaea* **PL. 3**

3–4¾ in. (7.5–12.1 cm); record 5 5/16 in. (13.5 cm). Adult whitish or pinkish, sometimes with faint traces of orange on lower sides, tail, and feet. *Eyes show as dark spots* beneath fused or partly fused eyelids. Larva brownish or purplish gray, with yellowish longitudinal flecks or dark streaks on sides, functional eyes, external gills, *high tail fin*. Costal grooves 16–19, usually 17. **SIMILAR SPECIES:** Only troglobitic salamander of the Ozark Plateau. Larvae of several brook salamanders might co-occur in cave waters or spring runs, but all have low tail fins and patterns with stippling or lichenlike patches. **HABITAT:** Adult inhabits caves; larva also in surface spring runs. **RANGE:** Ozark Plateau of sw. MO, extreme se. KS, and adjacent AR and OK. **REMARKS:** At least three lineages probably are deserving of recognition as distinct species. Previously assigned to the genus *Typhlotriton*.

OUACHITA STREAMBED SALAMANDER *Eurycea subfluvicola* **FIG. 23**

2–2 11/16 in. (5.1–6.8 cm). Neotenic. Very slender, head flattened, eyes small. Brown (dark brown blotches on amber to yellow ground color); lighter dorsolaterally. *Semitransparent undersurface* unpigmented, viscera or eggs show through the body wall, few dark specks under the tail. Dorsal and ventral coloration sharply demarcated. *External gills*; tail fin *low*. Costal grooves 21–22. **SIMILAR SPECIES:** Larval Many-ribbed Salamander has a longer, wider, less flattened head with larger eyes, a shorter body, and lateral black stripes on snout

FIG. 23. *The Ouachita Streambed Salamander* (Eurycea subfluvicola) *was until recently confused with the larval Many-ribbed Salamander* (E. multiplicata). *The eggs evident through the translucent ventral skin clearly show that this is a mature adult female despite its larval configuration.*

and head. **HABITAT:** Clear, cool streams with subsurface flow. **RANGE:** Slunger Creek Valley, se. Ouachita Mts., AR.

OKLAHOMA SALAMANDER *Eurycea tynerensis* **PL. 3**

1¾–3¼ in. (4.4–8.3 cm); record 4$^{1}/_{16}$ in. (10.2 cm). Some populations neotenic. Transformed adult gray or grayish or yellowish brown, essentially unicolored or with sides somewhat darker than back. Dark spots or V-shaped or chevronlike marks may form a broken, often faint middorsal line; often with lateral rows of faint light spots. *Undersurfaces gray to light gray*, usually with dark specks or patches. Larva and neotenic adult grayish (caused by heavy black stippling and streaking on cream ground color); dark pigment uniformly distributed over dorsum or densest on sides, leaving a broad light middorsal stripe; usually at least one lateral row of small light spots (as many as 3 in small individuals); belly pale and unmarked, except where viscera or eggs show through the body wall. Neotenic adult with *external gills*; tail fin *low*. Costal grooves 19–21, usually 20. **SIMILAR SPECIES:** (1) Grotto Salamander larva has a high tail fin and longitudinal lateral streaks. (2) Larvae of other brook salamanders with low tail fins occasionally co-occur with the Oklahoma Salamander but have 16 or fewer costal grooves. **HABITAT:** Cold, clear streams, often cave springs and runs. Neotenic populations usually in streams with gravel substrates, populations with transforming adults usually in streams with bedrock substrates. **RANGE:** Sw. MO and ne. OK south to the Arkansas R. **REMARK:** Transforming populations until recently were considered conspecific with the Many-ribbed Salamander.

FIG. 24. *As in many other cave-adapted salamanders, the eyes of the adult Georgia Blind Salamander* (Eurycea wallacei) *are covered by skin. This species was until recently thought to be the only species in the genus* Haideotriton.

GEORGIA BLIND SALAMANDER *Eurycea wallacei* **FIG. 24, P. 14**

1½–3 in. (4–7.6 cm). Neotenic. Pinkish white, slightly opalescent, with widely scattered dark spots on head and body. Belly unmarked, internal organs visible. External gills *long, slender, red*. Head long, broad, flattened, somewhat shovel-shaped. Eyes of juvenile tiny dark spots, those of adult barely visible under translucent skin. Tail fin *high*. **HABITAT:** Cave pools, streams, submerged passages. **RANGE:** Sw. GA and adjacent FL. **REMARK:** Previously assigned to the genus *Haideotriton*. **CONSERVATION:** Vulnerable.

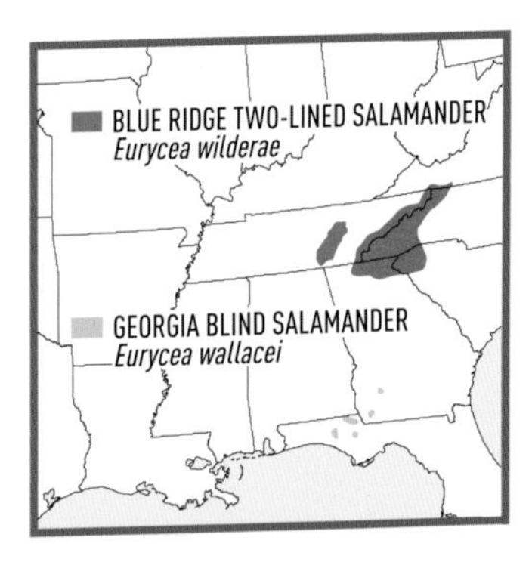

BLUE RIDGE TWO-LINED SALAMANDER *Eurycea wilderae* **PL. 3**

2¾–4¼ in. (7–10.7 cm); record 4¾ in. (12.1 cm). Bright yellow, yellowish orange, or orange. Relatively long-tailed (55–60 percent total length). Black dorsolateral lines usually *break into dots or blotches at about middle of tail*. Back often with scattered black spots. Belly yellow. Costal grooves 14–16 (generally 14 at lower elevations). **SIMILAR SPECIES:** See Northern and Southern Two-lined Salamanders. **HABITAT:** Rocky brooks, springs, seepages. **RANGE:** Sw. VA to n. GA; disjunct population on the Cumberland Plateau of TN.

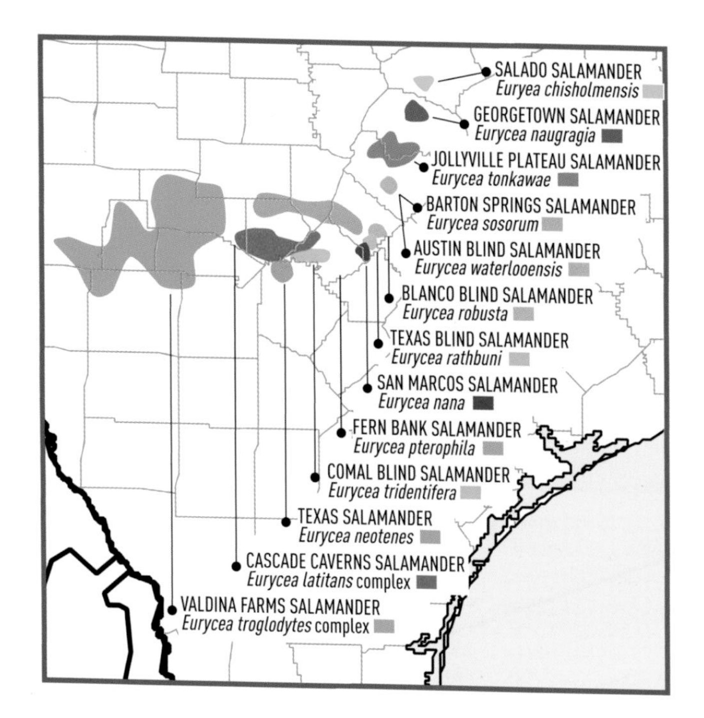

NEOTENIC BROOK SALAMANDERS FROM TEXAS

Both surface-dwelling and cave- or spring-adapted neotenic species from central Texas have small ranges, are often difficult to distinguish morphologically, and are best identified using geography and molecular analyses. Two species complexes contain populations in which at least some individuals metamorphose. All are threatened by water loss and contamination. Data are deficient for determining the conservation status of species without such listings.

SALADO SALAMANDER *Eurycea chisholmensis* **NOT ILLUS.**

1–1½ in. (2.5–3.8 cm). Grayish brown with light reticulations; cream-colored limbs speckled with brown above. Prominent golden yellow stripe along top of tail, tail fin well developed. Belly whitish. Costal grooves 15–16. **SIMILAR SPECIES:** See Georgetown Salamander. **HABITAT:** Vicinity of springs. **RANGE:** Big Boiling and Robertson Springs in Salado, Bell Co., TX. **REMARK:** Until recently considered conspecific with the Texas Salamander. **CONSERVATION:** Vulnerable (USFWS: Threatened).

FIG. 25. *Some populations of the Cascade Caverns Salamander* (Eurycea latitans complex), *as in this individual, have functional eyes and some pigmentation, but some other cave-dwelling populations have greatly reduced nonfunctional eyes and little or no pigment in the skin.*

CASCADE CAVERNS SALAMANDER *Eurycea latitans* complex **FIG. 25**

2–4 in. (5.1–10.2 cm). Surface form neotenic; head flattened, *dark bars from nostrils to eyes*; light grayish brown to yellowish, with dorsolateral rows of light spots, double rows of light flecks (iridophores) on sides; belly and lower sides white; tail low-finned. Cave form has less pigmentation, small eyes (sometimes partially covered by skin), somewhat shovel-nosed snout. Some individuals in some populations metamorphose. Costal grooves 14–15. **SIMILAR SPECIES:** See Texas Salamander. **HABITAT:** Springs, caves. **RANGE:** Comal, Kerr, Kendall, and Hays Cos., TX. **REMARK:** Undoubtedly a species complex. **CONSERVATION:** Vulnerable.

SAN MARCOS SALAMANDER *Eurycea nana* **PL. 4**

1½–2 in. (3.8–5.1 cm). Brown; *dark rings around small eyes*; dorsolateral *rows of yellowish flecks or spots.* Belly whitish or yellowish. Underside of tail pale yellow. Costal grooves 16–17. **SIMILAR SPECIES:** Texas Blind Salamander is pearly white, has a strongly shovel-nosed snout, 12 costal grooves, and eyespots covered by skin. **HABITAT:** Spring pool and upper reaches of the San Marcos R. **RANGE:** San Marcos R., Hays Co., TX. **CONSERVATION:** Vulnerable (USFWS: Threatened).

GEORGETOWN SALAMANDER *Eurycea naufragia* P. 1

1–1½ in. (2.5–3.8 cm). Brown; *dark rings around eyes*; light brown mid-dorsal areas (including a vaguely diamond-shaped mark on the head) extend to base of tail; scattered to somewhat intense dorsal blotches, 3 rows of light spots (iridophores) usually surrounded by *light rosettes or starburst-like areas*. Prominent golden yellow stripe along top of tail, tail fin weakly developed but with *complete dark lateral margins*. Belly whitish. Costal grooves 14–16. **SIMILAR SPECIES:** (1) Jollyville Plateau Salamander is best distinguished by range but has square or oblong light areas around dorsal light spots and dark margins of tail fin are irregular and broken dorsally. (2) Salado Salamander lacks dark eye rings. **HABITAT:** Spring flows, caves. **RANGE:** San Gabriel R. drainage in and near Georgetown, Williamson Co., TX. **REMARKS:** Until recently considered conspecific with the Texas Salamander. Also called the San Gabriel Springs Salamander. **CONSERVATION:** Endangered (USFWS: Threatened).

TEXAS SALAMANDER *Eurycea neotenes* PL. 4, FIG. 5

2–4 in. (5.1–10.2 cm). Head flattened (Fig. 5, p. 22). Light brown to yellowish with brown to dark brown mottling; *dark bars from nostrils to eyes*; double lateral rows of light flecks (iridophores). Tail low-finned. Belly white to pale cream, chin lightly pigmented. Costal grooves 15–17. **SIMILAR SPECIES:** Cascade Caverns, Fern Bank, and Valdina Farms Salamanders are best distinguished by geography and molecular markers. **HABITAT:** Springs. **RANGE:** Bexar and Kendall Cos., TX. **REMARK:** Many references to this species apply to populations now assigned to Salado, Georgetown, Barton Springs, Fern Bank, Jollyville Plateau, and Valdina Farms Salamanders. **CONSERVATION:** Vulnerable.

FERN BANK SALAMANDER *Eurycea pterophila* NOT ILLUS.

1¼–1½ in. (3.2–3.8 cm). Head flattened. Lightly mottled in brown and yellow (mottling darker on top of head and sides of body and tail); *dark bars from nostrils to eyes*; 2 lateral rows of light spots (iridophores). Dull orange stripe along top of tail. Chin and belly pale yellow. Costal grooves 14–16, usually 16. **SIMILAR SPECIES:** See Texas Salamander. **HABITAT:** Springs. **RANGE:** Blanco R. drainage in Blanco, Hays, and Kendall Cos., TX. **REMARKS:** Until recently considered conspecific with the Texas Salamander. Also called the Blanco River Springs Salamander.

TEXAS BLIND SALAMANDER *Eurycea rathbuni* PL. 4, FIGS. 5, 26

3¼–4¼ in. (8.3–10.7 cm); record 5⅜ in. (13.7 cm). Pearly white (pigment cells widely scattered). *Strongly shovel-nosed snout* (Fig. 5, p. 22; Fig. 26, p. 64), *small eyespots covered by skin*, head nearly as wide at level of eyespots as at its widest point in front of the gills. Legs *very*

FIG. 26. *This Texas Blind Salamander* (Eurycea rathbuni) *clearly shows the strongly shovel-nosed snout and eyespots covered by skin.*

thin. Tail fin *high*, extending full length of tail. Costal grooves 12; adpressed limbs *overlap by 5–9 costal folds*. **SIMILAR SPECIES:** (1) Blanco Blind Salamander is bulkier and head is distinctly narrower at level of eyespots when compared with the widest point in front of the gills. (2) Comal Blind Salamander is pale brown to brownish yellow and has 4–15 pairs of dorsolateral light spots on body and base of tail. (3) Austin Blind Salamander has short limbs (1–3 costal folds between adpressed limbs), at least some pigment in skin (often with a lavender tinge), and dorsolateral light spots on body and base of tail. (4) Other neotenic salamanders from central TX generally have 13 or more costal grooves, exposed eyes, normal limbs, and a head that might be flattened but is not shovel-nosed. **HABITAT:** Subterranean caverns. **RANGE:** Balcones Escarpment near San Marcos, Hays Co., TX. **REMARK:** Previously assigned to the genus *Typhlomolge*. **CONSERVATION:** Vulnerable (USFWS: Endangered).

BLANCO BLIND SALAMANDER *Eurycea robusta* **NOT ILLUS.**

4 in. (10.1 cm). Pearly white (pigment cells widely scattered). *Strongly shovel-nosed snout, tiny eyespots covered by skin*, head distinctly narrower at level of eyespots when compared with the widest point in front of the gills. Legs *very thin*. Tail fin *high*, extending full length of tail. Costal grooves 12; adpressed limbs *overlap by one costal fold*. **SIMILAR SPECIES:** See Texas Blind Salamander. **HABITAT:** Subterranean caverns. **RANGE:** Balcones Aquifer, Hays Co., TX. **REMARKS:** Known from only 4 specimens collected in 1951, all but one of which are lost. Previously assigned to the genus *Typhlomolge*.

FIG. 27. *The Barton Springs Salamander* (Eurycea sosorum), *like the Texas Salamander* (E. neotenes), *has a flattened but not shovel-nosed snout and a relatively low-finned tail.*

BARTON SPRINGS SALAMANDER *Eurycea sosorum* **FIG. 27**

1½–2¼ in. (3.8–5.7 cm). Snout flattened, small eyes sometimes covered at least partially by skin. Light brownish yellow to cream, with brown to olive brown mottling, double lateral rows of light flecks (iridophores) at least in smaller individuals (lower rows fade with age). Tail low-finned, with a narrow orange-yellow stripe. Belly and lower sides white or cream. Costal grooves 13–16, usually 14–15. **SIMILAR SPECIES:** Austin Blind Salamander is larger, has a shovel-nosed snout, eyespots covered by skin, thin legs, and 12 costal grooves. **HABITAT:** Springs, subterranean waters. **RANGE:** Barton Springs, Austin and Travis Cos., TX. **REMARK:** Until recently considered conspecific with the Texas Salamander. **CONSERVATION:** Vulnerable (USFWS: Endangered).

JOLLYVILLE PLATEAU SALAMANDER *Eurycea tonkawae* **NOT ILLUS.**

1–1½ in. (2.5–3.8 cm). Greenish brown, with yellowish front of face, *dark rings around eyes*, dark blotches concentrated along the dorsal midline, 3 rows of light spots (iridophores) usually surrounded by *square or oblong light areas*. Prominent orange-yellow stripe along top of tail, tail fin weakly developed with *irregular and broken dark lateral margins*, lower tail fin yellow. Belly whitish. Costal grooves 14–16. **SIMILAR SPECIES:** See Georgetown Salamander. **HABITAT:** Springs, spring runs; some cave populations tentatively assigned to this species. **RANGE:** Jollyville Plateau, Travis and Williamson Cos., TX.

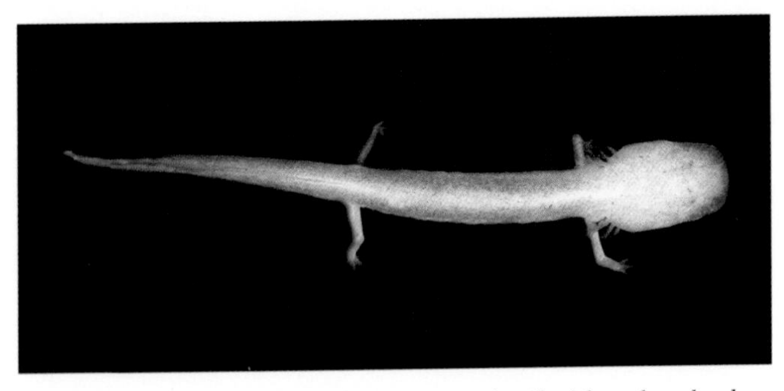

FIG. 28. *The Comal Blind Salamander* (Eurycea tridentifera) *has a large head, about one-third of the head-body length, and a shovel-nosed snout.*

REMARKS: Until recently considered conspecific with the Texas Salamander. Some cave populations might represent distinct species. **CONSERVATION:** Endangered (USFWS: Threatened).

COMAL BLIND SALAMANDER *Eurycea tridentifera* **FIGS. 5, 28**

1½–2⅞ in. (3.8–7.3 cm); record 3⅜ in. (8.5 cm). Very pale brown to brownish yellow. Head large, *shovel-nosed snout* depressed abruptly at about level of eyes (Fig. 5, p. 22), *relatively large eyespots usually covered by skin*. Has *4–15 pairs of dorsolateral light spots* (iridophores) on body and base of tail. Legs *thin*. Costal grooves 11–12; adpressed limbs separated by 1–2 costal folds, touch, or overlap by one costal fold. **SIMILAR SPECIES:** See Texas Blind Salamander. **HABITAT:** Caves. **RANGE:** Comal, Bexar, and possibly Kendall Cos., TX. **CONSERVATION:** Vulnerable.

VALDINA FARMS SALAMANDER *Eurycea troglodytes* complex **FIG. 29**

2–3 in. (5.1–7.6 cm). Surface form neotenic, snout flattened, *dark bars from nostrils to eyes*; light brown to yellowish, with brown to dark brown mottling, double lateral rows of light flecks (iridophores) at least in smaller individuals (lower rows fade with age), sometimes appearing as pale yellow lateral stripes; belly and lower sides white; tail low-finned, sometimes with a pale yellow stripe along the dorsal surface. Cave form has less pigmentation, small eyes sometimes at least partially covered by skin, and a somewhat shovel-nosed snout. Some individuals in some populations metamorphose. Costal grooves 13–14. **SIMILAR SPECIES:** See Texas Salamander. **HABITAT:** Springs, caves. **RANGE:** Discrete populations that might represent distinct species in Medina, Real, Kerr, Bandera, Edwards, Uvalde, and Gillespie Cos., TX. **REMARK:** Until recently considered conspecific with the Texas Salamander.

FIG. 29. *Spring- and cave-adapted neotenic salamanders, such as this Valdina Farms Salamander* (Eurycea troglodytes *complex*), *have reduced pigmentation, especially on the venter, allowing viscera and eggs to show through the translucent skin.*

AUSTIN BLIND SALAMANDER *Eurycea waterlooensis* **NOT ILLUS.**

1½–2⁹⁄₁₆ in. (3.8–6.6 cm). Pearly white, faintly pigmented skin tinged with lavender. *Flattened shovel-nosed snout, eyespots covered by skin*, few light spots (iridophores) along sides of body with a greater concentration on the tail. Legs *thin*. Tail fin *weakly developed* (ventral portion present only on distal half of tail, dorsal portion low or absent anteriorly). Costal grooves 12; adpressed limbs *separated by 1–3 costal folds*. **SIMILAR SPECIES:** See Texas Blind and Barton Springs Salamanders. **HABITAT:** Spring outlets, probably mostly in subterranean caverns. **RANGE:** Barton Springs, Austin, Travis Co., TX. **CONSERVATION:** Vulnerable (USFWS: Endangered).

SPRING, MUD, AND RED SALAMANDERS: *Gyrinophilus* and *Pseudotriton*

All are endemic to eastern North America.

BERRY CAVE SALAMANDER *Gyrinophilus gulolineatus* **FIG. 30**

3½–8¹⁵⁄₁₆ in. (8.9–22.7 cm). Neotenic cave-dweller, individuals rarely metamorphose; dark russet brown to deep brownish purple, dorsal spots roughly circular and blackish, extending from the jaw to the basal third of the tail. Venter pearl gray, but forward half of *throat with dark stripes or blotches*. Eyes small (diameter one-fourth or less of

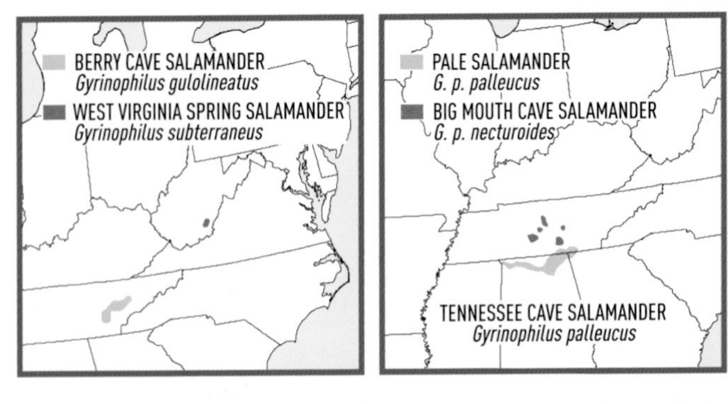

FIG. 30. *The Berry Cave Salamander* (Gyrinophilus gulolineatus) *is a neotenic cave-dweller. Listed as endangered on the IUCN Red List of Threatened Species, this species is known from only nine localities. Several populations are located in areas of urban development, and alterations of surface habitats can have substantive negative effects on the groundwater on which this species depends.*

distance from anterior corner of eye to tip of snout). **SIMILAR SPECIES:** See Tennessee Cave Salamander. **HABITAT:** Cave streams. **RANGE:** E. TN. **REMARK:** Until recently considered a subspecies of the Tennessee Cave Salamander. **CONSERVATION:** Endangered.

TENNESSEE CAVE SALAMANDER *Gyrinophilus palleucus* **PL. 4**

3–7¼ in. (7.5–18.4 cm). Neotenic cave-dweller. Head wide, snout spatulate. Color variable (see Subspecies). Eyes small (diameter one-

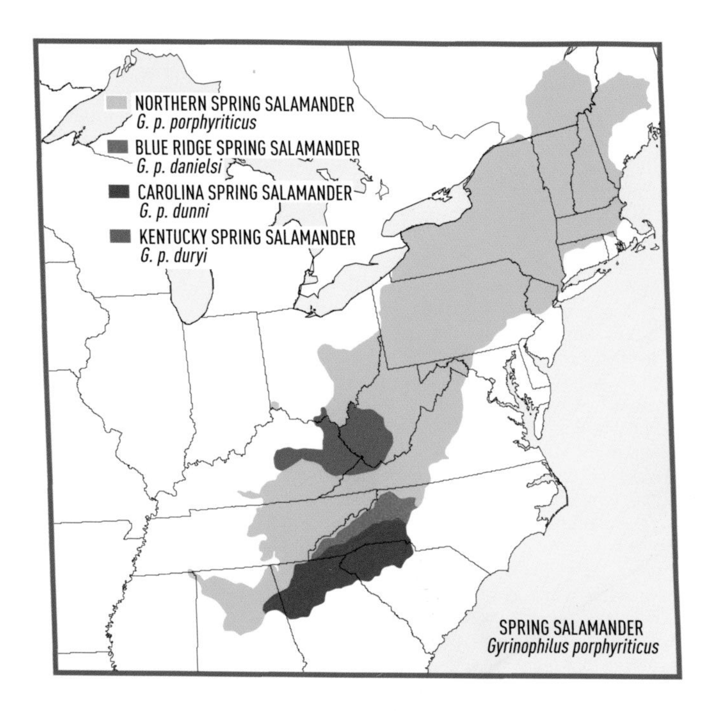

fourth or less of distance from anterior corner of eye to tip of snout). **SIMILAR SPECIES:** Berry Cave Salamander has dark stripes or blotches on forward half of throat. See also Spring Salamander. **HABITAT:** Cave streams. **RANGE:** Cen. TN and adjacent ne. GA and n. AL. Subspecies: (1) **Pale Salamander** (*G. p. palleucus*); pinkish above and below, without dark spots; external gills bright red. (2) **Big Mouth Cave Salamander** (*G. p. necturoides*); dark russet brown to deep brownish purple, dorsal spots blackish and roughly circular, extending from the jaw to the basal third of the tail; venter pearl gray. **REMARK:** Not all populations can be clearly assigned to subspecies. **CONSERVATION:** Vulnerable.

SPRING SALAMANDER *Gyrinophilus porphyriticus* **PL. 5**

4¾–7½ in. (12.1–19 cm); record 9⅛ in. (23.2 cm). Reddish, salmon, light brownish pink, or orange-yellow, often with black or brown spots or flecks (sometimes vague) producing a cloudy appearance. Pronounced ridge (canthus rostralis) with *light lines from eyes to nostrils* bordered below by black or gray (latter often inconspicuous). **SIMILAR SPECIES:** (1) Tennessee and Berry Cave Salamanders seldom transform and have smaller eyes. (2) West Virginia Spring Salamander has smaller eyes, indistinct lines from eyes to nostrils, and paler coloration. (3) Red and Mud Salamanders lack a canthus rostralis and light lines from eyes to nostrils. **HABITAT:** Cool springs, mountain streams, caves, swamps, lake margins, seepages, wet forest areas. **RANGE:** ME and s. QC southwest to n. AL and extreme ne. MS. **SUBSPECIES:** (1) **Northern Spring Salamander** (*G. p. porphyriticus*); light lines from eyes to nostrils bordered below by gray (often inconspicuous); dorsum salmon or light yellowish brown with reddish tinges and a mottled or clouded appearance; sides darker with a reticulate pattern enclosing light spots; venter pinkish; small, scattered black spots on belly, throat, and especially on margins of the lower jaw in old individuals; recently transformed young are salmon red with darker mottling poorly developed. (2) **Blue Ridge Spring Salamander** (*G. p. danielsi*); white lines from eyes to nostrils bordered below by conspicuous black or dark brown lines, often also with dark lines above white lines; dorsum reddish or rich salmon with scattered black spots; individuals from high elevations have heavy dark speckling on the chin. (3) **Carolina Spring Salamander** (*G. p. dunni*); white lines from eyes to nostrils bordered below by conspicuous black or dark brown lines, often also with dark lines above white lines; dorsum orange-yellow to light reddish, profusely flecked with dark pigment; venter salmon pink, usually immaculate except for margins of the jaw, which are mottled with black and white. (4) **Kentucky Spring Salamander** (*G. p. duryi*); light lines from eyes to nostrils bordered below by gray (often inconspicuous); dorsum salmon pink to light brownish pink with small black spots usually arranged in lateral rows, sometimes scattered along sides or over entire dorsal surface, but most strongly concentrated on sides; venter pinkish and unmarked except on lower lip and occasionally chin and throat.

WEST VIRGINIA SPRING SALAMANDER *Gyrinophilus subterraneus* FIG. 31

4–7 in. (10–17.7 cm). Pale cave-dweller. Brownish pink, with a light gray reticulate pattern (back somewhat darker than sides). Light lines from eyes to nostrils indistinct; *eyes very small*. **SIMILAR SPECIES:** See Spring Salamander. **HABITAT:** Cave stream. **RANGE:** General Davis Cave, Greenbrier Co., WV. **CONSERVATION:** Endangered.

MUD SALAMANDER *Pseudotriton montanus* PL. 4

3–6½ in. (7.5–16.5 cm); record 8⅛ in. (20.7 cm). Snout blunt, rounded. Reddish, usually with round, black spots (individuals from s. GA and n. FL virtually unmarked, vague mottling and a few spots restricted to tail). Young usually brightly colored, spots distinct, belly largely immaculate. Older adults darker, spots often inconspicuous, undersurfaces frequently spotted or flecked with brown or black. Irises *brown*. **SIMILAR SPECIES:** See Red Salamander. **HABITAT:** Mucky seepage areas, swamps, backwater ponds, muddy environs of springs and shallow streams in hammocks or mixed forests. **RANGE:** S. NJ to n. FL and west to MS and extreme e. LA. **SUBSPECIES:** (1) **Eastern Mud Salamander** (*P. m. montanus*); red to light reddish brown or chocolate back

FIG. 31. *The known range of the West Virginia Spring Salamander* (Gyrinophilus subterraneus) *is limited to a single cave. Whether this represents the historical range of the species or if it was once more widely distributed is unknown.*

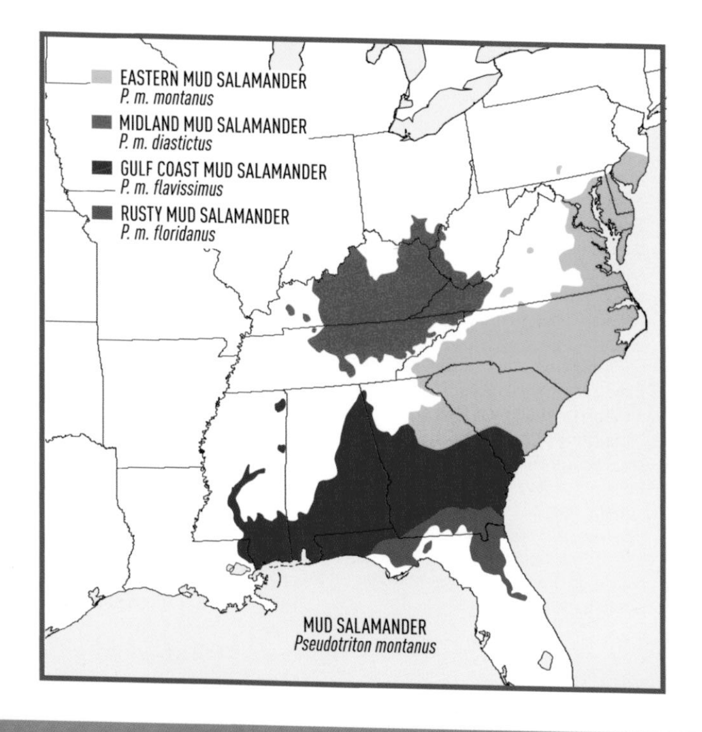

distinct from lighter sides and belly; well-separated black spots round; juvenile belly virtually unmarked, frequently flecked with brown or black in old adults. (2) **Midland Mud Salamander** (*P. m. diastictus*); bright coral pink or red to brown; black spots few and scattered; undersurfaces unmarked except occasional dark lines on rim of lower jaw. (3) **Gulf Coast Mud Salamander** (*P. m. flavissimus*); light brownish salmon with many small, well-separated round spots; undersides of head and trunk clear salmon pink. (4) **Rusty Mud Salamander** (*P. m. floridanus*); rusty dorsum sometimes slightly mottled with indistinct darker areas and few small, irregular, light pinkish spots; no dark dots on back, but a few may be scattered on top of the tail; sides pinkish buff and rust, sometimes with black streaks or specks; undersurfaces buffy, sparsely marked with small, irregular blackish spots. **REMARKS:** The Gulf Coast and Rusty Mud Salamanders sometimes are considered a separate species. Additional research is needed on this species.

RED SALAMANDER *Pseudotriton ruber* PL. 4

2¾–6 in. (7–15.2 cm); record 7⅛ in. (18.1 cm). Snout somewhat pointed. Red or reddish orange, profusely dotted with irregular, rounded black spots and scattered white spots (latter largely restricted to southern populations). Individuals in some populations and old adults dull purplish brown, spots larger and running together. Venter with variously sized black or brown spots; chin virtually unmarked, flecked with black along margins, or heavily pigmented. Irises usually *yellowish*. **SIMILAR SPECIES:** (1) Mud Salamander is more likely to occupy muddy habitats and has a rounded snout, usually brown irises, round and less numerous spots (when present), and dorsal and ventral ground colors usually more sharply divided. (2) Spring Salamander has a prominent canthus rostralis with light and dark lines extending from eyes to nostrils. **HABITAT:** Cold, clear, rocky or sandy streams and springs. **RANGE:** S. NY south to MS and extreme e. LA. **SUBSPECIES:** (1) **Northern Red Salamander** (*P. r. ruber*); red or reddish orange dorsum profusely dotted with irregular, rounded black spots; margin of chin flecked with black; old adults dull purplish brown, spots larger and running together, undersurfaces with black or brown spots. (2) **Blue Ridge Red Salamander** (*P. r. nitidus*); similar to Northern Red Salamander but without black pigment on distal half of tail and little or none on chin; even old adults brightly colored. (3) **Black-chinned Red Salamander** (*P. r. schencki*); red, black under chin heavier and broader than narrow black flecking of other subspecies; tail spotted almost to tip. (4) **Southern Red Salamander** (*P. r. vioscai*); reddish or salmon, fairly large blue-black blotches often producing a purplish effect; profusion of white flecks largely concentrated on snout and sides of head; undersurfaces light but with many small dark spots.

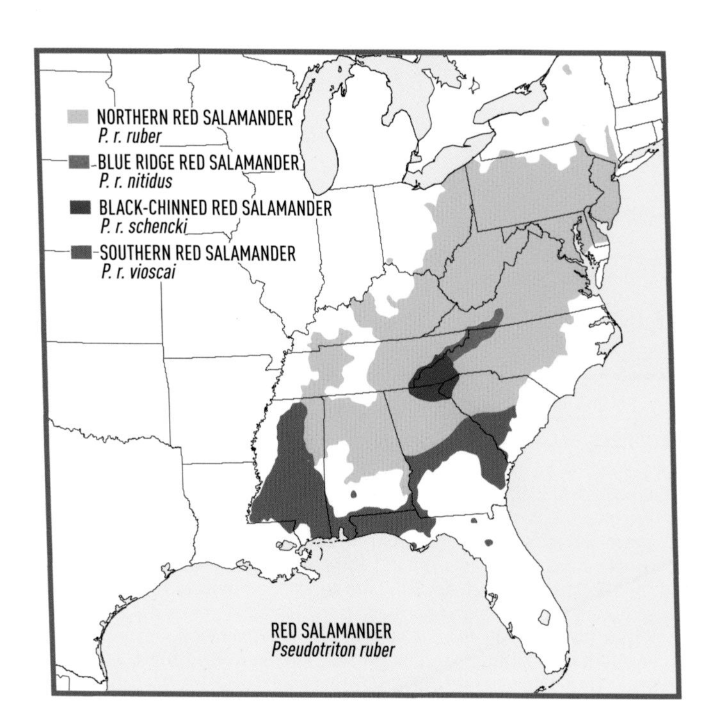

FOUR-TOED and RED HILLS SALAMANDERS: *Hemidactylium* and *Phaeognathus*

These two genera, each with only one species, occur solely in eastern North America.

FOUR-TOED SALAMANDER *Hemidactylium scutatum* **PL. 5, FIG. 37**

2–3½ in. (5.1–9 cm); record 4½ in. (11.3 cm). Dorsum rusty or gray-brown; sides grayish, often speckled with black and bluish spots. Belly white, boldly marked with *black spots* (Fig. 37, p. 80). *Four toes on all limbs. Marked constriction* at base of tail. **SIMILAR SPECIES:** Dwarf and Chamberlain's Dwarf Salamanders have 4 toes on all limbs but lack basal tail constriction and bold spots on a white belly. **HABITAT:** Frequently associated with sphagnum moss in swamps, boggy streams, wet wooded or open areas near ponds or pools. **RANGE:** Disjunct; NS west to MN and south to e. LA and the FL Panhandle.

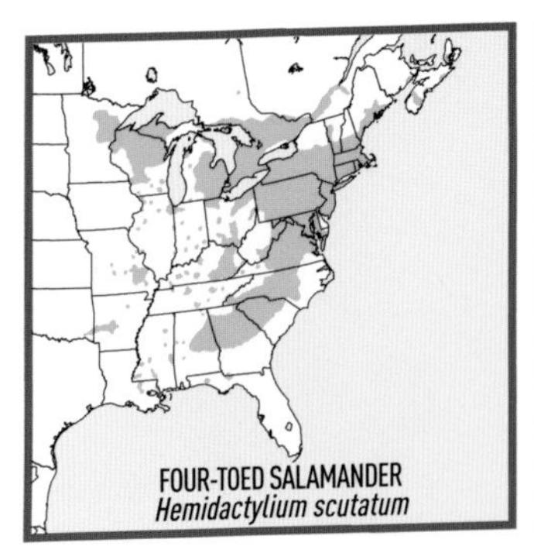

RED HILLS SALAMANDER *Phaeognathus hubrichti* **PL. 5**

4–8 in. (10–20.3 cm); record 10 9/16 in. (26.8 cm). Uniformly deep dark brown except for a slightly paler snout, jaws, and soles of limbs. Body *elongated*, legs short. Costal grooves 20–22, about 14 costal folds between adpressed limbs. **SIMILAR SPECIES:** (1) Dusky salamanders usually have pale lines extending from eyes to angles of the jaw and only 13–15 costal grooves. (2) Slimy salamanders usually are sprinkled with light spots and normally have 16 costal grooves. **HABITAT:** Fossorial; largely confined to cool, moist, forested ravines, especially northern exposures under full canopies. **RANGE:** Red Hills of AL. **CONSERVATION:** Endangered (USFWS: Threatened).

WOODLAND SALAMANDERS: *Plethodon*

Strictly North American; 45 of 55 currently recognized species are widely distributed throughout forested portions of our area, one presumably is extinct, and nine occur in the West. Eggs are deposited in moist areas; development occurs within eggs (no aquatic larval stage). Adult males of some species have a prominent circular mental gland. **SIMILAR SPECIES:** (1) Mole salamanders lack grooves from nostrils to lip. (2) Dusky salamanders usually have light lines from the eyes to the angles of the jaw, and their hindlimbs are larger and stouter than their forelimbs.

BLUE RIDGE GRAY-CHEEKED SALAMANDER *Plethodon amplus* **P. xi**

3½–5 11/16 in. (9–14.5 cm). Belly very dark. **RANGE:** Blue Ridge Mts. in Buncombe, Rutherford, Henderson Cos., NC. **CONSERVATION:** Vulnerable.

SOUTH MOUNTAIN GRAY-CHEEKED SALAMANDER **FIG. 32**
Plethodon meridianus

3½–5 15/16 in. (9–14.5 cm). Belly dark, chin light. **RANGE:** South Mts. in Burke, Cleveland, and Rutherford Cos., NC. **CONSERVATION:** Vulnerable.

FIG. 32. *All four species of gray-cheeked salamanders, including the South Mountain Gray-cheeked Salamander* (Plethodon meridianus) *shown here, were until recently considered color variants of the Red-cheeked Salamander* (P. jordani).

SOUTHERN GRAY-CHEEKED SALAMANDER *Plethodon metcalfi* **PL. 6**

3½–5⅝ in. (9–14.3 cm). Belly light in n. portions of range; belly dark, sometimes with brassy dorsal flecks and white spots on sides in s. portions of the range. **RANGE:** Sw. NC and adj. nw. SC and ne. GA.

NORTHERN GRAY-CHEEKED SALAMANDER *Plethodon montanus* **NOT ILLUS.**

3½–5½ in. (9–14 cm). Belly light, chin lighter. **RANGE:** Disjunct; w. VA south to w. NC and adj. TN.

All slender, dark gray, with lighter areas behind eyes. Costal grooves 16. Virtually indistinguishable, best identified using geography. **SIMILAR SPECIES:** (1) Red-cheeked Salamander almost always has red pigment on cheeks. (2) Cheoah Bald and Red-legged Salamanders generally have red pigment on legs. (3) Southern Appalachian Salamander often with small red spots on legs, tiny dorsal white spots, larger lateral white spots, and chin lighter than gray belly. (4) Northern and White-spotted Slimy Salamanders usually have dark bellies and abundant white spots on sides and back. (6) Yonahlossee Salamander has red or chestnut pigment on the back. (7) Eastern and Southern Red-backed Salamanders lacking a middorsal stripe superficially resemble gray-cheeked salamanders, but have a salt-and-pepper pattern on the belly and 19 costal grooves. **HABITAT:** Dense mesic forests. **REMARK:** Until recently considered color variants of the Red-cheeked Salamander.

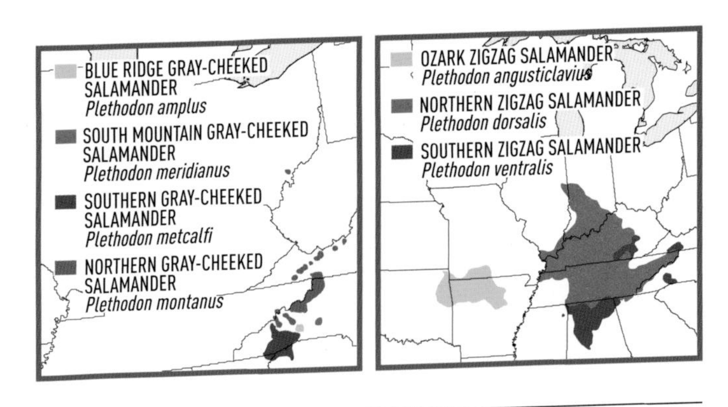

OZARK ZIGZAG SALAMANDER *Plethodon angusticlavius* **FIG. 33**

$2^3/_8$–$3^7/_8$ in. (6–9.8 cm). Slender, body elongated. Uniformly dark or with a red to orange or yellowish middorsal stripe; stripe narrow with straight edges for at least part of its length, sometimes wider with wavy or scalloped margins; edges of the stripe sometimes indistinct; if wide, usually widest at base of tail and continuing *broadly* onto the tail. Belly light with extensive dark mottling, often with some red pigment. Costal grooves 17–19 (usually 18); 6–7 costal folds between adpressed limbs. **SIMILAR SPECIES:** See Southern Red-backed Salamander. **HABITAT:** Ravines, canyons, rubble, seepage areas, caves,

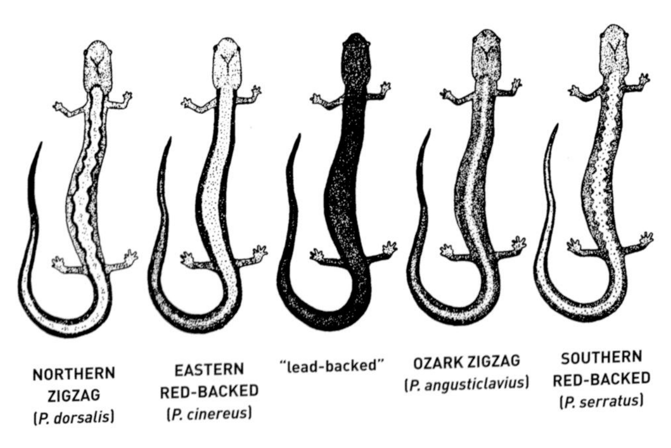

FIG. 33. *Dorsal patterns of some small woodland salamanders* (Plethodon). *Note that "lead-backed" variants occur in at least some populations of most of these species.*

FIG. 34. *The Tellico Salamander* (Plethodon aureolus) *cannot be reliably distinguished from the Northern Slimy Salamander* (P. glutinosus), *but the two species co-occur at only one site in extreme northern Polk County, Tennessee.*

wooded slopes. **RANGE:** Ozark Highlands of n. AR and adjacent MO and OK. **REMARK:** Until recently considered a subspecies of the Northern Zigzag Salamander.

TELLICO SALAMANDER *Plethodon aureolus* FIG. 34

4–5 15/16 in. (10–15.1 cm). Black, with *abundant dorsal brassy spotting and flecking* and *lateral white or yellow spots* (sometimes absent in populations at higher elevations in northeastern parts of the range). Chin usually lighter than the uniformly slate gray belly. Costal grooves usually 16. **SIMILAR SPECIES:** (1) Southern Appalachian Salamander is larger and has white dorsal spots. (2) Red-legged and Red-cheeked Salamanders lack brassy dorsal spots, have few white lateral spots, and usually have red legs or cheeks. (3) Northern Slimy Salamander cannot be reliably distinguished but co-occurs with the Tellico Salamander at only one site. **HABITAT:** Forested mountain slopes and lowlands. **RANGE:** Extreme se. TN and adjacent NC.

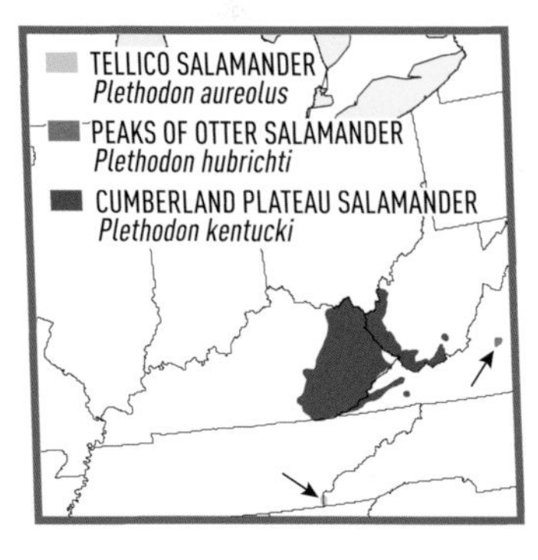

FIG. 35. *Like many other species in the* Plethodon jordani *species complex, the Cheoah Bald Salamander* (P. cheoah) *was until recently considered a color variant of the Red-cheeked Salamander* (P. jordani).

CADDO MOUNTAIN SALAMANDER *Plethodon caddoensis* **PL. 5**

3½–4 in. (9–10 cm); record 4⅜ in. (11.1 cm). Black back and sides profusely marked with small whitish spots; scattered brassy dorsal flecks. *Throat and chest light* (but speckled with black), belly dark. Costal grooves usually 16. **SIMILAR SPECIES:** See Rich Mountain Salamander. **HABITAT:** Talus slopes or rocky sites on forested north-facing slopes. **RANGE:** Caddo Mts., AR. **REMARK:** Four lineages might represent distinct species.

CHEOAH BALD SALAMANDER *Plethodon cheoah* **FIG. 35**

3½–5 in. (9–12.5 cm). Slender, black, with white or yellow lateral spots, usually red on upper surfaces of limbs, sometimes red on cheeks. Belly light gray with or without light whitish or yellowish spots. Costal grooves usually 16. **SIMILAR SPECIES:** See Red-cheeked Salamander. Dusky salamanders with red cheek patches (Imitator Salamander) or red legs (Ocoee Salamander) have an obvious dorsal pattern, lines from eyes to angles of the jaw, and hindlimbs larger than forelimbs. **HABITAT:** Humid forests. **RANGE:** Cheoah Bald, NC. **REMARK:** Interbreeds with the Southern Appalachian Salamander. **CONSERVATION:** Vulnerable.

EASTERN RED-BACKED SALAMANDER *Plethodon cinereus* **PL. 5, FIGS. 33, 36, 37**

2¼–4 in. (5.7–10 cm); record 5 in. (12.7 cm). Slender, elongated, long-tailed (over one-half total length). Uniformly dark gray to black (lead-backed), occasionally all red (especially in northern portions of the range), or with a red, orange, yellow, or even light gray *straight-edged*

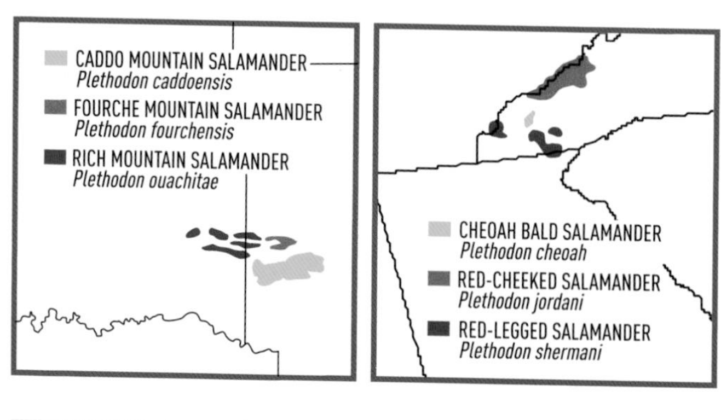

FIG. 36. *The Eastern Red-backed Salamander* (Plethodon cinereus) *has both red-backed and lead-backed color phases. Either type can predominate in any local population. Many other small woodland salamanders also have both striped and lead-backed variants.*

middorsal stripe extending from base of head to tail, usually *narrowing slightly* on base of tail, often bordered by dark pigment extending onto sides; rarely with a partial stripe. Belly light with extensive dark mottling, producing a *salt-and-pepper* effect (Fig. 37, p. 80). Costal grooves 17–22 (usually 19); 6–10 costal folds between adpressed limbs. **SIMILAR SPECIES:** (1) Big Levels Salamander is very similar but has a wider head, longer limbs (1–5 costal folds between adpressed limbs), and more white pigment on the belly. (2) Southern Red-backed Salamander has a shorter tail (less than one-half total length) and middorsal stripes, that when present, have saw-toothed margins. (3) Northern Zigzag Salamander has longer legs (6–7 costal folds between adpressed limbs), usually some ventral red pigment, and a middorsal stripe, when present, with scalloped or wavy edges, often widest on base of tail. (4) Northern Ravine, Valley and Ridge, Peaks of Otter, Cheat Mountain, Shenandoah, and Shenandoah Mountain Salamanders are superficially similar to the lead-backed variant, but their bellies are almost uniformly dark (may have some light mottling, but never a salt-and-pepper effect). **HABITAT:** Forested areas. **RANGE:** Atlantic Provinces and the Great Lakes Region south to NC and n. KY.

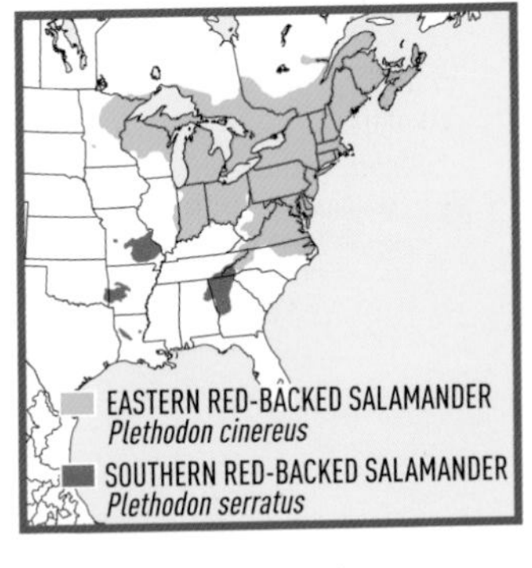

NORTHERN ZIGZAG SALAMANDER *Plethodon dorsalis* **PL. 5, FIG. 33**
SOUTHERN ZIGZAG SALAMANDER *Plethodon ventralis* **NOT ILLUS.**

2½–3½ in. (6.4–9 cm); record 4¾ in. (12 cm). Slender, elongated. Uniformly gray to dark gray or with a red to yellowish middorsal stripe generally narrow with straight edges for at least part of its length or wider with wavy or scalloped margins; edges of the stripe sometimes indistinct; if wide, usually widest at base of tail and continuing *broadly* onto the tail. In KY, where ranges overlap widely, the Northern Zigzag Salamander tends to either lack middorsal stripes (although some red pigment often is present) or have red middorsal stripes, whereas

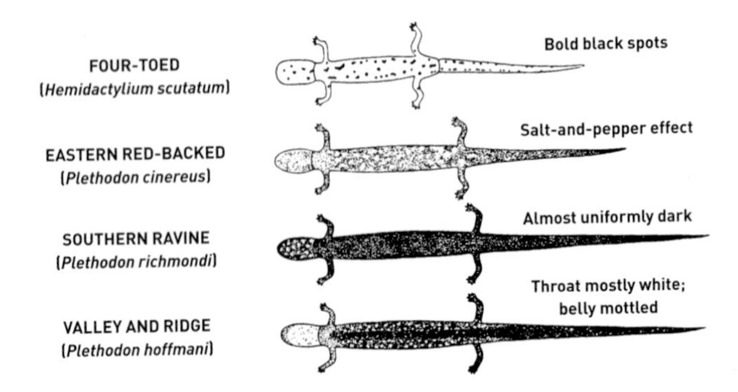

FIG. 37. *Ventral patterns of some small lungless salamanders* (Plethodontidae).

the Southern Zigzag Salamander tends to have 2 short, narrow, parallel dorsolateral stripes anteriorly. Some individuals have numerous light flecks, resulting in a frosted effect. Often with some red color at base of limbs. *Belly light, but with extensive black and orange or reddish mottling*. Costal grooves 17–19 (usually 18); 6–7 costal folds between adpressed limbs. **SIMILAR SPECIES:** Webster's Salamander cannot be distinguished using morphological characters, but ranges overlap only in Jefferson Co., AL, where Webster's Salamander usually is striped and the Southern Zigzag Salamander usually lacks stripes. See also Eastern Red-backed Salamander. **HABITAT:** Ravines, rubble, seepages, caves, wooded slopes. **RANGE:** Disjunct; extreme e.-cen. and s. IL and cen. IN south to n. AL and ne. MS and east to sw. VA and w. NC.

NORTHERN RAVINE SALAMANDER *Plethodon electromorphus* **NOT ILLUS.**
SOUTHERN RAVINE SALAMANDER *Plethodon richmondi* **PL. 6, FIG. 37**

3–4½ in. (7.5–11.5 cm); record 5⅝ in. (14.3 cm). Elongated, slender, short-legged. Brown to nearly black, sprinkled with minute silvery white and bronzy or brassy specks; very small, irregular white blotches on lower sides. Virtually *plain dark belly, lightly mottled chin* (Fig. 37, facing page). Costal grooves 19–23, often but not always 22 or more in the Northern Ravine Salamander and fewer than 22 in the Southern Ravine Salamander.

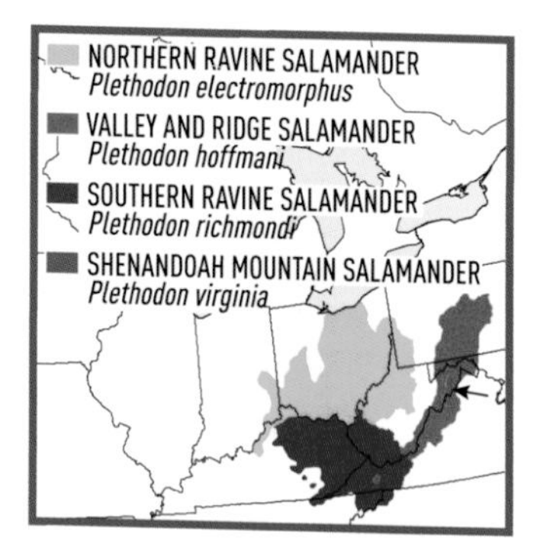

SIMILAR SPECIES: (1) Eastern Red-backed Salamander has a salt-and-pepper pattern on the belly. (2) Northern Zigzag Salamander usually has some reddish or orange pigment on the belly. (3) Wehrle's and Weller's Salamanders have proportionately longer legs and 15–18 costal grooves. (4) Northern Gray-cheeked Salamander has 16 costal grooves and gray cheeks. (5) Slimy salamanders have fewer costal grooves (average 16), a more robust body, and proportionately longer legs. **HABITAT:** Wooded slopes, forested talus; rarely ridges, hilltops, or valley floors. **RANGE:** Northern Ravine Salamander: sw. PA east to se. IN and n. KY; Southern Ravine Salamander: south of the Ohio R. and west of the New and Kanawha R. in VA and WV south to ne. TN and nw. NC. **REMARKS:** Until recently, these two species were considered conspecific; they interbreed in n. KY.

FIG. 38. *Dorsal patterns of two woodland salamanders* (Plethodon) *with restricted ranges.*

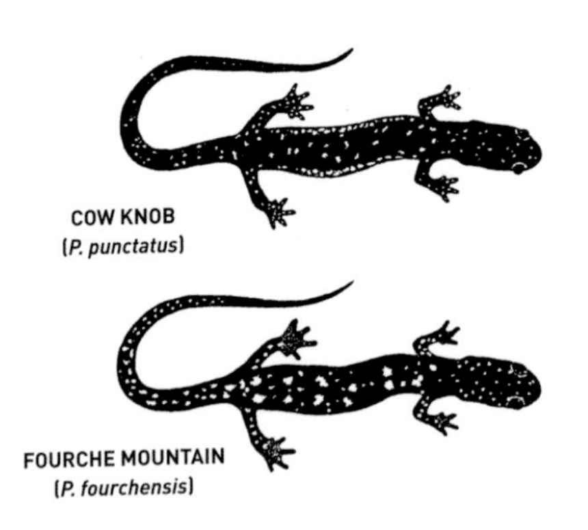

FOURCHE MOUNTAIN SALAMANDER *Plethodon fourchensis* **PL. 5, FIG. 38**

4½–6 in. (11.5–15.2 cm); record 7 in. (17.8 cm). Black, with *2 irregular dorsal rows of large whitish spots* and a variable number of small white spots and brassy flecks (Fig. 38, above). Yellowish white spots on cheeks, legs, sides, and dark belly; chin *light*, chest dark. Costal grooves usually 16. **SIMILAR SPECIES:** See Rich Mountain Salamander. **HABITAT:** High-elevation forested ridges and slopes. **RANGE:** Fourche and Irons Fork Mts., AR. **REMARK:** Four lineages might represent distinct species. **CONSERVATION:** Vulnerable.

SLIMY SALAMANDERS *Plethodon glutinosus* complex **PL. 6, FIG. 39**

4¾–6¾ in. (12.1–17.2 cm); record 8⅛ in. (20.6 cm). Species are best identified by range. Black, most variously sprinkled with silvery white spots or brassy flecks or both; belly usually lighter than dorsum. Costal grooves usually 16 (rarely 15 or 17).

WESTERN SLIMY SALAMANDER *Plethodon albagula* (Fig. 39)
CHATTAHOOCHEE SLIMY SALAMANDER *Plethodon chattahoochee*
ATLANTIC COAST SLIMY SALAMANDER *Plethodon chlorobryonis*
WHITE-SPOTTED SLIMY SALAMANDER *Plethodon cylindraceus*
NORTHERN SLIMY SALAMANDER *Plethodon glutinosus* (Pl. 6)
SOUTHEASTERN SLIMY SALAMANDER *Plethodon grobmani*
KIAMICHI SLIMY SALAMANDER *Plethodon kiamichi*
LOUISIANA SLIMY SALAMANDER *Plethodon kisatchie*
MISSISSIPPI SLIMY SALAMANDER *Plethodon mississippi*
OCMULGEE SLIMY SALAMANDER *Plethodon ocmulgee*
SAVANNAH SLIMY SALAMANDER *Plethodon savannah*
SEQUOYAH SLIMY SALAMANDER *Plethodon sequoyah*
SOUTH CAROLINA SLIMY SALAMANDER *Plethodon variolatus*

SIMILAR SPECIES: A black back with irregularly scattered white spots and/or brassy flecks will usually distinguish species in this complex

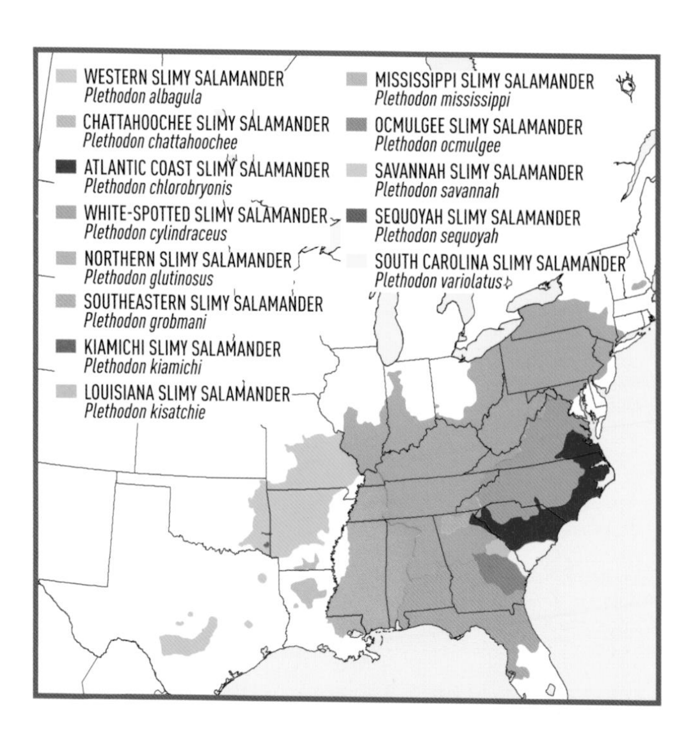

FIG. 39. *The Western Slimy Salamander* (Plethodon albagula) *is part of the wide-ranging* Plethodon glutinosus *species complex. The taxonomic status of some populations currently assigned to* P. albagula *is uncertain.*

from most other large, dark woodland salamanders with 15–17 costal grooves. However, see Cumberland Plateau and Southern Appalachian Salamanders. **HABITAT:** Woodland ravines or hillsides in hardwood or mixed forests, along streams and rivers, cave entrances. **RANGE:** S. CT and NY south to cen. FL and west to e. OK and TX.

VALLEY AND RIDGE SALAMANDER *Plethodon hoffmani* **FIG. 37**

SHENANDOAH MOUNTAIN SALAMANDER *Plethodon virginia* **FIG. 40**

3 9/16–4 3/8 in. (9.1–11.2 cm); record 5 3/8 in. (13.7 cm). Elongated, slender, short-legged, long-tailed (over one-half total length). Brown to nearly black back sprinkled with minute silvery white and bronzy or brassy specks; very small, irregular white blotches on lower sides. Red dorsal pigment sometimes present; some Shenandoah Mountain Salamanders on Reddish Knob and Shenandoah Mt. along the VA-WV border have a narrow reddish middorsal stripe. Throat *mostly white*; belly dark but with a moderate amount of white mottling (Fig. 37, p. 80). Costal grooves usually 20 or 21. **SIMILAR SPECIES:** Northern and Southern Ravine Salamanders are even more elongated (19–23 costal grooves) and have light flecks on a dark throat. See also Eastern Red-backed Salamander. **HABITAT:** Mature hardwood forests. **RANGE:** Valley and Ridge Salamander: cen. PA south to the New R. in sw. VA and adjacent WV; isolated population in w. VA; Shenandoah Mountain Salamander: restricted to Shenandoah Mt. in nw. VA and e. WV.

PEAKS OF OTTER SALAMANDER *Plethodon hubrichti* **FIG. 41**

3 3/16–4 13/16 in. (8.1–12.2 cm); record 5 1/8 in. (13.1 cm). Elongated, slender, long-tailed (about one-half total length). Dark brown to nearly black with abundant dorsal brassy to mosslike greenish flecking and occasional larger white spots. Sides with white spots; belly plain dark

FIG. 40. *Until recently, the Shenandoah Mountain Salamander* (Plethodon virginia) *was considered conspecific with the Valley and Ridge Salamander* (P. hoffmani). *The two species interbreed at some localities.*

FIG. 41. *Like many other woodland salamanders, the Peaks of Otter Salamander* (Plethodon hubrichti) *is territorial, but its home range is less than one square yard (0.8 square meters).*

brown to gray; chin with small white spots. Costal grooves 18–22 (usually 19 or 20). No striped phase, but hatchling often with small scattered dorsal red spots. **SIMILAR SPECIES:** Valley and Ridge Salamander has a light chin and 20–21 costal grooves. See also Eastern Red-backed Salamander. **HABITAT:** Hardwood forests, usually above 2,500 ft. (760 m). **RANGE:** Peaks of Otter Region, Bedford and Botetourt Cos., VA. **CONSERVATION:** Vulnerable.

RED-CHEEKED SALAMANDER *Plethodon jordani* **PL. 6**

3½–5 in. (9–12.5 cm); record 7¼ in. (18.4 cm). Slender; black with red or pink cheeks (rarely missing); very rarely with red on legs; few scattered white or yellow spots on back and sides. Belly gray, usually without light spots. Costal grooves 15–16. **SIMILAR SPECIES:** (1) Cheoah Bald and Red-legged Salamanders generally have red pigment on legs and little or none on cheeks. (2) Southern Appalachian Salamander often has small red spots on legs, tiny dorsal white spots, larger lateral white spots, and a chin lighter than the gray belly. (3) Northern and White-spotted Slimy Salamanders usually have abundant white spots on sides and back and a dark belly. (4) Tellico Salamander has brassy dorsal spots and many white spots on sides and legs. (5) Yonahlossee Salamander has chestnut pigment on the back. (6) Imitator Salamander has red cheek patches but also has an obvious dorsal pattern, lines from eyes to angles of the jaw, and hindlimbs larger than forelimbs. **HABITAT:** Dense forests. **RANGE:** Great Smoky Mountains National Park, NC and TN. **REMARKS:** Cheoah Bald, Red-legged, and 4 species of gray-cheeked salamanders until recently were considered pattern variants of *P. jordani* as then de-

FIG. 42. *The range of the Cumberland Plateau Salamander* (Plethodon kentucki) *overlaps completely with that of the Northern Slimy Salamander* (P. glutinosus). *Until genetic analyses determined that it was a distinct species, the Cumberland Plateau Salamander was thought to be part of the slimy salamander complex.*

fined. Although some pattern differences exist among these species, reliable identification often requires molecular analysis. Interbreeds with the Southern Appalachian Salamander.

CUMBERLAND PLATEAU SALAMANDER *Plethodon kentucki* FIG. 42

$3^7/_8$–$6^5/_8$ in. (9.8–16.8 cm). Black, with small, widely scattered dorsal white spots and very little brassy flecking; lateral white spots larger; usually few if any white spots on distal two-thirds of tail. *Chin lighter than the uniformly slate gray belly.* Breeding male has a large mental gland. Costal grooves usually 16. **SIMILAR SPECIES:** (1) Northern Slimy Salamander is larger, has a dark chin, generally more abundant dorsal spotting, and adult male has a smaller mental gland. (2) Wehrle's Salamander is dark gray or brown with large white or yellowish spots, and generally has light flecking and spotting confined to sides. **HABITAT:** Forested ridges of the Cumberland Plateau (also into the Valley and Ridge Region). **RANGE:** E. KY and adjacent TN, VA, and WV.

CHEAT MOUNTAIN SALAMANDER *Plethodon nettingi* PL. 5

3–4 in. (7.5–10 cm); record $4^3/_8$ in. (11.1 cm). Slender, elongated. Black or very dark brown, back usually sprinkled with *small brassy flecks* that extend onto the tail and are most numerous on or near the head; sometimes also with silvery flecks or spots or without light markings altogether. Flanks usually with larger white spots. Belly and throat plain dark slaty gray to black. Costal grooves 17–19 (usually 18). No

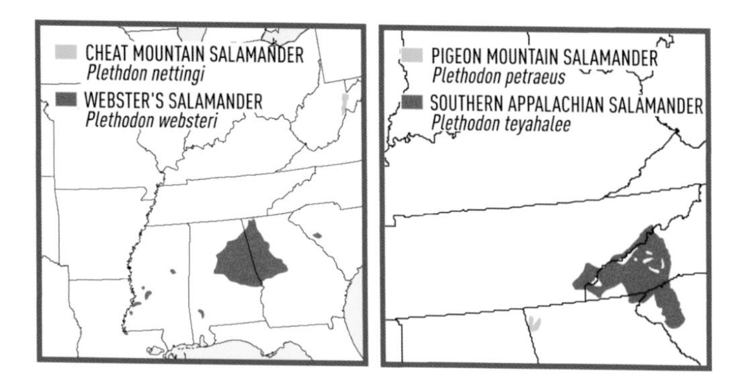

striped variant, but hatchling often with some dorsal red pigment. **SIMILAR SPECIES:** Valley and Ridge Salamander has a mostly white throat and 20–21 costal grooves. See also Eastern Red-backed Salamander. **HABITAT:** Spruce and mixed deciduous forests, generally above 2,950 ft. (900 m). **RANGE:** E.-cen. WV. **CONSERVATION:** (USFWS: Threatened).

RICH MOUNTAIN SALAMANDER *Plethodon ouachitae* **PL. 5**

4–5 in. (10–12.5 cm); record 6¼ in. (15.9 cm). Black, with chestnut pigment and numerous small white specks and metallic (brassy) flecks; some pattern elements sometimes missing, chestnut pigment reduced or lacking in populations not on Rich Mt. Chin *light*, chest and belly *dark*. Costal grooves usually 16. **SIMILAR SPECIES:** (1) Western Slimy Salamander is larger, has back liberally sprinkled with white spots and a dark chin. (2) Kiamichi Slimy Salamander is larger but otherwise difficult to distinguish (Rich Mountain Salamanders on Kiamichi Mt. almost always lack chestnut dorsal pigmentation); chin lighter than chest and belly, but usually not as light as that of the Rich Mountain Salamander. (3) Fourche Mountain Salamander has 2 irregular rows of large whitish dorsal spots and lacks dorsal chestnut coloration. (4) Caddo Mountain Salamander has a light throat and chest. **HABITAT:** Mostly north-facing forested ridges and slopes. **RANGE:** Ouachita Mts., e. OK and adjacent AR. **REMARKS:** Interbreeds with the Fourche Mountain Salamander at the extreme western end of Fourche Mt. Seven lineages might represent distinct species.

PIGEON MOUNTAIN SALAMANDER *Plethodon petraeus* **PL. 6**

4½–6½ in. (11.5–16.5 cm); record 7³⁄₁₆ in. (18.3 cm). Black but with *reddish brown, brown, or olive brown* on the head, back, and base of tail; smaller individuals with 3–12 opposing or alternating reddish brown dorsal spots. Head, body, and legs with sparsely scattered small light spots and brassy flecking. Sides with white or yellowish spots and some brassy flecking. Belly and underside of tail black;

chin with reddish brown pigment or yellowish spotting; throat and chest with yellowish mottling. Toes with *expanded tips* and *one elongated toe* on each limb. Costal grooves 15–17 (usually 16). **SIMILAR SPECIES:** Northern Slimy Salamander has a black back with a variable number of white spots and flecks and lacks dorsal brown areas. **HABITAT:** Limestone outcrops, boulder fields, twilight zones of caves in deciduous forests. **RANGE:** Eastern slope of Pigeon Mt., GA. **CONSERVATION:** Vulnerable.

COW KNOB SALAMANDER *Plethodon punctatus* FIG. 38

4–5 in. (10–12.5 cm); record 6¾ in. (17.1 cm). Grayish black, with numerous large white or cream spots on body, tail, and legs (Fig. 38, p. 82). Top of head often with a purplish cast. Eyes protrude more than in most woodland salamanders. Throat *light pinkish gray*, belly dark gray with small cream spots. Toes on hindfeet more extensively webbed than in slimy salamanders. Costal grooves 17–18. **SIMILAR SPECIES:** (1) Northern Slimy Salamander has brassy to silvery dorsal spots, 16 or fewer costal grooves, and considerably less webbing between toes on hindlimbs. (2) White-spotted Slimy Salamander has large irregular dorsal white spots and abundant lateral white or yellow spots, 16 or fewer costal grooves, and considerably less webbing between toes on hindlimbs. **HABITAT:** Higher-elevation forests with extensive rock outcrops, especially old-growth hemlock forests. **RANGE:** Shenandoah, North, and Great North Mts. along the VA-WV border. **REMARK:** Also called the White-spotted Salamander.

SOUTHERN RED-BACKED SALAMANDER *Plethodon serratus* PL. 5, FIG. 33

3³⁄₁₆–4⅛ in. (8.1–10.5 cm). Slender, relatively short-tailed (less than one-half total length). Either unicolored brownish to dark gray or nearly black above (lead-backed variant most abundant in AL and GA) or with a reddish or orangish middorsal stripe, occasionally straight-edged but usually with saw-toothed edges (lateral projections alternate with costal grooves); stripe often narrows slightly on base of tail. Sides brown with varying amounts of white, occasionally with some red on sides but almost never on belly; often with some red at base of legs. Costal grooves 18–22; 8–10 costal folds between adpressed limbs. **SIMILAR SPECIES:** (1) Ozark Zigzag and Southern Zigzag Salamanders have red pigment on the belly, a middorsal stripe (when present) with wavy or scalloped edges, longer limbs (6–7 costal folds between adpressed limbs), and usually 18 costal grooves. (2) Webster's Salamander usually has 18 costal grooves and a middorsal stripe (when present) with wavy edges at least anteriorly. See also Eastern Red-backed Salamander. **HABITAT:** Forested areas, often near seeps and springs during dry periods; occasionally twilight zones of caves. **RANGE:** Disjunct; mountains of w.-cen. AR and adjacent OK, cen. and se. MO, cen. LA, and ne. GA and adjacent AL, TN, and NC.

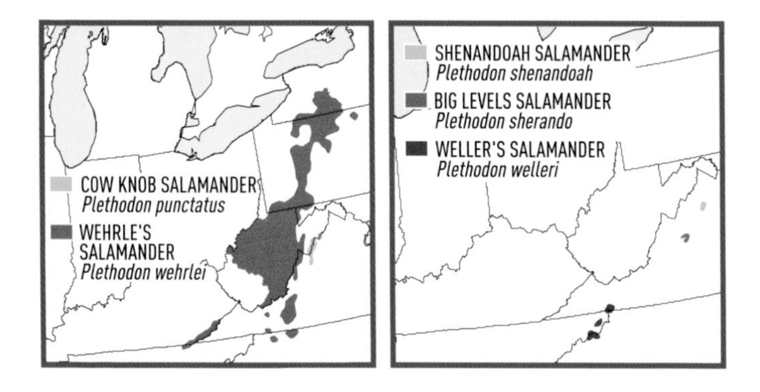

SHENANDOAH SALAMANDER *Plethodon shenandoah* **FIG. 43**

3–4$^{3}/_{8}$ in. (7.5–11 cm). Slender, elongated. Either uniformly dark brown to black with scattered brassy flecks or with a narrow red or yellow middorsal stripe. Sides with a variable number of small white spots. *Belly dark brown or black*, sometimes with scattered light, irregular markings; *chin dark* but usually lighter than belly and with some light flecks. Costal grooves 17–19 (usually 18). **SIMILAR SPECIES:**

FIG. 43. *The Shenandoah Salamander (*Plethodon shenandoah*) resembles the Eastern Red-backed Salamander (FIG. 36) and, like the latter, has both striped and unstriped variants. However, the striped variant of the Shenandoah Salamander has a much narrower middorsal stripe and the Eastern Red-backed Salamander has a ventral salt-and-pepper pattern, whereas the belly of the Shenandoah Salamander is uniformly dark.*

Valley and Ridge Salamander has a mostly white throat and belly with a moderate amount of white speckling. See also Eastern Red-backed Salamander. **HABITAT:** North-facing talus slopes in forested areas generally above 2,625 ft. (800 m). **RANGE:** Isolated populations on the 3 highest peaks in Shenandoah National Park, VA. **CONSERVATION:** Vulnerable (USFWS: Endangered).

BIG LEVELS SALAMANDER *Plethodon sherando* **FIG. 44**

2¼–4 in. (5.7–10 cm). Slender, elongated, long-tailed (over one-half total length). Either uniformly dark gray to black (lead-backed) or with a red, orange, yellow, or even light gray, *straight-edged* middorsal stripe extending from base of head to tail, usually *narrowing slightly* on base of tail, and bordered by dark pigment extending onto sides. Both striped and lead-backed variants often with numerous light flecks, resulting in a frosted effect. Belly light with extensive dark mottling, producing a *salt-and-pepper* effect. Costal grooves 17–20 (usually 18); 1–5 costal folds between adpressed limbs. **SIMILAR SPECIES:** See Eastern Red-backed Salamander. **HABITAT:** Forested areas. **RANGE:** Big Levels, VA. **CONSERVATION:** Vulnerable.

RED-LEGGED SALAMANDER *Plethodon shermani* **PL. 6**

3½–5½ in. (9–14 cm); record 6 11/16 in. (17 cm). Slender, black, usually with red blotches on limbs (limbs sometimes entirely red and sometimes red entirely absent), occasionally with red on cheeks. Few scattered white or yellow spots on back and sides. Belly light gray, usually

FIG. 44. *The light lichenlike mottling seen on many Big Levels Salamanders* (Plethodon sherando) *renders them highly cryptic. The range of this species consists of 15 localities, all within the range of the very similar Eastern Red-backed Salamander* (P. cinereus).

FIG. 45. *The Southern Appalachian Salamander* (Plethodon teyahalee) *is known to interbreed with the Red-cheeked Salamander* (P. jordani), *of which it was once thought to be a subspecies. Once recognized as being distinct, it was named* P. oconaluftee, *but that name did not follow the rules of scientific nomenclature and was subsequently changed to* P. teyahalee.

without light spots. Costal grooves 15–16. **SIMILAR SPECIES:** Cheoah Bald Salamander generally has more abundant white or yellow spots on sides. See also Red-cheeked Salamander. **HABITAT:** Moist forests. **RANGE:** Disjunct; Nantahala Mts., sw. NC and adjacent GA and TN. **REMARKS:** Until recently considered a color variant of the Red-cheeked Salamander. Interbreeds with the Southern Appalachian Salamander. **CONSERVATION:** Vulnerable.

SOUTHERN APPALACHIAN SALAMANDER *Plethodon teyahalee* **FIG. 45**

4¾–6¾ in. (12.1–17.2 cm); record 8$^{3}/_{16}$ in. (20.7 cm). Black, with very small white dorsal spots and larger lateral white spots; *small red spots often present on legs*; belly uniformly slate gray; chin usually lighter than belly. Costal grooves usually 16. **SIMILAR SPECIES:** See Tellico, Red-cheeked, and Red-legged Salamanders. **HABITAT:** Higher forested slopes of Blue Ridge physiographic province. **RANGE:** Sw. NC and adjacent TN, SC, and GA. **REMARK:** Interbreeds with Cheoah Bald, Red-cheeked, and Red-legged Salamanders.

WEBSTER'S SALAMANDER *Plethodon websteri* **NOT ILLUS.**

2¾–3$^{3}/_{16}$ in. (7–8.2 cm). Slender, elongated. Either uniformly gray to dark gray (lead-backed) or with a red to yellowish middorsal stripe generally narrow and with straight edges for at least part of its length, or wider with wavy or scalloped margins at least anteriorly; edges of stripe sometimes indistinct; if wide, usually widest at base of tail and continuing *broadly* onto tail. Some individuals from throughout the range have numerous light flecks, resulting in a frosted effect. *Belly light with extensive black and orange or reddish mottling.* Costal grooves 17–19, usually 18; 6–7 costal folds between adpressed limbs. **SIMILAR SPECIES:** See Southern Zigzag and Southern Red-backed Salamanders. **HABITAT:** Forested hillsides, especially north-facing slopes with rocky outcrops, often along rocky streambeds. **RANGE:** Disjunct; w. SC to sw. MS and adjacent LA.

WEHRLE'S SALAMANDER *Plethodon wehrlei* **PL. 5**

4–5¼ in. (10–13.3 cm); record 6⅝ in. (16.8 cm). Dark gray or dark brown, with irregular white, bluish white, or yellowish spots and dashes along sides, often with brassy dorsal flecks and few very small white dots; middorsal area may lack conspicuous markings or have small *red, orange-red,* or *yellow* spots often arranged in pairs, especially in southern portions of the range. Some individuals from TN, KY, and WV have 2 rows of large yellow dorsal spots; individuals from Dixie Caverns and Blankenship Cave near Roanoke, VA, have a purplish brown back profusely frosted with small light flecks and bronzy mottling. Throat *white or blotched with white,* with white spots frequently extending onto the chest. Belly and underside of tail uniformly gray. Webbing on hindfeet often extends almost to tips of first 2 toes. Costal grooves 16–18 (usually 17). **SIMILAR SPECIES:** See Cow Knob Salamander (closely related and similar in appearance, but ranges do not overlap). **HABITAT:** Moist upland forests, also drier rocky forested hillsides or twilight zones of caves. **RANGE:** Sw. NY south to ne. TN and nw. NC.

WELLER'S SALAMANDER *Plethodon welleri* **PL. 5**

2½–3⅛ in. (6.4–7.9 cm). Slender, with a short, thick tail (less than one-half total length). Black with profusion of dull *golden or brassy dorsal blotches* that often run together. Belly *plain black or black with white spots.* Costal grooves 15–17 (usually 16). **SIMILAR SPECIES:** Southern Ravine Salamander is larger, with 19–23 costal grooves and a longer tail (about one-half total length). See also Eastern Red-backed Salamander. **HABITAT:** Spruce-fir, birch-hemlock, and primarily deciduous forests, often in rocky areas. **RANGE:** Isolated mountaintops from sw. VA south to ne. TN and adjacent NC. **CONSERVATION:** Endangered.

YONAHLOSSEE SALAMANDER *Plethodon yonahlossee* **PL. 6**

4½–7 in. (11.5–18 cm); record 8$^{11}/_{16}$ in. (22.1 cm). Broad *red or chestnut dorsal stripe* extends from neck well onto base of tail, *bordered laterally by light gray or whitish pigment*; individuals in the Bat Cave area of Rutherford Co., NC, have a reduced reddish stripe often broken into scattered flecks or blotches. Head, limbs, and tail usually black with

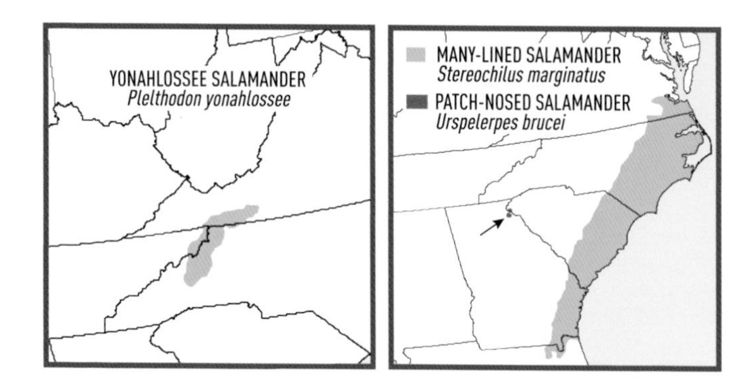

light specks. Belly dark gray, often mottled with light spots. Young with 4–6 pairs of red dorsal spots and a light belly. Usually 16 costal grooves. **SIMILAR SPECIES:** Other red- or chestnut-backed salamanders in range have red dorsal stripe bordered by dark instead of light pigment. **HABITAT:** Wooded hillsides and ravines, often in rockslides; areas with old windfalls, rock outcrops, or even grassy areas near woodlands. **RANGE:** Sw. VA, w. NC, and extreme ne. TN.

MANY-LINED and PATCH-NOSED SALAMANDERS: *Stereochilus* and *Urspelerpes*

These two genera, each with only one species, occur solely in eastern North America.

MANY-LINED SALAMANDER *Stereochilus marginatus* **PL. 6**

2½–3¾ in. (6.4–9.5 cm); record 4½ in. (11.4 cm). Head small; tail short, laterally compressed. Brown, occasionally dull yellow; sides with series of narrow, indistinct, dark *longitudinal lines*, sometimes reduced to a few dark spots; occasionally with indistinct dark (or light) markings on back. Belly yellow with scattered dark specks. **HABITAT:** Cypress and gum swamps, small ponds in pine forests, large drainage ditches, sluggish streams. **RANGE:** Coastal Plain from se. VA to n. FL.

PATCH-NOSED SALAMANDER *Urspelerpes brucei* **FIG. 46**

2–2⅜ in. (5–6 cm). *Yellow patch* on top of snout, thin yellow line along top of tail, belly yellow. Adult male yellow or yellowish or greenish brown, with black dorsolateral stripes and a vague black middorsal line often broken into a series of spots. Young male and female uniformly yellowish or greenish brown. **SIMILAR SPECIES:** (1) Dwarf Salamander has only 4 digits on all limbs and lacks light patches on snout. (2) Two-lined salamanders are sometimes similar but lack

FIG. 46. *The Patch-nosed Salamander* (Urspelerpes brucei) *is unique among plethodontid salamanders in our area in that the male and female have different patterns. The male has black dorsolateral stripes and a vague black middorsal line often broken into a series of spots; the female is uniformly yellowish or greenish brown.*

light patches on the snout and are much larger (newly metamorphosed individuals of comparable size will lack prominent cirri of an adult male Patch-nosed Salamander). **HABITAT:** Small streams in steep-walled ravines. **RANGE:** Foot of the Blue Ridge Escarpment in ne. GA and adjacent SC.

MUDPUPPIES and WATERDOGS: Proteidae

All but one species occur in eastern North America (the other in southern Europe). These aquatic salamanders retain external gills throughout life, with the size and condition of the gills usually reflecting environmental conditions: large and bushy in stagnant water low in dissolved oxygen, smaller in flowing water. Usually called mudpuppies in the North and waterdogs in the South. All have *four* toes on all limbs.

ALABAMA WATERDOG *Necturus alabamensis* **FIG. 47**

6–8½ in. (15.2–21.6 cm). Reddish to dark brown or black, nearly uniformly dark or spotted or mottled with sides paler than back. Dark spots fairly conspicuous in young but coalesce and blend with ground color in older individuals. Small scattered light spots often present. Center of belly white, *unmarked* (Fig. 47, facing page). Tips of digits whitish. Tail relatively short, dorsal fin low. **SIMILAR SPECIES:** See Common Mudpuppy. **HABITAT:** Medium to large streams. **RANGE:** N. AL. **REMARK:** Also called the Black Warrior River Waterdog. **CONSERVATION:** Endangered.

GULF COAST WATERDOG *Necturus beyeri* **PL. 6, FIGS. 6, 47**

6¼–8¾ in. (15.9–22.2 cm). Dark brown, but many tan spots join to form a reticulate pattern; dark brown to almost black round or oval spots scattered or in irregular rows. Belly invaded by spots and dorsal ground color (Fig. 47, facing page). Dark stripes through eyes some-

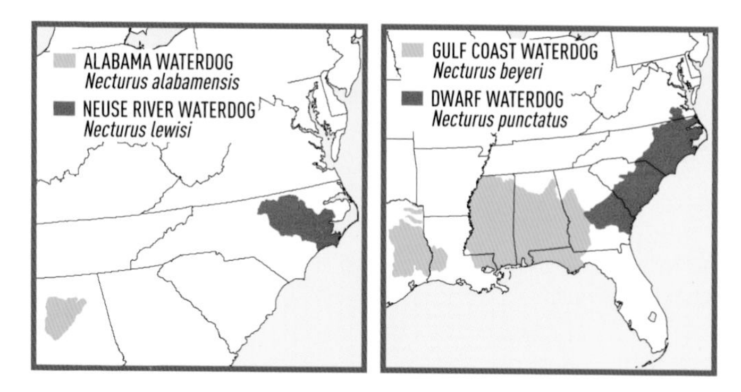

FIG. 47. *Ventral patterns of waterdogs and mudpuppies* (Necturus).

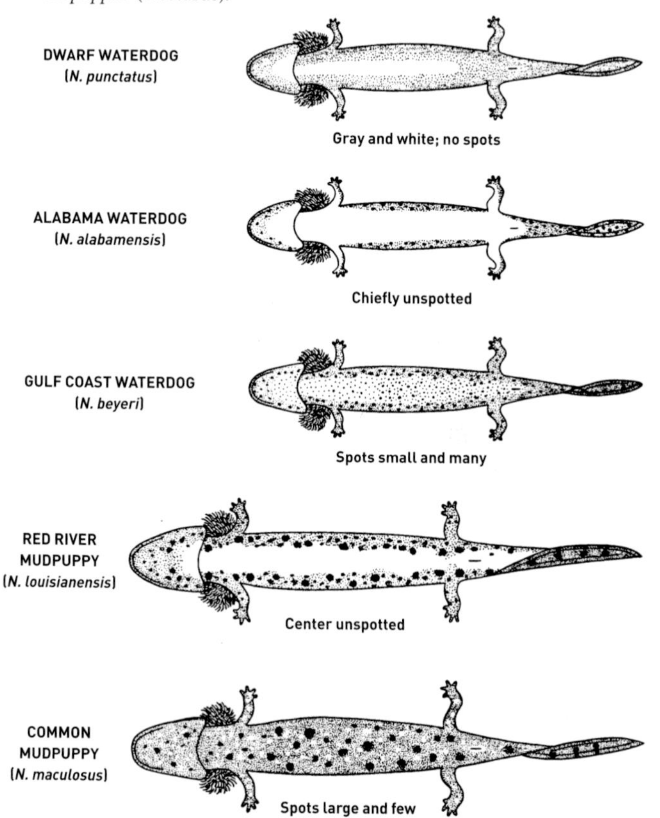

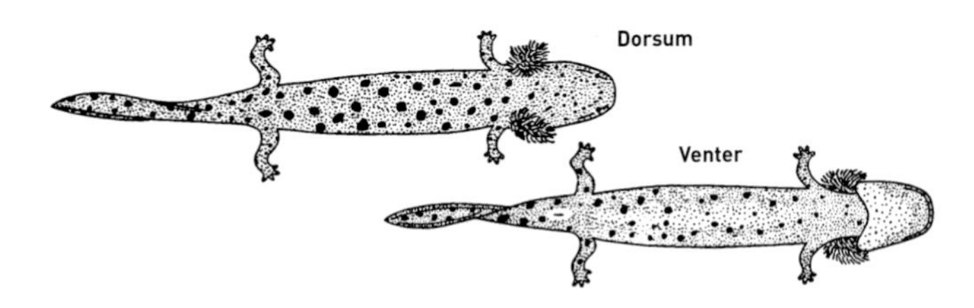

FIG. 48. *The back and belly of the Neuse River Waterdog* (Necturus lewisi) *are similarly patterned.*

times present. Larva (Fig. 6, p. 26) and juvenile with dull yellow spots on head and body and in rows on edges of tail fin; dark spots develop with age. **SIMILAR SPECIES:** See Common Mudpuppy. **HABITAT:** Spring-fed streams with sandy bottoms. **RANGE:** Disjunct; e. GA to the Mississippi R.; w. LA to e. TX. **REMARK:** Probably consists of 2 species.

NEUSE RIVER WATERDOG *Necturus lewisi* **FIG. 48**

6–9 in. (15.2–23 cm); record 10 7/8 in. (27.6 cm). Rusty brown, with dark brown or bluish black spots above and below (fewer and smaller on belly; Fig. 48, above). Dark lines through eyes. Belly dull brown or slate. Juvenile spotted; very small individuals sometimes with a light middorsal stripe and flecks on dark sides. **SIMILAR SPECIES:** See Dwarf Waterdog. **HABITAT:** Clean, moderate to swift-flowing water, especially larger streams. **RANGE:** Neuse and Tar R. systems, NC.

RED RIVER MUDPUPPY *Necturus louisianensis* **FIGS. 6, 47, 49**

7–9 in. (18–23 cm); record 12 in. (30.7 cm). Light yellowish brown or tan, with broad, fairly distinct dorsolateral light stripes bordering a darker middorsal area; dorsum usually with distinct scattered, rounded, dark brown to blue-black spots. Dark stripes through eyes extend from nostrils to base of gills. Center of belly grayish white, unmarked but tinged with pink (Fig. 47, p. 95). Larva (Fig. 6, p. 26) and juvenile almost always striped, with a broad middorsal dark stripe flanked by yellowish stripes and conspicuous dark lateral stripes extending from gills to tailtip. **SIMILAR SPECIES:** See Common Mudpuppy. **HABITAT:** Lakes, ponds, rivers, streams, other permanent bodies of water. **RANGE:** Arkansas R. and adjacent drainages from s. MO and se. KS to

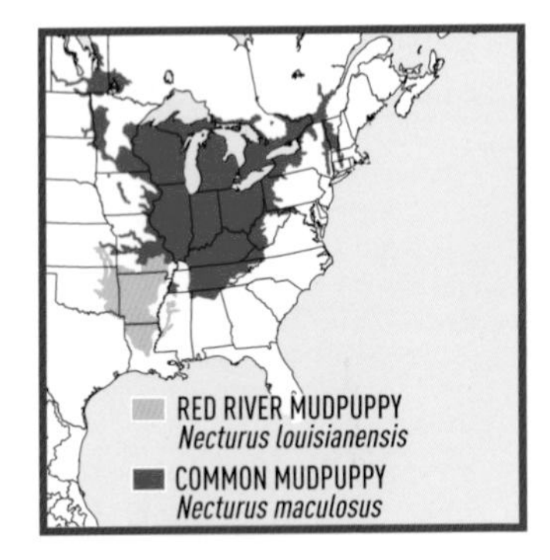

FIG. 49. *The Red River Mudpuppy* (Necturus louisianensis) *frequently is considered a subspecies of the Common Mudpuppy* (N. maculosus).

n. LA. **REMARK:** Until recently considered a subspecies of the Common Mudpuppy.

COMMON MUDPUPPY *Necturus maculosus* **PL. 6, FIGS. 6, 47**

8–13 in. (20–33 cm); record 19⅛ in. (48.6 cm). Gray or rusty brown to almost black; usually with rather indistinct, scattered, rounded, blue-black spots, rarely absent and occasionally fused to form dorsolateral stripes. Dark stripes through eyes. Tail fin often with an orange or reddish tinge. Belly grayish *with dark spots* (Fig. 47, p. 95). Larva (Fig. 6, p. 26) and juvenile almost always striped, with a broad middorsal dark stripe flanked by yellowish stripes and conspicuous dark lateral stripes extending from gills to tailtip; young rarely uniformly gray. **SIMILAR SPECIES:** (1) Red River Mudpuppy has unmarked midventral region and often has light lateral stripes. (2) Alabama Waterdog has an unmarked midventral region. (3) Gulf Coast Waterdog has numerous small dorsal spots, small spots on belly, and dark stripes through eyes are absent or poorly developed. (4) Hellbenders have a flat head, lateral skinfolds, and adults lack external gills. (5) Mole salamander larvae have 5 toes on hindlimbs. **HABITAT:** Lakes, ponds, rivers, streams, other permanent bodies of water. **RANGE:** S. QC west to se. MB and e. KS, south to the Tennessee R. system; presumably introduced in several places in New England.

DWARF WATERDOG *Necturus punctatus* **PL. 6, FIG. 47**

4½–6¼ in. (11.5–15.9 cm); record 7 7/16 in. (18.9 cm). Slate gray to dark brown or black (sometimes purplish black) with at most a few small pale spots. Throat whitish; central portion of belly bluish white, plain or partly invaded by dorsal color (Fig. 47, p. 95). Juvenile gray, with or without spots. **SIMILAR SPECIES:** (1) Neuse River Waterdog is conspicuously spotted. (2) Mole salamander larvae have 5 toes on hindlimbs. **HABITAT:** Slow, sand- or mud-bottomed streams and connected ditches, cypress swamps; stream-fed rice fields and millponds. **RANGE:** Coastal Plain from se. VA to GA.

NEWTS: Salamandridae

Twenty-one genera and more than 100 species occur in eastern and western North America, Europe, North Africa, and Asia. Newts have rougher skin than most salamanders and their costal grooves are barely discernible. Most are essentially aquatic, but two species in our area have a terrestrial eft stage between the aquatic larval and adult stages. The tail of efts is round in cross section and the skin is quite rough; the tail of adults is laterally compressed and the skin is smoother. Skin-gland secretions are toxic and most potent in bright red efts in eastern populations of the Eastern Newt.

BLACK-SPOTTED NEWT *Notophthalmus meridionalis* **PL. 7**

3–4 5/16 in. (7.5–11 cm). *Large* black spots on back and belly, wavy or uneven *yellowish dorsolateral stripes*, often with a suggestion of a brown or russet middorsal stripe; no red spots. Venter bright orange to yellow-orange. **SIMILAR SPECIES:** See Eastern Newt. **HABITAT:** Ponds, lagoons, swampy areas, roadside ditches, quiet stream pools. **RANGE:** S. TX south into Mex. **SUBSPECIES:** **TEXAS BLACK-SPOTTED NEWT** (*N. m. meridionalis*). **CONSERVATION:** Endangered.

STRIPED NEWT *Notophthalmus perstriatus* **PL. 7**

2–4 1/8 in. (5.1–10.5 cm). Frequently neotenic. Olive green to dark brown. *Red, narrowly and unevenly black-bordered dorsolateral stripes* continuous on trunk but often broken on head and tail; stripes bright to dull red, sometimes partly obscured by dusky pigment. Sometimes with lateral rows of red spots and a faint light middorsal stripe. Venter yellow, usually sparsely marked with black specks. Eft orange-red, with red stripes like those of adult. **SIMILAR SPECIES:** See Eastern Newt. **HABITAT:** Ephemeral wetlands, almost any body of shallow, standing water in sandhills, scrub, and flatwoods. **RANGE:** GA and n. FL.

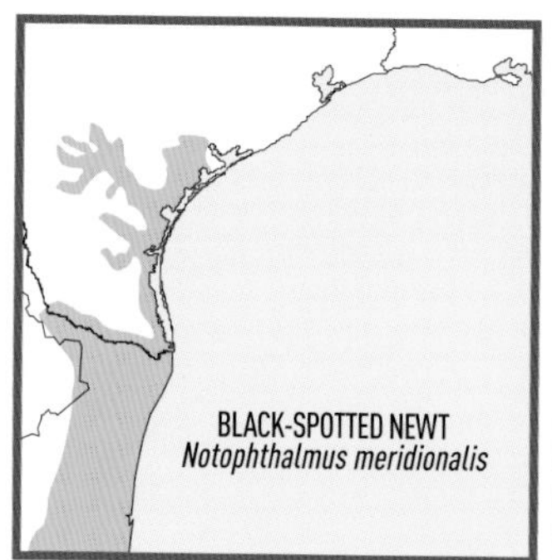

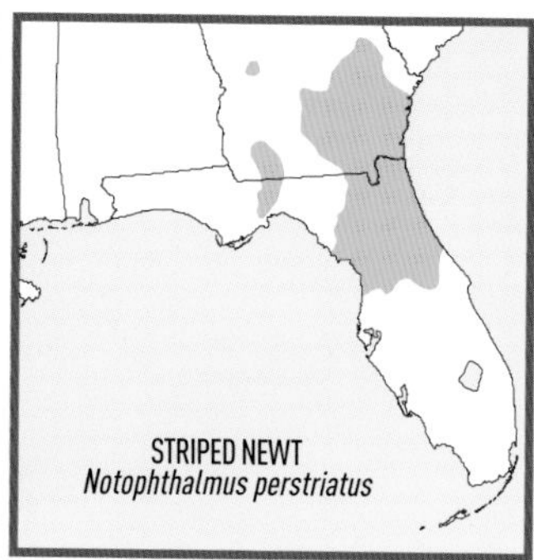

EASTERN NEWT *Notophthalmus viridescens* **PL. 7, FIG. 50**

2¼–4 13/16 in. (5.7–12.2 cm); record 5½ in. (14 cm). Aquatic adult normally olive green (also yellowish or olive brown to dark greenish brown or almost black); red spots or dorsolateral stripes variously outlined in black; stripes, if present, broken at least once on head or body and rarely extend onto tail. Dorsal ground color vaguely or sometimes sharply demarcated from yellow or orange-yellow belly, which bears small black spots or a peppering of dark specks. Male with high tail fin and black excrescences on hindlimbs during spring breeding season (may develop again in autumn). Eft 1⅜–3⅜ in. (3.5–8.6 cm), bright orange-red, dull red, orange, or brown with a faint reddish cast (usually brightest in moist forested mountains or other uplands, especially in eastern portions of the range). Recently transformed efts yellowish brown or dull reddish brown, individuals transforming into adults often very dark or almost black. **SIMILAR SPECIES:** (1) Striped Newt has narrowly and unevenly bordered red dorsolateral stripes usually continuous at least on body. (2) Black-spotted Newt has large black spots on back and belly and yellowish dorsolateral stripes. **HABITAT:** Ponds, small lakes, marshes, ditches, quiet portions of streams, other permanent or semipermanent bodies of water; efts often on forest floor. **RANGE:** Atlantic Provinces west to the Great Lakes and south through FL and much of TX. **SUBSPECIES:** (1) **Red-spotted Newt** (*N. v. viridescens*); as many as 21 red spots present at all life stages; belly yellow with small black spots. (2) **Broken-striped Newt** (*N. v. dorsalis*); black-bordered red dorsolateral stripes broken at least once or twice on head and trunk, rarely extending onto tail; sometimes with lateral rows of small red spots and a light middorsal line; eft reddish brown, with red stripes not strongly black-bordered. (3) **Central Newt** (*N. v. louisianensis*); usually without red spots (if present, small or only partly outlined by black); dorsal ground color sharply demarcated from yellow belly; efts uncommon, neoteny frequent on the Coastal Plain of se. states. (4) **Peninsula Newt** (*N. v. piaropicola*); dorsum olive, dark brown, or almost black; belly peppered with black specks on

FIG. 50. *This brightly colored red eft is the terrestrial form of the Red-spotted Newt* (Notophthalmus viridescens viridescens; *a subspecies of the Eastern Newt*). *The eft bridges the developmental period between the aquatic larval and adult stages. Its skin-gland secretions are toxic, causing many potential predators to avoid efts.*

yellow or orange-yellow; efts rare, neoteny common. **REMARK:** Eft stage is sometimes omitted, especially in Coastal Plain populations.

SIRENS: Sirenidae

Two genera are endemic to the southeastern United States. These aquatic salamanders *lack hindlimbs* and retain external gills throughout life, with the size and condition of the gills usually reflecting environmental conditions: large and bushy in stagnant water low in dissolved oxygen, smaller in flowing water. Dwarf sirens have *three digits* on each limb, sirens *four*.

SOUTHERN DWARF SIREN *Pseudobranchus axanthus* **PL. 7, FIG. 7**

4¾–7½ in. (12.1–19 cm); record 9⅞ in. (25.1 cm). Broad, buff-colored lateral stripes; broad, dark middorsal stripe contains 3 narrow light stripes or is reduced to narrow dark lines. Thirty-two chromosome pairs. **SIMILAR SPECIES:** See Northern Dwarf Siren. **HABITAT:** Open marsh and prairie ponds, low-gradient streams, lake tributaries. **RANGE:** Peninsular FL. **SUBSPECIES:** (1) **NARROW-STRIPED DWARF SIREN** (*P. a. axanthus*); snout blunt; dorsal dark lines narrow, light stripes with vague margins, pattern details obscure; looks "muddy." (2) **EVERGLADES DWARF SIREN** (*P. a. belli*); small, slender; broad dark middorsal stripe contains 3 narrow light stripes; lateral stripes sharply defined; belly gray.

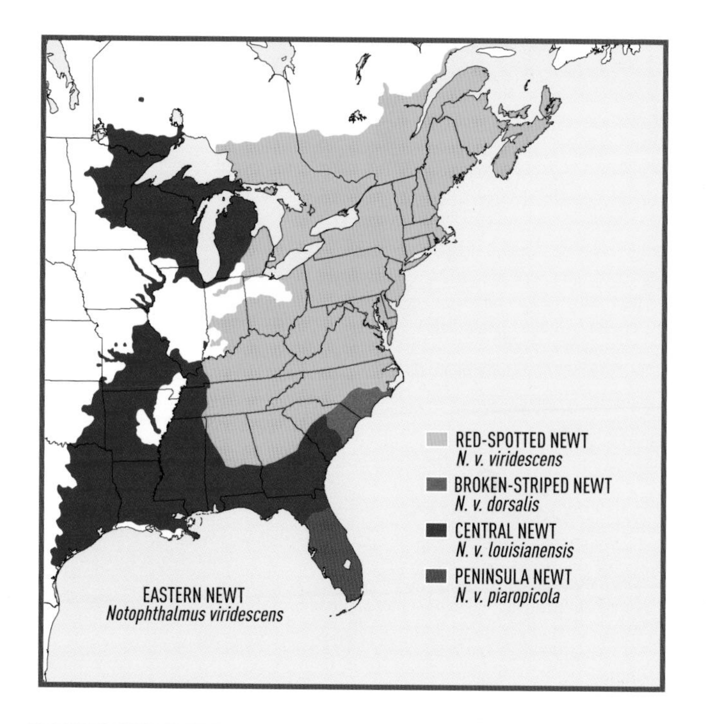

NORTHERN DWARF SIREN *Pseudobranchus striatus* **PL. 7, FIG. 7**

4–8½ in. (10–21.6 cm); record 8⅝ in. (22 cm). Stripes distinct, including wide buff, tan, or orange-brown lateral stripes. Belly gray to black with scattered light spots. Twenty-four chromosome pairs. **SIMILAR SPECIES:** (1) Southern Dwarf Siren has 32 chromosome pairs, and individuals in the contact zone between the two species tend to have vague patterns and occupy open marshes and prairie ponds instead of cypress ponds. (2) Greater and Lesser Sirens have 4 digits on each limb. **HABITAT:** Cypress domes and strands, marshes, ponds, ditches, Carolina bays, other permanent and temporary shallow freshwater habitats. **RANGE:** S. SC to n. FL. **SUBSPECIES:** (1) **Broad-striped Dwarf Siren** (*P. s. striatus*); body short and stocky; broad dark brown middorsal stripe contains a vague light middorsal line and is bordered by broad yellow stripes; belly dark but heavily mottled with yellow. (2) **Gulf Hammock Dwarf Siren** (*P. s. lustricolus*); body stout, snout blunt; broad dark middorsal stripe contains 3 narrow light stripes; 2 broad, sharply defined light lateral stripes, upper stripes orange-brown, lower stripes silvery white; belly black with light mottling. (3) **Slender Dwarf Siren** (*P. s. spheniscus*); snout narrow, wedge-shaped; dark middorsal stripe contains 3 vague light lines; 2 lateral stripes are bright tan or yellow.

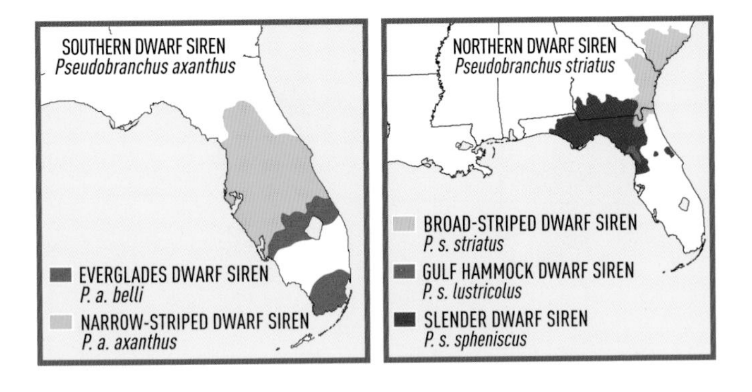

LESSER SIREN *Siren intermedia* **PL. 7, FIG. 51**

7–27 in. (18–68.6 cm). Dark brown to bluish black, sometimes olive green; darker individuals virtually without markings, lighter animals with irregularly scattered black dots. Generally 31–35 costal grooves. Juvenile with a red band across snout and alongside head; older juvenile olive green with tiny brown spots. **SIMILAR SPECIES:** (1) Greater Siren generally has at least some pronounced light markings and 36–40 costal grooves. (2) Dwarf sirens have only 3 digits on each limb. (3) Amphiumas lack external gills and have 4 small limbs. (4) True eels (fish, not amphibians) have lateral fins behind the head and no limbs.

FIG. 51. *Sirens, like this Eastern Lesser Siren* (Siren intermedia intermedia), *are long, slender, and look superficially like eels (which are fish, not amphibians). The tiny forelimbs are very close to the gills and hindlimbs are missing. Distinguishing Lesser Sirens and Greater Sirens* (S. lacertina) *is often difficult.*

HABITAT: Swamps, sloughs, ponds, lakes, ditches, less frequently rivers and streams. **RANGE:** E. VA into FL, west to s. TX and adjacent Mex., and north along the Mississippi River Valley to s. MI. **SUBSPECIES:** (1) **Eastern Lesser Siren** (*S. i. intermedia*); small, max. length 15 in. (38.1 cm); plain black or brown above or with minute black dots sprinkled over the dorsal surface and tail; venter uniformly dark but paler than dorsum; costal grooves 31–35. (2) **Western Lesser Siren** (*S. i. nettingi*); olive to gray or brownish gray above with scattered, minute black spots; venter dark with numerous light spots or, especially in animals from the Lower Rio Grande Valley, light gray but even paler behind angles of the jaw, under gills and limbs, and around vent; costal grooves 34–38. **REMARK:** Lineages possibly representing distinct species do not correspond to traditionally recognized subspecies.

GREATER SIREN *Siren lacertina* **PL. 7**

20–30 in. (51–76 cm); record 38½ in. (97.8 cm). Olive to light gray, back darker than sides, sides with faint greenish or yellowish dots and dashes; some individuals with defined circular black spots on top of head, back, and sides. Costal grooves 36–40. Belly with numerous small greenish or yellowish flecks. Juvenile has prominent light lateral stripes and a light dorsal fin but becomes almost uniformly dark with age. **SIMILAR SPECIES:** See Lesser Siren. **HABITAT:** Ditches, ponds, lakes, streams. **RANGE:** Vicinity of Washington, DC, to s. FL and s. AL.

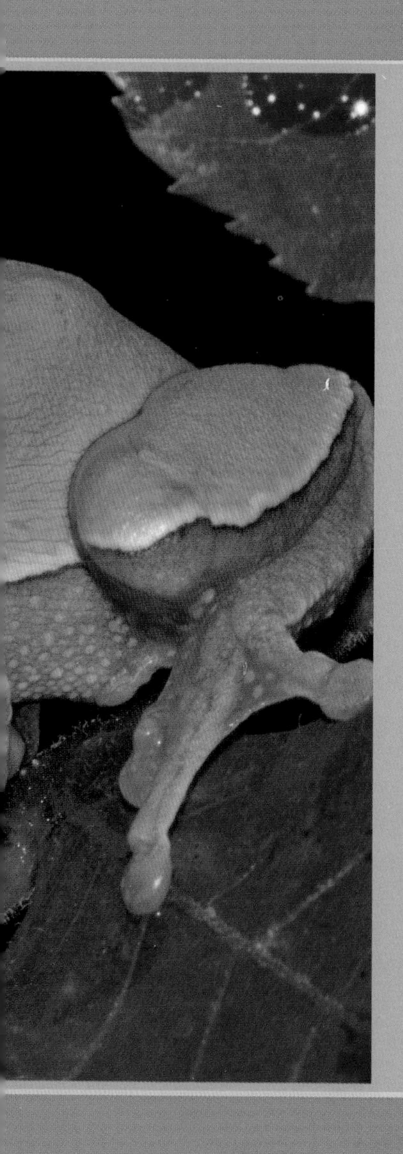

FROGS

NORTH AMERICAN TOADS (*Anaxyrus*)

Two tubercles under hindfoot—spadelike in some species. See also Figs. 53 and 54 (p. 117).

SOUTHERN TOAD *A. terrestris* **P. 123**
Prominent cranial knobs.

AMERICAN TOAD *A. americanus* **P. 116**
One or two large warts in each dark spot.

CANADIAN TOAD *A. hemiophrys* **P. 120**
Boss between eyes.

CHIHUAHUAN GREEN TOAD *A. debilis* **P. 119**
WESTERN CHIHUAHUAN GREEN TOAD (*A. d. insidior*): Black reticulate pattern on at least part of body.

EASTERN CHIHUAHUAN GREEN TOAD (*A. d. debilis*): Spotted with black and yellow.

FOWLER'S TOAD *A. fowleri* **P. 120**
Three or more warts in most dark spots.

WOODHOUSE'S TOAD *A. woodhousii* **P. 123**
Warts in dark spots variable in number.

RED-SPOTTED TOAD *A. punctatus* **P. 121**
Parotoid glands round; warts often reddish; no middorsal light stripe.

TEXAS TOAD *A. speciosus* **P. 122**
Parotoid glands oval; no light middorsal stripe; cranial crests indistinct or absent.

GREAT PLAINS TOAD *A. cognatus* **P. 118**
Large light-edged dark dorsal blotches; boss on snout.

OAK TOAD *A. quercicus* **P. 122**
Paired black spots along light middorsal line.

CENTRAL AMERICAN TOADS (*Incilius*)

GULF COAST TOAD *I. nebulifer* **P. 125**
Broad dark lateral stripes bordered above by light stripes; deep "valley" between eyes.

SOUTH AMERICAN TOADS (*Rhinella*)

CANE TOAD *R. marina* **P. 125**
Parotoid glands enormously enlarged and extending onto sides.

NORTHERN RAINFROGS (*Craugastor*)

BARKING FROG *C. augusti* **P. 126**
Toadlike but no warts; stout forearms; dorsolateral skinfolds.

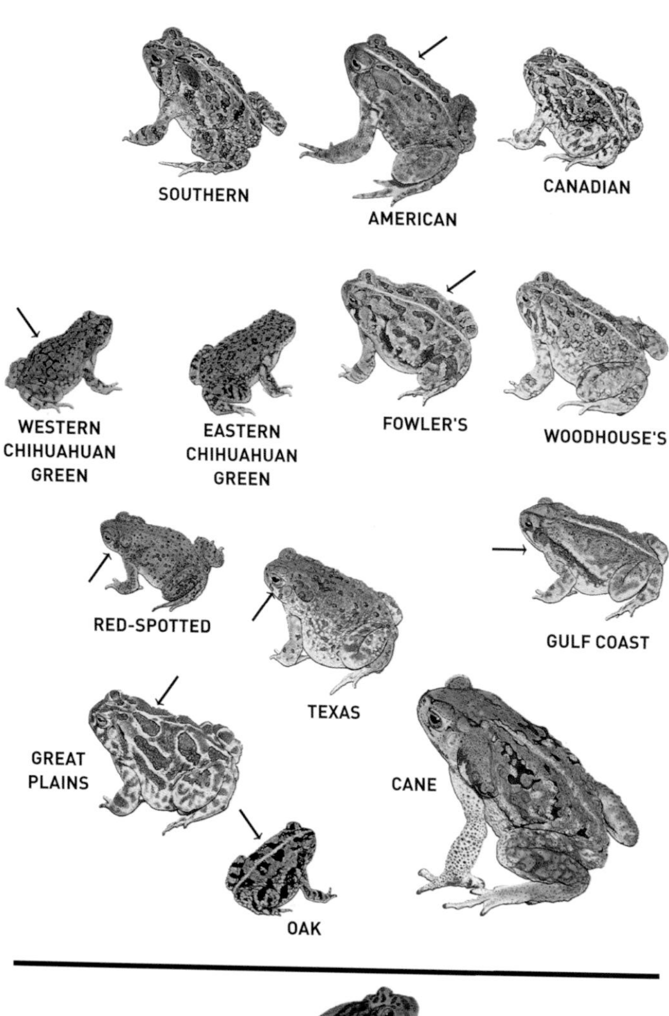
SOUTHERN
AMERICAN
CANADIAN
WESTERN CHIHUAHUAN GREEN
EASTERN CHIHUAHUAN GREEN
FOWLER'S
WOODHOUSE'S
RED-SPOTTED
TEXAS
GULF COAST
GREAT PLAINS
CANE
OAK

BARKING

PLATE 9

RAINFROGS (*Eleutherodactylus*)

SPOTTED CHIRPING FROG *E. guttilatus* **P. 127**
Dark bar between eyes; dark vermiculate pattern.

CLIFF CHIRPING FROG *E. marnockii* **P. 128**
Greenish; dark markings but no definite pattern; skin smooth.

RIO GRANDE CHIRPING FROG *E. cystignathoides* **P. 127**
Usually grayish brown or grayish olive; dark shadow from eyes to nostrils; legs with crossbars.

CUBAN FLAT-HEADED FROG *E. planirostris* **Non-native** **P. 128**
Striped or mottled.

CRICKET FROGS (*Acris*)

EASTERN CRICKET FROG *A. crepitans* **P. 130**
Short hindlimbs; dark triangle between eyes.

SOUTHERN CRICKET FROG *A. gryllus* **P. 131**
Long hindlimbs; dark triangle between eyes.

BLANCHARD'S CRICKET FROG *A. blanchardi* **P. 129**
Plump, stocky body; dark triangle between eyes.

HOLARCTIC TREEFROGS (*Hyla*)

PINE WOODS TREEFROG *H. femoralis* **P. 135**
Dark dorsal blotches; light spots on concealed surface of thighs (Fig. 62, p. 135).

SQUIRREL TREEFROG *H. squirella* **P. 136**
Green or brown or combinations of both; highly variable.

PINE BARRENS TREEFROG *H. andersonii* **P. 132**
Whitish (or yellowish) and purplish lateral stripes.

GREEN TREEFROG *H. cinerea* **P. 134**
Light lateral stripes variable in length, occasionally absent; small scattered yellowish dorsal spots.

COPE'S GRAY TREEFROG *H. chrysoscelis* **P. 133**
GRAY TREEFROG *H. versicolor* **P. 133**
Light spots below eyes; concealed surface of hindlimbs yellowish orange.

BIRD-VOICED TREEFROG *H. avivoca* **P. 133**
Light spots below eyes; concealed surface of hindlimbs greenish or yellowish white.

CANYON TREEFROG *H. arenicolor* **P. 132**
Dark bars in front of light spots below eyes.

BARKING TREEFROG *H. gratiosa* **P. 135**
Stout; profusion of dark rounded spots.

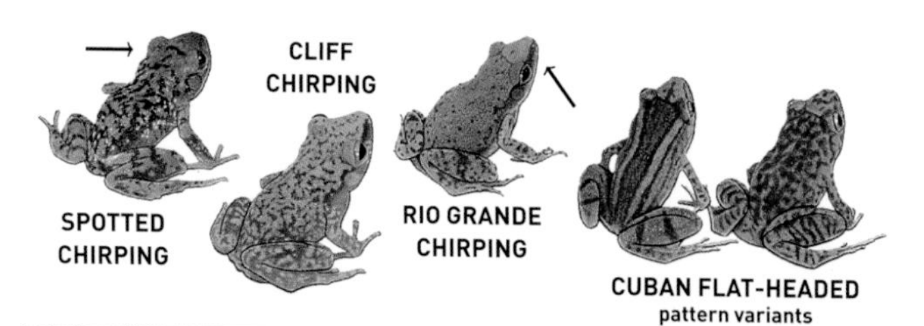
CLIFF CHIRPING
SPOTTED CHIRPING
RIO GRANDE CHIRPING
CUBAN FLAT-HEADED
pattern variants
CRICKET FROGS
EASTERN
SOUTHERN
BLANCHARD'S

PINE WOODS
SQUIRREL
color variants
GREEN
PINE BARRENS
pattern variants
GRAY
BIRD-VOICED
CANYON
BARKING
pattern variants

PLATE 10

WEST INDIAN and MEXICAN TREEFROGS
(*Osteopilus* and *Smilisca*)

CUBAN TREEFROG *O. septentrionalis* **Non-native** **P. 136**
Large toepads; warty skin; skin on head fused to skull.

MEXICAN TREEFROG *S. baudinii* **P. 137**
Light spots below eyes; black patches on shoulders behind eardrums; light spots at base of forelimbs.

CHORUS FROGS (*Pseudacris*)

BOREAL CHORUS FROG *P. maculata* **P. 143**
Three broad longitudinal dark stripes; hindlimbs short.

MIDLAND CHORUS FROG *P. triseriata* **P. 145**
Three broad longitudinal dark stripes (Fig. 52); hindlimbs long.

BRIMLEY'S CHORUS FROG *P. brimleyi* **P. 138**
Black lateral stripes; dorsal stripes brown or gray; dark spots on chest.

SPRING PEEPER *P. crucifer* **P. 140**
Dark X-shaped mark on back.

UPLAND CHORUS FROG *P. feriarum* **P. 140**
Pattern spotted or weakly striped (Fig. 52).

SOUTHERN CHORUS FROG *P. nigrita* **P. 144**
White lines on lips, lips black-and-white spotted on frogs from FL; dorsal stripes usually broken, frogs in FL with three rows of black spots (Fig. 52).

LITTLE GRASS FROG *P. ocularis* **P. 144**
Dark lines through eyes; pattern and coloration variable.

MOUNTAIN CHORUS FROG *P. brachyphona* **P. 138**
Dorsal stripes curved (Fig. 52); triangle between eyes.

ORNATE CHORUS FROG *P. ornata* **P. 145**
Bold black spots on sides and near groin.

SPOTTED CHORUS FROG *P. clarkii* **P. 139**
Black-bordered green dorsal markings; triangle between eyes.

STRECKER'S CHORUS FROG *P. streckeri* **P. 142**
Dark spots below eyes; forearms stout.

FIG. 52. *Dorsal patterns of some chorus frogs* (Pseudacris).

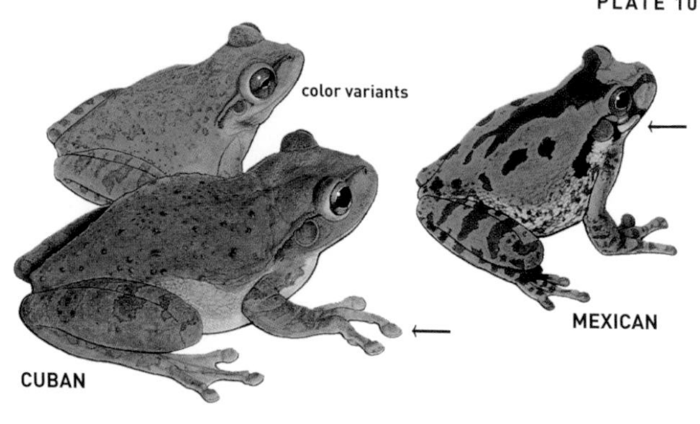
color variants
MEXICAN
CUBAN

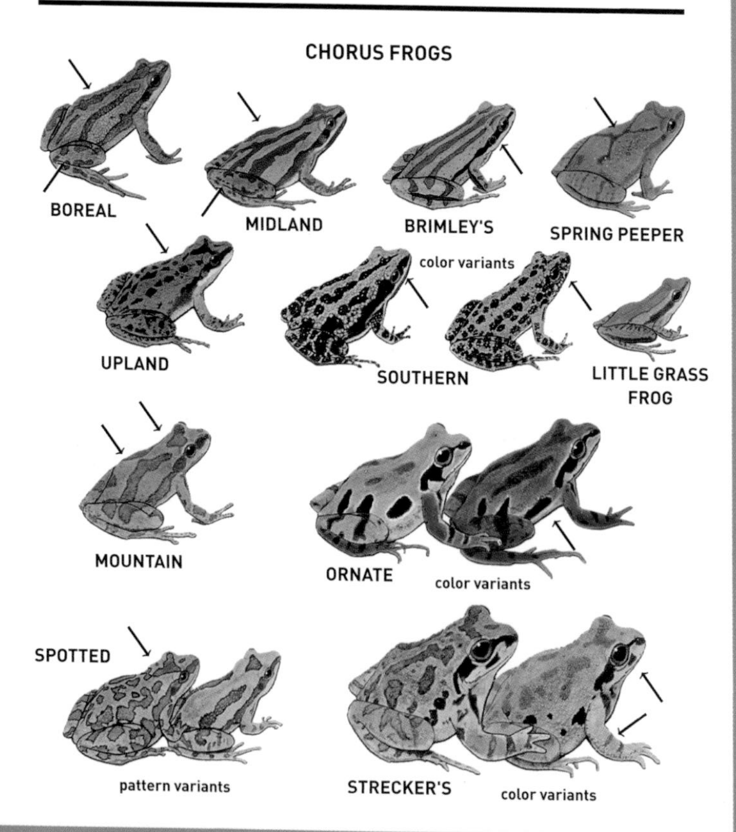
CHORUS FROGS
BOREAL
MIDLAND
BRIMLEY'S
SPRING PEEPER
UPLAND
color variants
SOUTHERN
LITTLE GRASS FROG
MOUNTAIN
ORNATE
color variants
SPOTTED
pattern variants
STRECKER'S
color variants

PLATE 11

NEOTROPICAL THIN-TOED FROGS (*Leptodactylus*)

MEXICAN WHITE-LIPPED FROG *L. fragilis* P. 146
Light lines on lips; ventral disc (Fig. 57, p. 126).

NARROW-MOUTHED FROGS (*Gastrophryne* and *Hypopachus*)

WESTERN NARROW-MOUTHED FROG *G. olivacea* P. 147
Plain gray, tan, or greenish, with or without small black spots.

EASTERN NARROW-MOUTHED FROG *G. carolinensis* P. 146
Broad light dorsolateral stripes; stripes reddish in Key West variant.

SHEEP FROG *H. variolosus* P. 148
Narrow light middorsal line.

AMERICAN WATER FROGS (*Lithobates*)

Most have dorsolateral ridges (Fig. 70, p. 150). Males of species marked with an asterisk (*) have eardrums larger than eyes.

FLORIDA BOG FROG *L. okaloosae* P. 156
Small size; dorsum unspotted; fourth toe extends beyond webbing.

MINK FROG *L. septentrionalis** P. 158
Mottled or spotted (Fig. 75, p. 158); often without dorsolateral ridges.

CARPENTER FROG *L. virgatipes** P. 160
Golden brown dorsolateral and lateral stripes; no dorsolateral ridges.

WOOD FROG *L. sylvaticus* P. 160
Dark mask through eyes.

NORTHERN LEOPARD FROG *L. pipiens* P. 157
Light-bordered dark spots; snout rounded; dorsolateral ridges unbroken.

SOUTHERN LEOPARD FROG *L. sphenocephalus* P. 159
Light spots on eardrums; pointed snout; dorsolateral ridges unbroken.

RIO GRANDE LEOPARD FROG *L. berlandieri* P. 149
Pallid; no spots in front of eyes; dorsolateral ridges inset before groin.

PLAINS LEOPARD FROG *L. blairi* P. 151
Brown (rarely green); spot on snout; dorsolateral ridges inset before groin.

PICKEREL FROG *L. palustris* P. 156
Squarish dark spots in parallel rows; bright yellow or orange on hindlimbs.

GOPHER FROG *L. capito* P. 151
Stubby; irregular markings on light ground color; belly spotted (Fig. 69, p. 150).

DUSKY GOPHER FROG *L. sevosus* P. 158
Stubby; dark; belly spotted (Fig. 69, p. 150).

CRAWFISH FROG *L. areolatus* P. 148
Stubby; light-bordered spots rounded; belly mostly plain (Fig. 69, p. 150).

MEXICAN
WHITE-
LIPPED
SHEEP FROG

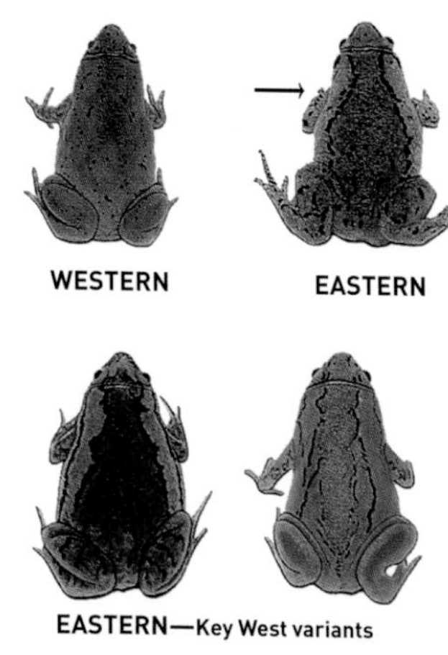
WESTERN
EASTERN
EASTERN—Key West variants

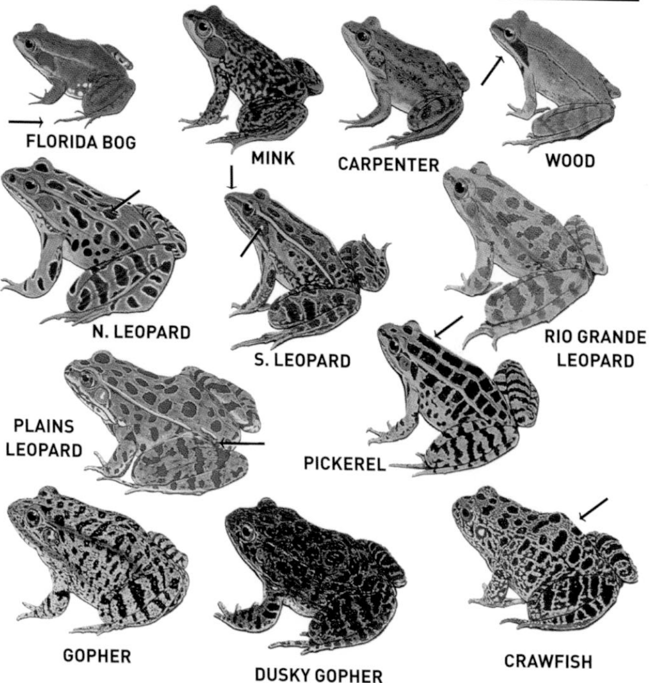
FLORIDA BOG
MINK
CARPENTER
WOOD
N. LEOPARD
S. LEOPARD
RIO GRANDE
LEOPARD
PLAINS
LEOPARD
PICKEREL
GOPHER
DUSKY GOPHER
CRAWFISH

AMERICAN WATER FROGS (*Lithobates*) (cont.)

Males of species marked with an asterisk (*) have eardrums larger than eyes. See also Fig. 70 (p. 150).

GREEN FROG *L. clamitans** **P. 153**
Green, brown, or bronzy; dorsolateral ridges extend about two-thirds length of body.

RIVER FROG *L. heckscheri** **P. 155**
Dark; white spots on lips; no dorsolateral ridges. See also Fig. 74 (p. 155).

PIG FROG *L. grylio** **P. 154**
Pointed snout; fourth toe extends only slightly beyond webbing (Fig. 72, p. 152); no dorsolateral ridges.

AMERICAN BULLFROG *L. catesbeianus** **P. 152**
Blunt snout; fourth toe extends well beyond webbing (Fig. 72, p. 152); no dorsolateral ridges; pattern variable.

BURROWING TOADS (*Rhinophrynus*)

BURROWING TOAD *R. dorsalis* **P. 161**
Rotund; cone-shaped snout; orange middorsal stripe.

NORTH AMERICAN SPADEFOOTS (*Scaphiopus* and *Spea*)

Single sharp-edged spade on hindfeet (Fig. 76, p. 162).

COUCH'S SPADEFOOT *Scaphiopus couchii* **P. 162**
Dark irregular pattern on yellowish or greenish ground color; spades elongate, sickle-shaped.

HURTER'S SPADEFOOT *S. hurterii* **P. 164**
Light dorsal lines; boss between eyes (Fig. 78, p. 163); spades elongate, sickle-shaped; pectoral glands (Fig. 77, p. 163).

EASTERN SPADEFOOT *S. holbrookii* **P. 162**
Light dorsal lines in form of a lyre; no boss between eyes; spades elongate, sickle-shaped; pectoral glands (Fig. 77, p. 163).

PLAINS SPADEFOOT *Spea bombifrons* **P. 164**
Faint longitudinal light stripes; boss between eyes (Fig. 78, p. 163); spades rounded, wedge-shaped.

MEXICAN SPADEFOOT *S. multiplicata* **P. 165**
Gray, green, or brown with small dark markings; spades rounded, wedge-shaped.

GREEN
color variants
RIVER
PIG
AMERICAN
BULLFROG

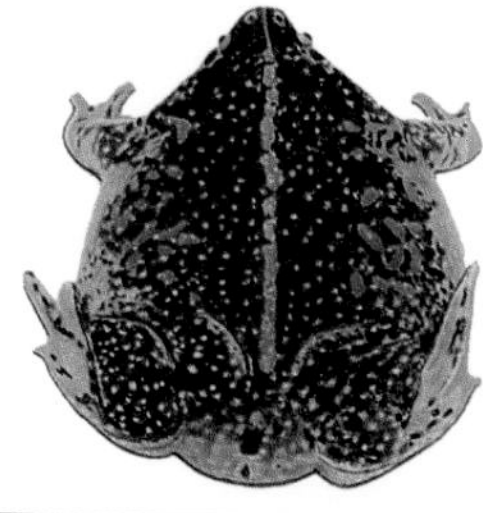
BURROWING
TOAD

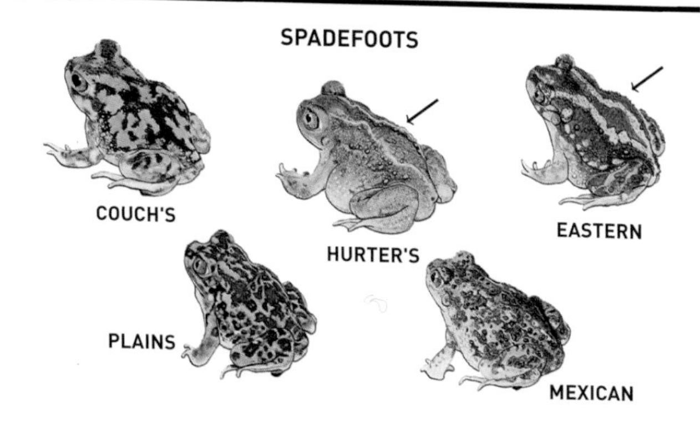
SPADEFOOTS
COUCH'S
HURTER'S
EASTERN
PLAINS
MEXICAN

FROGS (ANURA)

Nearly 6,500 species range from above the Arctic Circle to almost the southern tips of Africa, Australia, and South America. No hard-and-fast criteria distinguish frogs and toads. Typical toads (Bufonidae) have warty skin and short legs, whereas typical frogs (Ranidae) have relatively smooth skin and long legs. However, species in many families exhibit numerous variations that defy placement in either category.

Males of most species (all in our area) advertise their presence with distinctive mating calls. Frogs also may scream when seized by predators, utter territorial calls, vocalize during wet weather, and issue various warning chirps or croaks (including release calls of males or unresponsive females to discourage males that indiscriminately grasp anything during the excitement of courtship). We include brief descriptions in the species accounts, but listening to recordings is much more informative (see listings on p. 12).

Most species lay eggs that typically hatch into aquatic larvae called tadpoles. We have made no effort to identify tadpoles in this guide and refer interested readers to the book on larval amphibians by Altig and McDiarmid (see References, p. 456). Additional information is available in several online guides, including www.pwrc.usgs.gov/tadpole/.

TOADS: Bufonidae

Fifty genera and about 575 species have a nearly worldwide distribution. Three genera and 13 native species occur in our area. All until recently were included in the genus *Bufo*, a name now restricted to species in the Eastern Hemisphere. Toads in our area have distinct warts, horizontally oval pupils, two tubercles on the hindfeet, well-developed parotoid glands, and most have prominent cranial crests (Figs. 53 and 54, facing page). Touching toads does not cause warts. However, skin-gland secretions are irritating to mucous membranes, so wash your hands after handling these animals.

NORTH AMERICAN TOADS: *Anaxyrus*

AMERICAN TOAD *Anaxyrus americanus* **PL. 8, FIG. 54**

2–3½ in. (5.1–9 cm); record 6⅛ in. (15.5 cm). Brown to gray, olive, or brick red, sometimes almost uniformly colored but often (especially female) with patches of yellow, buff, or other light colors. Light middorsal line sometimes present. Dark spots brown or black, warts yellow, orange, or red to dark brown. Usually *1–2 large warts* in largest dark spots. Chest and front of abdomen *usually spotted. Enlarged warts on tibia.* Parotoid glands usually kidney-shaped, either *separated from ridges behind eyes or connected to them by short spurs* (Fig. 54, facing page). **SIMILAR SPECIES:** See Woodhouse's Toad. **HABITAT:** Forests, prairies, suburban backyards, anywhere with shallow water for breeding. **VOICE:** Long musical trills lasting 6–30 sec.; trill rate about

FIG. 53. *Characteristics of North, Central, and South American Toads* (Anaxyrus, Incilius, Rhinella).

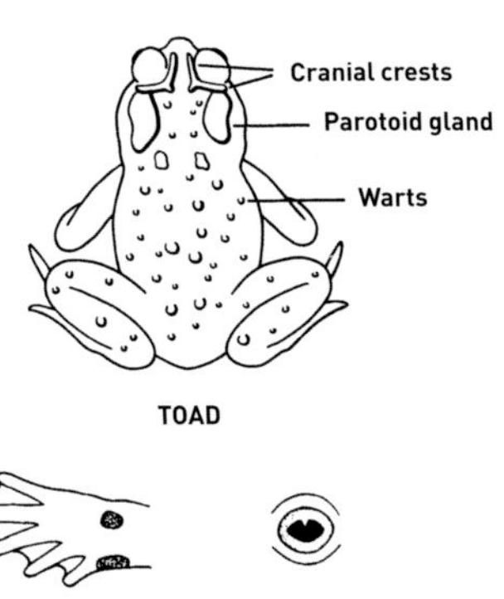

TOAD

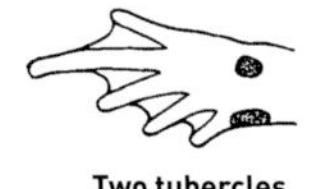

Two tubercles

Horizontal pupil

FIG. 54. *Cranial crests and parotoid glands of various toads* (Anaxyrus *and* Incilius).

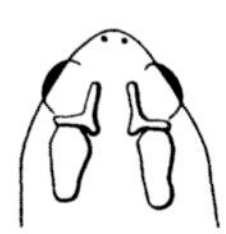

FOWLER'S
(*A. fowleri*)
Parotoid touches postorbital ridge

AMERICAN
(*A. americanus*)
Parotoid separate or connected to ridge by a spur

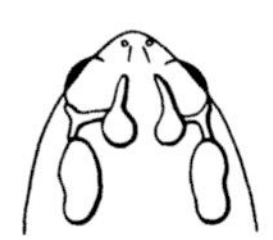

SOUTHERN
(*A. terrestris*)
Pronounced knobs at rear of interorbital crests

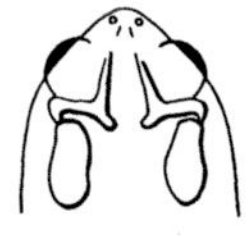

HOUSTON
(*A. houstonensis*)
Postorbital ridges thickened

CANADIAN
(*A. hemiophrys*)
A boss between eyes

GREAT PLAINS
(*A. cognatus*)
Interorbital ridges converge to meet a boss on the snout

RED-SPOTTED
(*A. punctatus*)
Parotoid round

GULF COAST
(*I. nebulifer*)
Deep "valley" on center of head; parotoid triangular

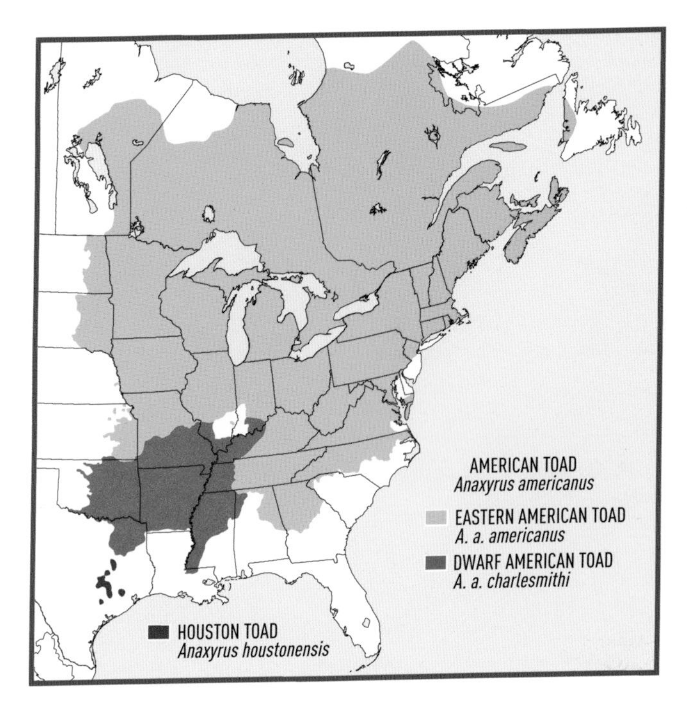

30–40/sec. **RANGE:** Atlantic Provinces west to MB and south to GA and ne. TX. **SUBSPECIES:** (1) **Eastern American Toad** (*A. a. americanus*); described above. (2) **Dwarf American Toad** (*A. a. charlesmithi*); smaller, record 4 in. (10.2 cm); often reddish; dorsal spots, when present, small and with single warts; venter faintly spotted if at all. **REMARK:** Interbreeds with Fowler's, Canadian, and Woodhouse's Toads.

GREAT PLAINS TOAD *Anaxyrus cognatus* **PL. 8, FIGS. 54, 55**

1⅞–3½ in. (4.8–9 cm); record 4½ in. (11.5 cm). Gray, brown, greenish, or yellowish, with *large, paired, light-bordered green, olive, or dark gray blotches* enclosing many warts; sometimes with a narrow light middorsal line. Cranial crests apart posteriorly but converge to form a *boss on the snout*; parotoid glands far apart and *diverging posteriorly* (Fig. 54, p. 117). Cranial crests in young form a V between the eyes, dorsum often dotted with small red tubercles. Vocal sac of male *sausage-shaped* (Fig. 55, facing page). **HABITAT:** Deserts, grasslands, semidesert shrublands, open floodplains, agricultural areas. **VOICE:** Shrill, piercing metallic trill often lasting 20 sec. or more (rarely more than 50 sec.); trill rate 13–20/sec. **RANGE:** S. AB south through the Great Plains to TX, west to CA and south into Mex.

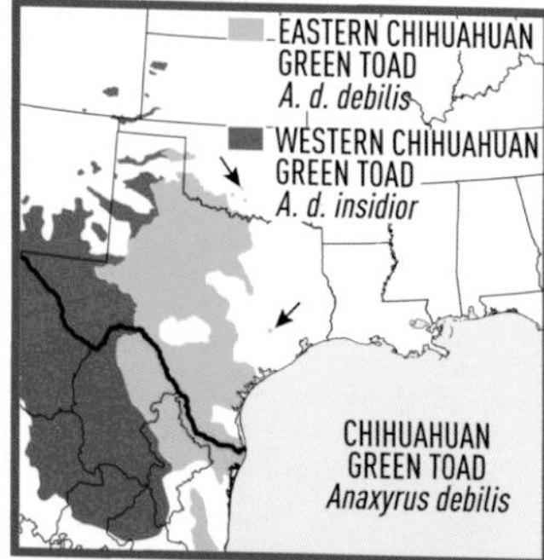

FIG. 55. *The inflated vocal sac of a male Great Plains Toad* (Anaxyrus cognatus) *is large and sausage-shaped. Male Oak Toads* (A. quercicus) *and Texas Toads* (A. speciosus) *also have a sausage-shaped vocal sac.*

CHIHUAHUAN GREEN TOAD *Anaxyrus debilis* **PL. 8**

1¼–2⅛ in. (3.2–5.4 cm). *Green*, head and body flattened, scattered black dorsal dots sometimes connected by black lines. *No prominent cranial crests* between eyes; *parotoid glands elongated, extending onto body*. Warts numerous on dorsum but less conspicuous than in most toads. Throat black or dusky in male, yellow or white in female. **HABITAT:** Arid and semiarid plains, desert shrublands. **VOICE:** See Subspecies. **RANGE:** W. KS and se. CO southwest to se. AZ and south into Mex. **SUBSPECIES:** (1) **Eastern Chihuahuan Green Toad** (*A. d. debilis*); bright green, black dorsal spots scattered; warts on parotoid glands and upper eyelids broad and flattened; voice shrill trills, duration 5–6 sec. at intervals of about same length. (2) **Western Chihuahuan Green Toad** (*A. d. insidior*); pale green, black dorsal spots connected by black lines; warts on parotoid glands and upper eyelids with black points; voice cricketlike trill lasting 3–7 sec. **REMARK:** Subspecies probably represent only clinal variation.

FOWLER'S TOAD *Anaxyrus fowleri* **PL. 8, FIG. 54**

2–3 in. (5.1–7.5 cm); record 3¾ in. (9.5 cm). Grayish, yellowish, or greenish brown to olive or very dark (almost black), often with a reddish wash; dark dorsal blotches usually with 3 or more warts. Light middorsal line usually present; occasionally with vague light lateral stripes. Belly white or yellowish, sometimes unmarked but frequently with dark areas on chest breaking into small spots on belly. Cranial crests prominent, *oval parotoid glands usually in contact with cranial crest* (Fig. 54, p. 117); *cranial crests extend down over eardrums* (especially in western populations). Warts on shanks no larger than those on thighs and usually without spines. Not all individuals have all features. **SIMILAR SPECIES:** Gulf Coast Toad has pronounced dark stripes along sides and a deep valley between the eyes. See also Woodhouse's Toad. **HABITAT:** Wooded areas, river valleys, floodplains, agricultural areas, urban and suburban yards and gardens. **VOICE:** Short bleats, nasal *w-a-a-a-h* lasting 1–4 sec. **RANGE:** New England west to MI and IA and south to the Gulf. **REMARKS:** Interbreeds with American and Woodhouse's Toads. Populations in e. TX within the extensive hybrid zone with Woodhouse's Toad sometimes are considered a separate species.

CANADIAN TOAD *Anaxyrus hemiophrys* **PL. 8, FIG. 54**

2–3 in. (5.1–7.5 cm); record 3⅝ in. (9.1 cm). Brownish, greenish, sometimes reddish, usually with a light middorsal line; dark brown to nearly black dorsal spots contain 1–3 brown or reddish warts. Cranial crests between eyes form a *pronounced boss*, often grooved, extending from snout to just behind eyes (Fig. 54, p. 117; Fig. 73, p. 163). Borders of parotoid glands indistinct and blend with skin. Young without cranial crests, which develop later, eventually uniting to form a boss. **SIMILAR SPECIES:** See Woodhouse's Toad. **HABITAT:** Prairies, plains, aspen parklands, frequently near water. **VOICE:** Rather soft, low-pitched trills lasting 2–5 sec., repeated 2–3 times/min.; trill rate about 90/sec. **RANGE:** Nw. MN and ne. SD northwest to e. AB and extreme s. NT. **REMARK:** Interbreeds with American and Woodhouse's Toads.

HOUSTON TOAD *Anaxyrus houstonensis* **FIGS. 54, 56**

2–2⅝ in. (5.1–6.7 cm); record 3$\frac{5}{16}$ in. (8.4 cm). Cream or light brown to purplish gray, sometimes with dark green patches. Dark brown to almost black dorsal spots often form a mottled pattern frequently arranged in a vague herringbone fashion; middorsal light line usually

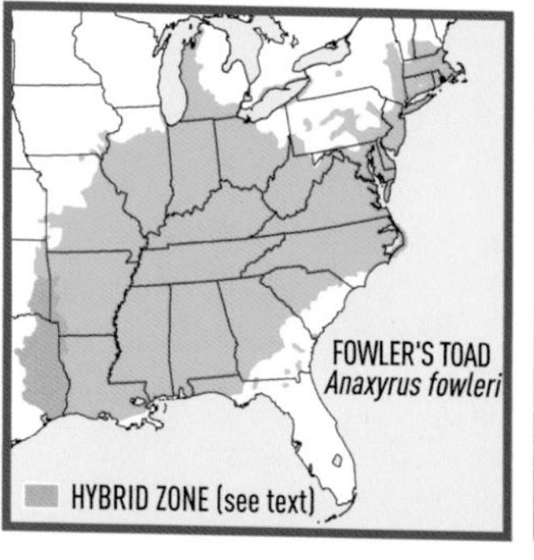

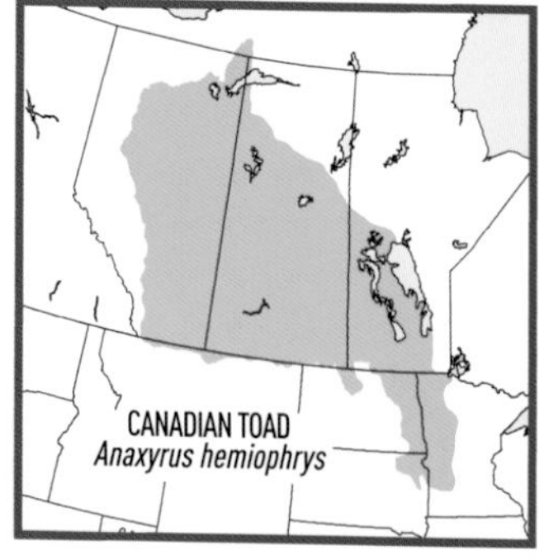

FIG. 56. *The inflated vocal sac of the Houston Toad* (Anaxyrus houstonensis) *is rounded, resembling that of most of the toads in our area.*

present. Belly with numerous small dark spots. *Cranial ridges thickened,* especially behind eyes (Fig. 54, p. 117). **SIMILAR SPECIES:** See Woodhouse's Toad. **HABITAT:** Dry forests, coastal prairies. **VOICE:** Piercing, rather musical trill (rate 32/sec.), duration 4–17 sec. **RANGE:** Se. TX; many populations extirpated. **CONSERVATION:** Endangered (USFWS: Endangered).

RED-SPOTTED TOAD *Anaxyrus punctatus* **PL. 8, FIG. 54**

1½–2½ in. (3.8–6.4 cm); record 3 in. (7.6 cm). Somewhat flattened; gray, light to medium brown, or pale olive, with buff or reddish warts sometimes contained in small dark blotches (individuals from the

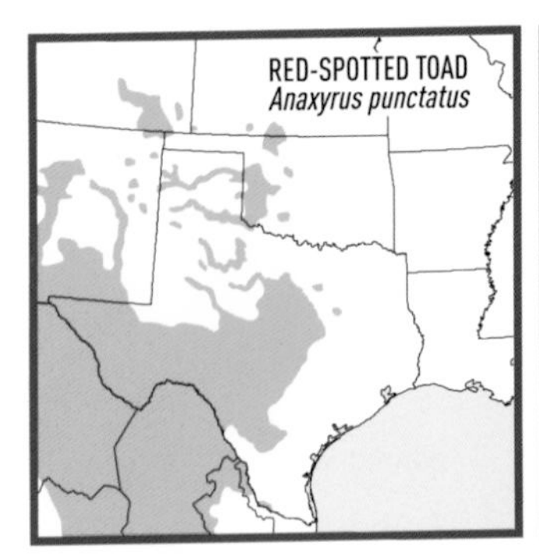

Edwards Plateau in cen. TX pale gray, virtually unmarked). *Cranial crests absent* or slightly developed; *parotoid glands round* and small, no larger than eyes (Fig. 54, p. 117). **HABITAT:** Rocky canyons and gullies in deserts, grasslands, and dry woodlands, usually near springs, seepages, persistent pools along streams, cattle tanks. **VOICE:** Clear, musical trill, duration 4–10 sec. **RANGE:** Sw. KS to se. CA and south into Mex.

OAK TOAD *Anaxyrus quercicus* **PL. 8**

¾–1⅜ in. (1.9–3.5 cm). Pearl gray to almost black, with a *conspicuous white, cream, yellow, or orange middorsal line* (occasionally black with very little pattern beyond the light middorsal line) and 4–5 pairs of black or brown dorsal spots; warts red, orange, or reddish brown. *Cranial crests inconspicuous*. Vocal sac of male *sausage-shaped*. **HABITAT:** Sandy pine flatwoods, savannas, oak scrub, open woodlands, maritime forests. **VOICE:** Like the peeping of a newly hatched chick. **RANGE:** Coastal Plain from se. VA south through FL and some Keys and west to e. LA.

TEXAS TOAD *Anaxyrus speciosus* **PL. 8**

2–3¼ in. (5.1–8.3 cm); record 3¹⁵⁄₁₆ in. (10 cm). "Chubby." Gray, with yellowish green or brown spots and pink, orange, or greenish warts; no light middorsal stripe. *Cranial crests indistinct or absent*; parotoid glands oval. Two tubercles on hindfeet sharp-edged, often black, inner one *sickle-shaped*. Vocal sac of male *sausage-shaped*. **SIMILAR SPECIES:** (1) Red-spotted Toad has small round parotoid glands and buff or reddish warts. (2) Chihuahuan Green Toad has elongated parotoid glands that extend onto the body. (3) Other toads in range have prominent cranial crests. **HABITAT:** Grasslands, cultivated areas, mesquite savannas. **VOICE:** Series of loud, explosive trills, each ½ sec. or longer; trill rate 39–57/sec. **RANGE:** W. OK through TX and se. NM south into Mex.

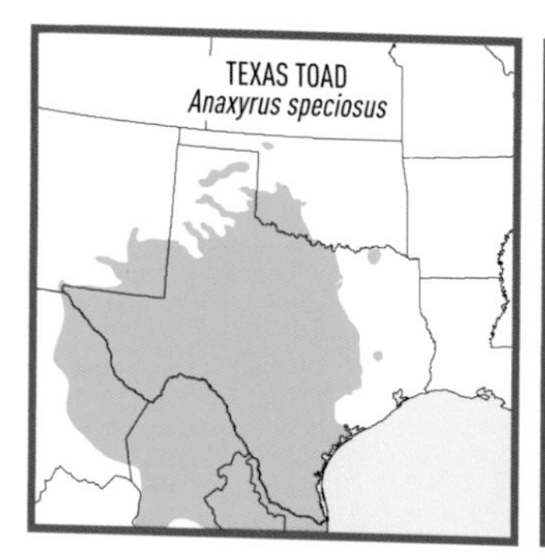

SOUTHERN TOAD *Anaxyrus terrestris* **PL. 8, FIG. 54**

1⅝–3 in. (4.1–7.5 cm); record 4⅞ in. (12.3 cm). Generally brown, sometimes reddish, occasionally very dark, almost black. Dark spots, when present, usually with 1–2 warts (occasionally more). Light middorsal line sometimes present, often obscure, especially posteriorly. Cranial crests high, usually with *pronounced knobs*; ridges running forward from knobs converge toward snout (Fig. 54, p. 117). Knobs not well developed in young, but locations indicated by backward extensions of cranial crests. **SIMILAR SPECIES:** No other toads in range have pronounced cranial knobs. For young: (1) Fowler's Toad has smaller warts, usually 3 or more in each dark spot. (2) Oak Toad has inconspicuous cranial crests and a prominent light middorsal line. (3) American Toad lacks distinct backward extensions from interorbital crests. **HABITAT:** Woodlands and open areas. **VOICE:** Shrill musical trills; duration 2–9 sec.; trill rate 52–78/sec. **RANGE:** Coastal Plain from se. VA south through FL and the Keys and west to e. LA.

WOODHOUSE'S TOAD *Anaxyrus woodhousii* **PL. 8, FIG. 54**

2½–4 in. (6.4–10 cm); record 5 in. (12.7 cm). Grayish, yellowish, or greenish brown to olive or very dark (almost black), irregular dark blotches with one to several warts. Light middorsal line usually present. Belly white or yellowish, unmarked or with few dark markings restricted to chest. Cranial crests prominent, *oval parotoid glands usually in contact with the cranial crest* (Fig. 54, p. 117). Warts on shanks no larger than those on thighs. Not all individuals have all features. **SIMILAR SPECIES:** (1) Fowler's Toad has large, dark, well-defined dorsal spots (often 6), each containing 3 or more warts; where species co-occur, Fowler's Toad usually has extensions of cranial crests extending down over eardrums. (2) American Toad has numerous dark markings on chest, enlarged warts on tibia, kidney-shaped parotoid glands separated from cranial crests behind eyes or connected to them by spurs. (3) Houston Toad has thick cranial crests, especially behind eyes, that contact parotoid glands via spurs. (4) Texas Toad

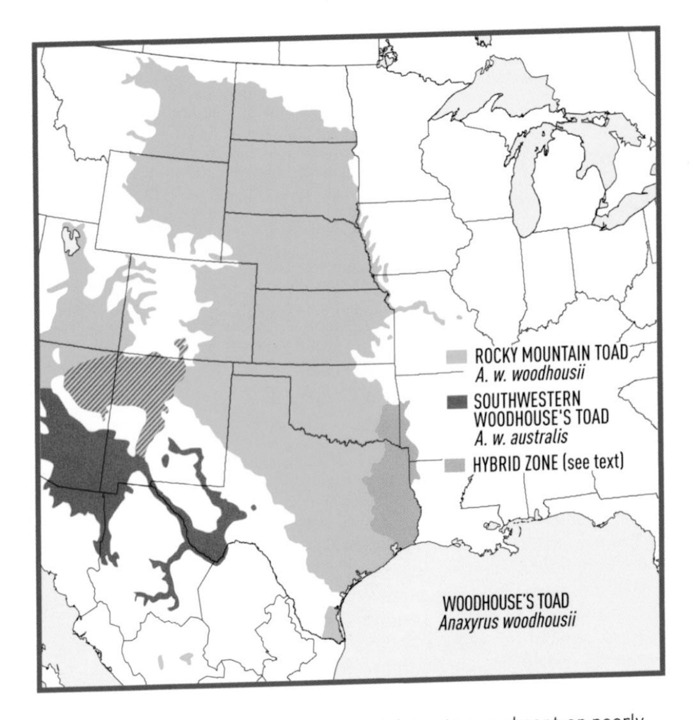

lacks a light middorsal line and cranial crests are absent or poorly developed. (5) Red-spotted Toad has round parotoid glands, lacks cranial crests and dark dorsal spots (although reddish warts sometimes surrounded by small spots). (6) Canadian Toad has a prominent boss between eyes. **HABITAT:** Grasslands, deserts, semidesert shrublands, river valleys, floodplains, agricultural areas. **VOICE:** Nasal *w-a-a-a-h*, lasting 1–2½ sec., similar to Fowler's Toad but lower in pitch. **RANGE:** MT and ND south to the Gulf, west to se. CA and south into Mex.; isolated populations in the Northwest. **SUBSPECIES:** (1) **Rocky Mountain Toad** (*A. w. woodhousii*); middorsal line often extends onto snout, belly white or yellowish, usually unmarked, but sometimes with a dark breast spot and flecks between forelimbs. (2) **Southwestern Woodhouse's Toad** (*A. w. australis*); light middorsal lines usually not involving snout, chest with extensive dark markings. **REMARKS:** The Southwestern Woodhouse's Toad might represent a distinct species. Interbreeds with American, Canadian, and Fowler's Toads. Hybrid zone with the latter is extensive.

CENTRAL AMERICAN TOADS: *Incilius*

About 40 species occur from the southern United States to Ecuador.

GULF COAST TOAD *Incilius nebulifer* **PL. 8, FIG. 54**

2–4 in. (5.1–10 cm); record 5³⁄₁₆ in. (13.2 cm). Body flattened, very dark, almost black, with touches of rich orange to yellow-brown with whitish spots. Light middorsal line almost always present; usually with prominent *broad dark lateral stripes* bordered above by light stripes. Cranial crests strongly developed, bordering a *deep valley* down center of head. Parotoid glands often triangular (Fig. 54, p. 117). Throat of male clear yellowish green, vocal sac large and round. **SIMILAR SPECIES:** No other toads in range have pronounced dark lateral stripes and a deep valley between the eyes. **HABITAT:** Coastal prairies, barrier beaches, yards and gardens. **VOICE:** Short trill, 2–6 sec., repeated several times at intervals of about 1–4 sec. **RANGE:** S. MS east through TX and south into Mex. **REMARK:** Until recently considered a subspecies of *I. valliceps*, a name now restricted to populations south of our area.

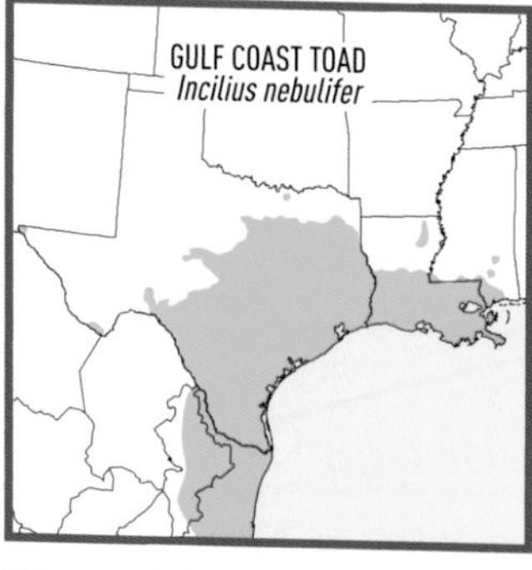

SOUTH AMERICAN TOADS: *Rhinella*

Nearly 90 species occur from the southern United States to southern South America.

CANE TOAD *Rhinella marina* **PL. 8**

4–6 in. (10–15.2 cm); record 9⅜ in. (23.8 cm); probably not over 6⅞ in. (17.5 cm) in the U.S.; female up to 3.3 lb. (1.5 kg). Huge, with somewhat flattened head and body; light to dark brown, occasionally with yellowish, reddish, or olive tinges. Light middorsal line and irregular patterns of darker spots sometimes present. Cranial crests prominent. *Large, deeply pitted parotoid glands extend onto sides of body.* Paramedian rows of enlarged warts most evident in breeding male. **SIMILAR SPECIES:** No other toads in our area have both prominent cranial crests and enlarged parotoid glands. **HABITAT:** Cane fields, savannas, open forests, yards and gardens, locally in dry forests; frequently in degraded habitats, often most abundant in disturbed settings in and near human settlements. In TX, often near pools in the Rio Grande Valley; in FL, usually in yards and gardens, irrigated agricultural fields, and along canals. **VOICE:** Slow, low-pitched trill. **RANGE:** Extreme s. TX south through the Amazon Basin. Widely introduced, including FL and HI. **REMARKS:** Comprises multiple lineages. Because the current name is associated with S. American populations, that applied to U.S. populations will likely change. Skin-gland secretions highly toxic, can be lethal to dogs and other animals.

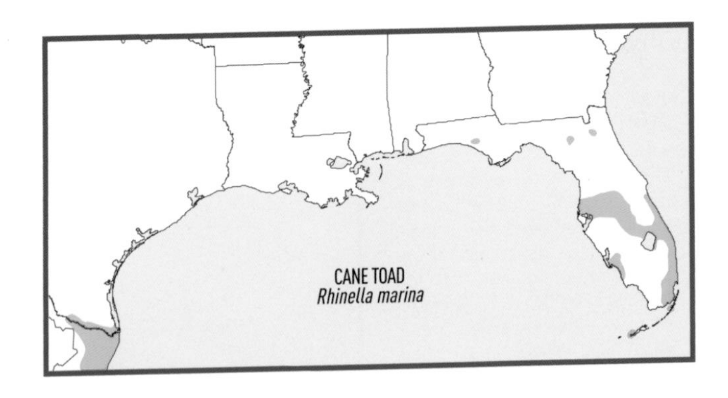

NORTHERN RAINFROGS: Craugastoridae

Nine genera and about 740 species range from central Texas south to northeastern Argentina. Eggs are laid in moist places on land. Tiny froglets emerge directly from eggs.

BARKING FROG *Craugastor augusti* **PL. 8, FIG. 57**

2½–3 in. (6.4–7.5 cm); record 3¾ in. (9.4 cm). Squat, smooth-skinned, with a distinct ventral disc (Fig. 57, below). *Skinfolds behind head continue as dorsolateral ridges.* Forearms stout. Tan or brown to greenish, sometimes tinged pink or reddish brown; dark dorsal blotches have narrow light borders. Hindlimbs banded. Young often greenish with whitish to black bands across middle of back. **SIMILAR SPECIES:** Cliff Chirping Frog is smaller and lacks prominent skinfolds and bands across back. **HABITAT:** Low-elevation shrublands and deserts, brushy or open woodlands, often associated with clumps of cactus and rocky limestone areas; in TX, usually in canyons. **VOICE:** Explosive, from a distance like a dog barking, at close range like a guttural *whurr*; single note repeated at regular intervals of 2–3 sec. **RANGE:** Cen. TX west to se. NM and south into Mex. (western populations range north into AZ). **SUBSPECIES:** **Balcones Barking Frog** (*C. a. latrans*). **REMARKS:** Until recently assigned to the genus *Eleutherodactylus*. The Balcones Barking Frog might represent a distinct species.

FIG. 57. *The ventral disc of the Barking Frog* (Craugastor augusti). *The Mexican White-lipped Frog* (Leptodactylus fragilis) *also has a ventral disc.*

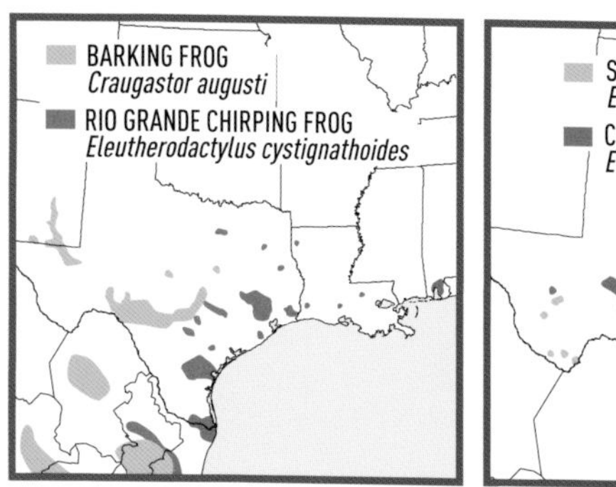

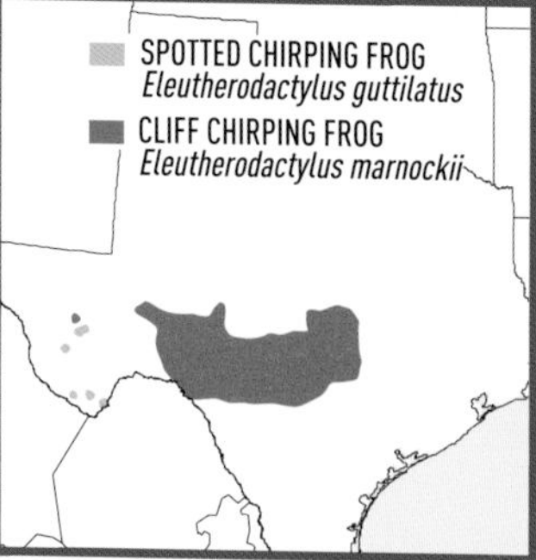

RAINFROGS: Eleutherodactylidae

Sometimes called free-toed or robber frogs; four genera and more than 200 species range from central Texas south into much of South America and north through the Antilles. Eggs are laid in moist places on land and tiny froglets emerge directly from eggs. Three native species until recently were placed in the genus *Syrrhophus*. One West Indian species occurs in Florida; another (the Puerto Rican Coquí) apparently is restricted to artificial environments in greenhouses and is not included in this guide. All native species have *smooth skin*, a *proportionately large head*, and a *flattened head and body*.

RIO GRANDE CHIRPING FROG *Eleutherodactylus cystignathoides* **PL. 9**

5/8–1 in. (1.6–2.6 cm). Brown to grayish or yellowish olive with scattered dark spots especially on back. Dark lines from nostrils through eyes and on top of head between eyes usually not prominent. **SIMILAR SPECIES:** See Cliff Chirping Frog. **HABITAT:** Moist shaded vegetation, palm groves, thickets, ditches, lawns, gardens. **VOICE:** Erratic cricket-like chirps. **RANGE:** Extreme s. TX south into Mex. Widely introduced along the Gulf Coast and as far north as Dallas. **SUBSPECIES:** **Rio Grande Chirping Frog** (*E. c. campi*).

SPOTTED CHIRPING FROG *Eleutherodactylus guttilatus* **PL. 9**

3/4–1 1/4 in. (1.9–3.2 cm). Yellowish to yellowish brown with a very dark brown vermiculate pattern, scattered light flecks, vague dark bar between eyes, and darkly banded hindlimbs. **SIMILAR SPECIES:** See Cliff Chirping Frog. **HABITAT:** Springs, canyons, caves, ravines in dry forests. **VOICE:** Sharp, relatively short whistle. **RANGE:** Sw. TX and disjunct areas in Mex. (not mapped).

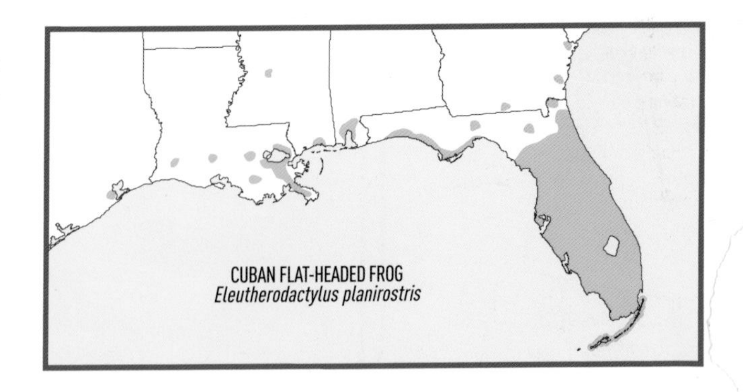

CLIFF CHIRPING FROG *Eleutherodactylus marnockii* **PL. 9**

¾–1½ in. (1.9–3.8 cm). Greenish to greenish brown with brown mottling, scattered light flecks, and banded hindlimbs. **SIMILAR SPECIES:** (1) Rio Grande Chirping Frog is difficult to distinguish and best identified by range, but tends to be darker and has a less distinct dorsal pattern. (2) Spotted Chirping Frog also is very similar and best distinguished by range, but is darker, less likely to be greenish, has a more pronounced vermiculate pattern, a vague dark bar between eyes, and darker hindlimb banding. (3) Green Toad is greener and has warty skin. **HABITAT:** Cracks, caves, crevices in cliffs and limestone hills in woodland, scrubland, grassland, desert; also in human-generated debris, occasionally watered lawns. **VOICE:** Cricketlike chirps or trills throughout the year; mating calls clearer and sharper. **RANGE:** Cen. TX west to the Pecos R. Valley; presumably isolated populations farther west.

CUBAN FLAT-HEADED FROG *Eleutherodactylus planirostris* Non-native **PL. 9**

5/8–1¼ in. (1.6–3.2 cm); record 3 7/16 in. (8.7 cm). Brown (sometimes with a reddish or greenish cast), with irregular dark and light mottling or longitudinal light stripes. Belly white; irises dark with gold flecks; *toepads barely wider than toes*; *no webbing between toes*. **SIMILAR SPECIES:** All other frogs in range have some webbing between toes. **HABITAT:** Gardens, greenhouses, dumps, hardwood hammocks, Gopher Tortoise burrows, small stream valleys. **VOICE:** Short, melodious birdlike chirps, usually 4–6 in series. **RANGE:** FL, additional populations in GA, AL, MS, LA, TX. **NATURAL RANGE:** Cuba, Bahamas, Cayman Is. Widely introduced, including HI. **SUBSPECIES: CUBAN FLAT-HEADED FROG** (*E. p. planirostris*). **REMARK:** Introduced populations often called the Greenhouse Frog.

TREEFROGS: Hylidae

Comprising 48 genera and about 950 species, treefrogs are distributed on all habitable continents but are absent from sub-Saharan Africa. Five genera and 27 species occur in our area. Treefrogs generally are slim-waisted, long-limbed, and most are small. Females are larger than males. Although in the treefrog family, cricket frogs and chorus frogs are not climbers, but all have intercalary cartilages that serve as an extra joint between the toes and toepads.

CRICKET FROGS: *Acris*

All three small, *warty*, nonclimbing species occur in our area, ranging from southeastern New York, extreme southern Ontario, the southern peninsula of Michigan, and southeastern South Dakota south to the southern tip of Florida and northern Mexico. Although they are variable in color and pattern, consistent elements include a *dark triangle or V-shaped mark between the eyes, longitudinal dark stripes on the rear surface of the thighs*, and often white lines extending from beneath the eyes to the mouth. Males have a single vocal sac under the chin. Webs on the feet are at least half the length of the longest (fourth) toe, and *toepads are barely wider than toes* (Fig. 60, p. 130).

BLANCHARD'S CRICKET FROG *Acris blanchardi* **PL. 9, FIGS. 58, 59, 60**

5/8–1½ in. (1.6–3.8 cm). Snout blunt, dark (occasionally light) *triangle between eyes, irregular dark stripes on back surface of thighs* often blending with dark pigment above stripes and near the vent (Fig. 58, below). Brown or gray, often with gray to green or reddish brown markings varying from vivid to pale; middorsal stripes sometimes interrupted and frequently bordered by small black spots. Belly white. Throat white (occasionally pale yellow), male often with a yellow or gray area on the vocal sac. Anal warts prominent. *Hindlimbs relatively*

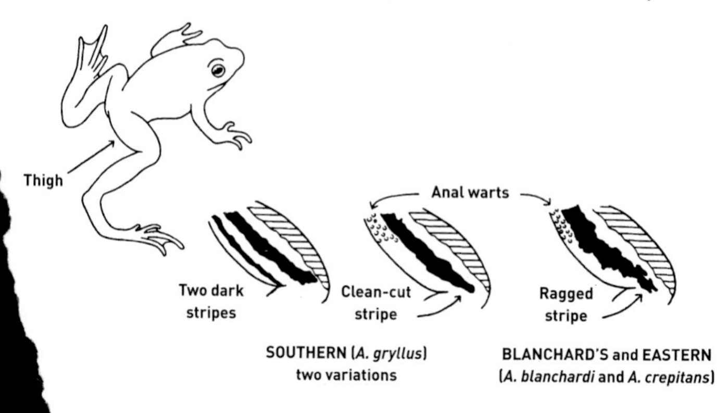

58. *Thigh patterns of cricket frogs* (Acris).

FIG. 59. *Like other cricket frogs, Blanchard's Cricket Frog* (Acris blanchardi) *has warty skin and terminal toepads barely wider than the digits. Although the ground color is a subdued brown or gray, many individuals are vividly marked with reddish brown or bright green. Such spots or streaks apparently break up the outline of the frog, causing it to blend into the substrate.*

FIG. 60. *Webbing and toepads of cricket frogs* (Acris).

BLANCHARD'S and EASTERN
(*A. blanchardi* and *A. crepitans*)
Webbing extensive

SOUTHERN
(*A. gryllus*)
Webbing scant

short (adpressed heel does not reach snout). Usually only *one joint of longest (fourth) toe is free of webbing*; *webbing of innermost toes extends to toepads* (Fig. 60, above). **SIMILAR SPECIES:** (1) Eastern Cricket Frog is essentially similar and best distinguished by range and molecular markers. (2) Southern Cricket Frog has a more pointed snout, longer hindlimbs (when adpressed, heel extends beyond snout), innermost toes fully webbed, and usually 2 dark stripes on back surface of thighs, with at least one sharply defined. (3) Chorus frogs have smoother skin, toes only slightly webbed. (4) American water frogs, the young of which sometimes resemble cricket frogs, lack a triangle between the eyes, dark stripes on back surface of thighs, and light lines from eyes to mouth. **HABITAT:** Ponds, lakes, marshes, streams, less frequently impoundments and rivers, potentially any permanent body of water. **VOICE:** Like clicking pebbles in rapid succession; duration $\frac{1}{2}$–$1\frac{1}{2}$ sec. **RANGE:** WV, s. MI, and adjacent se. ON west to se. SD south to TX (status of frogs in w. MS uncertain); also in the Rio Grande Valley. **REMARK:** Until recently considered a subspecies of the Eastern Cricket Frog.

EASTERN CRICKET FROG *Acris crepitans* **PL. 9, FIGS. 58, 60**

$\frac{5}{8}$–$1\frac{3}{8}$ in. (1.6–3.5 cm). Essentially identical to Blanchard's Cricket Frog. **SIMILAR SPECIES:** See Blanchard's Cricket Frog. **HABITAT:** Permanent and seasonal ponds, lakes, marshes, other wetlands in open forests and meadows, slow-moving streams and ditches. **VOICE:** *Gick*,

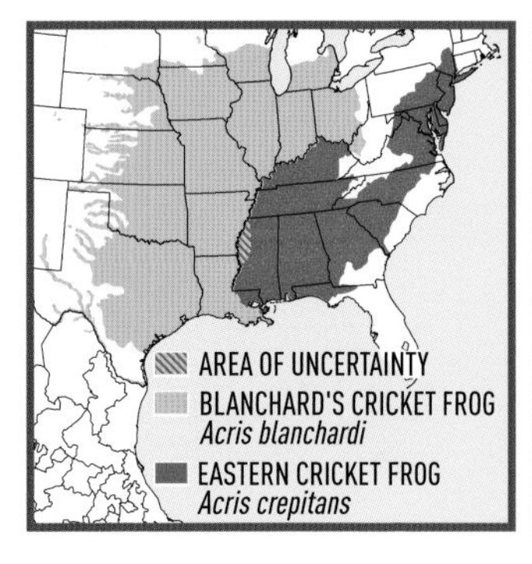

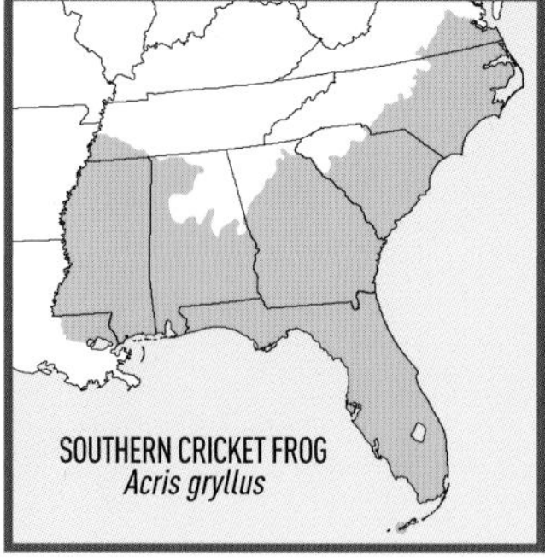

gick-gick; like pebbles clicked together, slowly at first but picking up speed and continuing for 20–30 or more beats. **RANGE:** Se. NY to the FL Panhandle and west to the Mississippi R. Apparently extirpated on Long Island, NY.

SOUTHERN CRICKET FROG *Acris gryllus* **PL. 9, FIGS. 58, 60**

5/8–1¼ in. (1.6–3.3 cm). Snout pointed, dark *triangle between eyes* (occasionally absent), often *2 dark stripes on back surface of thighs* (Fig. 58, p. 129), upper stripe narrow and usually sharply defined, lower one ranging from sharply defined to vague stippling. Gray, brown, almost black, or olive, usually with a gray, yellow, green, or reddish brown middorsal stripe; colors vivid to pale. Sides dark, often blotched. Undersurfaces white, but throat of male dark during breeding season. Anal warts often poorly developed or nearly absent. *Hindlimbs relatively long* (adpressed heel extends beyond snout). At least *2 joints of longest (fourth) toe free of webbing*, at least *one joint of innermost toes free of webbing* (Fig. 60, facing page). **SIMILAR SPECIES:** See Blanchard's Cricket Frog. **HABITAT:** Grassy margins of swamps, marshes, lakes, ponds, streams, ditches, bogs, nearby temporary pools, essentially any body of water; chiefly in lowlands, but will follow river valleys into uplands. **VOICE:** Similar to the Eastern Cricket Frog but less percussive; duration 12–28 sec. **RANGE:** Se. VA south through FL and some Keys and west to the Mississippi R.

HOLARCTIC TREEFROGS: *Hyla*

Thirty-five currently recognized species occur in central and southern Europe, eastern Asia, northwestern Africa, and North America. Nine species occur in our area. Depending on light, moisture, temperature, stress, or activity level, the same frog may be patterned or uniformly colored, gray or green or brown. Some species have distinctive markings or coloration on the concealed surfaces of the hindlimbs and must be captured to ensure accurate identification. Juveniles of several species are uniformly bright green,

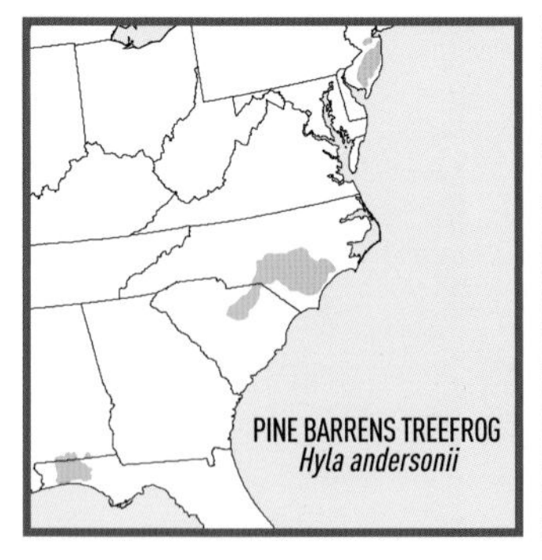

FIG. 61. *Extensive webbing and large toepads of treefrogs* (Hyla, Osteopilus, and Smilisca).

rendering distinctions difficult. Most males have a vocal sac that looks like a small balloon when inflated; however, the Canyon Treefrog has a slightly bi-lobed vocal sac (and Cuban and Mexican Treefrogs, which are not in the genus *Hyla*, have paired vocal sacs). All of these treefrogs have extensive webbing and large toepads (Fig. 61, above).

PINE BARRENS TREEFROG *Hyla andersonii* **PL. 9, P. 104**

1⅛–1¾ in. (2.8–4.4 cm); record 2 in. (5.1 cm). Bright green, with *broad lavender brown bands* with narrow white borders extending from nostrils through the eyes and eardrums to the groin. Concealed surfaces of legs and groin yellow-orange. **SIMILAR SPECIES:** See Green Treefrog. **HABITAT:** Pine-oak forests near swamps and bogs. **VOICE:** Rapid nasal *quonk-quonk-quonk* repeated as many as 3 times/sec. (slower in cool weather). **RANGE:** Disjunct; s. NJ, se. NC and adjacent SC, FL Panhandle and adjacent s.-cen. AL.

CANYON TREEFROG *Hyla arenicolor* **PL. 9**

1¼–2 in. (3.2–5.1 cm); record 2¼ in. (5.7 cm). Camouflaged gray to brownish gray or olive green, sometimes tinted with pink, with *dark-bordered whitish patches beneath eyes* (occasionally just dark bars). Usually with medium to dark brown or almost black or greenish dorsal spots, occasionally almost unicolored. Concealed surfaces of fore- and hindlimbs orange-yellow. Skin moderately rough. Male with

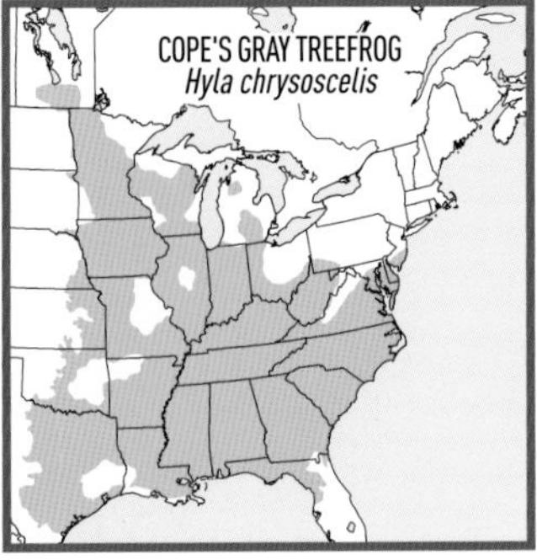

gray, brown, or black throat and slightly bilobed vocal sac. **SIMILAR SPECIES:** No other frogs in range have dark-bordered light patches beneath eyes. **HABITAT:** Rocky canyons in forested uplands and arroyos in semiarid grasslands; usually along intermittent or permanent streams. **VOICE:** Harsh, rattling, mechanical trills, *brrrrr* or *ah-ah-ah-ah*, duration about 1 sec., repeated every few seconds. **RANGE:** Disjunct; s. CO and UT south into Mex.

BIRD-VOICED TREEFROG *Hyla avivoca* **PL. 9**

1⅛–1¾ in. (2.8–4.4 cm), female much larger than male; record 2$^{1}/_{16}$ in. (5.2 cm). Camouflaged gray, brown, or green, with *dark-bordered yellowish, cream, or whitish patches beneath eyes*; often with irregular dark spots or mottling on back. Concealed portions of hindlimbs *yellowish green to greenish or yellowish white*. Skin somewhat rough. **SIMILAR SPECIES:** See Cope's Gray Treefrog. **HABITAT:** Wooded swamps bordering rivers and streams. **VOICE:** Rapid series of 10–20 musical, birdlike whistles lasting several seconds, sometimes varying in tempo. **RANGE:** S. IL and w. KY south to LA and east to sw. SC; isolated populations in LA, AR, and se. OK. **REMARK:** Interbreeds with Cope's Gray Treefrog.

COPE'S GRAY TREEFROG *Hyla chrysoscelis* **PL. 9**

GRAY TREEFROG *Hyla versicolor* **PL. 9**

1¼–2⅜ in. (3.2–6 cm); record 2$^{7}/_{16}$ in. (6.2 cm). Essentially identical. Gray Treefrog slightly less likely to be green, has somewhat rougher skin, and has twice as many chromosomes (tetraploid) as Cope's Gray Treefrog. Both camouflaged gray or green, sometimes grayish brown or green or even almost white, *dark-bordered whitish (rarely cream) patches beneath eyes*; often with irregular dark spots or mottling on back. Concealed portions of hindlimbs *yellowish orange*. Belly white. Skin somewhat rough. **SIMILAR SPECIES:** (1) Bird-voiced Treefrog is smaller, light patches beneath eyes are more likely yellowish or cream, concealed surfaces of hindlimbs washed with green or yellow-

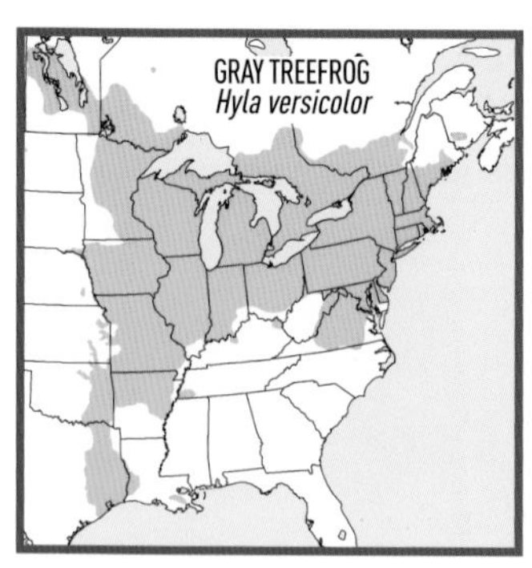

ish white, call is a series of musical whistles instead of a trill. (2) Squirrel Treefrog sometimes similar but has smoother skin and lacks dark-bordered light patches beneath eyes and an orange wash on concealed surfaces of hindlimbs. **HABITAT:** Wooded areas and woodland edges (including small wooded areas in prairies), usually near water. **VOICE:** Cope's Gray Treefrog: buzzlike trill (44–75 notes/sec.) lasting 3–4 sec.; Gray Treefrog: slower trill (usually fewer than 35 notes/sec.) lasting about ½ sec. and repeated every few seconds. Trill rates vary with temperature. Distinctions are possible when males of both species call at the same place and time. **RANGE:** Cope's Gray Treefrog: s. NJ to n. FL west to cen. TX and north to MB. Gray Treefrog: NB west through the Great Lakes to s. MB south to extreme n.-cen. NC in the East and to the Gulf in the West.

GREEN TREEFROG *Hyla cinerea* **PL. 9**

1¼–2¼ in. (3.2–5.7 cm); record 2⁹⁄₁₆ in. (6.6 cm). Usually bright green, with *white or cream stripes* extending from lips onto sides and to the groin, frequently with scattered *small yellowish spots* on back, but coloration variable; ground color sometimes yellowish or dull greenish olive or slate gray (latter most likely when inactive during cool weather). White or yellowish stripes frequently have narrow black borders, vary in length (may be short or lacking entirely); occasional individuals lack yellow dorsal spots. Skin smooth. **SIMILAR SPECIES:** (1) Pine Barrens Treefrog has broad light-bordered lavender brown stripes and concealed surfaces of legs and groin are yellow-orange. (2) Squirrel Treefrog is more likely to be brown (but can be bright green), often has dark spots between eyes, and almost always lacks small dorsal spots; sometimes has light lateral stripes, but lower borders indistinct, not sharply defined, and stripes rarely extend to the groin. (3) Barking Treefrog is often green and has white stripes on lips, but dorsum is almost always spotted and skin is rough. (4) Treefrogs of many species, especially young, may be bright green, but most lack light lateral stripes. **HABITAT:** Swamps, marshes, banks of ponds, lakes, streams, especially in areas with emergent or floating

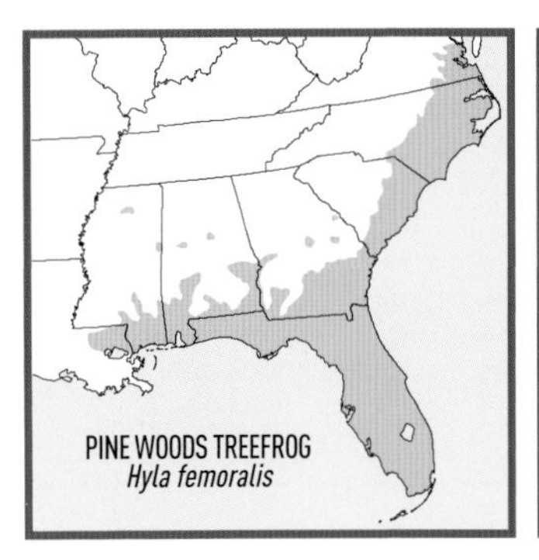

vegetation. **VOICE:** Short, nasal *quonk* repeated about 2 times/sec. **RANGE:** Delmarva Peninsula south through FL and the Keys and west to TX and north in the Mississippi River Valley to IL and IN; isolated populations in MO, Big Bend Region of TX (not mapped), and elsewhere are introduced; also introduced in PR. **REMARK:** Occasionally interbreeds with the Barking Treefrog.

PINE WOODS TREEFROG *Hyla femoralis* PL. 9, FIG. 62

1–1½ in. (2.5–3.8 cm); record 1¾ in. (4.4 cm). Gray or greenish gray to brown, most frequently reddish brown; usually with irregular dark lines through eyes and often onto sides. *Rows of small orange, yellow, or whitish spots on back of thighs* (Fig. 62, below). **SIMILAR SPECIES:** (1) Squirrel Treefrog is very similar when brown but lacks rows of light spots on back of thighs and has a very different call. (2) Bird-voiced and Cope's Gray Treefrogs have dark-bordered light patches below eyes. (3) Spring Peeper has a dark X on the back. **HABITAT:** Pine flatwoods, savannas, pine-oak forests, usually near bogs or ponds, occasionally in hardwood forests and swamps. **VOICE:** Nearly continuous but irregular strings of raspy notes resembling someone tapping a telegraph key. **RANGE:** Se. VA to s. FL and west to e. LA. **REMARK:** Sometimes called the dot-and-dash frog or Morse-code frog.

BARKING TREEFROG *Hyla gratiosa* PL. 9

2–2⅝ in. (5.1–6.7 cm); record 2¾ in. (7 cm). *Stout-bodied*, dark brown and bright green to pale gray or yellowish, but nearly always some green evident. *Round or squarish dark dorsal spots* usually obvious,

FIG. 62. *Thigh pattern of the Pine Woods Treefrog* (Hyla femoralis).

but sometimes missing, especially when frogs are dark brown or very pale. Sharply defined *light lines on upper lip* sometimes extend onto shoulders or even sides, where they break into dark and light mottling on lower sides. Skin somewhat *rough*. **SIMILAR SPECIES:** See Green Treefrog. **HABITAT:** Sandy areas in pine savannas, low wet woods and swamps. **VOICE:** Breeding call a single explosive *donk* repeated every 1–2 sec.; barking calls consist of 9–10 raucous syllables repeated more slowly and irregularly than the breeding call. **RANGE:** Disjunct; DE to s. FL and west to e. LA, chiefly in the Coastal Plain. Introduced population in s. NJ probably extirpated. **REMARK:** Interbreeds with the Green Treefrog.

SQUIRREL TREEFROG *Hyla squirella* **PL. 9**

7/8–1¾ in. (2.2–4.5 cm). Color variable; brown or green, plain or spotted. Usually with dark spots or a bar between the eyes and light stripes extending from lips onto shoulders and sides; however, spots and bars on the head and body can range from clearly defined to vague or absent; stripes similarly vary in definition and length, sometimes fading out below the eardrums and occasionally extending to the groin. Skin smooth. **SIMILAR SPECIES:** See Green Treefrog. **HABITAT:** Open woods, urban areas, especially in thick vegetation near water. **VOICE:** Nasal, buzzing, ducklike quack repeated 1–2 times/sec., producing a harsh trill or rasp. "Rain calls," usually voiced away from water, are scolding squirrel-like rasps. **RANGE:** Se. VA south through FL and the Keys and west to TX. Isolated populations in TX are introduced.

WEST INDIAN TREEFROGS: *Osteopilus*

Eight currently recognized species occur in the Bahamas and Greater Antilles.

CUBAN TREEFROG *Osteopilus septentrionalis* Non-native **PL. 10**

Male 1½–3½ in. (3.8–9 cm); female 2–5 in. (5.1–12.5 cm); record 6 1/8 in. (15.5 cm), in captivity 6½ in. (16.5 cm). Largest treefrog in our area. Gray to tan or brown, sometimes essentially unicolored but usually with brown or green to olive dorsal spots or mottling or a lichenlike pattern. *Skin fused to skull*, skin elsewhere warty. *Toepads very large* (as large as eardrums). Male has paired vocal sacs. Juvenile pale

green to tan, often with broad cream to yellow dorsolateral stripes; many young have red eyes, and all have *greenish blue bones* visible in hindlimbs. **SIMILAR SPECIES:** No other treefrogs in range have skin fused to skull or toepads as large as eardrums. **HABITAT:** Moist forests and buildings. **VOICE:** Grating squawk, sometimes followed by series of clicks. **RANGE:** FL and the Keys, scattered records in other states (most not mapped). **NATURAL RANGE:** Cuba, Bahamas, Cayman Is. Widely introduced in the W. Indies, also in HI (apparently not established) and Costa Rica. **REMARKS:** Although considered non-native, this species might have reached the FL Keys by natural overwater dispersal from Cuba. Highly invasive; adults prey on and compete with native frogs and other small vertebrates; tadpoles compete with native species. Skin secretions are highly irritating if they contact eyes or mouth.

MEXICAN TREEFROGS: *Smilisca*

Eight currently recognized species range from southern Arizona and extreme southern Texas to northwestern South America.

MEXICAN TREEFROG *Smilisca baudinii* **PL. 10**

2–2¾ in. (5.1–7 cm); record 3⁹⁄₁₆ in. (9 cm). Coloration can change rapidly; light to dark brown or gray to brownish green, yellowish, or reddish, often with dark dorsal spots that vary in size and number. Consistent markings include *light patches beneath the eyes, dark spots or stripes from eardrums onto the shoulders*, and *light spots at the base of each forelimb*. Undersides white to cream or yellow, with darker areas marking paired vocal sacs of male. *Eardrums very large*, nearly as large as eyes. **HABITAT:** Ponds, pools, canals, flooded fields, yards and gardens. **VOICE:** Series of short, explosive notes, *wonk-wonk-wonk*, or blurred notes sounding like *heck*, sometimes interspersed with chuckles. **RANGE:** S. TX south to Costa Rica.

FIG. 63. *Reduced webbing and small toepads of chorus frogs* (Pseudacris). *Note, however, that the Spring Peeper* (P. crucifer) *has toepads similar to those of treefrogs* (Hyla, Osteopilus, *and* Smilisca).

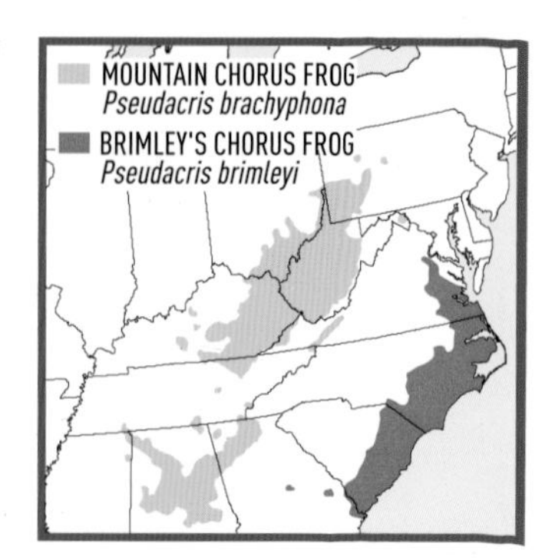

CHORUS FROGS: *Pseudacris*

Eighteen currently recognized North American species (14 in our area) range from the Gulf of Mexico to Hudson Bay in the East and from southern Baja California to southern Alaska in the West. *Toepads are small*, and *toes are only slightly webbed* (Fig. 63, above). Most species have light lines along the upper lip. Males have a single round vocal sac that, when collapsed, is gray or brown against cream or yellowish ground color.

MOUNTAIN CHORUS FROG *Pseudacris brachyphona* **PL. 10, FIG. 52**

1–1¼ in. (2.5–3.2 cm); record 1½ in. (3.7 cm). Stocky, brown to gray brown or olive, with white lines on lips, dark masks through eyes, usually with a *dark triangle* between eyes, dorsal marking like *reversed parentheses or crescents* (Fig. 52, p. 110) that sometimes touch to produce a crude dark X on the back (crescents occasionally broken into spots). *Concealed surfaces of hindlimbs yellow*. **SIMILAR SPECIES:** Upland Chorus Frog has dorsal pattern basically consisting of 3 longitudinal stripes; if stripes are broken, spots usually are arranged in 3 rows. See also Spring Peeper. **HABITAT:** Forested slopes and hilltops, often far from water; generally absent from valleys and flat areas occupied by other species of chorus frogs. **VOICE:** Harsh, raspy trills repeated about 2 times/sec. **RANGE:** Appalachian Mts. from sw. PA southwest to AL and extreme ne. MS.

BRIMLEY'S CHORUS FROG *Pseudacris brimleyi* **PL. 10**

1–1³/₈ in. (2.5–3.5 cm). Slender, light brown to brownish yellow or gray, with white lines on lips, no dark triangle between eyes, *bold black stripes extending uninterrupted from snout through eyes to the groin*. Three considerably less pronounced brown or gray middorsal stripes. Undersides yellow with *dark spots on chest*; *dark longitudinal bars or stripes* instead of crossbands on hindlimbs. **SIMILAR SPECIES:** See Upland Chorus Frog. **HABITAT:** Low areas in hardwood forests,

swamps near rivers and streams, marshes, wet open woods. **VOICE:** Short, raspy trills lasting less than 1 sec. and repeated 12 times or more. **RANGE:** Coastal Plain from VA south to GA.

OTTED CHORUS FROG *Pseudacris clarkii* **PL. 10**

¾–1¼ in. (1.9–3.1 cm). Pale gray or grayish brown to olive or greenish, white lines on lips, a dark but sometimes irregular *triangle between eyes*, dark bands or stripes extend from snout through eyes onto the body and sometimes to the groin (often interrupted), *black-bordered green dorsal spots* occasionally fused into stripes. Belly white. **SIMILAR SPECIES:** (1) Cajun and Boreal Chorus Frogs have 3 gray or brown (rarely greenish) stripes or dark spots arranged in 3 longitudinal rows. (2) Strecker's Chorus Frog is larger, somewhat toadlike, with black stripes from snout to shoulders, almost always with dark spots below eyes. **HABITAT:** Prairie grasslands, pastures, meadows, shrubby areas, lawns, woodland edges, most commonly near shallow semipermanent to permanent ponds, irrigation canals, cattle tanks. **VOICE:** Rasping trill, *wrrank-wrrank-wrrank*, rapidly repeated 20–30 or more times with intervals between notes about equal to duration of notes. **RANGE:** Cen. KS south to s. TX and extreme ne. Mex. **REMARK:** Interbreeds with the Boreal Chorus Frog.

SPRING PEEPER *Pseudacris crucifer* **PL. 10**

¾–1¼ in. (1.9–3.2 cm); record 1½ in. (3.7 cm). Straw, light brown, or rusty orange to gray or olive. No white lines on lips (lips might be slightly lighter than head). Dark dorsal streaks or lines include *lines between eyes*, crossbands on hindlimbs, and *X on center of back* (often imperfect or uneven). *Toepads distinctly wider than toes*. Belly light, virtually unmarked or with dark spots (latter most prevalent in s. GA and n. FL). **SIMILAR SPECIES:** Other chorus frogs tend to be striped, mottled, or spotted, and most have sharply defined light lines along upper lip and toepads barely wider than toes. (1) Mountain Chorus Frog sometimes has dorsal markings resembling an X, but also has a well-defined dark triangle between eyes and distinct white lines on lip. (2) Bird-voiced, Cope's Gray, and Gray Treefrogs have dark-bordered light patches beneath eyes and rough skin. (3) Pine Woods Treefrog has rows of small orange, yellow, or whitish spots on concealed surfaces of thighs. **HABITAT:** Moist wooded areas, especially near breeding pools. **VOICE:** High *peep*, single clear note rising slightly in pitch from beginning to end and repeated at intervals of about 1 sec. **RANGE:** Atlantic Provinces south to n. FL and west to SK and e. TX. **REMARK:** Because toepads are wider than toes, the Spring Peeper was long thought to be in the genus *Hyla* instead of *Pseudacris*.

UPLAND CHORUS FROG *Pseudacris feriarum* **PL. 10, FIG. 52**

¾–1⅜ in. (1.9–3.5 cm); record 1½ in. (3.9 cm). Dark brown to light tan, with white lines on lips often extending to shoulders. Sometimes with indications of a dark triangle or other figure between eyes; dark bands or lines extend from the snout through eyes and eardrums onto shoulders and sometimes continuing (often broken) to the groin. Dorsal pattern of 3 *narrow* longitudinal dark stripes *often* broken into streaks or rows of small spots (Fig. 52, p. 110) usually lighter than lateral stripes and sometimes not in strong contrast to ground color; stripes sometimes reduced to a few scattered spots or entirely absent. Tibial bands narrow, often indistinct. Belly whitish to cream with dark stippling on chest. Length of tibia about one-half head-body length. **SIMILAR SPECIES:** (1) Cajun Chorus Frog cannot be reliably distinguished, but its call has a slightly longer duration and slower pulse rate; best distinguished using geography and molecular markers. (2) New Jersey, Midland, and Boreal Chorus Frogs tend to have broader, usually unbroken dorsal stripes (if broken, usually only the middorsal stripe), and the Boreal Chorus Frog has shorter legs; however, all are variable and best distinguished using geography and molecular markers. (3) Southern Chorus Frog has a more pointed snout, very dark dorsal markings in strong contrast with ground color, broad tibial bands with narrow light interspaces, and the call has a slower pulse rate. (4) Brimley's Chorus Frog has bold, black lateral stripes much more sharply defined than dorsal stripes, dark longitudinal bars or stripes instead of crossbands on hindlimbs, and a spotted chest. (5) Ornate Chorus Frog has conspicuous light-bordered black

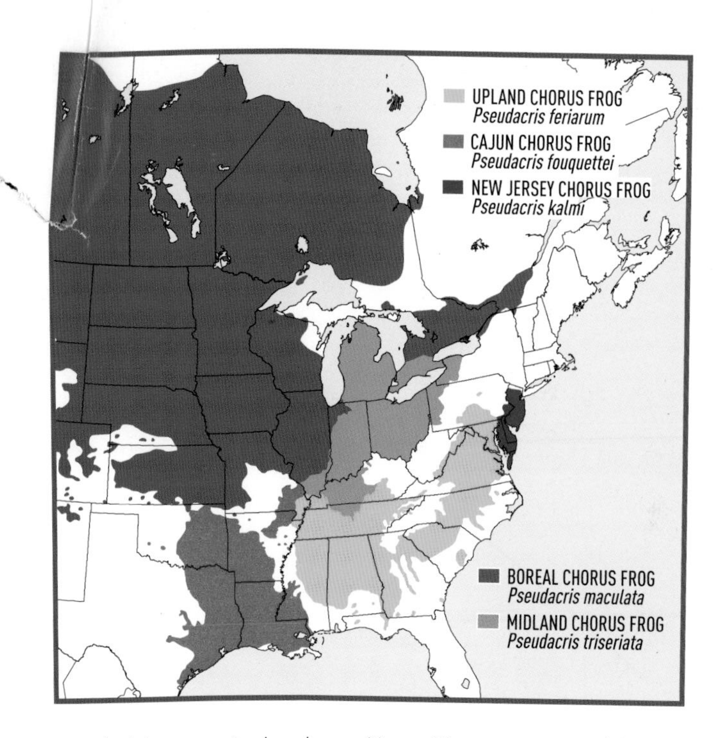

or dark brown spots along lower sides and 2 more near groin. (6) Little Grass Frog is smaller and lacks distinct white lines on lips. **HABITAT:** Meadows, moist forests, bottomland swamps, vicinity of ponds, bogs, marshes; mainly in uplands in northern parts of range, in lowlands farther south. **VOICE:** Clicking trills with rising inflections, *crrreeeeek*; similar to trills of Midland and Boreal Chorus Frogs, but with a faster pulse rate (within trills) than Cajun and Southern Chorus Frogs. **RANGE:** Cen. PA south to the FL Panhandle, west to MS and se. MO. **REMARK:** Species in the *P. triseriata* complex (Upland, Cajun, New Jersey, Boreal, and Midland Chorus Frogs) until recently were considered subspecies of what was called the Northern Chorus Frog.

CAJUN CHORUS FROG *Pseudacris fouquettei* **NOT ILLUS.**

¾–1 in. (1.9–2.5 cm); record 1³⁄₁₆ in. (3 cm). Essentially indistinguishable from the Upland Chorus Frog. **SIMILAR SPECIES:** See Upland Chorus Frog. **HABITAT:** Forests, fields, swamps, marshes, irrigation ditches, temporarily flooded areas. **VOICE:** See Upland Chorus Frog. **RANGE:** Se. MO and e. OK south to the Gulf. **REMARK:** See Upland Chorus Frog.

FIG. 64. *The Illinois Chorus Frog* (Pseudacris illinoensis) *occurs in the Mississippi River Valley but often is considered conspecific with Strecker's Chorus Frog* (P. streckeri), *which occupies a disjunct range to the west.*

ILLINOIS CHORUS FROG *Pseudacris illinoensis* **FIG. 64**
STRECKER'S CHORUS FROG *Pseudacris streckeri* **PL. 10**

1–1⅝ in. (2.5–4.1 cm); record 1⅞ in. (4.8 cm). Stocky, almost toadlike, with *stout forelimbs*. Gray, hazel, and brown to olive or green, with *dark brown or black masklike stripes* extending from the snout through the eyes (behind which stripes are wider) and eardrums onto shoulders and frequently continuing as series of dark spots along sides. Additional, usually less prominent, dorsal spots or streaks sometimes present. Upper lip with narrow dark lines. *Dark spots almost always present below eyes* vary in size and sometimes may be little more than upward bulges from narrow dark liplines. **SIMILAR SPECIES:** Spring Peeper is small and slender, toepads are distinctly wider than toes, and usually has a dark dorsal X and toepads distinctly wider than toes. Other chorus frogs in range have continuous light lines along the upper lip. **HABITAT:** Moist, shady woods, rocky ravines, sand prairies, cultivated fields, usually in vicinity of streams, lagoons, cypress swamps. **VOICE:** Clear, metallic single note quickly repeated. **RANGE:** Illinois Chorus Frog: Mississippi River Valley of AR, MO, IL; Strecker's Chorus Frog: extreme s.-cen. KS south

FIG. 65. *The New Jersey Chorus Frog* (Pseudacris kalmi) *is in the* Pseudacris triseriata *species complex, which includes the Upland* (P. feriarum), *Cajun* (P. fouquettei), *Boreal* (P. maculata), *and Midland* (P. triseriata) *Chorus Frogs. Species in the complex frequently interbreed where ranges overlap.*

to the Gulf. **REMARK:** These two species are sometimes considered conspecific.

NEW JERSEY CHORUS FROG *Pseudacris kalmi* **FIG. 65**

¾–1⅜ in. (1.9–3.5 cm); record 1 7/16 in. (3.6 cm). Gray to brown, sometimes greenish or olive. Very similar to the Upland Chorus Frog, except the dorsal pattern of 3 *broad* longitudinal dark stripes is *rarely* broken into streaks or rows of small spots or entirely absent. **SIMILAR SPECIES:** See Upland Chorus Frog. **HABITAT:** Grassy floodplains, wet woodlands with shallow wetlands such as ephemeral pools, ditches, wooded swamps, freshwater marshes. **VOICE:** See Upland Chorus Frog. **RANGE:** NJ south through the Delmarva Peninsula. **REMARK:** See Upland Chorus Frog.

BOREAL CHORUS FROG *Pseudacris maculata* **PL. 10**

¾–1⅜ in. (1.9–3.5 cm); record 1½ in. (3.9 cm). Dark brown to light tan, occasionally greenish. Similar to the Upland Chorus Frog, except the dorsal pattern of 3 *broad* longitudinal dark stripes is *sometimes*, especially the middle stripe, broken into streaks or rows of spots. *Legs short* (length of tibia less than one-half head-body length). **SIMILAR SPECIES:** See Upland Chorus Frog. **HABITAT:** Open and wooded areas, often near still bodies of water and associated wetlands and meadows. **VOICE:** See Upland Chorus Frog. **RANGE:** IL west to AZ and north to Hudson Bay and NT; disjunct in se. ON, s. QC, and extreme n. NY.

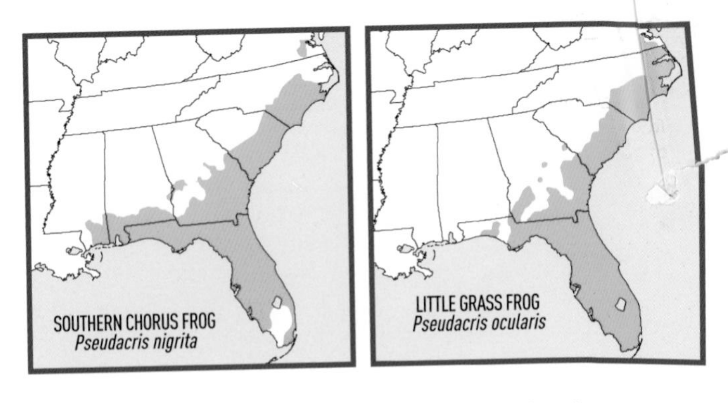

REMARKS: Interbreeds with the Spotted Chorus Frog. See also Upland Chorus Frog.

SOUTHERN CHORUS FROG *Pseudacris nigrita* **PL. 10, FIG. 52**

¾–1¼ in. (1.9–3.3 cm). Darkest of the chorus frogs. Light gray, tan, or silvery with *very dark, usually black markings*; sometimes indications of a dark triangle or other figure between eyes; dark bands or lines extend from the *pointed snout* through eyes and eardrums onto shoulders and usually continue to the groin. Dorsal pattern of 3 *narrow* longitudinal dark stripes *often broken into streaks or rows of spots* (Fig. 52, p. 110); dorsal stripes frequently darker than lateral stripes and always in strong contrast to ground color; stripes occasionally reduced to scattered spots or entirely absent. Distinct white lines on lips in most populations often extend to shoulders, but frogs in peninsular FL have black interrupting white liplines so some individuals have primarily black lips. *Tibial bands broad* with narrow light interspaces. Belly whitish to cream with dark stippling on chest. **SIMILAR SPECIES:** See Upland Chorus Frog. **HABITAT:** Pine flatwoods and forests, wet meadows, moist woodlands, prairies, often near small ponds, potholes, or ditches with grassy margins or emergent vegetation. **VOICE:** Slow, clicking trill, *crrreeeek*, with about 8–10 pulses/trill, rising in pitch, lasting about 1 sec., and repeated at regular intervals of about 2–3 sec. **RANGE:** E. NC south through FL and west to s. MS.

LITTLE GRASS FROG *Pseudacris ocularis* **PL. 10**

$^{7}/_{16}$–$^{5}/_{8}$ in. (1.1–1.6 cm); record $^{7}/_{8}$ in. (2 cm). Smallest N. American frog, often mistaken for a juvenile of another species. Narrow, somewhat pointed head. Tan, brown, greenish, pink, or reddish. No white lines on lips. A vague dark triangle between eyes (sometimes continuing onto body as a faint, often interrupted middorsal line); *dark lines of variable length pass through eyes and onto sides of the body*, additional narrow dorsolateral stripes (often poorly defined). Crossbands on hindlimbs. Undersides white or yellowish, no spots on chest. **SIMILAR**

SPECIES: See Upland Chorus Frog. **HABITAT:** Grass or sedgy areas near bogs or ponds in pine flatwoods or along pools and streams in hardwood forests and cypress swamps. **VOICE:** Very high pitched, insectlike chirps, *set-see*, each consisting of 2 parts, introductory note followed by a trill and repeated about 1 time/sec. Choruses sound like chirping crickets. **RANGE:** Se. VA through FL.

ATE CHORUS FROG *Pseudacris ornata* PL. 10

1–1¼ in. (2.5–3.2 cm); record 1⁹⁄₁₆ in. (4 cm). Chunky. Brown or reddish brown or nearly black to silvery gray or green; can rapidly change from light to dark. *Black masklike stripes* extend from the snout through eyes and eardrums onto shoulders. Light-bordered *dark brown to black spots on lower sides* (like discontinuous extensions of mask) and usually 2 more *spots near the groin*; additional dorsal marks, including a triangle between eyes, sometimes present. Yellow in groin and numerous small yellow spots on concealed portions of hindlimbs. Pattern details poorly developed in juvenile. **SIMILAR SPECIES:** Spring Peeper is small and slender, toepads are distinctly wider than toes, and usually has a dark dorsal X. Other chorus frogs in range have a pattern consisting primarily of longitudinal dark stripes or rows of spots and lack conspicuous light-bordered black or dark brown spots along sides. **HABITAT:** Cypress ponds, pine barren ponds, flooded meadows, flatwoods ditches. **VOICE:** Series of shrill, metallic notes repeated 2–3 times/sec. **RANGE:** Coastal Plain from NC south to cen. FL and west to se. MS. Populations in sw. MS and adjacent se. LA extirpated (not mapped).

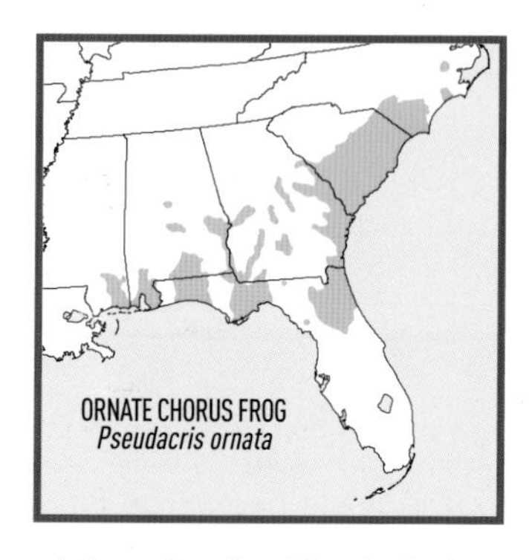

MIDLAND CHORUS FROG *Pseudacris triseriata* PL. 10, FIG. 52

¾–1³⁄₁₆ in. (1.9–3 cm); record 1¼ in. (3.2 cm). Light or dark brown to gray or greenish. Similar to the Upland Chorus Frog, except the dorsal pattern of 3 *broad* longitudinal dark stripes *rarely* broken into streaks or rows of small spots (Fig. 52, p. 110) or entirely absent. **SIMILAR SPECIES:** See Upland Chorus Frog. **HABITAT:** Largely open areas, damp meadows, marshes, forest edges, bottomland swamps, floodplains. **VOICE:** See Upland Chorus Frog. **RANGE:** W. NY west to MI and south to TN. **REMARKS:** Often called the Western Chorus Frog, but the name was changed to reflect recent studies that show a more restricted range east of the Mississippi R. See also Upland Chorus Frog.

NEOTROPICAL THIN-TOED FROGS: Leptodactylidae

Recently revised, this family now includes 13 genera and about 200 species. All occur in the Americas. Many species, including the one that ranges into our area, build foam nests near water. Tadpoles live in the liquefied center of the nest until rains enable them to swim into nearby pools.

MEXICAN WHITE-LIPPED FROG *Leptodactylus fragilis* **PL.**

1 1/16–1 3/4 in. (2.7–4.4 cm). Gray to chocolate brown, with a variable number of dark dorsal and lateral spots, *distinct white or cream-colored lines on upper lip*, dorsolateral ridges, a *ventral disc* (Fig. 57, p. 126), and essentially *no webbing* between toes. **SIMILAR SPECIES:** (1) Spotted Chorus Frog has some webbing between toes and lacks dorsolateral ridges and a ventral disc. (2) Rio Grande Leopard Frog grows much larger, has extensive webbing between toes, and lacks a ventral disc. **HABITAT:** Cultivated fields, roadside ditches, irrigated fields, moist meadows, drains, potholes, oxbow lakes, low grasslands, runoff areas. **VOICE:** *Throw-up, throw-up* repeated continually and with rising inflections at end. **RANGE:** Extreme s. TX south to n. S. America. **REMARK:** Until recently known as *Leptodactylus labialis* (many references use that name).

NARROW-MOUTHED FROGS: Microhylidae

Sixty genera and more than 560 species range across North and South America, sub-Saharan Africa, and from the Indian subcontinent and Korea south through northern Australia. Two closely related genera found only in North and Central America are represented by three species in our area. These are small, plump frogs with *short limbs*, a *pointed head*, and a *skinfold across the back of the head* (Fig. 66, below). Males almost always have a dark throat, whereas that of females is light. Skin-gland secretions are distasteful to predators and an irritant to humans.

EASTERN NARROW-MOUTHED FROG *Gastrophryne carolinensis* **PL. 11, FIG. 67**

7/8–1 1/4 in. (2.2–3.2 cm); record 1 1/2 in. (3.9 cm). Gray or silvery to brown or reddish, dark or light at various times; broad dark middorsal area flanked by broad light stripes, stripes frequently obscured (often

FIG. 66. *Narrow-mouthed and sheep frogs* (Gastrophryne *and* Hypopachus) *are small, plump frogs with short limbs, a pointed head, and a skinfold across the back of the head.*

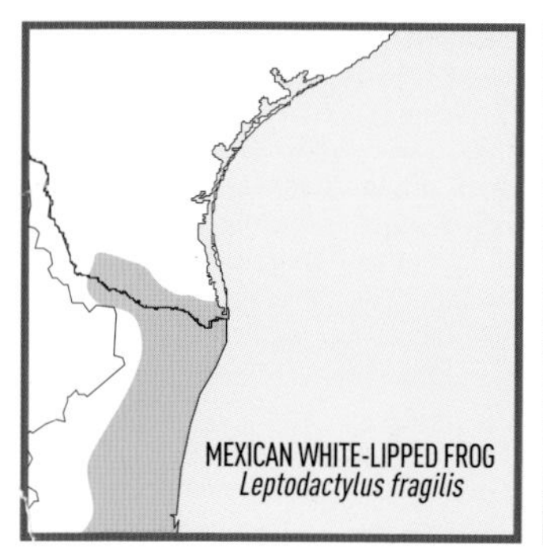

FIG. 67. *Venters of the Sheep Frog* (Hypopachus variolosus), *Western Narrow-mouthed Frog* (Gastrophryne olivacea), *and Eastern Narrow-mouthed Frog* (G. carolinensis).

completely) by patches, spots, and mottling of dark or light pigment. *Belly strongly mottled* (Fig. 67, above). About half of individuals on the Lower FL Keys have a middorsal area only slightly darker than the light dorsolateral stripes and separated from them by irregular dark lines; about one-fourth are tan with virtually no pattern; the rest are similar to frogs on the mainland, but usually more reddish. *Single spade* on hindlimbs. **SIMILAR SPECIES:** (1) Western Narrow-mouthed Frog has a light, generally unmarked belly, lacks any dorsal pattern or has only a few dark dorsal spots. (2) Sheep Frog has a thin, light middorsal line and 2 spades on each hindfoot. **HABITAT:** Moist, shaded areas, often near water. **VOICE:** Like the bleat of a lamb, occasionally with a very short preliminary *peep*; calls vibrant, something like an electric buzzer, duration ½–4 sec. **RANGE:** S. MD south through FL and the Keys and west to e. TX.

WESTERN NARROW-MOUTHED FROG *Gastrophryne olivacea*

PL. 11, FIG. 67

⅞–1½ in. (2.2–3.8 cm); record 1$^{11}/_{16}$ in. (4.3 cm). Gray or olive green, dark or light at various times; dorsum often with widely scattered small black spots, otherwise patternless. *Belly light, unmarked* or virtually so (Fig. 67, p. 147). Young dark brown with a conspicuous dark dorsal leaflike pattern that disappears with age. **SIMILAR SPECIES:** See Eastern Narrow-mouthed Frog. **HABITAT:** Semiarid and arid lowlands, grasslands, rocky wooded hills, cultivated fields, river floodplains, near springs, streams, rain pools, along marshes; sometimes in tarantula burrows. **VOICE:** Very short, distinct *whit* or *peep* followed by a low-pitched, insectlike buzz lasting 1–4 sec. **RANGE:** Extreme s. NE and w. MO, where it follows the Missouri R. eastward, south into Mex. and west through most of TX and n. Mex. into sw. NM; also s. AZ south into Mex. **REMARK:** Also called the Great Plains Narrow-mouthed Frog or ant-eating toad.

SHEEP FROG *Hypopachus variolosus* **PL. 11, FIG. 67**

1–1½ in. (2.5–3.8 cm); record 1 7/8 in. (4.7 cm). Tan or brown to olive, dorsum with small scattered dark spots or streaks and a *narrow, yellow middorsal line* from the snout along full length of body (line sometimes broken, occasionally branched, rarely missing). Belly gray with dark mottling and a *thin white midventral line* running full length of body with branches from chest to forelimbs (Fig. 67, p. 147). *Two prominent spades* on hindfeet. **SIMILAR SPECIES:** See Eastern Narrow-mouthed Frog. **HABITAT:** Disturbed and open habitats, moist sites in arid areas, often near water. **VOICE:** Sheeplike bleat lasting 1½–2½ sec., repeated every 10–20 sec. **RANGE:** Disjunct in s. TX and n. Mex.; continuous from n.-cen. Mex. to Costa Rica. **REMARKS:** Comprises multiple lineages. Because the current name applies to Costa Rican populations, the name for U.S. populations is likely to change.

TRUE FROGS: Ranidae

Sixteen genera and mo[illegible] species have an almost worldwide distribution. Seventeen [illegible] species of American water frogs (*Lithobates*) occur in our a[illegible] ently, these species were assigned to the genus *Rana*, a nar[illegible] cted to frogs in the Eastern Hemisphere and western North A[illegible] ese frogs generally are long-legged, narrow-waisted, and rather sm[illegible] skinned, with fingers free and toes joined by webs. Vocal sacs of males may be single or paired (Fig. 68, facing page).

CRAWFISH FROG *Lithobates areolatus* **PL. 11, FIGS. 68, 69**

2¼–3 in. (5.7–7.5 cm); record 4 13/16 in. (12.2 cm). Stout, toadlike. Relatively *smooth* skin. Dark, densely spaced, *light-bordered dorsal spots* on variable ground color. Belly whitish, unmarked (Fig. 69, p. 150); chin and throat unmarked or with some dark pigment laterally. Dorsolateral ridges prominent to barely noticeable. Male has lateral vocal sacs (Fig. 68, facing page). Sometimes yellow or greenish yellow on dorsolateral ridges and concealed surfaces of limbs. **SIMILAR SPECIES:** See Gopher and Plains Leopard Frogs. **HABITAT:** Prairie, pasturelands, floodplains, pine scrub. **VOICE:** Low-pitched, nasal, often

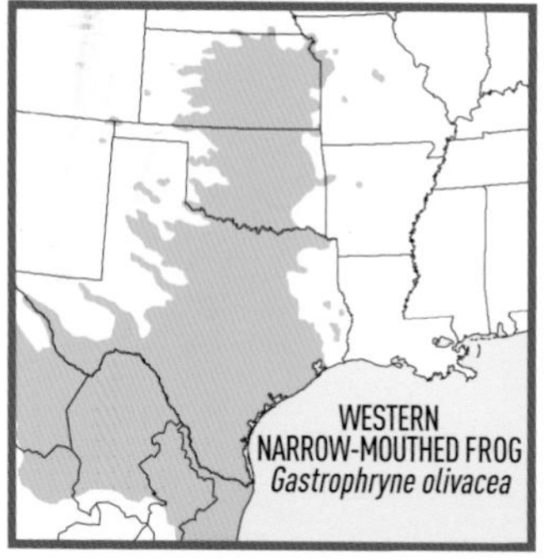

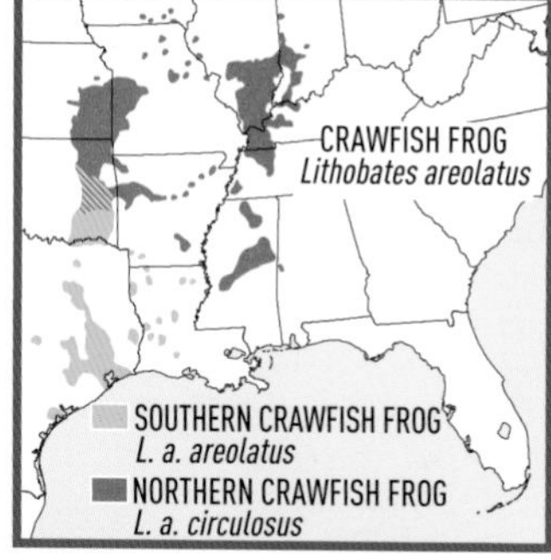

FIG. 68. *Paired vocal sacs are clearly evident in this Crawfish Frog* (Lithobates areolatus).

chuckling snores lasting about 1 sec. **RANGE:** Disjunct; IN and IA south to the Gulf. **SUBSPECIES:** (1) **Southern Crawfish Frog** (*L. a. areolatus*); smaller, max. length 3⅝ in. (9.2 cm); head short, relatively narrow, skin smooth or nearly so, dorsolateral ridges not prominent. (2) **Northern Crawfish Frog** (*L. a. circulosus*); head short, broad, skin rougher, dorsolateral ridges more prominent.

RIO GRANDE LEOPARD FROG *Lithobates berlandieri* **PL. 11, FIG. 70**

2¼–4 in. (5.7–10 cm); record 4½ in. (11.4 cm). Tan, beige, or light gray (usually pale, occasionally green or olive or somewhat darker) with prominent dark spots; usually no spots on head in front of eyes; light central spots on eardrums sometimes present. Light lines on jaw fade and are indistinct in front of eyes. Back of thighs with a dark reticulate pattern. *Dorsolateral ridges interrupted anterior to groin and inset medially* (Fig. 70, p. 150). Male has paired vocal sacs and vestigial

FIG. 69. *Venter of the Crawfish Frog* (Lithobates areolatus) *compared with that of the Gopher Frog* (L. capito) *and Dusky Gopher Frog* (L. sevosus). *Some Gopher Frogs, especially in Florida, have dark pigment restricted to the chin and throat (the belly usually is unmarked posteriorly).*

CRAWFISH
(*L. areolatus*)
Largely unmarked

GOPHER and DUSKY GOPHER
(*L. capito* and *L. sevosus*)
Pigmented, often heavily

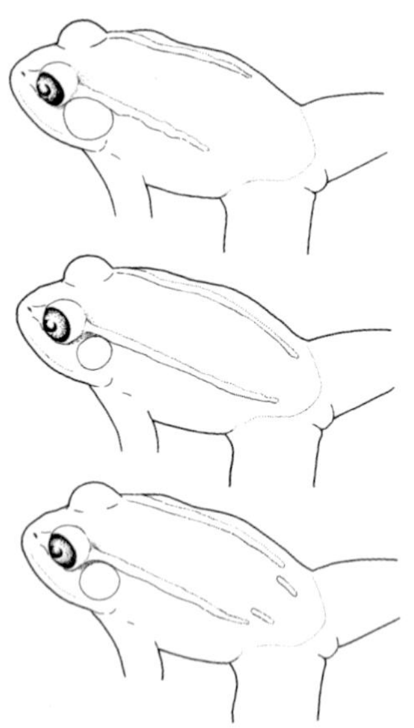

FIG. 70. *Dorsolateral ridges in six species of American water frogs* (Lithobates).

oviducts (visible only upon dissection), except in frogs from Trans-Pecos TX. **SIMILAR SPECIES:** See Plains Leopard Frog. **HABITAT:** Streams, rivers, springs, stock ponds, backwaters, canals, drainage ditches, arroyo pools in grasslands, shrublands, savannas, desert, woodlands. **VOICE:** Guttural, rattling trill (13 or more pulses/sec.); pulses last about ½ sec., uttered singly or in groups of 2–3, often followed by soft garbled notes; occasionally a stark *chuck*. **RANGE:** Cen. TX and extreme s. NM south into Mex. Introduced in several western states and in Mex. (not mapped).

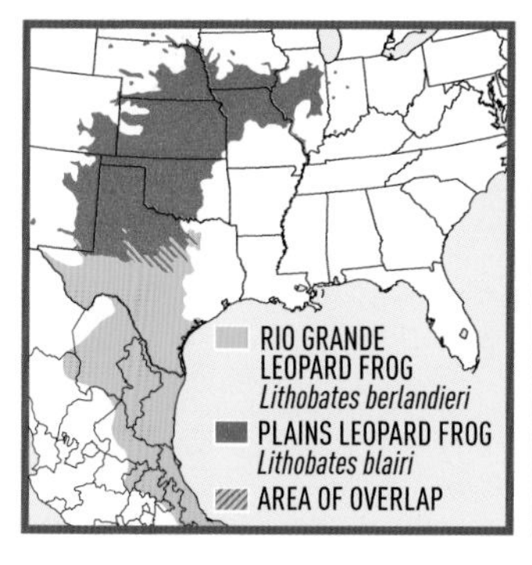

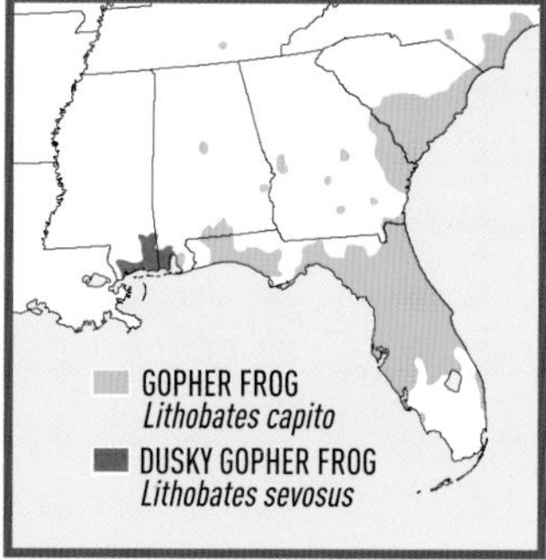

PLAINS LEOPARD FROG *Lithobates blairi* **PL. 11, FIGS. 70, 204 (P. 455)**

2–3¾ in. (5.1–9.5 cm); record 4½ in. (11.4 cm). Stocky, relatively short-headed. Tan to brown (occasionally green) with dark dorsal spots (not light-bordered), often with a spot on top of snout. Light spots on eardrums almost always present; light lines along upper lip *distinct*; considerable yellow in groin and to varying degrees on ventral surfaces of thighs. *Dorsolateral ridges interrupted anterior to groin and inset medially* (Fig. 70, facing page). Male has paired vocal sacs; lacks vestigial oviducts. **SIMILAR SPECIES:** (1) Rio Grande Leopard Frog usually lacks spot on snout, only sometimes has light spots on eardrums, and light lines on upper lip tend to fade in front of eyes. (2) Northern, Southern, and Kauffeld's Leopard Frogs have dorsolateral ridges that extend continuously to the groin. (3) Pickerel Frog has squarish or rectangular dorsal spots arranged in 2 rows between dorsolateral ridges. (4) Crawfish Frog is squat and stubby, with light-bordered, closely spaced dark dorsal spots. **HABITAT:** Prairie, desert grasslands, farmland, usually near water. **VOICE:** Two to 4 abrupt guttural notes repeated several times in rapid succession, *chuck-chuck-chuck*, rising slightly in pitch and often ending with soft grunt. **RANGE:** W. IN and s. SD south to e. NM and cen. TX; isolated populations as far west as se. AZ (not mapped).

GOPHER FROG *Lithobates capito* **PL. 11, FIG. 69**

2½–3¾ in. (7–9.5 cm); record 4⅜ in. (11.2 cm). Stout, toadlike. Light to dark gray to grayish brown or brown with a pattern of irregular very dark brown to black spots. Skin *warty*, circular or oval to elongated warts prominent; dorsolateral ridges strongly developed. Occasional individuals almost solid black; those from FL much lighter, warts smaller, often with some yellow on dorsolateral ridges, some warts along upper jaw and hidden surfaces of limbs. Ventral surfaces heavily mottled (producing a clouded or marbled pattern) or with dark spots limited to chin and throat. Male has lateral vocal sacs. Juvenile with less spotting on ventral surfaces, especially on abdomen, where often lacking. **SIMILAR SPECIES:** (1) Dusky Gopher Frog is similar, but usually darker with a more heavily spotted belly, at least from chin to midbody. (2) Crawfish Frog has smoother skin, a predominately light, unspotted belly, and rounded dark, light-bordered dorsal spots. (3) River Frog is less stocky, has conspicuous light spots on lips, and dorsolateral ridges curve around eardrums (not extending onto body). **HABITAT:** Relatively dry upland areas, dry to moist flatwoods, sand pine scrub, oak hammocks; frequently in Gopher Tortoise or rodent burrows. **VOICE:** Deep, rattling snores lasting about 3 sec. **RANGE:** Coastal Plain of the Carolinas south through FL and west to s. AL; isolated populations inland. **REMARK:** Until recently considered a subspecies of the Crawfish Frog.

AMERICAN BULLFROG *Lithobates catesbeianus* PL. 12, FIGS. 71, 72

3½–6 in. (9–15.2 cm); record 8⅝ in. (22 cm). Largest frog in our area (only the Cane Toad is larger). Plain or nearly plain green above or with a reticulate pattern of gray or brown on a green background. Venter whitish, often mottled with gray and with a yellowish wash (Fig. 71, below). Occasionally (especially in southeastern parts of the range) heavily patterned with dark gray or brown, sometimes (especially in FL) almost black above and heavily mottled below. *No dorso-*

FIG. 71. *Venter of the American Bullfrog* (Lithobates catesbeianus).

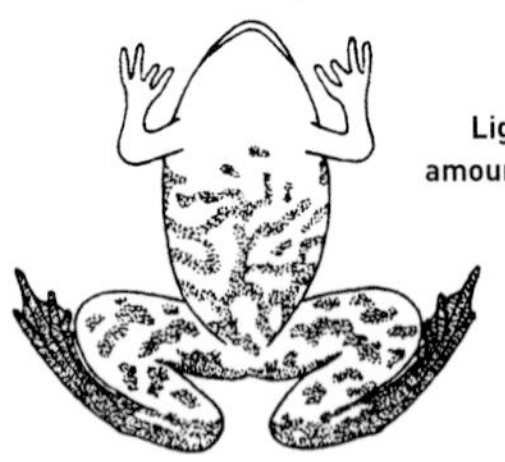

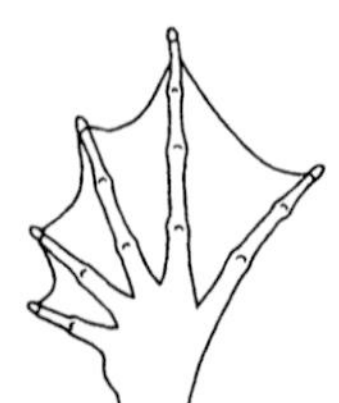

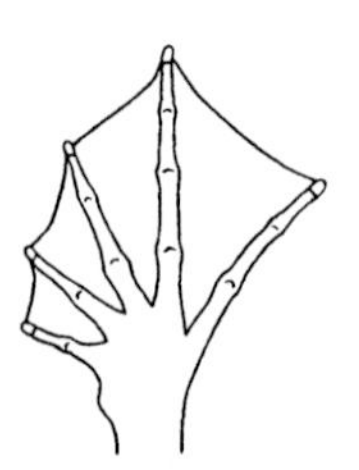

FIG. 72. *Toe webbing in the American Bullfrog* (Lithobates catesbeianus) *and the Pig Frog* (L. grylio). *Webbing in the River Frog* (L. heckscheri) *and the Carpenter Frog* (L. virgatipes) *is similar to that of the American Bullfrog.*

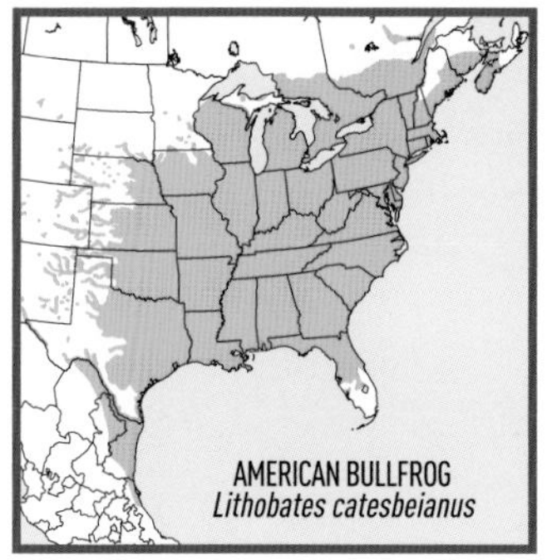

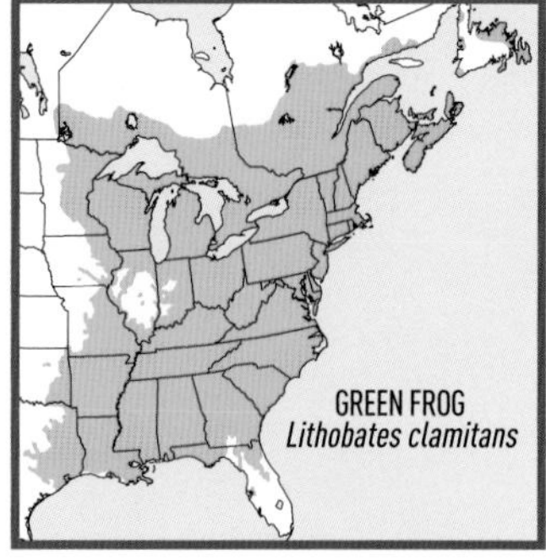

lateral ridges on body (ridges curve around eardrums). Hindtoe webs indented on either side of longest (fourth) toe, which extends beyond the web by at least one full joint (Fig. 72, facing page). Male has eardrums larger than eyes; internal vocal sac single. **SIMILAR SPECIES:** (1) Pig Frog has fully webbed hindfeet (webbing of fourth toe extends virtually to toetip). (2) Green Frog has dorsolateral ridges that extend onto the body. (3) River Frog has light spots on lips, rougher skin, and a dark venter marked with light spots or wavy lines. (4) Mink Frog may lack dorsolateral ridges (sometimes present) but is smaller and has more fully webbed hindlimbs. (5) Carpenter Frog has light longitudinal dorsolateral and lateral lines and more fully webbed hindlimbs. **HABITAT:** Lakes, reservoirs, ponds, swamps, marshes, bogs, sluggish portions of streams, cattle tanks, irrigation canals, occasionally temporary pools or flooded fields; usually near water. **VOICE:** Series of sonorous bass notes, *rumm . . . rumm . . . rumm* or *jug-o'-rum*. When startled, especially juveniles utter a loud *eeek* when leaping into water. **RANGE:** Atlantic Provinces and s. QC south to s. FL, west to the Rocky Mts., and south into n. Mex. Western limits of the range are confused because of many introductions into western states and s. BC. Widely introduced elsewhere (including HI). **REMARK:** Populations in FL might represent a distinct species.

GREEN FROG *Lithobates clamitans* **PL. 12, FIGS. 70, 73**

2⅛–3 in. (5.4–7.5 cm); record 4¼ in. (10.8 cm). Green or greenish brown to brown or bronzy (very rarely blue) with a scattering of dark brown or blackish spots on back and often with a light green upper lip. Venter white with dark mottling or vermiculate markings; throat of some males washed with yellow (Fig. 73, p. 154). Dorsolateral ridges *extend about two-thirds length of body* (do not reach groin; Fig. 70, p. 150). Center of eardrums elevated. Hindlimbs crossbanded; webbing fails to reach tip of fifth toe and barely passes beyond second joint of fourth (longest) toe. Male has eardrums larger than eyes and 2 (internal) vocal sacs. Juvenile has numerous dark dorsal spots and a venter with heavy brown or black vermiculate markings. **SIMILAR**

FIG. 73. *The eardrum of a male Green Frog* (Lithobates clamitans) *is larger than the eye and has a raised center. The throat is often yellow. An actively calling individual also inflates two lateral vocal pouches that amplify the call, which frequently is likened to an explosive banjolike throaty* gunk.

SPECIES: See American Bullfrog. **HABITAT:** Permanent or semipermanent water (young may use temporary pools), frequently swamps and small streams. **VOICE:** Explosive banjolike throaty *gunk*, usually in short series that drop in pitch and volume. Frightened individuals often squeak or chirp when leaping into water. **RANGE:** NS west to se. MB south to cen. FL and e. TX. Introduced in NL, also BC and some western states (not mapped).

PIG FROG *Lithobates grylio* **PL. 12, FIG. 72**

3¼–5½ in. (8.3–14 cm); record 6½ in. (16.5 cm). Head pointed, rather narrow; olive to dark brown with numerous dark dorsal spots, frequently with light lines or rows of light spots across rear of thighs. Venter white or pale yellow; dark brown or gray or even black reticu-

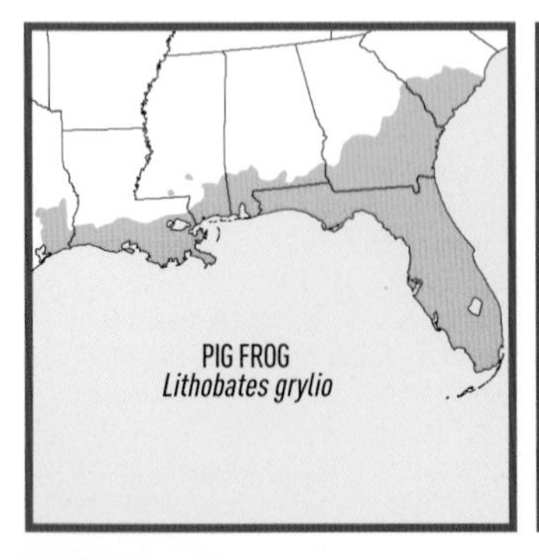

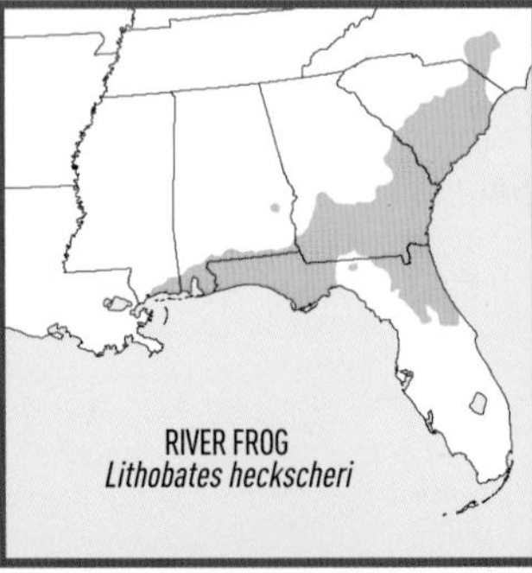

late pattern on thighs sometimes extends onto abdomen. *No dorsolateral ridges on body* (ridges curve around eardrums). Hindfeet fully webbed, with *fourth (longest) toe webbed virtually to tip* (Fig. 72, p. 152). Male has eardrums larger than eyes. Single internal vocal sac appears to have 3 sections when inflated. **SIMILAR SPECIES:** See American Bullfrog. **HABITAT:** Strongly aquatic; lakes, large ponds, marshes, swamps, cypress bays. **VOICE:** Low-pitched guttural piglike grunts, usually repeated 2–3 (occasionally more) times. **RANGE:** Coastal Plain from s. SC through FL and west to e. TX.

RIVER FROG *Lithobates heckscheri* **PL. 12, FIG. 74**

3¼–4⅝ in. (8.3–11.7 cm); record 6⅛ in. (15.5 cm). Dark brownish green to gray; skin *rough* with ragged black dorsal spots, streaks, or blotches. *Conspicuous light spots on lips* extend to eardrums and usually are largest on the lower jaw, often producing a scalloped effect along edges of the upper lip. Venter medium to dark gray (sometimes almost black), *marked with light spots or short wavy lines*, usually with pale crescents in groin (Fig. 74, below). *No dorsolateral ridges on body* (ridges curve around eardrums). Male has eardrums larger than eyes; single internal vocal sac. Longest (fourth) toes extends beyond webbing by at least one full joint (Fig. 72, p. 152). Juvenile often has reddish eyes. **SIMILAR SPECIES:** See American Bullfrog. **HABITAT:** Swamps and swampy shores of streams, shallow ponds, impoundments (including beaver ponds), bayous. **VOICE:** Deep, sonorous rolling snores lasting about 2 sec.; also snarling, explosive grunts. **RANGE:** Coastal Plain from s. NC and n.-cen. FL west to se. MS.

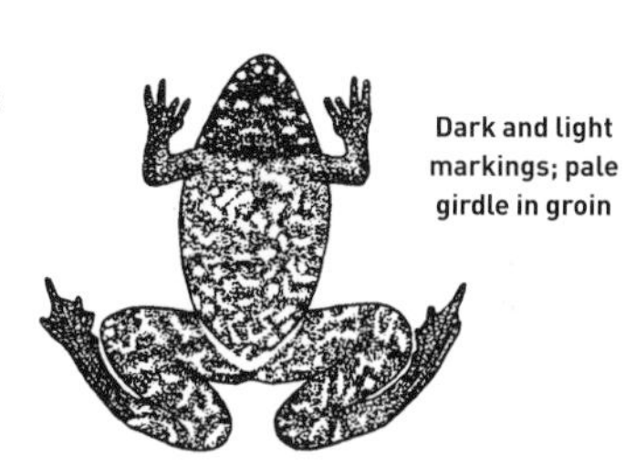

FIG. 74. *Venter of the River Frog* (Lithobates heckscheri).

KAUFFELD'S LEOPARD FROG *Lithobates kauffeldi* **FIG. 70**

2–3 in. (5.1–7.6 cm); record 3 5/16 in. (8.5 cm). Essentially indistinguishable from the Southern Leopard Frog but with light marks on a dark field on back of thighs (Southern Leopard Frogs in the Northeast have dark marks on a light field). *Dorsolateral ridges extend continuously to groin* (Fig. 70, p. 150). Vocal sacs of male paired. Similar species: See Plains and Southern Leopard Frogs. **HABITAT:** Essentially any body of fresh water, even slightly brackish marshes. **VOICE:** Single *chuck* occasionally followed by secondary "groans." **RANGE:** S. CT to s. NJ

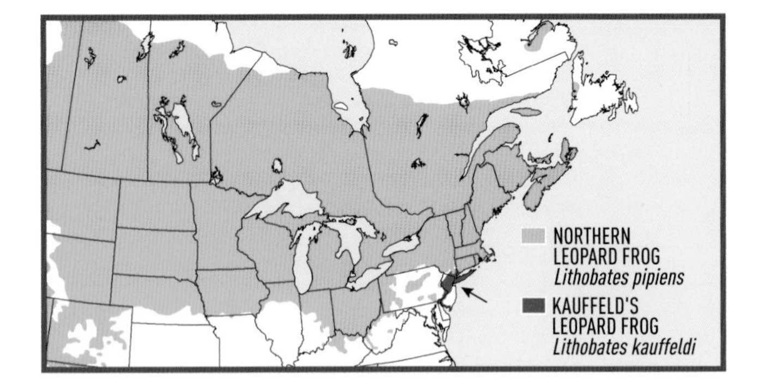

(mapped), possibly from cen. CT to ne. NC; at least occasionally shares ponds with the Northern Leopard Frog. **REMARK:** Originally named the Atlantic Coast Leopard Frog.

FLORIDA BOG FROG *Lithobates okaloosae* **PL. 11**

1³/₈–1¾ in. (3.5–4.4 cm); record 2¼ in. (5.7 cm). Smallest frog in the genus. Brown or brownish green to green, with light green lips, brown eardrums, unspotted above or with vague darkish dorsal spots, often with lighter spots on lower sides. *Light dorsolateral ridges stop short of groin.* Throat yellow, contrasting with gray belly bearing darker vermiculate markings. *Toes of hindfeet extend well beyond extremely reduced webbing.* **SIMILAR SPECIES:** (1) Green Frog is larger, dorsolateral ridges are usually the same color as dorsum, centers of eardrums are elevated, and hindfeet are extensively webbed. (2) American Bullfrog, Pig Frog, and River Frog are much larger and dorsolateral ridges curve behind eardrums (do not extend onto body). (3) Southern Leopard and Pickerel Frogs have distinct dorsal spots. **HABITAT:** Clear acidic streams and shallow freshwater seeps, especially shallow boggy overflows dominated by sphagnum moss. **VOICE:** Series of 5–15 or more guttural notes repeated 4–5 times/sec., dropping in volume at end, somewhat like *Currt-Currt-Currt-currt-curt.* Responses to nearby males may elicit soft throaty notes or a sharp *pit.* **RANGE:** W. FL Panhandle. **CONSERVATION:** Vulnerable.

PICKEREL FROG *Lithobates palustris* **PL. 11**

1¾–3 in. (4.4–7.5 cm); record 3⁷/₁₆ in. (8.7 cm). Light brown (occasionally with a greenish cast), light lines on upper lip, *prominent squarish dorsal spots arranged in 2 (rarely 3) parallel rows* between prominent dorsolateral ridges; adjacent squares occasionally fused to form rectangles or long, longitudinal bars. *Bright yellow or orange on concealed surfaces of hindlimbs.* Venter plain whitish in northern populations but usually mottled or marbled with dark pigment in frogs from the Coastal Plain. Male has paired vocal sacs. Juvenile lacks bright

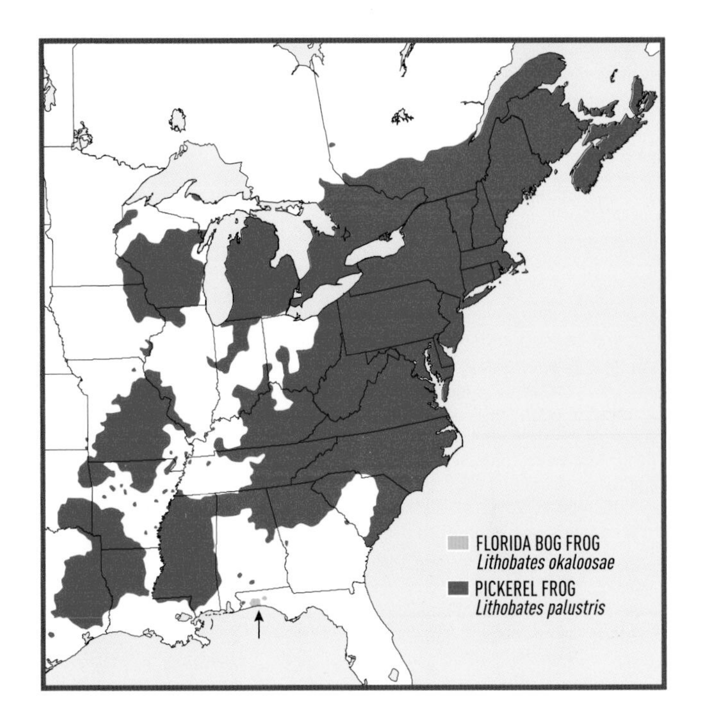

colors under legs and has a lower lip clouded with dark pigment. **SIMILAR SPECIES:** (1) Rio Grande, Plains, Northern, and Southern Leopard Frogs have circular or oval and irregularly distributed dark dorsal spots, and all but the Plains Leopard Frog lack bright yellow or orange on concealed surfaces of hindlimbs. (2) Crawfish Frog is squat, toadlike, with closely spaced light-bordered dark dorsal spots. **HABITAT:** Cool, clear water in the North (sphagnum bogs, springs, rocky ravines, meadow streams, many other aquatic situations); in the South, primarily warm, turbid, often tea-colored waters of the Coastal Plain and floodplain swamps; frequently in twilight zones of caves. **VOICE:** Soft, low-pitched, grating snores lasting about 2 sec.; also garbled, throaty notes along with a sharp *guck*. **RANGE:** Atlantic Provinces south to the Carolinas and west to se. MN and e. TX. **REMARK:** Skin-gland secretions render this species distasteful to predators.

NORTHERN LEOPARD FROG *Lithobates pipiens* PL. 11, FIG. 70

2–3½ in. (5.1–9 cm); record 4⅜ in. (11.1 cm). Brown or green. Snout *rounded*, light lines on upper lip, usually with a spot on top of the snout. Usually no distinct light spots on eardrums; 2 or 3 rows of rounded and irregularly placed, dark, light-bordered dorsal spots

FIG. 75. *Variations (diagrammatic) in dorsal patterns of the Mink Frog* (Lithobates septentrionalis), *showing spotted and mottled types.*

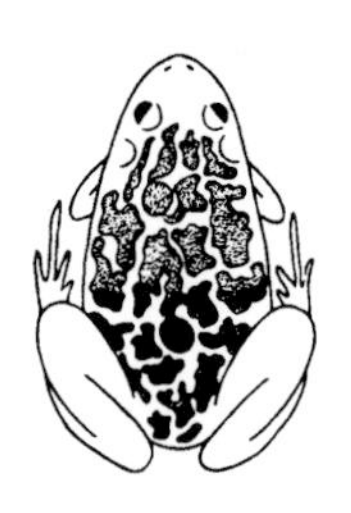

between conspicuous dorsolateral ridges; spots occasionally run together (some individuals in parts of MN and ND have dark mottling largely obscuring spots or lack spots altogether); numerous additional rounded dark spots on sides. *Dorsolateral ridges extend continuously to groin* (Fig. 70, p. 150). Male usually has vestigial oviducts (visible only upon dissection); paired vocal sacs visible only when calling. **SIMILAR SPECIES:** See Plains Leopard Frog. **HABITAT:** Springs, slow streams, marshes, bogs, ponds, canals, floodplains, reservoirs, lakes. **VOICE:** Deep rattling snores start softly and rise in volume before tapering off at end, duration 3 sec. or longer, pulse rate about 20/sec.; interspersed with clucking grunts consisting of 1, 2, or more notes. **RANGE:** Atlantic Provinces west to extreme s. NT and south to WV and KY in the East and NM and AZ in the West; many populations in western portions of the natural range are extirpated. Introduced in NL, BC, and many western states, so the western extent of the natural range is difficult to discern.

MINK FROG *Lithobates septentrionalis* **PL. 11, FIG. 75, 206 (P. 461)**

1⅞–2¾ in. (4.8–7 cm); record 3 in. (7.6 cm). Dark green to olive or brown, usually with green on the upper jaw (sometimes extending onto back). Dorsum with dark spots, blotches, or a reticulate pattern (Fig. 75, above); spots often round, size variable; *dark markings on dorsal surfaces of hindlimbs roughly rectangular*, with long axes more or less parallel to limbs. Dorsolateral ridges absent, partially developed, or prominent. Webbing on hindfeet *extends to the last joint of fourth (longest) toe and to tip of fifth toe*. Male has eardrums larger than eyes; vocal sacs paired. **SIMILAR SPECIES:** See American Bullfrog. **HABITAT:** Bogs, cold lakes and ponds. **VOICE:** Series of about 4 sharp, relatively deep-pitched taps, *cut-cut-cut-cut*. **RANGE:** S. NL and se. MB south to n. NY, s. ON, and MN.

DUSKY GOPHER FROG *Lithobates sevosus* **PL. 11, FIG. 69**

2½–3½ in. (6.4–9 cm); record 3⅞ in. (9.8 cm). Stout, toadlike. Dark gray or grayish brown to very dark brown or almost black (very dark individuals almost solid black), with a pattern of irregular very dark brown or reddish brown to black spots. Skin *warty*, with prominent

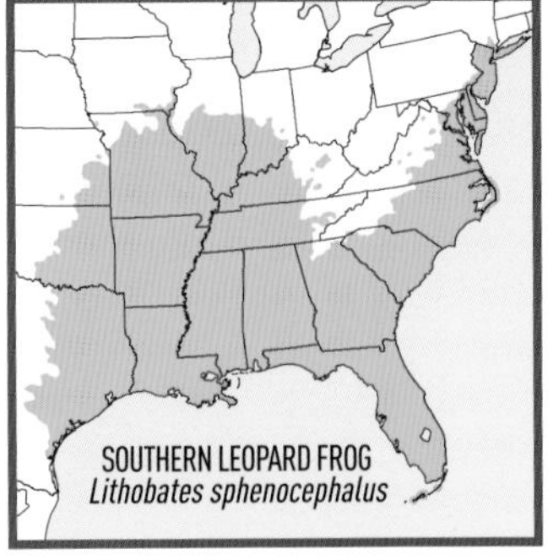

circular or oval warts; dorsolateral ridges strongly developed. Venter heavily spotted, at least from chin to midbody (Fig. 69, p. 150). Male has lateral vocal sacs. Juvenile often with less spotting on venter, especially on abdomen, where often lacking entirely. **SIMILAR SPECIES:** See Gopher Frog. **HABITAT:** Upland sandy areas with isolated temporary wetlands historically forested with longleaf pine; frequently in Gopher Tortoise or rodent burrows. **VOICE:** Deep, rattling snores lasting about 3 sec. **RANGE:** Gulf Coast from AL west to s. MS; populations in e. LA extirpated (not mapped). **REMARKS:** Until recently considered a subspecies of the Crawfish Frog. Isolated populations in AL previously assigned to this species are now referred to the Gopher Frog. **CONSERVATION:** Critically Endangered (USFWS: Endangered).

SOUTHERN LEOPARD FROG *Lithobates sphenocephalus* PL. 11, FIG. 70

2–3½ in. (5.1–9 cm); record 5 in. (12.7 cm). Head elongated, snout *pointed*. Brown to gray or green, light lines on upper lip, usually without spots on top of the snout, usually with distinct light spots on eardrums. Variable number of rounded, often longitudinally elongated dark dorsal spots without light margins, few dark spots on sides; occasional individuals lack spots. *Dorsolateral ridges extend continuously to groin* (Fig. 70, p. 150). Male has paired vocal sacs; vestigial oviducts absent (except in peninsular FL). **SIMILAR SPECIES:** See Plains and Kauffeld's Leopard Frogs. **HABITAT:** Essentially any body of fresh water, even slightly brackish marshes. **VOICE:** Five to 10 or more abrupt guttural notes repeated several times in rapid succession, *chuck-chuck-chuck*, often followed by grunting or scraping sounds. Alarm call a piercing scream. **RANGE:** Se. NY south through FL and some Keys west to e.-cen. TX; introduced in the Bahamas. **REMARK:** FL populations, in which males have vestigial oviducts and vocal sacs that fold inward and out of sight when not inflated, are sometimes recognized as a distinct subspecies.

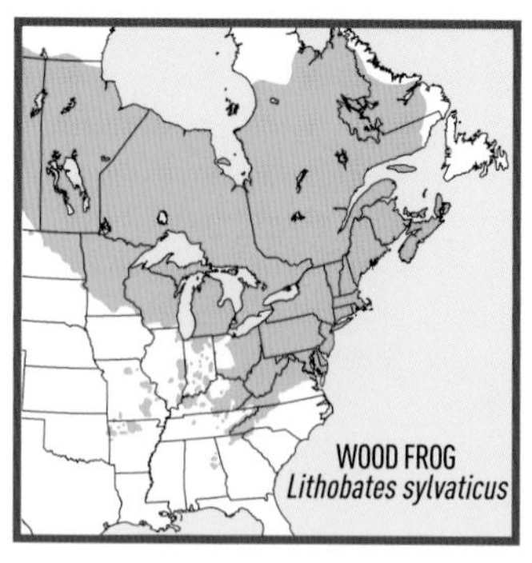

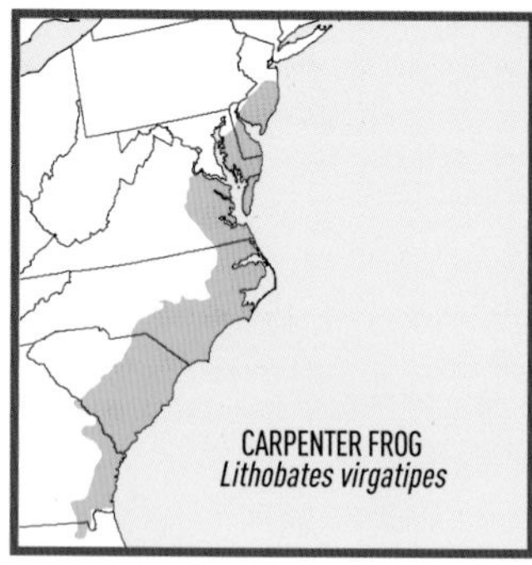

WOOD FROG *Lithobates sylvaticus* **PL. 11**

1⅜–2¾ in. (3.5–7 cm); record 3¼ in. (8.3 cm). Brown or pinkish brown to coppery or tan, with a *dark mask* that begins as thin lines from the snout to eyes but extends back from the eyes as broad bands over the eardrums to shoulders (sometimes indistinct in very dark individuals); some frogs from northern portions of the range have a light mid-dorsal stripe. Dorsolateral ridges prominent; often yellow in groin. Hindlimbs proportionately shorter in northern populations (short-legged frogs often hop like toads instead of leaping). **HABITAT:** Edges of ponds and streams in woodlands, willow thickets, and grass/willow associations; wherever shallow water is available in the far North, even tundra ponds. **VOICE:** Hoarse ducklike cackling with little carrying power, sometimes issued in series that almost blend together. **RANGE:** NL south through the Appalachian Mts. to n. GA and northwest to Alaska; many isolated populations in central and western states (some presumably introduced). **REMARK:** Eastern and western populations might be distinctive and worthy of taxonomic recognition.

CARPENTER FROG *Lithobates virgatipes* **PL. 11, P. viii**

1⅝–2⅝ in. (4.1–6.7 cm); record 2⅞ in. (7.3 cm). Brown and reddish brown to olive; lips light; *4 light longitudinal stripes* (dorsolateral stripes often less pronounced than lateral stripes, which are essentially extensions of light lips), usually with scattered dark spots between stripes. *No dorsolateral ridges on body* (ridges curve around eardrums). Fourth toe extends well beyond webbing (Fig. 72, p. 152). Male has eardrums larger than eyes and paired vocal sacs. **SIMILAR SPECIES:** See American Bullfrog. **HABITAT:** Bogs, swamps, edges of lakes and ponds. **VOICE:** Series of sharp, doubled, rapping notes, *pu-tunk´, pu-tunk´, pu-tunk´*, almost like two carpenters hitting nails a fraction of a second apart. **RANGE:** Coastal Plain from s. NJ to n. FL.

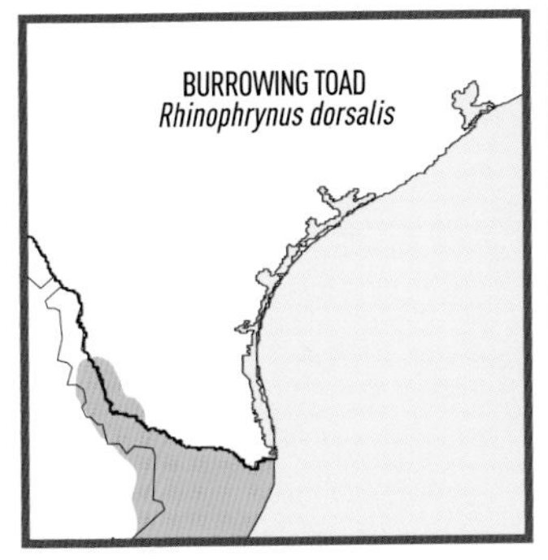

BURROWING TOADS: Rhinophrynidae

Only one currently recognized species is assigned to this family.

BURROWING TOAD *Rhinophrynus dorsalis* **PL. 12**

2–2¾ in. (5.1–7 cm); record 3½ in. (9 cm). Body *round*, head small, snout *cone-shaped*, small widely spaced eyes with *vertical pupils*, relatively short limbs, prominent spades on heavily webbed hindfeet. When calling or alarmed, the body is inflated and resembles a tennis ball with a snout. Dark gray or brown to black dorsum, venter somewhat lighter; *broad reddish or orange middorsal stripe* (buff in young), sometimes with similarly colored spots on sides, nearly always with scattered light spots on head, back, and belly. **SIMILAR SPECIES:** Sheep Frog is much smaller, lighter in color, and has very narrow yellow middorsal and midventral lines. **HABITAT:** Dry and moist forested lowlands, thorn scrub, savannas, cultivated areas; often seasonally flooded locations. **VOICE:** Loud, low-pitched *wh-o-o-o-a*. **RANGE:** Extreme s. TX to Cen. America. **REMARKS:** Often called the Mexican Burrowing Toad. Likely a complex of species.

NORTH AMERICAN SPADEFOOTS: Scaphiopodidae

Two genera and seven currently recognized species range from southwestern Canada to temperate regions in southern Mexico. Five species occur in our area. North American spadefoots until recently were placed in the same family as superficially similar Eurasian spadefoots (Pelobatidae). All have a *single, sharp-edged, black spade* on each hindfoot; spades are elongated and sickle-shaped in eastern spadefoots (*Scaphiopus*) and wedge-shaped in western spadefoots (*Spea*) (Fig. 76, p. 162). All lack obvious parotoid glands and have rather *smooth* skin and *vertically elliptical pupils*. In contrast, toads (Bufonidae) in our area have two spades on each hindfoot, warty skin, parotoid glands, horizontally oval pupils, and most have prominent cranial crests. Many people experience strong allergic reactions from handling spadefoots.

FOOT

HEAD

Sickle-shaped spade

Eyelids as wide as space between them

COUCH'S SPADEFOOT
(*Scaphiopus couchii*)

Wedge-shaped spade

Eyelids wider than space between them

PLAINS and MEXICAN SPADEFOOTS
(*Spea bombifrons* and *S. multiplicata*)

FIG. 76. *Characteristics of spadefoots* (Scaphiopus *and* Spea).

COUCH'S SPADEFOOT *Scaphiopus couchii* **PL. 12, FIG. 76**

2¼–2⅞ in. (5.7–7.3 cm); record 3½ in. (9 cm). Spades elongate (dark edge no more than twice as long as wide), *sickle-shaped*; no boss between eyes; diameter of eyelids about equal to distance between eyes (Fig. 76, above); no pectoral glands; fingers webbed. Bright greenish yellow to dull brownish yellow, mottled or marbled with black, green, or dark brown (dark pattern may fade during the breeding season). **SIMILAR SPECIES:** See Hurter's Spadefoot. **HABITAT:** Arid and semiarid shrublands, shortgrass plains, mesquite savannas, creosote bush desert, thorn forest, cultivated areas (tropical deciduous forests in Mex.). **VOICE:** Groaning goat- or sheeplike bleat lasting ½–1 sec. **RANGE:** Sw. OK and CO to se. CA and south into Mex.

EASTERN SPADEFOOT *Scaphiopus holbrookii* **PL. 12, FIG. 77**

1¾–2¼ in. (4.4–5.7 cm); record 3¹⁄₁₆ in. (7.8 cm). Spades elongate (dark edge 3 times longer than wide), *sickle-shaped*; no boss between eyes; pectoral glands present (Fig. 77, facing page); essentially no webs between fingers. Brown (grayish or blackish brown or sepia), almost always with 2 yellowish lines, one originating at each eye and running down the back to form inverted parentheses, a lyre-shaped pattern, or an outline of a somewhat misshapen hourglass; frequently with additional, sometimes broken light lines along sides. Occasionally almost uniformly dark gray to nearly black. **SIMILAR SPECIES:** See

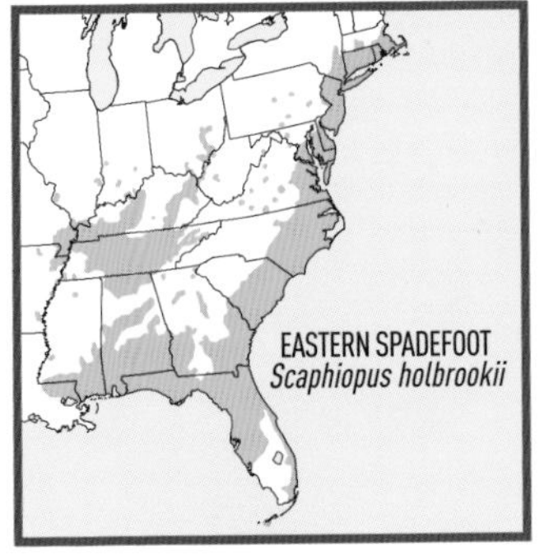

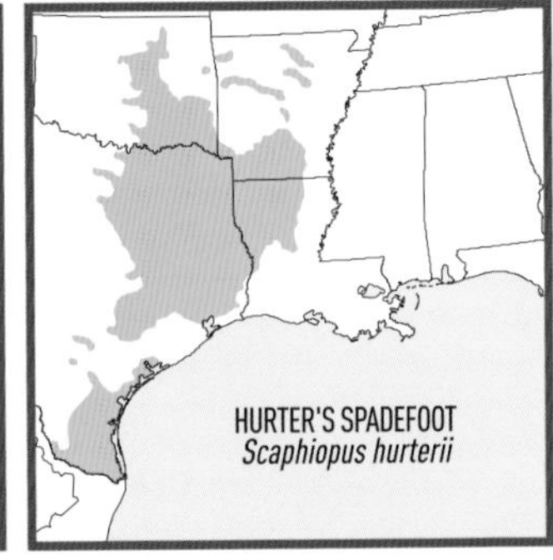

FIG. 77. *Pectoral glands of the Eastern Spadefoot* (Scaphiopus holbrookii) *and Hurter's Spadefoot* (S. hurterii).

FIG. 78. *Heads of Hurter's Spadefoot* (Scaphiopus hurterii) *and the Plains Spadefoot* (Spea bombifrons), *showing the location of the boss. Note that the Canadian Toad* (Anaxyrus hemiophrys) *has a similar boss formed by the cranial crests between the eyes and extending to the snout.*

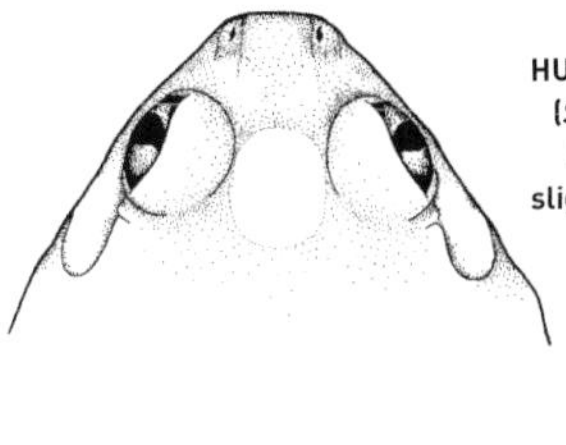

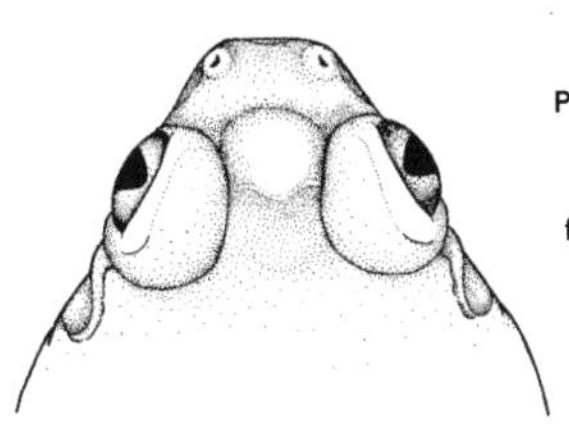

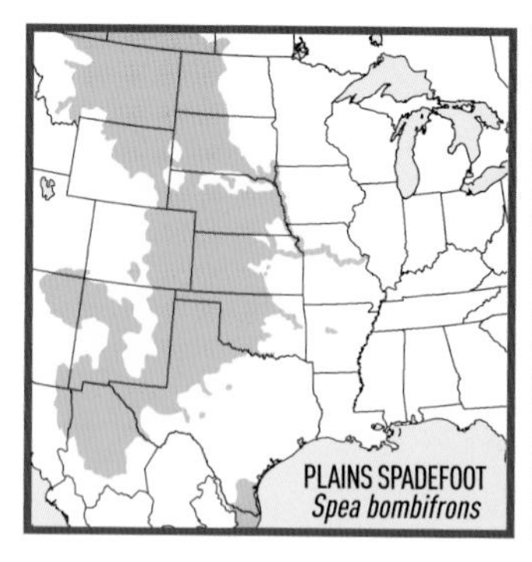

Hurter's Spadefoot. **HABITAT:** Sandy, gravelly, or soft, light soils in wooded or open terrain. **VOICE:** Explosive grunt, rather low-pitched, short in duration, repeated at brief intervals. **RANGE:** S. New England through FL west to the Mississippi R. and across the river to se. MO and e. AR. Populations in the FL Keys probably extirpated.

HURTER'S SPADEFOOT *Scaphiopus hurterii* **PL. 12, FIGS. 77, 78**

1¾–2¼ in. (4.4–5.7 cm); record 3¼ in. (8.3 cm). Spades elongate (dark edge 3 times longer than wide), *sickle-shaped*; *boss between and slightly behind eyes* (Fig. 78, p. 163); pectoral glands present (Fig. 77, p. 163); essentially no webs between fingers. Grayish green to chocolate or greenish brown to almost black, almost always with 2 curved light dorsal stripes (as in the Eastern Spadefoot). **SIMILAR SPECIES:** (1) Couch's and Eastern Spadefoots have no boss between eyes. (2) Plains and Mexican Spadefoots have wedge-shaped spades. **HABITAT:** Areas with sandy, gravelly, or soft, light soils in wooded or open terrain, savannas, arid mesquite scrub. **VOICE:** Bleating note, slightly explosive, short in duration (less than ½ sec.). **RANGE:** E. OK and w. AR south to s. TX. **REMARK:** Until recently considered a subspecies of the Eastern Spadefoot.

PLAINS SPADEFOOT *Spea bombifrons* **PL. 12, FIGS. 76, 78, 79**

1½–2 in. (3.8–5.1 cm); record 2½ in. (6.4 cm). Spades rounded, *wedge-shaped*; *high, bony boss extends forward of eyes* (Fig. 78, p. 163); eyelids wider than distance between eyes (Fig. 76, p. 162). Grayish or brownish, often with a greenish tinge; dark markings brown or gray; small dorsal tubercles yellowish or reddish; often with 4 vague longitudinal light dorsal lines. **SIMILAR SPECIES:** (1) Mexican Spadefoot lacks boss between eyes. (2) Couch's, Eastern, and Hurter's Spadefoots have sickle-shaped spades. **HABITAT:** Shrublands, grasslands, semidesert areas, almost always near temporary pools, usually not in wooded areas. **VOICE:** Short rasping bleat repeated at intervals of ½–1 sec. or a

FIG. 79. *This adult Plains Spadefoot* (Spea bombifrons) *is in a defensive posture, during which it inflates its body to make it look larger. This also makes it more difficult for a predator to swallow it.*

rasping, rather low-pitched snore, each trill lasting ½–¾ sec. **RANGE:** Sw. MB and s. AB south to n. Mex.; follows the Missouri River Valley eastward across MO; disjunct in s. TX and adjacent Mex. **REMARK:** Interbreeds with the Mexican Spadefoot.

MEXICAN SPADEFOOT *Spea multiplicata* **PL. 12, FIG. 76**

1½–2 in. (3.8–5.1 cm); record 2⁹⁄₁₆ in. (6.5 cm). Spades rounded and *wedge-shaped* (slightly longer than wide); no boss between eyes; eyelids wider than distance between eyes (Fig. 76, p. 162). Dusky grayish brown to brown or dusky green, sometimes almost black; scattered darker spots and blotches; small dorsal tubercles often reddish; occasional individuals with a vague suggestion of longitudinal, light-colored lines. **SIMILAR SPECIES:** See Plains Spadefoot. **HABITAT:** Desert grasslands, shortgrass plains, playas and alkali flats of arid and semiarid regions; creosote bush, sagebrush, and semidesert shrublands, mixed grassland and chaparral, open evergreen forests, but absent from extreme deserts. **VOICE:** Vibrant metallic trill lasting about ¾–1½ sec. **RANGE:** W. OK and cen. TX west to se. UT and south into Mex. **REMARKS:** Until recently considered a subspecies of the Western Spadefoot, which does not occur in our area. Interbreeds with the Plains Spadefoot.

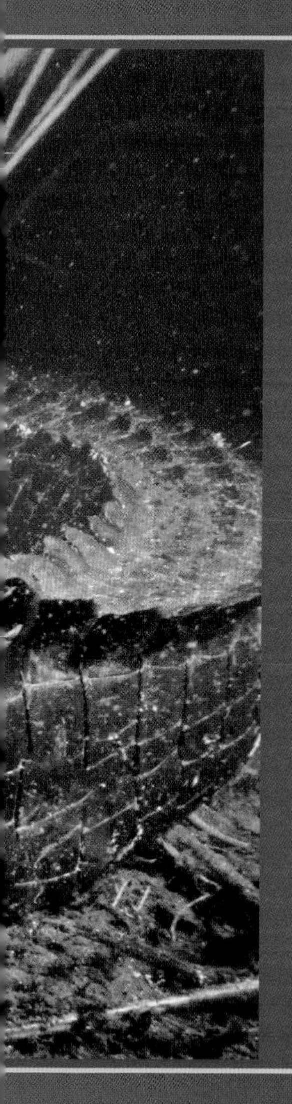

CROCODILIANS

PLATE 13

CROCODILIANS

YOUNG

AMERICAN ALLIGATOR *Alligator mississippiensis* **P. 170**
Black with yellowish crossbands.

SPECTACLED CAIMAN *Caiman crocodilus* Non-native **P. 171**
Gray with dark brown crossbands; curved bony ridge in front of eyes (Fig. 80).

AMERICAN CROCODILE *Crocodylus acutus* **P. 171**
Gray or greenish gray with black crossbands or rows of spots; snout narrow.

ADULTS

AMERICAN CROCODILE *Crocodylus acutus* **P. 171**
Snout narrow; light-colored; fourth teeth of lower jaw fit into grooves in upper jaw and remain visible when mouth is closed (Fig. 80).

AMERICAN ALLIGATOR *Alligator mississippiensis* **P. 170**
Broadly rounded snout.

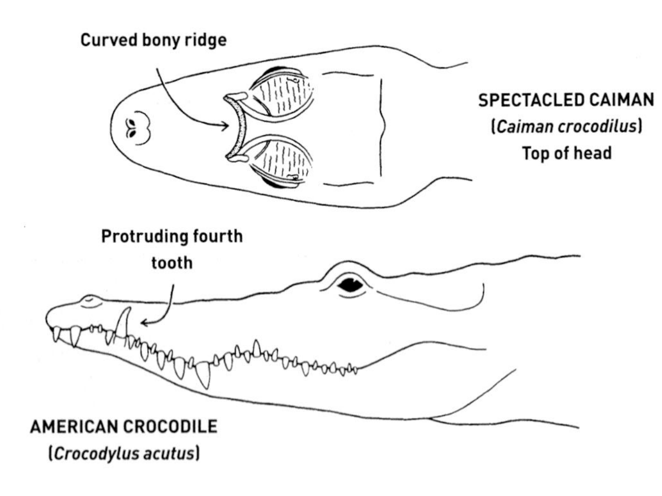

FIG. 80. *Head characteristics of some crocodilians.*

YOUNG
AMERICAN ALLIGATOR
SPECTACLED CAIMAN
AMERICAN CROCODILE

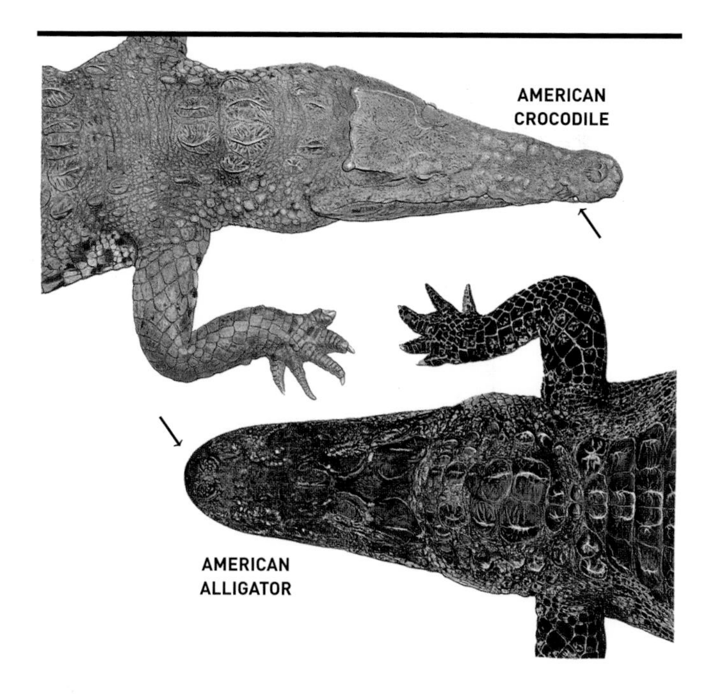
AMERICAN CROCODILE
AMERICAN ALLIGATOR

CROCODILIANS (CROCODYLIA)

More than 25 species in three families occur in tropical and subtropical regions of the world. Two species in two families are native to North America and a third species is established in Florida.

ALLIGATORS and CAIMANS: Alligatoridae

Seven species are endemic to the Americas, an eighth occurs in China.

AMERICAN ALLIGATOR *Alligator mississippiensis* **PL. 13, P. 166, P. 464**

6–16½ ft. (1.8–5 m); record 19 ft. 2 in. (5.84 m). *Broadly rounded snout.* Dark gray, brown, or olive to black, but bold yellowish crossbands of young sometimes persist in adult. Hatchling 8½–9 in. (21.5–23 cm). **SIMILAR SPECIES:** (1) American Crocodile has a tapering snout, and the fourth tooth on either side of lower jaw is visible when mouth is closed in all but very small individuals. (2) Spectacled Caiman has a curved, bony ridge in front of the eyes. **HABITAT:** Swamps, lakes, bayous, marshes, other bodies of water. **VOICE:** Adult male emits throaty, bellowing roars, adult female bellows less loudly; female grunts like a pig when calling young; juvenile emits a moaning grunt. Alligators of all sizes hiss. **RANGE:** Coastal Plain from NC south through FL and some Keys and west to TX. Disjunct in the Tennessee R. of n. AL. Lower Rio Grande Valley population probably introduced.

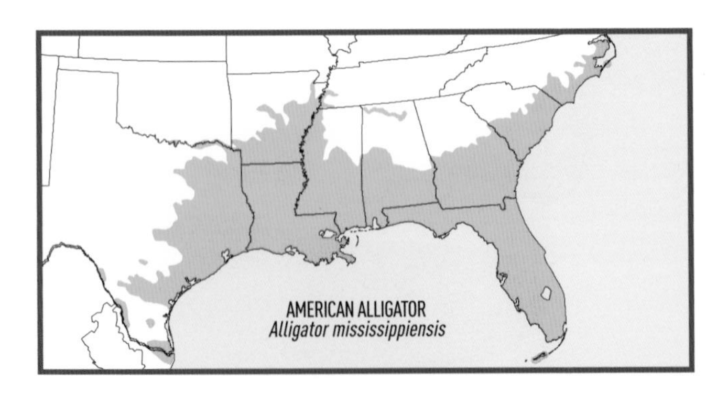

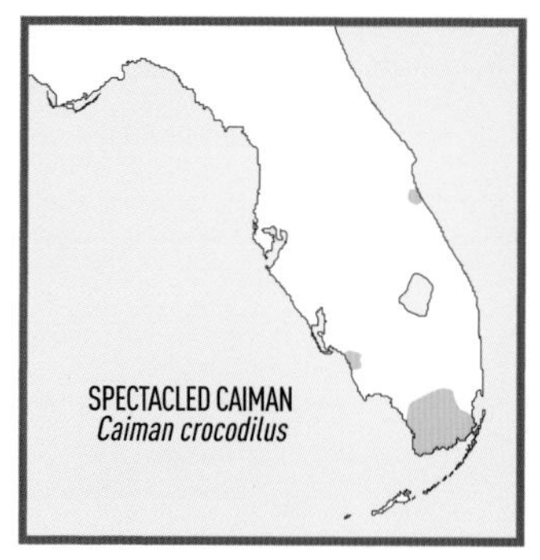

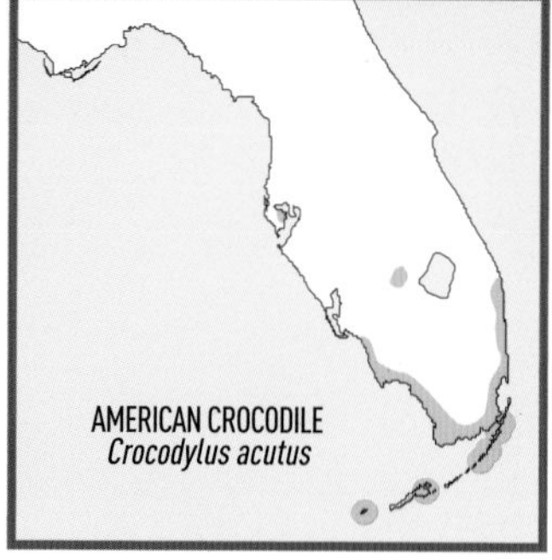

SPECTACLED CAIMAN *Caiman crocodilus* Non-native **PL. 13, FIG. 80, P. 452**

3½–6 ft. (1.1–1.8 m); record 8 ft. 8 in. (2.64 m). Broad snout with a *bony ridge in front of eyes* (Fig. 80, p. 168), ridge sometimes broken or irregular. Greenish, yellowish, or brownish gray with dark brown crossbands. Hatchling 8 in. (20.3 cm). **SIMILAR SPECIES:** See American Alligator. **HABITAT:** Natural and artificial waterways. **VOICE:** Adult male emits a throaty roar likened to a soft bark; female grunts to call young; young emits a soft high-pitched cough to call the mother. Caimans of all sizes moan or screech when distressed. **RANGE:** S. FL; also introduced elsewhere in FL, several other states (not mapped), and the Greater Antilles. **NATURAL RANGE:** S. Mex. to n. Argentina.

CROCODILES: Crocodylidae

Sixteen species occur in tropical America, Africa, Asia, and Australia.

AMERICAN CROCODILE *Crocodylus acutus* **PL. 13, FIG. 80, P. ii–iii**

7½–12 ft. (2.3–3.7 m); record 15 ft. (4.6 m) in U.S., to 23 ft. (7 m) in S. America. *Long tapering snout* with large fourth tooth on either side of the lower jaw visible when mouth is closed (Fig. 80, p. 168). Gray to tannish gray or dark greenish gray with dusky markings. Young gray or greenish gray with narrow black crossbands or rows of spots. Hatchling 8½–10 in. (21.5–25.4 cm). **SIMILAR SPECIES:** See American Alligator. **HABITAT:** Estuaries. **VOICE:** Male growls or emits a low rumble; young issues a high-pitched grunt. **RANGE:** Extreme s. FL and the Keys, occasionally farther north; also Greater Antilles and Mex. through n. S. America. **CONSERVATION:** Vulnerable (USFWS: Threatened).

TURTLES

SEA TURTLES

LOGGERHEAD SEA TURTLE *Caretta caretta* **P. 192**
Reddish brown; five (or more) pleurals on each side, first touching cervical; usually three scutes on bridge (Fig. 81).

KEMP'S RIDLEY SEA TURTLE *Lepidochelys kempii* **P. 194**
Gray; five pleurals on each side, first touching cervical; interanal scute present; usually four scutes on bridge (Fig. 81).

GREEN SEA TURTLE *Chelonia mydas* **P. 192**
Four pleurals on each side, first not touching cervical; one pair of prefontals between eyes (Fig. 89, p. 193).

LEATHERBACK SEA TURTLE *Dermochelys coriacea* **P. 194**
Prominent ridges along back; no scutes; skin smooth.

HAWKSBILL SEA TURTLE *Eretmochelys imbricata* **P. 194**
Tortoiseshell pattern; scutes overlap (except in very large individuals); two pairs of prefrontals between eyes (Fig. 89, p. 193).

FIG. 81. *Scutes of sea turtles.*

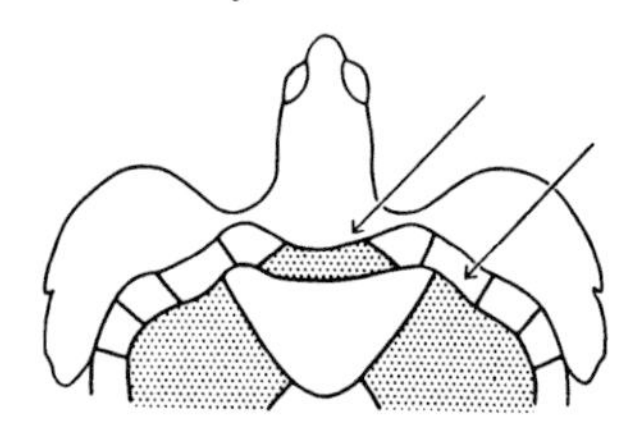

GREEN and HAWKSBILL
(*Chelonia mydas* and *Eretmochelys imbricata*)
Cervical separated from pleurals

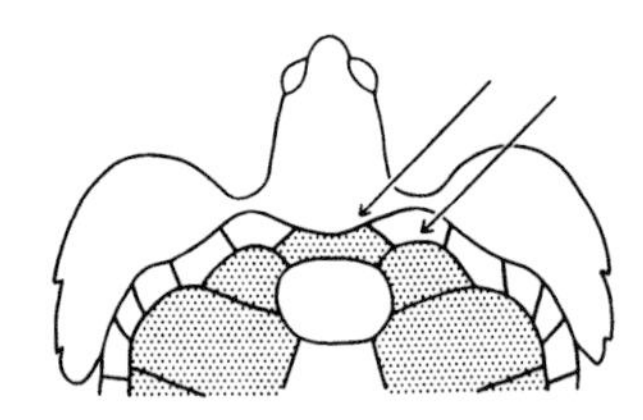

LOGGERHEAD and KEMP'S RIDLEY
(*Caretta caretta* and *Lepidochelys kempii*)
Cervical touches first pleurals

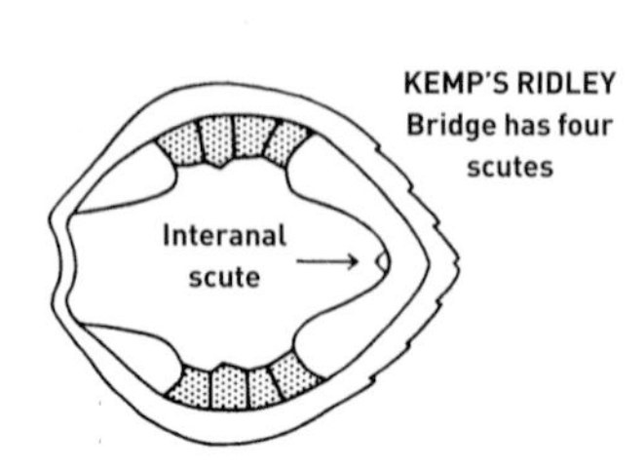

KEMP'S RIDLEY
Bridge has four scutes

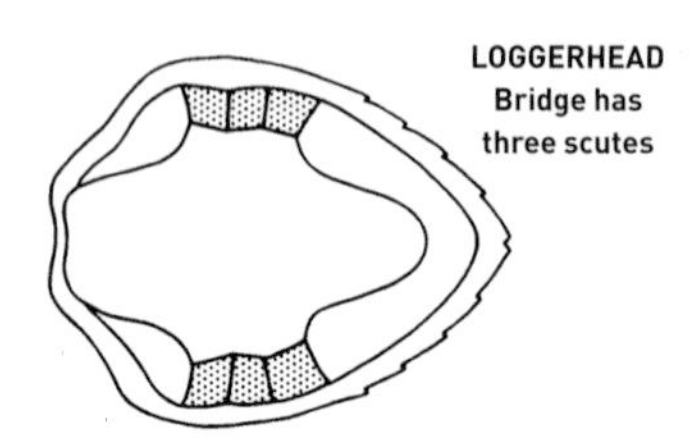

LOGGERHEAD
Bridge has three scutes

SNAPPING TURTLES

ALLIGATOR SNAPPING TURTLE *Macrochelys temminckii* **P. 196**
Extra rows of scutes on sides of carapace; three prominent dorsal keels; head very large; beak strongly hooked.

EASTERN SNAPPING TURTLE *Chelydra serpentina* **P. 195**
Long, saw-toothed tail.

LOGGERHEAD
GREEN
KEMP'S RIDLEY
HAWKSBILL
LEATHERBACK

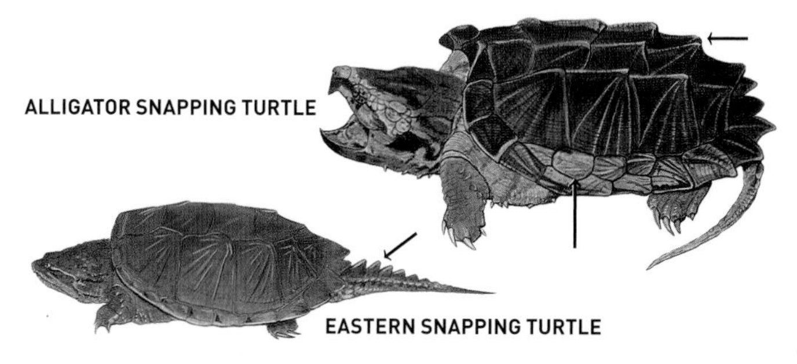
ALLIGATOR SNAPPING TURTLE
EASTERN SNAPPING TURTLE

PLATE 15

PAINTED TURTLES (*Chrysemys*)

PAINTED TURTLE *C. picta* **P. 198**

MIDLAND PAINTED TURTLE (*C. p. marginata*): Pleurals arranged in alternating fashion across back.

EASTERN PAINTED TURTLE (*C. p. picta*): Pleurals with broad olive edges arranged in more or less straight rows across back.

WESTERN PAINTED TURTLE (*C. p. bellii*): Light reticulate lines on carapace; bars on marginals.

SOUTHERN PAINTED TURTLE *C. dorsalis* **P. 197**

Broad red or orange middorsal stripe (sometimes yellowish).

SPOTTED TURTLES (*Clemmys*)

SPOTTED TURTLE *C. guttata* **P. 198**

Scattered yellow spots on carapace; orange or yellow spots on head.

CHICKEN and BLANDING'S TURTLES (*Deirochelys* and *Emydoidea*)

CHICKEN TURTLE *D. reticularia* **P. 199**

Long, striped neck; light reticulate pattern on shell; forelimbs broadly striped; striped "pants" (Fig. 91, p. 200).

BLANDING'S TURTLE *E. blandingii* **P. 200**

Carapace with many light spots; plastron with transverse hinge; bright yellow throat.

SCULPTED TURTLES (*Glyptemys*)

BOG TURTLE *G. muhlenbergii* **P. 201**

Orange head patches; small size.

WOOD TURTLE *G. insculpta* **P. 201**

Orange on neck and legs; shell rough, sculptured.

DIAMOND-BACKED TERRAPINS (*Malaclemys*)

DIAMOND-BACKED TERRAPIN *M. terrapin* **P. 210**

ORNATE DIAMOND-BACKED TERRAPIN (*M. t. macrospilota*): Concentric dark and light rings with yellow centers in each large scute.

NORTHERN DIAMOND-BACKED TERRAPIN (*M. t. terrapin*): Concentric rings on each large scute.

MIDLAND PAINTED
EASTERN PAINTED
WESTERN PAINTED
SOUTHERN PAINTED
SPOTTED
CHICKEN
BLANDING'S
BOG
plastron
WOOD
ORNATE DIAMOND-BACKED TERRAPIN
NORTHERN DIAMOND-BACKED TERRAPIN

MAP TURTLES (*Graptemys*)

Females of all species grow larger than males. See also FIGS. 92 (p. 202) and 93 (p. 209).

TEXAS MAP TURTLE *G. versa* **P. 209**
Horizontal or J-shaped lines behind eyes; anterior scutes of carapace distinctly convex.

BLACK-KNOBBED MAP TURTLE *G. nigrinoda* **P. 206**
Rounded black middorsal knobs; narrow light rings on pleurals.

RINGED MAP TURTLE *G. oculifera* **P. 206**
Broad light rings on pleurals.

ALABAMA MAP TURTLE *G. pulchra* **P. 208**
Broad light bars on marginals; longitudinal light bar under chin (Fig. 82). ♀ large, large-headed.

YELLOW-BLOTCHED MAP TURTLE *G. flavimaculata* **P. 204**
Solid orange or yellow spots on large scutes.

BARBOUR'S MAP TURTLE *G. barbouri* **P. 202**
Narrow light markings on marginals; curved or transverse bar under chin (Fig. 82). Mature ♀ very large, head enormous, pattern obscure.

FALSE MAP TURTLE *G. pseudogeographica* **P. 207**
MISSISSIPPI MAP TURTLE (*G. p. kohnii*): Yellow crescents prevent neck stripes from reaching eyes.

NORTHERN FALSE MAP TURTLE (*G. p. pseudogeographica*): Yellow spots behind eyes; some neck stripes reach eyes; middorsal spines conspicuous.

NORTHERN MAP TURTLE *G. geographica* **P. 204**
Yellowish spots behind eyes; maplike pattern on carapace; middorsal spines not prominent.

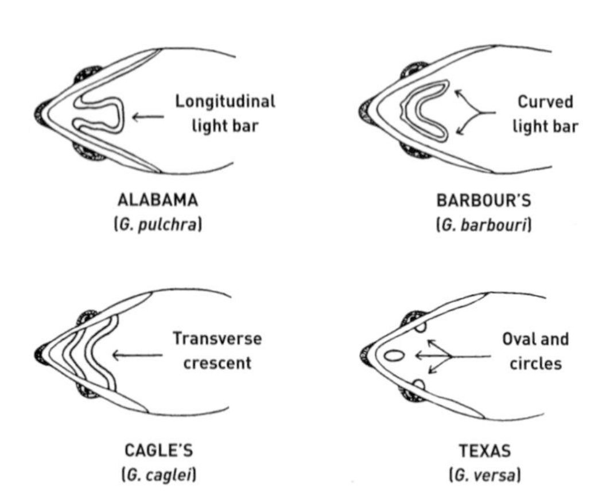

FIG. 82. *Chin patterns of some map turtles* (Graptemys).

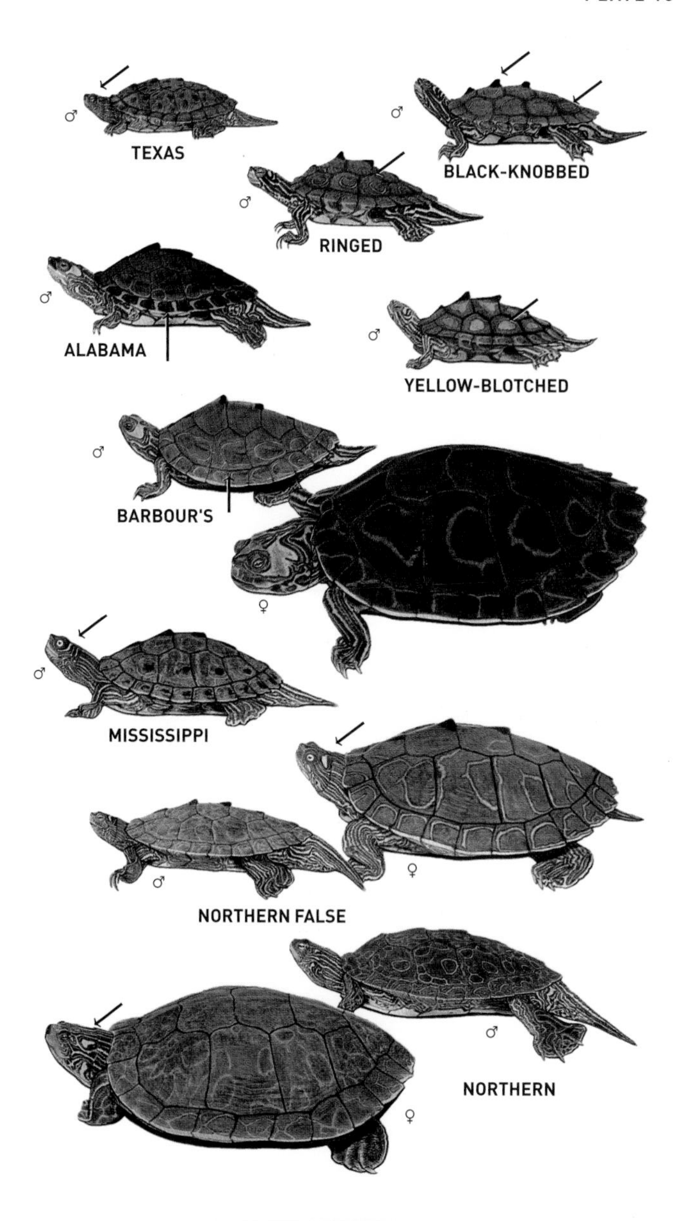
TEXAS
BLACK-KNOBBED
RINGED
ALABAMA
YELLOW-BLOTCHED
BARBOUR'S
MISSISSIPPI
NORTHERN FALSE
NORTHERN

PLATE 17

COOTERS (*Pseudemys*)

See also Pl. 18.

PENINSULA COOTER *P. peninsularis* **P. 215**
"Hairpins" on head (Fig. 83); dark smudges on marginals; plastron unmarked.

FLORIDA RED-BELLIED COOTER *P. nelsoni* **P. 214**
Light vertical bands on pleurals; few stripes on head (Fig. 83).

RIVER COOTER *P. concinna* **P. 212**
Figure C on second pleurals; dark spots (often doughnut-shaped) on undersides of marginals; shell pinched inward in front of hindlimbs.

COASTAL PLAIN COOTER *P. floridana* **P. 213**
No "hairpins"; hollow ovals on undersides of marginals; plastron unmarked.

NORTHERN RED-BELLIED COOTER *P. rubriventris* **P. 216**
♀ with vertical reddish markings on carapace. ♂ dark, markings irregular.

Note: Patterns, often obscure in large individuals, are best seen in submerged turtles. Distinctive markings often fade or disappear in old adults.

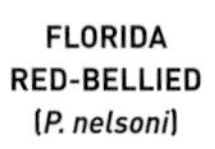

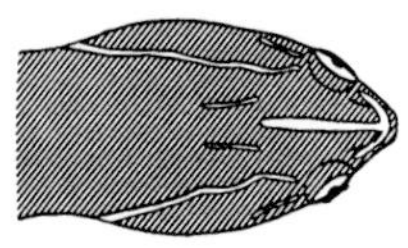

PENINSULA
(*P. peninsularis*)

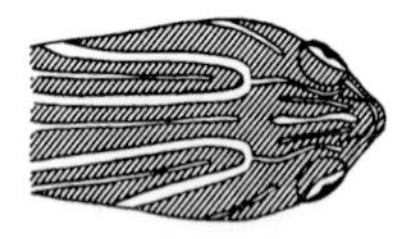

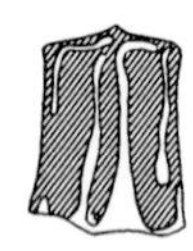

SUWANNEE
(*P. suwanniensis*)

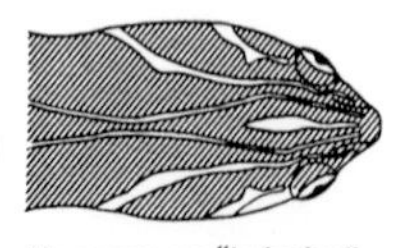

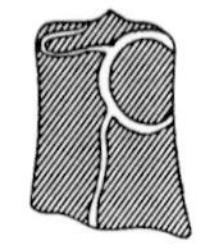

FIG. 83. *Head and scute patterns in some cooters* (Pseudemys). *Some of these same pattern elements occur in other species of cooters and are noted in the text by citing this figure. Note that patterns often are obscured in large individuals.*

♀
FLORIDA RED-BELLIED
♂
PENINSULA
plastron
♂
RIVER
plastron
young individual
♀
COASTAL PLAIN
♀
♂
NORTHERN RED-BELLIED

COOTERS AND SLIDERS (*Pseudemys* and *Trachemys*)

See also Pl. 17.

TEXAS COOTER *P. texana* **P. 217**

Broad head markings; pattern on plastron reduced to narrow dark lines.

SUWANNEE COOTER *P. suwanniensis* **P. 216**

Figure *C* on second pleurals; carapace dark; plastron with lines along seams.

POND SLIDER *T. scripta* **P. 217**

YELLOW-BELLIED SLIDER (*T. s. scripta*): Yellow head blotches; vertical yellowish bars on carapace; leg stripes narrow; striped "pants" (Fig. 91, p. 200).

RED-EARED SLIDER (*T. s. elegans*): ♀ and young with reddish stripes behind eyes; ♂ with reddish stripe reduced, sometimes completely obscured in old individuals.

Note: Patterns, often obscure in large individuals, are best seen in submerged turtles. Distinctive markings often fade or disappear in old adults.

FIG. 84. *Undersides of jaws of sliders* (Trachemys) *and cooters* (Pseudemys).

BOX TURTLES (*Terrapene*)

Plastron with a transverse hinge.

ORNATE BOX TURTLE *T. ornata* **P. 221**

Carapace flattened on top, with radiating light lines; plastron with bold light lines.

EASTERN BOX TURTLE *T. carolina* **P. 220**

Carapace high, domelike, with yellow, orange, or olive markings; plastron usually with some pattern.

FLORIDA BOX TURTLE *T. bauri* **P. 220**

Shell arched, highest toward rear, with radiating light lines; two yellow lines on head.

THREE-TOED BOX TURTLE *T. triunguis* **P. 222**

Orange on head (often on forelimbs); three toes on hindfeet; shell pattern reduced or absent.

TEXAS COOTER
SUWANNEE COOTER
young individual
YELLOW-BELLIED SLIDER
♂
♀
young ♂
RED-EARED SLIDER
ORNATE BOX
EASTERN BOX
plastron
plastron
FLORIDA BOX
THREE-TOED BOX

MUSK TURTLES (*Sternotherus*)

Plastron small. See also Figs. 101 (p. 224) and 104 (p. 227).

EASTERN MUSK TURTLE *S. odoratus* **P. 229**
Two light stripes on head; ♂ with large areas of skin between plastral scutes and large, stout tail; ♀ with plastral scutes close together.

LOGGERHEAD MUSK TURTLE *S. minor* **P. 228**
EASTERN LOGGERHEAD MUSK TURTLE (*S. m. minor*): ♂ with very large head; ♀ with moderate head and shell streaked, spotted, or blotched; young with streaked shell; three keels.

STRIPE-NECKED MUSK TURTLE (*S. m. peltifer*): Head and neck striped; middorsal keel.

RAZOR-BACKED MUSK TURTLE *S. carinatus* **P. 227**
Head spotted; strong middorsal keel.

FLATTENED MUSK TURTLE *S. depressus* **P. 228**
Reticulate head pattern; shell flattened.

MUD TURTLES (*Kinosternon*)

Plastron large. See also Figs. 101 (p. 224) and 102 (p. 225).

STRIPED MUD TURTLE *K. baurii* **P. 223**
Shell and head usually striped.

EASTERN MUD TURTLE *K. subrubrum* **P. 226**
Rounded shells.
MISSISSIPPI MUD TURTLE (*K. s. hippocrepis*): Two light stripes on head.
SOUTHEASTERN MUD TURTLE (*K. s. subrubrum*): No light stripes on head.

YELLOW MUD TURTLE *K. flavescens* **P. 224**
Throat plain yellow; ninth marginals higher than eighth; pectoral scutes narrowly in contact.

ROUGH-FOOTED MUD TURTLE *K. hirtipes* **P. 225**
Head spotted; tenth marginals higher than other marginals; pectoral scutes broadly in contact.

GOPHER TORTOISES (*Gopherus*)

GOPHER TORTOISE *G. polyphemus* **P. 230**
Hindfeet elephant-like; shell relatively long.

TEXAS TORTOISE *G. berlandieri* **P. 230**
Hindfeet elephant-like; shell relatively short.

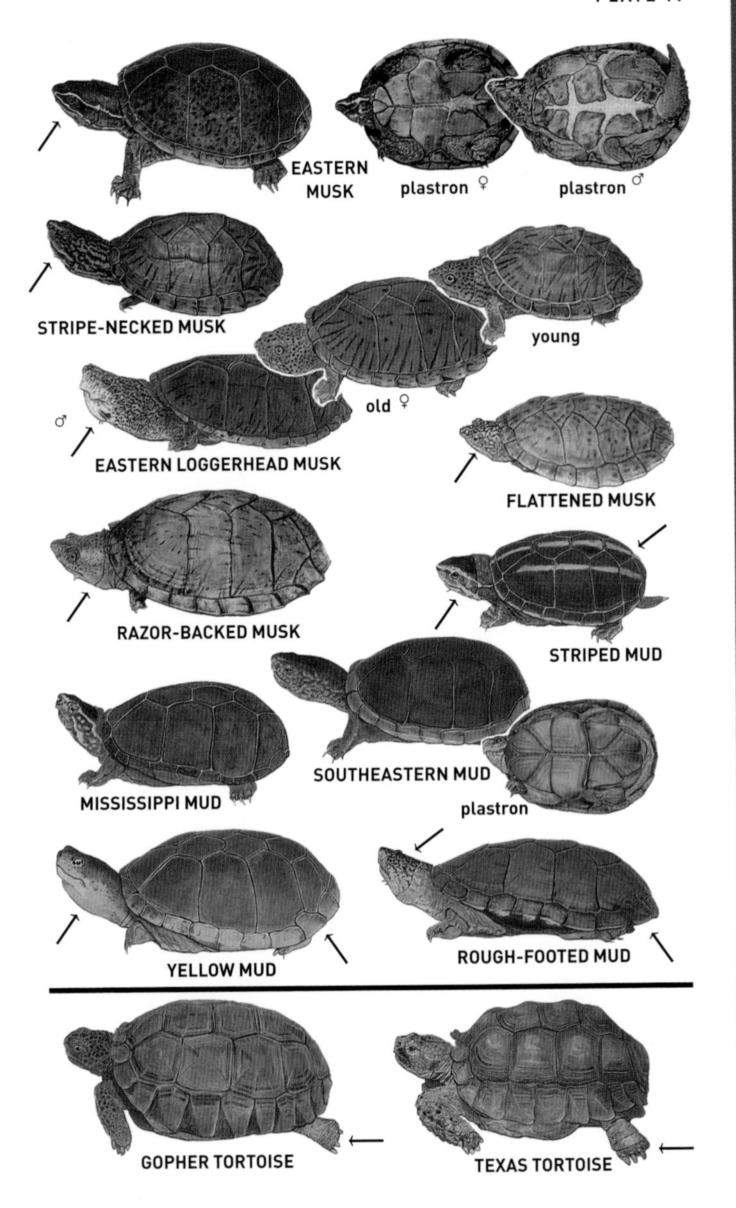
EASTERN
MUSK
plastron ♀
plastron ♂
STRIPE-NECKED MUSK
young
old ♀
♂
EASTERN LOGGERHEAD MUSK
FLATTENED MUSK
RAZOR-BACKED MUSK
STRIPED MUD
SOUTHEASTERN MUD
MISSISSIPPI MUD
plastron
YELLOW MUD
ROUGH-FOOTED MUD
GOPHER TORTOISE
TEXAS TORTOISE

SOFTSHELLS (*Apalone*)

MIDLAND SMOOTH SOFTSHELL *A. mutica* **P. 231**

Feet not strongly patterned; carapace smooth, no spines or bumps; no ridges in nostrils (Fig. 85); ♂ with vague dots and dashes; ♀ with indefinite mottling.

SPINY SOFTSHELL *A. spinifera* **P. 232**

Spines along front of carapace; ridges in nostrils (Fig. 85).

EASTERN SPINY SOFTSHELL (*A. s. spinifera*): Feet strongly patterned; ♂ with large eyelike spots or dark spots and small eyelike marks in western portions of the range, carapace rough like sandpaper at least posteriorly; ♀ pattern vague.

GULF COAST SPINY SOFTSHELL (*A. s. aspera*): ♂ with two or more rows of curved black lines bordering rear edge of carapace; light lines on head usually meet (Fig. 106, p. 233).

GUADALUPE SPINY SOFTSHELL (*A. s. guadalupensis*): ♂ with numerous light dots on carapace.

FLORIDA SOFTSHELL *A. ferox* **P. 232**

Shell proportionately longer than in other species; anterior surface of carapace with numerous small bumps; ridges in nostrils (Fig. 85).

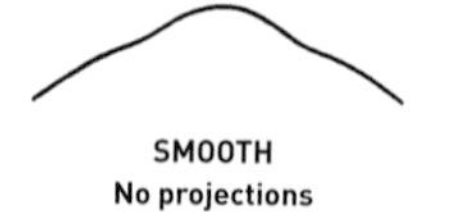

FIG. 85. *Nostrils and carapaces (upper shells) of softshells* (Apalone).

PLATE 20
SMOOTH
EASTERN SPINY
males variable
GULF COAST SPINY
FLORIDA
GUADALUPE SPINY

PLATE 21

YOUNG TURTLES (1)

EASTERN MUSK TURTLE *Sternotherus odoratus* **P. 229**
Light head stripes; edges of shell with light spots.

EASTERN LOGGERHEAD MUSK TURTLE *Sternotherus minor minor* **P. 228**
Three middorsal keels; plastron reddish.

SOUTHEASTERN MUD TURTLE *Kinosternon subrubrum subrubrum* **P. 227**
No stripes on head; plastron orange to pale yellow, center dark.

NORTHERN MAP TURTLE *Graptemys geographica* **P. 204**
Yellow spots behind eyes; maplike lines on shell; plastron with dark lines along seams.

MISSISSIPPI MAP TURTLE *Graptemys pseudogeographica kohnii* **P. 208**
Yellowish crescents behind eyes; plastron with broad dark markings with open centers.

BARBOUR'S MAP TURTLE *Graptemys barbouri* **P. 202**
Saw-toothed carapace; broad light areas behind eyes and across chin.

WOOD TURTLE *Glyptemys insculpta* **P. 201**
Carapace rough; head dark, unmarked; tail long.

BLANDING'S TURTLE *Emydoidea blandingii* **P. 200**
Light marks on head; chin yellow; tail long.

FLORIDA BOX TURTLE *Terrapene bauri* **P. 220**
Yellow middorsal stripe; mottled.

EASTERN BOX TURTLE *Terrapene carolina* **P. 220**
Light spot in each large scute; plastron with dark pigment concentrated centrally.

SPOTTED TURTLE *Clemmys guttata* **P. 198**
Light spot in each large scute and spots on head; plastron with dark pigment toward center.

ALLIGATOR SNAPPING TURTLE *Macrochelys temminckii* **P. 196**
Carapace extremely rough; beak strongly hooked; tail long.

EASTERN SNAPPING TURTLE *Chelydra serpentina* **P. 195**
Carapace rough, edged with light spots; tail long.

FIG. 86. *Carapace (upper shell) of a hatchling Razor-backed Musk Turtle* (Sternotherus carinatus).

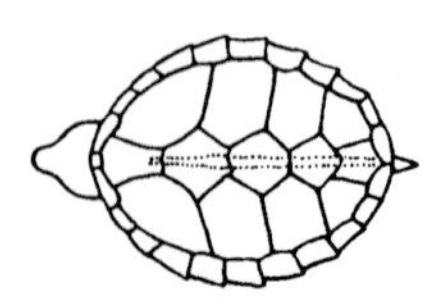

RAZOR-BACKED MUSK
(*S. carinatus*)
A middorsal keel; shell toothed along each side

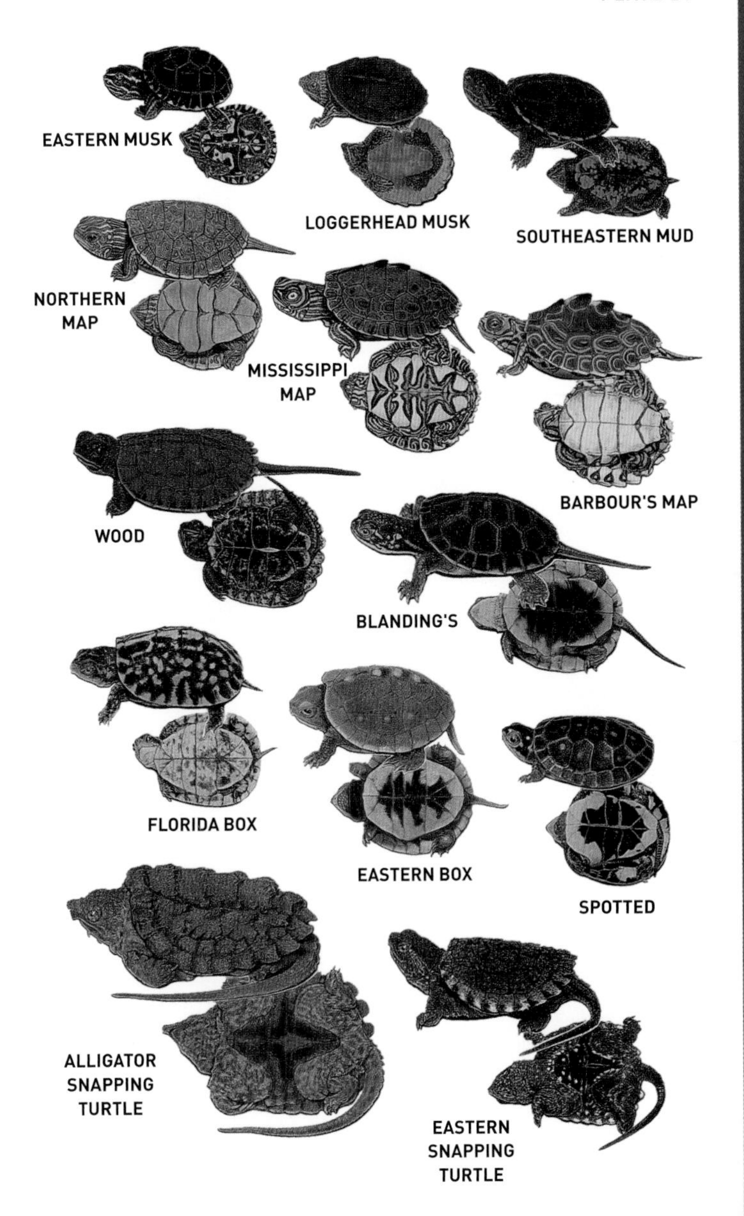
EASTERN MUSK
LOGGERHEAD MUSK
SOUTHEASTERN MUD
NORTHERN MAP
MISSISSIPPI MAP
BARBOUR'S MAP
WOOD
BLANDING'S
FLORIDA BOX
EASTERN BOX
SPOTTED
ALLIGATOR SNAPPING TURTLE
EASTERN SNAPPING TURTLE

YOUNG TURTLES (2)

SOUTHERN PAINTED TURTLE *Chrysemys dorsalis* **P. 197**
Broad red, orange, or yellow middorsal stripe.

PAINTED TURTLE *C. picta* **P. 198**
WESTERN PAINTED TURTLE (*C. p. bellii*): Pale reticulate markings; plastron with large dark area with outward extensions.

MIDLAND PAINTED TURTLE (*C. p. marginata*): No bold markings; plastron with dark central blotch.

EASTERN PAINTED TURTLE (*C. p. picta*): Pleurals with light borders; plastron usually unmarked.

NORTHERN DIAMOND-BACKED TERRAPIN
Malaclemys terrapin terrapin **P. 210**
Dark lines parallel to edges of scutes above and below.

POND SLIDER *Trachemys scripta* **P. 217**
RED-EARED SLIDER (*T. s. elegans*): Red patches or stripes behind eyes; large circular markings on plastron.

YELLOW-BELLIED SLIDER (*T. s. scripta*): Large yellow patches behind eyes; small circular markings restricted to forepart of plastron.

PENINSULA COOTER *Pseudemys peninsularis* **P. 215**
Curved lines on carapace; plastron unmarked; dark spots on anterior marginals.

RIVER COOTER *Pseudemys concinna* **P. 212**
Markings circular, especially on marginals; plastron with markings chiefly along seams.

CHICKEN TURTLE *Deirochelys reticularia* **P. 199**
Light reticulate pattern on carapace; striped "pants" (Fig. 91, p. 200).

FLORIDA SOFTSHELL *Apalone ferox* **P. 232**
Head striped; large round dorsal spots; plastron dark.

EASTERN SPINY SOFTSHELL *Apalone spinifera spinifera* **P. 232**
Small circular dorsal spots; plastron light, nearly matching underside of carapace.

MIDLAND SMOOTH SOFTSHELL *Apalone mutica* **P. 231**
Indistinct dorsal dots and dashes; plastron light; underside of carapace brown.

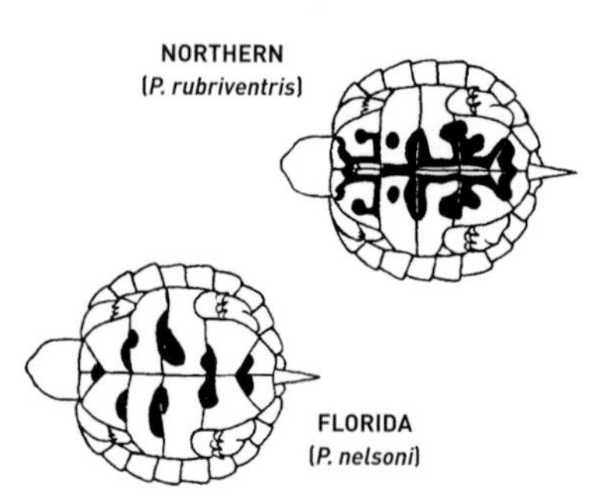

FIG. 87. *Plastral (lower shell) patterns of young red-bellied cooters* (Pseudemys). *The Florida Red-bellied Cooter* (P. nelsoni) *has solid dark plastral markings with flat sides along the seams, whereas the Northern Red-bellied Cooter* (P. rubriventris) *tends to have a more complex plastral pattern superficially resembling that of a Western Painted Turtle* (Chrysemys picta bellii).

WESTERN PAINTED
MIDLAND PAINTED
SOUTHERN PAINTED
EASTERN PAINTED
NORTHERN DIAMOND-
BACKED TERRAPIN
RED-EARED SLIDER
YELLOW-BELLIED SLIDER
PENINSULA COOTER
RIVER COOTER
CHICKEN
FLORIDA SOFTSHELL
EASTERN SPINY SOFTSHELL
SMOOTH
SOFTSHELL

TURTLES (TESTUDINES)

Turtles occur on all continents except Antarctica and are particularly abundant in eastern North America, where about 60 of more than 340 species occur. Shells are composed of an upper carapace and a lower plastron (Fig. 88, facing page). All of the turtles in our area belong to the Cryptodira (which means "hidden neck"). They can withdraw the head into the shell by bending the neck into the shape of an S. Pleurodira (or side-necked turtles) bend the head to the side but cannot draw it into the shell. They are restricted to the Southern Hemisphere.

SEA TURTLES: Cheloniidae and Dermochelyidae

Limbs are modified into flippers. Females usually leave water only to nest on beaches; males might never leave water, as mating occurs offshore. Many species have worldwide distributions or relatives in other oceans; the ranges indicated below pertain only to populations in the western Atlantic (not mapped). Males often have large curved claws on the forelimbs that help hold females during mating. All species are protected and should not be handled. Four species in the family Cheloniidae and one species in the family Dermochelyidae occur in our area.

LOGGERHEAD SEA TURTLE *Caretta caretta* **PL. 14, FIG. 81**

31–45 in. (79–114 cm); record 85+ in. (213 cm); 175–440 lb. (80–200 kg); record 1,200+ lb. (545 kg); records questionable, most individuals are much smaller. Male smaller than female, with a wider shell tapering posteriorly and a longer, thicker tail. Five or more pleurals on each side of carapace, *the first always in contact* with cervical. Usually 3, occasionally 4 large scutes on bridge (Fig. 81, p. 174). Middorsal keel low and inconspicuous in large individuals. Carapace, head, and limbs reddish brown. Young brown above, whitish, yellowish, or tan below; 3 dorsal keels, 2 on plastron. Hatchling 1⅝–1⅞ in. (4.1–4.8 cm). **SIMILAR SPECIES:** (1) Kemp's Ridley Sea Turtle is smaller, olive green carapace (gray in young) is almost circular, usually has interanal scute and almost always 4 large scutes on bridge. (2) Hawksbill and Green Sea Turtles have only 4 pairs of pleurals, the first not touching cervical. **HABITAT:** Hatchlings often in sargassum before moving into shallow coastal waters and estuaries. **RANGE:** Atlantic Provinces south to Argentina; nests as far north as the Carolinas and (rarely) MD and NJ. **CONSERVATION:** Endangered (USFWS: Threatened).

GREEN SEA TURTLE *Chelonia mydas* **PL. 14, FIGS. 81, 89, P. 456**

36–48 in. (90–122 cm); record 60⅜ in. (153 cm); 250–450 lb. (113–204 kg); record 835+ lb. (380 kg). Male slightly longer than female, with a flatter shell and much longer tail with a terminal "nail." Carapace light or dark brown, sometimes shaded with olive, frequently with ra-

FIG. 88. *Turtle shells showing the names of the scutes.*

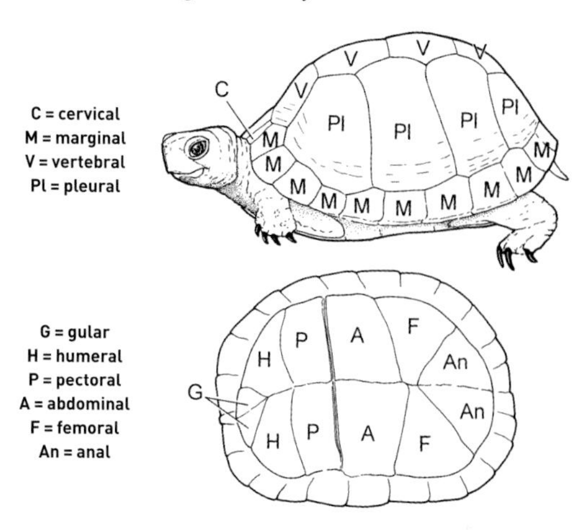

diating mottled or wavy dark markings or large dark brown blotches. Four pleurals on each side of carapace, the first not touching cervical (Fig. 81, p. 174). *One pair of prefrontals* between eyes (Fig. 89, below). Large scutes of carapace do not overlap except in very small individuals. Carapace of young dark brown (black in hatchling); venter white except for tips of flippers, which are black but edged with white; hatchling with a middorsal keel and 2 keels on plastron. Hatchling $1\frac{5}{8}$–$2\frac{3}{8}$ in. (4.1–6 cm). **SIMILAR SPECIES:** (1) Hawksbill Sea Turtle has 2 pairs of prefrontals, and large scutes on carapace often overlap. (2) Loggerhead and Atlantic Kemp's Ridley Sea Turtles have first pleural touching cervical. **HABITAT:** Hatchlings float in offshore currents before moving to coastal areas rich in sea grass and/or algae. **RANGE:** MA to n. Argentina. **REMARK:** Pacific populations sometimes considered subspecifically distinct. **CONSERVATION:** Endangered (USFWS: Endangered in FL, Threatened elsewhere).

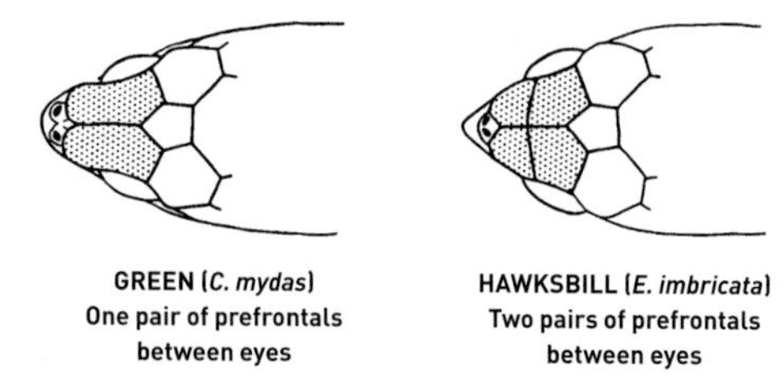

FIG. 89. *Head plates of the Green Sea Turtle* (Chelonia mydas) *and the Hawksbill Sea Turtle* (Eretmochelys imbricata).

HAWKSBILL SEA TURTLE *Eretmochelys imbricata* **PL. 14, FIGS. 81, 89**

30–35 in. (76–89 cm); record 45 in. (114 cm); 95–165 lb. (43–75 kg); record 280 lb. (127 kg). Male and female about same size, but male has a concave plastron and long, thick tail. Large scutes on carapace overlap except in very old individuals. Middorsal keel present. Four pleurals on each side of carapace, the first not touching cervical (Fig. 81, p. 174); *2 pairs of prefrontals* between eyes (Fig. 89, p. 193). Brown, especially smaller individuals with tortoiseshell pattern. Young black or very dark brown but light brown on raised ridges, edges of shell, and areas on neck and flippers; one middorsal keel, 2 keels on plastron. Hatchling 1½–1⅞ in. (3.8–4.8 cm). **SIMILAR SPECIES:** See Green Sea Turtle. **HABITAT:** Hatchlings float in offshore currents before moving to coastal coral reefs, sea grass, algal beds, mangrove bays and creeks, or mud flats. **RANGE:** S. New England to s. Brazil; most N. American records from FL. **SUBSPECIES:** **ATLANTIC HAWKSBILL SEA TURTLE** (*E. i. imbricata*). **CONSERVATION:** Critically Endangered (USFWS: Endangered).

KEMP'S RIDLEY SEA TURTLE *Lepidochelys kempii* **PL. 14, FIG. 81**

23–27½ in. (58–70 cm); record 29⅞ in. (76 cm); 80–100 lb. (36–45 kg); record 110 lb. (49.9 kg). Male slightly smaller than female and has a concave plastron and long prehensile tail. Carapace almost circular; olive green above and yellow below (gray above in smaller individuals). Five pleurals on each side of carapace, *the first touching cervical*; middle marginals mostly wider than long; almost always *4* (rarely 5) enlarged scutes on bridge, each pierced by a pore near posterior edge; interanal scute (Fig. 81, p. 174) usually present at posterior tip of plastron. Young with short light gray streaks along rear edges of front flippers; 3 tuberculate dorsal ridges and 4 ridges on plastron. Hatchling 1½–1¾ in. (3.8–4.4 cm). **SIMILAR SPECIES:** See Loggerhead Sea Turtle. **HABITAT:** Primarily coastal areas with muddy or sandy bottoms. **RANGE:** Chiefly Gulf of Mexico, but immature turtles as far north as New England and NS. **REMARKS:** Pacific Ridley Sea Turtle (*Lepidochelys olivacea*) might eventually appear in FL waters. This species usually has 6–7 pleurals (sometimes more) on each side of carapace, and middle marginal scutes are longer than wide. **CONSERVATION:** Critically Endangered (USFWS: Endangered).

LEATHERBACK SEA TURTLE *Dermochelys coriacea* **PL. 14**

53–70 in. (135–178 cm); record 96 in. (243.8 cm); 650–1,200 lb. (295–544 kg); record ±1,500 lb. (±680 kg), but reliable historical accounts record turtles exceeding 2,000 lb. (907 kg). Largest living turtle. Male larger than female and with a narrower, posteriorly tapered carapace, concave plastron, and much longer tail. *Seven prominent longitudinal ridges on carapace* and 5 similar ridges on plastron. Carapace and plastron *lack scutes*, covered instead by smooth, slate black, dark bluish, or dark bluish black skin. Irregular patches of white or pink, especially in female; white predominates on plastron. Young more conspicuously marked than adult; covered with many small, beadlike scales (soon shed); tail keeled above. Hatchling 2⅜–3 in. (6–7.5 cm). **HABITAT:** Juveniles generally in tropical coastal waters until more than 39 in. (100 cm) in carapace length; adults in open ocean. **RANGE:** NL to Argentina, frequently in New England waters in summer. Nests north to NC and on beaches around the Gulf of Mexico. **CONSERVATION:** Critically Endangered (USFWS: Endangered).

SNAPPING TURTLES: Chelydridae

Large freshwater turtles with a large head, powerful jaws, long neck and tail, and short tempers. Often exploited for their meat, snapping turtles are considered game animals in some states and provinces. Six currently recognized species range from Canada to South America.

EASTERN SNAPPING TURTLE *Chelydra serpentina* **PLS. 14, 21**

8–14 in. (20.3–36 cm); record 19⅜ in. (49.4 cm); 10–35 lb. (4.5–16 kg), but to 75 lb. (34 kg) for wild-caught and 86 lb. (39 kg) for fattened captive turtles. Male usually larger than female. Head large, plastron small and cross-shaped, tail long and *saw-toothed* above. Adult carapace almost black to light horn brown, often covered with algae. Young blackish or dark brown with light spots at edges of marginals. Carapace rough, with 3 fairly well-defined longitudinal keels. Rugosity less prominent as turtle ages, with large adults almost smooth but usually with traces of keels. Tail as long as carapace or longer, with vent in adult male usually posterior to rim of carapace. Hatchling ¾–1¼ in. (1.9–3.2 cm). **SIMILAR SPECIES:** (1) Alligator snapping turtles have extra rows of scutes between marginals and pleurals. (2) Mud and musk turtles have a shorter tail, smooth, domed shell (although juvenile may have keels), and plastron with 1 or 2 hinges. **HABITAT:** Permanent bodies of fresh water; often follows intermittent streams far from outlets; will enter brackish water. Often makes long overland journeys (especially males, although females will move far from water to find nesting sites). **RANGE:** S. Can. to the Gulf of Mexico and Atlantic Ocean to the Rocky Mts. Introduced farther west and possibly on some FL Keys. **REMARK:** Populations in peninsular FL and extreme s. GA once were considered subspecifically distinct.

ALLIGATOR SNAPPING TURTLE *Macrochelys temminckii* **PLS. 14, 21, FIG. 90**

APALACHICOLA ALLIGATOR SNAPPING TURTLE
Macrochelys apalachicolae **FIG. 90**

SUWANNEE ALLIGATOR SNAPPING TURTLE
Macrochelys suwanniensis **FIG. 90**

15–20 in. (38–50.8 cm); record 31½ in. (80 cm); 35–115 lb. (16–52.2 kg); record 251 lb. (113.9 kg) for captive, 126 lb. (57.1 kg) for wild-caught animal. Largest freshwater turtles in our area. Huge head with a *strongly hooked beak*, three prominent dorsal keels, and *extra rows of scutes* between marginals and pleurals. Young brown, shell exceedingly rough, tail very long. Caudal notch very wide in Suwannee Alligator Snapping Turtle, narrow and triangular or U-shaped in others (Fig. 90, below); species best distinguished by ranges. Hatchlings 1¼–1¾ in. (3–4.4 cm). **SIMILAR SPECIES:** See Eastern Snapping Turtle. **HABITAT:** Deep water in rivers, canals, lakes, swamps, streams; hatchlings and juveniles usually in smaller streams. Will lie on bottom with mouth open and "lure" fish with a pink, wormlike appendage in the mouth. **RANGE:** Alligator Snapping Turtle: W. FL Panhandle to e. TX and north in the Mississippi River Valley to IL (records from IA are probably invalid); Suwannee Alligator Snapping Turtle: Suwannee R. drainage in FL and GA; Apalachicola Alligator Snapping Turtle: FL Panhandle, w. GA, and e. AL. **REMARK:** These three species until recently were considered conspecific. **CONSERVATION:** Vulnerable (applies to all three species).

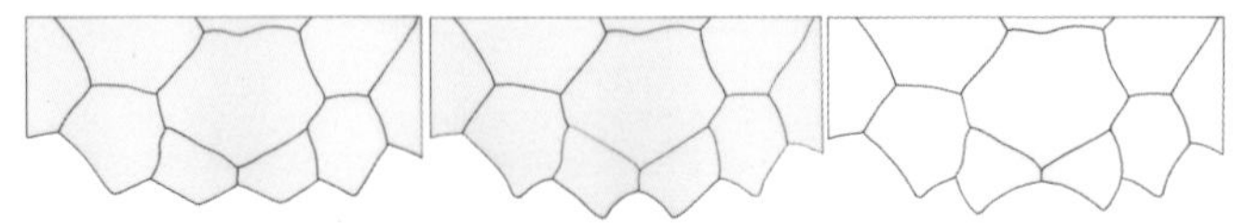

FIG. 90. *Caudal notches of alligator snapping turtles* (Macrochelys).

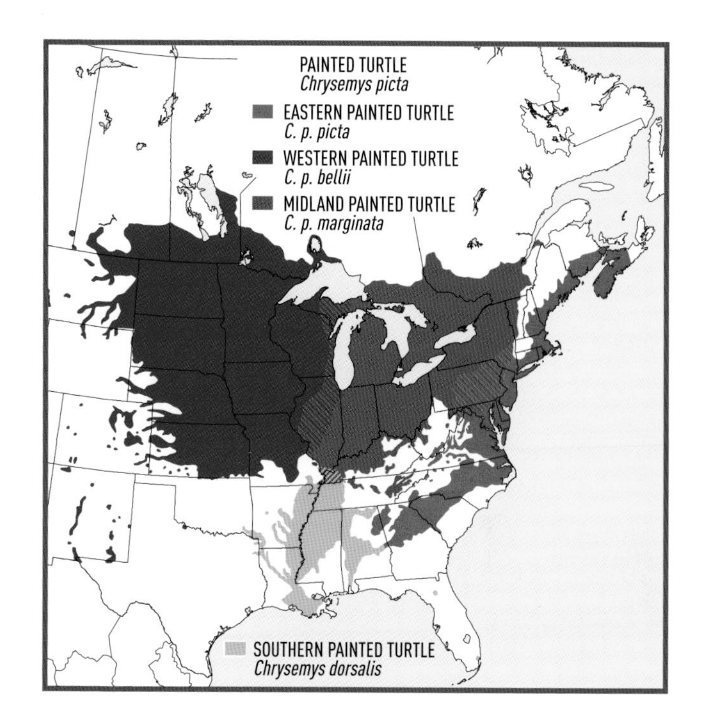

BOX AND WATER TURTLES: Emydidae

The most species-rich turtle family in North America, also with species in South America and Europe.

PAINTED TURTLES: *Chrysemys*

Females average larger than males. The carapace is smooth, without keels, and the posterior margin is essentially without indentations. Very long nails on the forefeet of males are used in courtship. Hatchlings $^7/_8$–$1^1/_8$ in. (2.2–2.8 cm), with faint medial keels lost as animals grow.

SOUTHERN PAINTED TURTLE *Chrysemys dorsalis* **PLS. 15, 22**

4–5 in. (10–12.5 cm); record $6^1/_8$ in. (15.6 cm). *Red or orange middorsal stripe* (sometimes yellow), plastron plain yellow (sometimes lightly spotted). **SIMILAR SPECIES:** (1) Painted Turtle lacks a colored stripe on the carapace. (2) Map turtles, cooters, and sliders have notches or indentations along the posterior margin of the carapace. (3) Chicken Turtle has broad yellow stripes along front of forelimbs and vertical stripes on back of thighs. **HABITAT:** Shallow ponds, marshes, ditches, edges of lakes, backwaters of streams. **RANGE:** Extreme s. IL to the Gulf; introduced in FL. **REMARK:** Until recently considered a subspecies of the Painted Turtle.

PAINTED TURTLE *Chrysemys picta* **PLS. 15, 22, P. 461**

4½–8 in. (11.5–20 cm); record 10 in. (25.4 cm). Red largely restricted to lower marginals. Yellow spots, lines, or stripes on head. Plastron plain yellow, lightly spotted, or with dark blotches along seams; pattern sometimes obscured in large adults. **SIMILAR SPECIES:** See Southern Painted Turtle. **HABITAT:** Shallow ponds, marshes, ditches, edges of lakes; westward in prairie sloughs, cattle tanks, river pools. **RANGE:** NS west to the Pacific Northwest and south to GA and AL; isolated populations in the Southwest. **SUBSPECIES:** (1) **EASTERN PAINTED TURTLE** (*C. p. picta*); large pleurals in more or less straight rows across back (scutes in other turtles alternate); olive front edges of pleurals form light bands across carapace; plastron plain yellow or lightly spotted. (2) **WESTERN PAINTED TURTLE** (*C. p. bellii*); large, often to 8 in. (20 cm) or more; large pleurals alternate; less red on marginals; often with light irregular lines forming a reticulate pattern on carapace; plastron with large dark figures extending along seams. (3) **MIDLAND PAINTED TURTLE** (*C. p. marginata*); similar to Eastern Painted Turtle, but large pleurals alternate; dark plastral blotch variable, typically oval, one-half or less the width of plastron, usually without extensions along seams.

SPOTTED TURTLES: *Clemmys*

The one species in this genus is endemic to eastern North America.

SPOTTED TURTLE *Clemmys guttata* **PLS. 15, 21**

3½–4½ in. (9–11.5 cm); record 5⅝ in. (14.3 cm). Male shorter than female and with a slightly concave plastron and longer tail. *Yellow spots* on black carapace variable in number. Hatchling usually has one spot per large scute, older turtles may have 100 or more; rare individuals have few or no spots. Head and neck with yellow or orange spots. Horny portion of jaws almost completely dark in male, yellowish and virtually unmarked in female. Hatchling 1⅛ in. (2.8 cm). **SIMILAR SPECIES:** (1) Bog Turtle has large orange head patches. (2) Blanding's Turtle has great number of yellow spots, yellow

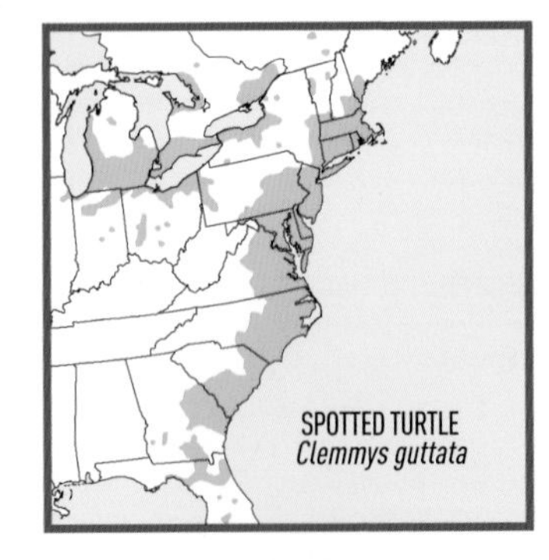

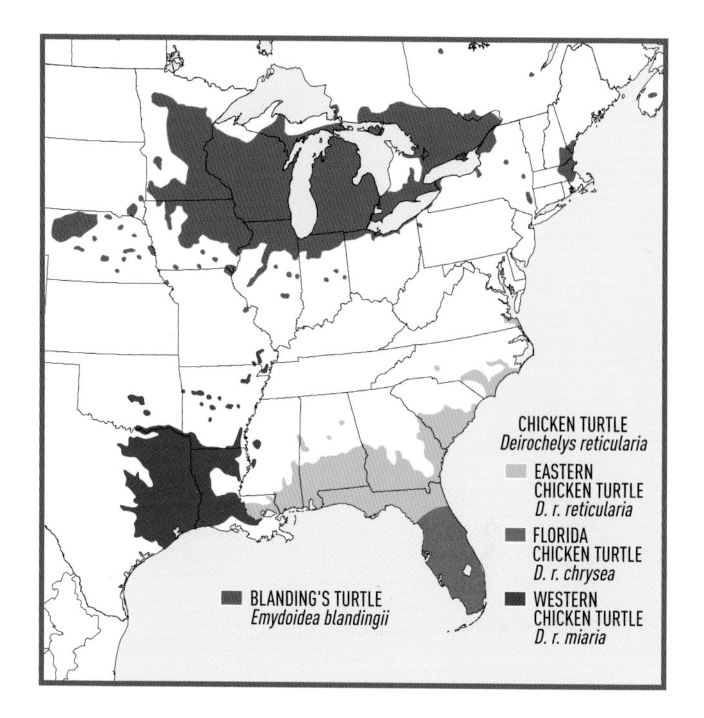

throat, hinged plastron. **HABITAT:** Shallow marshy meadows, bogs, swamps, small ponds, ditches. **RANGE:** Disjunct; s. Can. to IL, IN, and OH, and south in the East to cen. FL. Many isolated populations; those in s. QC possibly extirpated. **CONSERVATION:** Endangered.

CHICKEN and BLANDING'S TURTLES: *Deirochelys* and *Emydoidea*

Longest-necked turtles in our area except softshells. First vertebral is in contact with four marginals and the cervical (only two marginals and the cervical are normally in contact with the first vertebral in other emydid turtles in our area). Both of these eastern North American endemic genera contain only one species.

CHICKEN TURTLE *Deirochelys reticularia* **PLS. 15, 22**

4–6 in. (10–15.2 cm); record 10 in. (25.4 cm). Carapace sculptured with small linelike ridges, much longer than wide, widest over hindlimbs; light reticulate pattern, often with longitudinal dark bars or spots on bridge; plastron yellow or with dark pigment along seams (especially in western populations). *Neck long and strongly striped*; forelimbs with *broad yellow stripes*; back of hindlimbs *vertically striped* (Fig. 91, below). Hatchling 1⅛–1¼ in. (2.8–3.2 cm), carapace with slight keels. **SIMILAR SPECIES:** Some sliders have striped "pants," but yellow lines on forelimbs are narrow, carapace is rounder, and neck is much shorter. **HABITAT:** Still water, including ponds, marshes, sloughs, ditches; frequently on land. **RANGE:** Coastal Plain from se. VA through FL to e. TX and north to se. MO and cen. OK. **SUBSPECIES:** (1) **EASTERN CHICKEN TURTLE** (*D. r. reticularia*); narrow reticulate lines on carapace greenish or brownish; coloration subdued. (2) **FLORIDA CHICKEN TURTLE** (*D. r. chrysea*); reticulate pattern orange or yellow, bold and broad in younger turtles, less conspicuous in larger individuals; rim of carapace boldly edged with orange; plastron unpatterned orange or bright yellow. (3) **WESTERN CHICKEN TURTLE** (*D. r. miaria*); broad reticulate lines only slightly lighter than ground color; plastron with dark markings along seams; underside of neck unpatterned in adult.

BLANDING'S TURTLE *Emydoidea blandingii* **PLS. 15, 21**

6–9 in. (15–23 cm); record 11⅛ in. (28.4 cm). Profuse light spots on carapace often run together, forming bars or streaks; *bright yellow chin and throat* and *notched upper jaw*; hinge on plastron, but shell does not close completely (see Fig. 98, p. 220). Carapace of young plain gray or grayish brown; plastron blackish with yellow edges; tail proportionately much longer than in adult. Hatchling 1–1½ in. (2.5–3.8 cm). **SIMILAR SPECIES:** (1) Spotted Turtle has fewer, well-separated spots and no hinge on plastron. (2) Box turtles have a shell that closes tightly and a hooklike beak. **HABITAT:** Marshes, bogs, lakes, small streams, but often on land, moving between wetlands; nests in open grasslands, often far from water. **RANGE:** Disjunct; NS to NE; apparently extirpated in PA. **CONSERVATION:** Endangered.

SCULPTED TURTLES: *Glyptemys*

Both species in this genus are endemic to eastern North America.

YELLOW-BELLIED SLIDER
(*T. s. scripta*)
Narrow leg stripes

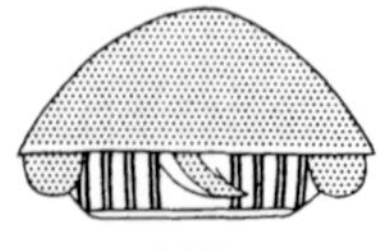

BOTH
Striped "pants"

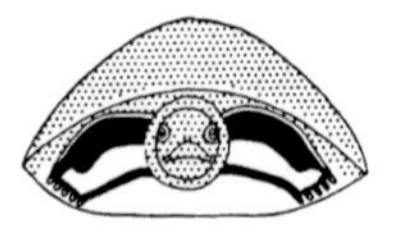

CHICKEN TURTLE
(*D. reticularia*)
Broad leg stripes

FIG. 91. *Pattern characteristics of the Chicken Turtle* (Deirochelys reticularia) *and the Yellow-bellied Slider* (Trachemys scripta scripta).

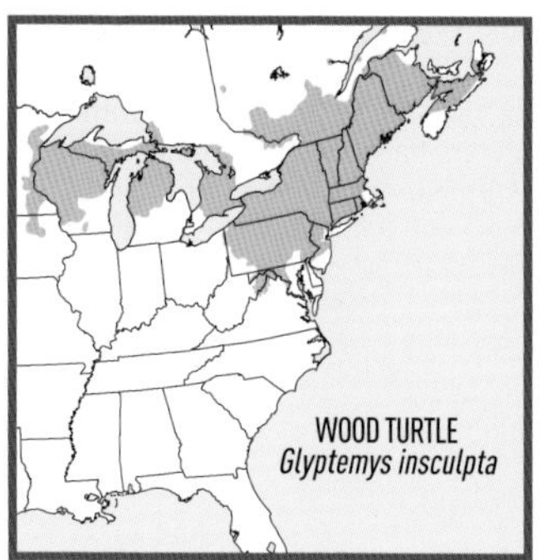

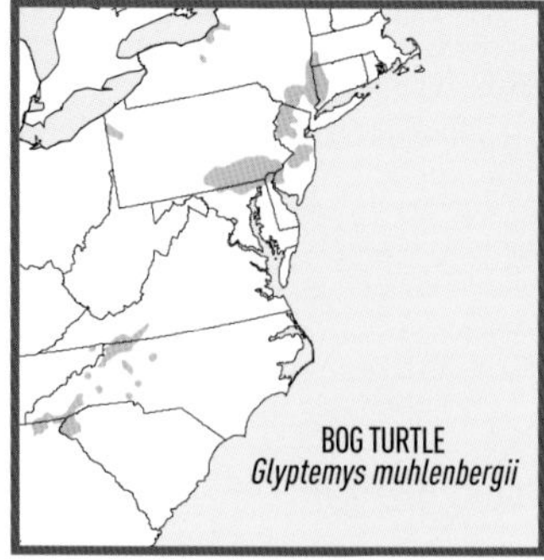

WOOD TURTLE *Glyptemys insculpta* **PLS. 15, 21**

5½–8 in. (14–20 cm); record 9⅞ in. (25.1 cm). Male larger than female, with a concave plastron, longer, thicker tail, and prominent scales on anterior surface of forelimbs. *Carapace very rough*; large scutes with irregular pyramids of concentric grooves and ridges. Plastron with large dark blotches on each scute. Bright *orange or yellow on neck and limbs*. Hatchling 1⅛–1⅝ in. (2.8–4.1 cm); shell broad and low, brown or grayish brown; no orange on neck or legs; tail almost as long as carapace. **SIMILAR SPECIES:** (1) Blanding's and Eastern Box Turtles lack concentric ridges on carapace and have a hinged plastron. (2) Diamond-backed Terrapin has concentric ridges on carapace but is restricted to coastal maritime marshes and hatchling is strongly patterned. **HABITAT:** Largely terrestrial, but overwinters in deep pools in streams. **RANGE:** NS to e. MN and south in the East to n. VA; many populations extirpated. **REMARK:** Until recently placed in the genus *Clemmys*. **CONSERVATION:** Endangered.

BOG TURTLE *Glyptemys muhlenbergii* **PL. 15**

3–3½ in. (7.5–9 cm); record 4½ in. (11.5 cm). *Orange patches on sides of head* occasionally yellow or split in two; carapace brown to black, large pleurals may have yellowish or reddish centers; plastron dark, may have a few light marks. Hatchling 1–1¼ in. (2.5–3.2 cm). **SIMILAR SPECIES:** Spotted Turtle has yellow dots on shell and many separate yellow or orange spots on head and neck. **HABITAT:** Sphagnum bogs, swamps, clear, slow-moving meadow streams with muddy bottoms. **RANGE:** Disjunct; nw. NY, w. PA, e. NY, w. MA south to MD and DE, and s. VA to extreme ne. GA; many populations extirpated. **REMARK:** Until recently placed in the genus *Clemmys*. **CONSERVATION:** Critically Endangered (USFWS: Threatened).

MAP TURTLES: *Graptemys*

All 14 currently recognized species occur in our area, but several are confined to single river systems. All have dorsal keels, some with sawlike projections. Rear margins of shells in juveniles and males are serrated (saw-toothed), less so in adult females. Females are larger than males, to more than twice the length and 10 times the mass in some species. Females of many species have a very large head and strong jaws with crushing surfaces for eating snails and clams. Juveniles, males, and females in species without enlarged heads feed primarily on aquatic insects. Hatchlings 1–1½ in. (2.5–3.8 cm).

BARBOUR'S MAP TURTLE *Graptemys barbouri*

PLS. 16, 21, FIGS. 82, 92, P. x

Female 7–10 13/16 in. (18–27.5 cm); male 3½–5¼ in. (9–13.5 cm). Head of female enormously enlarged. Male retains most juvenile markings, but large females become smudged and blotched with dark pigment that hides any pattern. Juvenile has well-developed "sawtooth" on carapace, small longitudinal keels on each pleural, *broad light areas between and behind eyes* (Fig. 92, below). *Postorbital blotches curve un-*

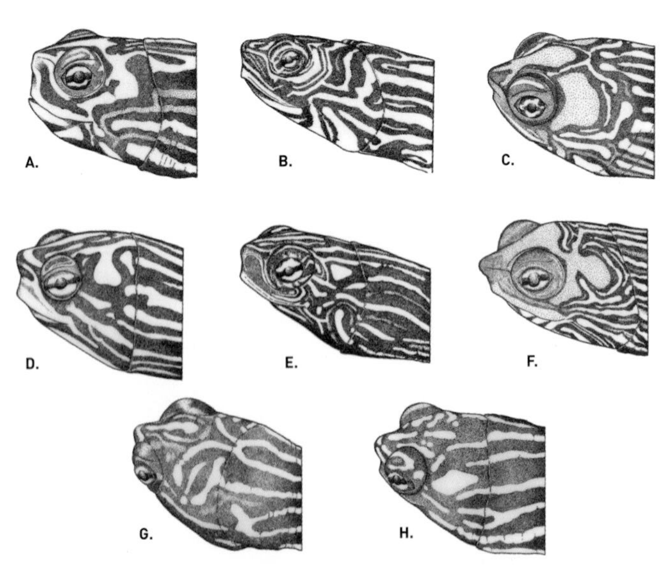

FIG. 92. *Head patterns of map turtles* (Graptemys): **A.** *Barbour's Map Turtle* (G. barbouri); **B.** *Cagle's Map Turtle* (G. caglei); **C.** *Escambia Map Turtle* (G. ernsti); **D.** *Yellow-blotched Map Turtle* (G. flavimaculata); **E.** *Northern Map Turtle* (G. geographica); **F.** *Pascagoula Map Turtle* (G. gibbonsi); **G.** *Black-knobbed Map Turtle* (G. nigrinoda); **H.** *Ringed Map Turtle* (G. oculifera).

der eyes. Light bar *across* or paralleling curve of chin (Fig. 82, p. 178) and heart- or Y-shaped light marks on top of head. *Pleurals with C-shaped markings*. Light markings on marginals, if present, *narrow*; light curved markings on carapace of juvenile often reduced or absent. Smaller individuals have *deep* shells when viewed in profile and spines on plastron. **SIMILAR SPECIES:** Escambia, Pascagoula, Pearl River, and Alabama Map Turtles lack C-shaped markings on pleurals, transverse bars under chin, and heart- or Y-shaped marks on head. **HABITAT:** Rivers with limestone outcrops, less frequently silty channels. **RANGE:** FL Panhandle, adjacent GA, se. AL. **REMARK:** Interbreeds with the Escambia Map Turtle in the Choctawhatchee R. system. **CONSERVATION:** Vulnerable.

CAGLE'S MAP TURTLE *Graptemys caglei* FIGS. 82, 92

Female to 8⅜ in. (21.3 cm); male 2¾–5 in. (7–12.6 cm). Bold, light, V-shaped mark on top of head, almost always with *triple crescents* behind eyes, the most anterior of which branches off arms of V-shaped marks on head (Fig. 92, facing page). Dark-edged, cream-colored *crescent or band across chin* (Fig. 82, p. 178). Carapace predominantly green, with vertebrals and pleurals bearing concentric light lines. Male usually with black flecking on light parts of plastron. **SIMILAR SPECIES:** Texas Map Turtle lacks dark-edged, cream-colored crescents or bands across chin, carapace predominantly olive. **HABITAT:** Rivers with riffles and pools. **RANGE:** Guadalupe R. system, s.-cen. TX. **CONSERVATION:** Endangered.

ESCAMBIA MAP TURTLE *Graptemys ernsti* FIG. 92

Female to 11¼ in. (28.5 cm); male 2¾–5⅛ in. (7–13.1 cm). Light blotch between eyes *not connected* to blotches behind eyes (Fig. 92, facing page); light blotches beneath eyes sometimes present. *Trident-shaped mark* on snout. Carapace high-domed with a series of laterally compressed knobs (most prominent on vertebrals 2 and 3) forming a middorsal keel; broken black middorsal stripe on olive ground color. Tops of marginals with prominent yellow bars. **SIMILAR SPECIES:** See Barbour's Map Turtle. **HABITAT:** Medium-sized to large rivers and creeks, especially with sand or gravel bottoms; rarely in estuaries, backwaters, floodplain swamps. **RANGE:** Escambia, Yellow, and Choctawhatchee R. systems in w. FL and adjoining AL. **REMARK:** Interbreeds with Barbour's Map Turtle in the Choctawhatchee R. system.

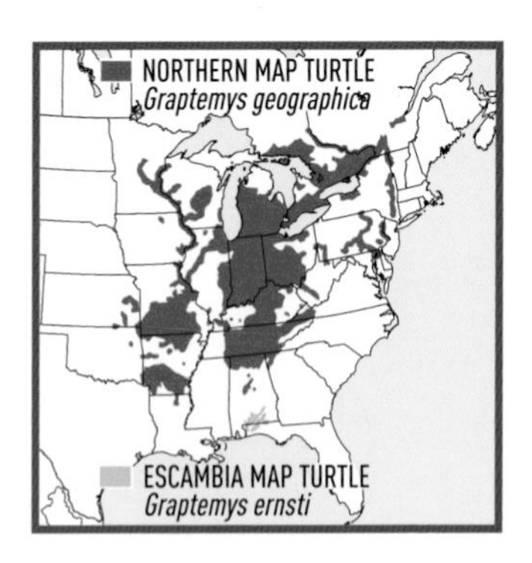

YELLOW-BLOTCHED MAP TURTLE *Graptemys flavimaculata*

PL. 16, FIG. 92

Female 4–7 in. (10–18 cm); male 3–4 5/16 in. (7.5–11 cm). *Solid areas* of yellow or orange in each large scute sometimes invaded by dark ground color (especially in female), but do not form series of symmetrical rings. Mandible light; postocular yellow spots usually present; 2 light neck stripes enter eyes (Fig. 92, p. 202); *curving yellow bar* on chin. Dorsal spines conspicuous in juvenile and adult male. Rear corners of marginals project outward in juvenile, producing a sawtoothed appearance. **SIMILAR SPECIES:** Black-knobbed and Ringed Map Turtles also have prominent dorsal spines but lack solid yellow or orange areas in large scutes. **HABITAT:** Streams with moderate current and sand, gravel, or clay bottoms. **RANGE:** Pascagoula and lower Escatawpa R., se. MS. **CONSERVATION:** Vulnerable (USFWS: Threatened).

NORTHERN MAP TURTLE *Graptemys geographica* **PLS. 16, 21, FIG. 92**

Female 7–10¾ in. (18–27.3 cm); male 3½–6¼ in. (9–16 cm). Carapace with a network of light markings (often obscure in female). Shell moderately low; keel with very small knobs. Roughly triangular *yellow postorbital spots* with rounded points, longer than tall, one point directed downward, and separated from eyes by 2–3 diagonal lines (Fig. 92, p. 202), usually largest in individuals from southern parts of the range. Almost always with lower lines on neck *curling upward anteriorly* toward postorbital spots. Head of female enlarged. Adult with virtually plain plastron. Juvenile with dorsal keels and dark lines bordering plastral seams. **SIMILAR SPECIES:** See False Map Turtle. **HABITAT:** Rivers and lakes. **RANGE:** S. QC west to cen. MN and south to n. LA and cen. AL.

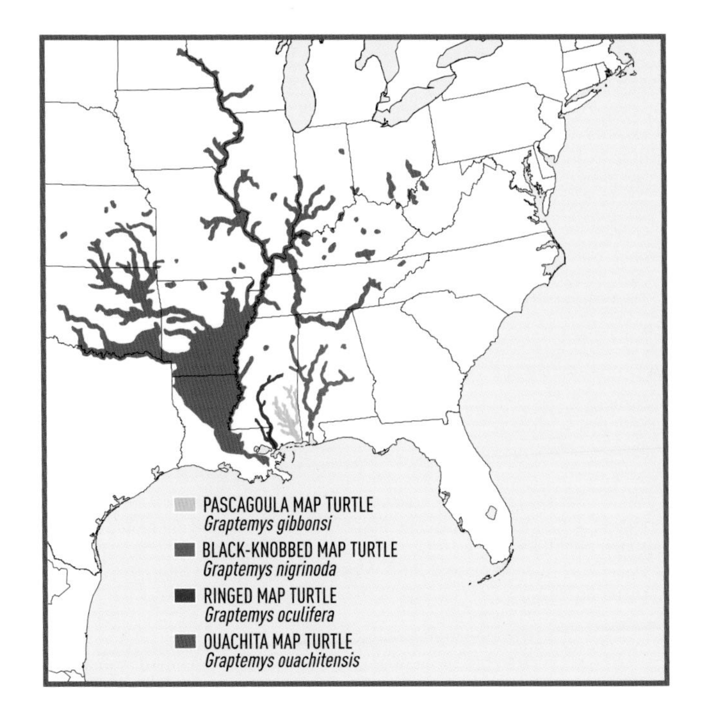

PASCAGOULA MAP TURTLE *Graptemys gibbonsi* **FIG. 92**

Female to 10⅝ in. (27.1 cm); male 2¾–5½ in. (7–14.1 cm). Female with enlarged head and jaws with broad crushing surfaces. Large blotch between eyes *connected* by thin stripes to blotches behind eyes (Fig. 92, p. 202). Three-pronged yellow blotch (nasal trident) often present behind nostrils. Carapace with a series of laterally compressed knobs (most prominent on vertebrals 2 and 3) forming a middorsal keel; black middorsal stripe on olive brown ground color often *broken*; yellow rings on posterior portions of first 3 pleurals. *Wide* yellow bars on upper sides of marginals, undersides of marginals with *narrow* dark bands. Plastron yellow with some dark pigment along seams. **SIMILAR SPECIES:** (1) Pearl River Map Turtle usually has unbroken middorsal stripes, narrower yellow bars generally without concentric rings on upper sides of marginals, and is more likely to have a 3-pronged nasal trident behind nostrils. (2) Yellow-blotched and Ringed Map Turtles have a narrower head and more extensive yellow pigmentation on carapace. **HABITAT:** Large to medium-sized clear-water rivers. **RANGE:** Pascagoula and lower Escatawpa R., se. MS. **REMARK:** Until recently thought to include populations in the Pearl R. system now assigned to the Pearl River Map Turtle. **CONSERVATION:** Endangered.

BLACK-KNOBBED MAP TURTLE *Graptemys nigrinoda* **PL. 16, FIG. 92**

Female 4–8$\frac{11}{16}$ in. (10–22.1 cm); male 3–4¾ in. (7.5–12.2 cm). Black spines on keel *bluntly knobbed*. Strongly serrated carapace margin extends to anterior marginals. Light rings on pleurals *narrow*, extent of dark figure on plastron about 30–60 percent of plastron. Head and legs range from mostly light to mostly black. Light postorbital marks linear, angular, or strongly recurved and J-shaped (Fig. 92, p. 202); *curving yellow bar* on chin. **SIMILAR SPECIES:** See Yellow-blotched Map Turtle. **HABITAT:** Streams with moderate currents and sand or clay bottoms. **RANGE:** Alabama, Tombigbee, Black, Warrior, Coosa, Tallapoosa, and Cahaba R. systems of AL and w. MS.

RINGED MAP TURTLE *Graptemys oculifera* **PL. 16, FIG. 92**

Female 5–8$\frac{11}{16}$ in. (12.5–22 cm); male 3–4$\frac{5}{16}$ in. (7.5–11 cm). *Broad light yellow to orange rings* on pleurals (sometimes obscure in large female). Mandible light, postorbital yellow spots large, 2 broad light neck stripes enter eyes (Fig. 92, p. 202); *curving yellow bar* on chin. Dorsal spines conspicuous in juvenile and male. **SIMILAR SPECIES:** See Yellow-blotched Map Turtle. **HABITAT:** Rivers and streams with moderate current and sand or clay bottoms. **RANGE:** Pearl R. system of s.-cen. MS and extreme e. LA. **CONSERVATION:** Vulnerable (USFWS: Threatened).

OUACHITA MAP TURTLE *Graptemys ouachitensis* **FIG. 93**

Female to 9$\frac{7}{16}$ in. (24 cm); male 2¾–5½ in. (7–14 cm). Bold light *blotches behind eyes usually continuous above with narrower neck stripes*; *1–6 neck stripes reach eyes* (Fig. 93, p. 209). Light spot under each eye, another on chin; spots behind and under eyes sometimes fused to form a crescent (especially in northern populations) that prevents neck stripes from reaching eyes. *Three light spots* under chin. Olive to brown carapace with low black knobs forming a middorsal keel; yellow vermiculations and dark blotches on pleurals. Plastral pattern extensive in juvenile but fades to seams in adult. Whitish irises often bisected by black horizontal lines. **SIMILAR SPECIES:** See False Map Turtle. **HABITAT:** Swift rivers, lakes, oxbows, swamps; sand and silt bottoms more frequently than mud or rock. **RANGE:** Mississippi R. and tributaries from MN and WI south to LA.

PEARL RIVER MAP TURTLE *Graptemys pearlensis* **FIG. 93**

Female to 11⅝ in. (29.5 cm); male 2¾–4¾ in. (7–12.1 cm). Female has enlarged head and jaws with broad crushing surfaces. Large blotch between eyes *connected* by thin stripes to blotches behind eyes (Fig. 93, p. 209). Three-pronged yellow blotch (nasal trident) almost always present behind nostrils. Series of laterally compressed knobs (most prominent on vertebrals 2 and 3) form a middorsal keel; black mid-

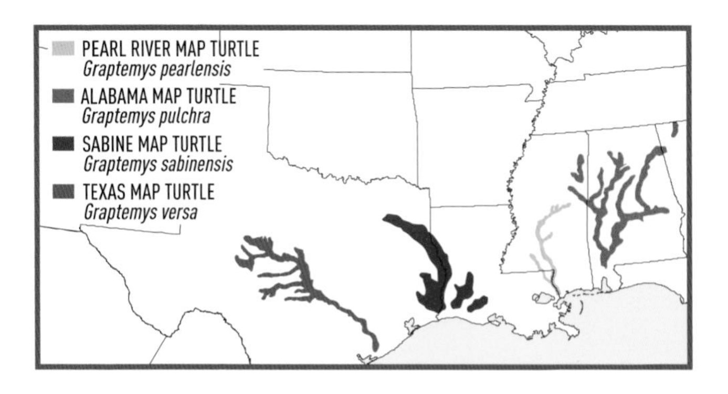

dorsal stripe on olive brown ground color usually *unbroken*; yellow rings and vermiculations on posterior portions of first 3 pleurals. Yellow bars, generally without concentric rings, on upper sides of marginals; undersides of marginals with *narrow* dark bands. Plastron yellow with some dark pigment along seams. **SIMILAR SPECIES:** See Pascagoula Map Turtle. **HABITAT:** Medium to large clear-water rivers. **RANGE:** Pearl R. system of s.-cen. MS and extreme e. LA. **REMARK:** Until recently considered conspecific with the Pascagoula Map Turtle. **CONSERVATION:** Endangered.

FALSE MAP TURTLE *Graptemys pseudogeographica* **PLS. 16, 21, FIG. 93**

Female 6–10⅝ in. (15–27 cm); male 3½–5⅞ in. (9–15 cm). Marks behind eyes relatively small, often downward extensions of a neck stripe when 4–7 neck stripes reach eyes; however, sometimes no neck stripes reach eyes (Fig. 93, p. 209; see Subspecies). Carapace olive to brown with a middorsal keel of small, distinct knobs; yellow ocelli on upper and lower marginal surfaces. Juvenile plastron with dark pigment along seams. Head relatively small. Irises yellow, white, greenish, or dark brown, often bisected by horizontal black lines (rarely when irises white). **SIMILAR SPECIES:** (1) Ouachita and Sabine Map Turtles often with light spots under eyes and on lower jaw; irises always white and usually bisected by horizontal black lines. (2) Northern Map Turtle has a broader head, less prominent keel with weakly developed or absent knobs, and reduced or absent plastral pattern. (3) Texas Map Turtle has light postocular stripes that extend posteriorly from lower edges of eyes. (4) Pascagoula and Pearl River Map Turtles have wide light marks between eyes connected by thin stripes to extensive blotches behind eyes. **HABITAT:** Rivers, lakes, ponds, oxbows, occasionally marshlands, usually with mud bottoms. **RANGE:** Mississippi-Missouri R. and tributaries from WI, MN, and ND south to LA; Mermenteau and Calcasieu R. drainages, LA; Sabine, Trinity, and Brazos R. systems, TX. Records from OH and n. IN are likely incorrect; records in FL and elsewhere outside the natural range (not mapped)

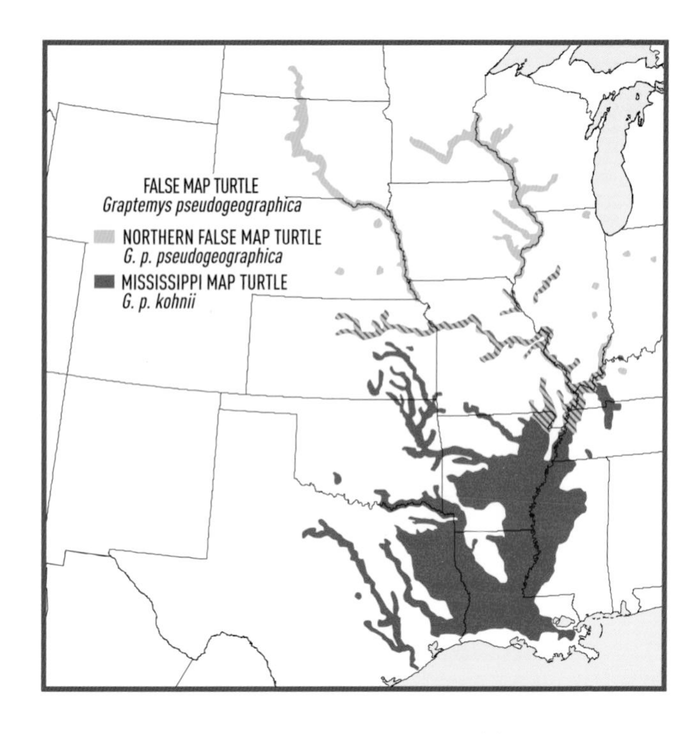

are attributable to released pets. **SUBSPECIES:** (1) **Northern False Map Turtle** (*G. p. pseudogeographica*); yellow postocular lines narrow, usually 4–7 neck stripes reach eyes; no enlarged spots on mandible; fewer lines on legs. (2) **Mississippi Map Turtle** (*G. p. kohnii*); crescent-shaped marks behind eyes prevent neck stripes from reaching eyes (Fig. 93, facing page). **REMARK:** The Mississippi Map Turtle is sometimes considered a separate species.

ALABAMA MAP TURTLE *Graptemys pulchra* PL. 16, FIGS. 82, 93

Female 7–11½ in. (18–29.2 cm); male 3½–5⅛ in. (9–13 cm). Juvenile has middorsal black lines involving spines, *broad* light areas between eyes broadly connected to blotches behind eyes (Fig. 93, facing page), and a *longitudinal light bar* running back from point of chin (Fig. 82, p. 178); male retains juvenile features but vertebral spines essentially absent in female. *Prominent* light markings on marginals. Female has enlarged head but slightly more pointed snout and jaws less wide than in other species in which female has an enlarged head. Half-grown individual has a *shallow* plastron when viewed in profile. **SIMILAR SPECIES:** Northern Map Turtle lacks broadly fused blotches between and behind eyes. See also Barbour's Map Turtle. **HABITAT:**

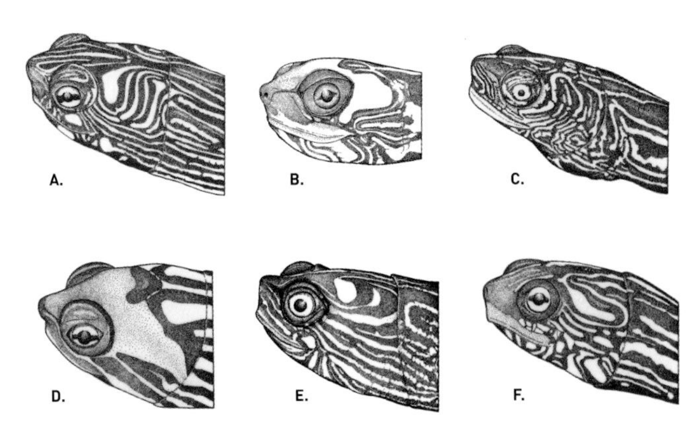

FIG. 93. *Head patterns of map turtles* (Graptemys): **A.** *Ouachita Map Turtle* (G. ouachitensis); **B.** *Pearl River Map Turtle* (G. pearlensis); **C.** *Mississippi Map Turtle* (G. pseudogeographica kohnii); **D.** *Alabama Map Turtle* (G. pulchra); **E.** *Sabine Map Turtle* (G. sabinensis); **F.** *Texas Map Turtle* (G. versa).

Swift-flowing creeks and rivers, females often in deep pools or impoundments. **RANGE:** Mobile Bay drainage of AL, e. MS, and nw. GA.

SABINE MAP TURTLE *Graptemys sabinensis* **FIG. 93**

Female to 9 7/16 in. (24 cm); male 2¾–5.5 in. (7–14 cm). *Oval or elongate blotches* behind eyes usually continuous with narrower neck stripes; *5–9 neck stripes reach eyes* (Fig. 93, above). Often with small spots under eyes that usually are part of prominent necklines that extend forward under eyes; light transverse bar under chin. Olive to brown carapace with low black knobs forming a middorsal keel; yellow vermiculations and dark blotches on pleurals. Plastron with many fine lines in juvenile, but fading to lines along seams in adult. Whitish irises often bisected by black horizontal lines. **SIMILAR SPECIES:** See False Map Turtle. **HABITAT:** Swift rivers, also lakes, oxbows, swamps. **RANGE:** Mermentau and Calcasieu R. systems, LA; Sabine R. system of TX and LA. **REMARK:** Until recently considered a subspecies of the Ouachita Map Turtle.

TEXAS MAP TURTLE *Graptemys versa* **PL. 16, FIGS. 82, 93**

Female 4–8⅜ in. (10–21.4 cm); male 2¾–4½ in. (7–11.5 cm). Light yellow or orange lines (often J-shaped) extend *backward from lower end* of postorbital markings (Fig. 93, above). Chin marked with a light orange oval and smaller, round, light orange spots farther back on each side (Fig. 82, p. 178). Carapace olive, with vertebrals and pleurals bearing sets of concentric light lines; more anterior carapacial scutes high and rounded, and sutures form distinct grooves; vertebral keel with low horn-colored knobs and yellow areas in front of each knob. **SIMILAR SPECIES:** See Cagle's Map Turtle. **HABITAT:** Upstream in clear, spring-fed rivers with pools and riffles; downstream in larger streams and sloughs. **RANGE:** Colorado R. system from cen. TX nearly to the Gulf.

DIAMOND-BACKED TERRAPINS: *Malaclemys*

The only species in this genus occurs in our area. Market hunting once seriously reduced populations, but many have recovered as a consequence of conservation efforts.

DIAMOND-BACKED TERRAPIN *Malaclemys terrapin* **PLS. 15, 22, P. 458**

Female 6–9$^3/_8$ in. (15.2–23.8 cm); male 4–5½ in. (10–14 cm). *Concentric grooves and ridges or concentric dark and light markings* on each carapacial scute. *Flecked or spotted head and legs.* Skin gray. Carapace with a central keel, low and inconspicuous along the Atlantic Coast but prominent and often knobbed along the Gulf Coast. Plastron with or without bold dark markings. Hatchling 1–1¼ in. (2.5–3.2 cm); more brightly patterned than adult. **SIMILAR SPECIES:** (1) Wood Turtle has concentric carapacial ring but has orange on neck and limbs and avoids brackish water. (2) Snapping and mud turtles enter brackish water, but the Eastern Snapping Turtle has a long, saw-toothed tail, and mud turtles in e. N. America have a double-hinged plastron and smooth shell. **HABITAT:** Coastal salt marshes, tidal flats, coves, estuaries, brackish coastal streams, especially those bordered by mangroves, inner edges of barrier beaches, rarely far offshore. **RANGE:** Coastal areas from Cape Cod to s. TX. **SUBSPECIES:** Geographically from north to south: (1) **NORTHERN DIAMOND-BACKED TERRAPIN** (*M. t. terrapin*); coloration variable, some with carapace boldly patterned with dark rings on light gray or brown ground color, others uniformly black or dark brown; plastron orange or yellowish to greenish gray, with or without bold dark markings; carapace wedge-shaped when viewed from above, widest posteriorly; plastron with nearly parallel sides. (2) **CAROLINA DIAMOND-BACKED TERRAPIN** (*M. t. centrata*); similar to Northern Diamond-backed Terrapin, but sides of carapace more nearly parallel and sides of plastron tend to curve inward posteriorly. (3) **EASTERN FLORIDA DIAMOND-BACKED TERRAPIN** (*M. t. tequesta*); carapace dark or horn colored without a pattern of concentric circles; centers of large scutes only slightly lighter than surrounding areas. (4) **MANGROVE DIAMOND-BACKED TERRAPIN** (*M. t. rhizophorarum*); dark spots on neck fused, producing bold streaks; bulbous bumps on the dorsal keel; stripes on rear of thighs. (5) **ORNATE DIAMOND-BACKED TERRAPIN** (*M. t. macrospilota*); centers of large pleurals orange or yellow, bulbous bumps or tubercles on the middorsal keel most evident in juvenile and male; juvenile shell is light horn color, except tubercles and edges of dorsal scutes black. (6) **MISSISSIPPI DIAMOND-BACKED TERRAPIN** (*M. t. pileata*); very dark, carapace usually uniformly black or

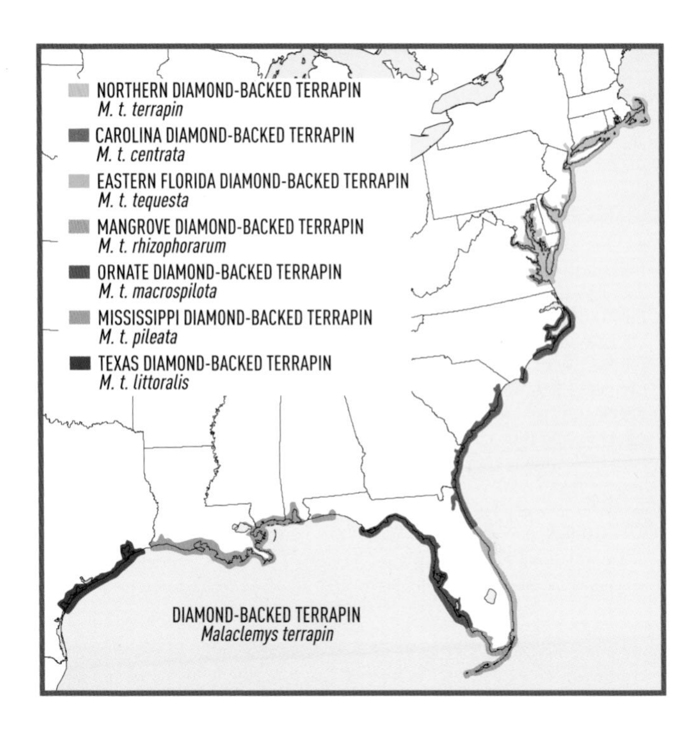

brown; skin very dark; plastron yellow, often clouded; strongly tuberculate central keel; edges of shell orange or yellow and turned upward; most males and some females with a black "mustache" on upper jaw. (7) **Texas Diamond-backed Terrapin** (*M. t. littoralis*); similar to Mississippi Diamond-backed Terrapin but with a deeper shell with highest point posteriorly; skin greenish gray, heavily marked with black spots; plastron nearly white; "mustache" usually missing. **REMARK:** Traditionally recognized subspecies intergrade extensively and are not well supported by genetic data.

COOTERS and SLIDERS: *Pseudemys* and *Trachemys*

Cooters (6 to 9 species) and sliders (perhaps as many as 16 species, 2 in our area) range from the northeastern United States to Argentina and Uruguay. Shells of adults usually are rugose (wrinkled) and the rear margin of the carapace is saw-toothed, with indentations in or between the posterior marginals. Except in the Big Bend Slider (a subspecies of the Mexican Plateau Slider), adult males have greatly elongated "fingernails" used in courtship. The shell of males is flat compared with the arched shell of larger females. Hatchlings $^7/_8$–1½ in. (2.2–3.8 cm). Cooters (*Pseudemys*) tend to have flattened undersides of jaws, whereas those of sliders (*Trachemys*) tend to be rounded (Fig. 84, p. 182). The classification of many cooters remains controversial. Morphological and genetic data are sometimes in conflict, and interbreeding and introgression are extensive. Detailed studies are needed to resolve these issues.

ALABAMA RED-BELLIED COOTER *Pseudemys alabamensis* **FIG. 94**

8–12 in. (20.3–30.6 cm); record 15 in. (38.3 cm). Carapace brown to olive (adult) or greenish (young) with yellow or red bars on second pleurals; *plastron reddish*, usually with dark bars and a light-centered dark figure; head olive to black with few yellow stripes, but with a *slender "arrow"* between eyes pointing to the snout; upper jaw saw-toothed, with a notch flanked by cusps (as in Fig. 83, p. 180). **SIMILAR SPECIES:** See River Cooter. **HABITAT:** Shallow fresh to moderately brackish water. **RANGE:** Mobile Bay drainage in AL and lower Pascagoula R. and Biloxi Bay drainages in MS. **CONSERVATION:** Endangered (USFWS: Endangered).

RIVER COOTER *Pseudemys concinna* **PLS. 17, 22**

9–13 in. (23–33 cm); record 14¾ in. (37.5 cm). *Light C-shaped marks on second pleurals* (as in Fig. 83, p. 180) (*C* may face forward or back-

FIG. 94. *The Alabama Red-bellied Cooter* (Pseudemys alabamensis) *is the official state reptile of Alabama. The Alliance for Zero Extinction has listed it as one of the world's most vulnerable species as a consequence of human-mediated habitat alterations and a range now restricted to the lower Mobile Bay drainage basin.*

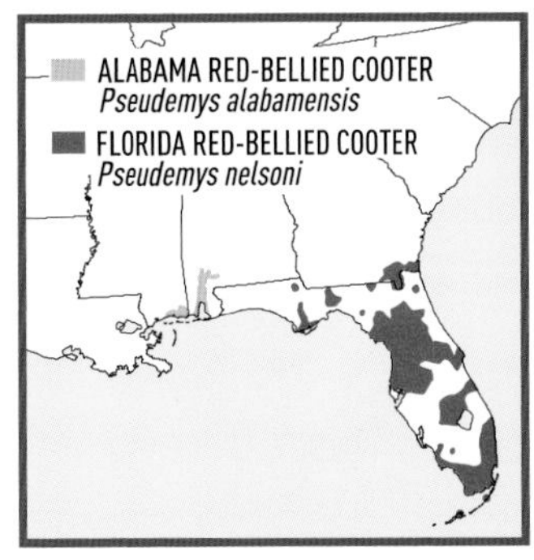

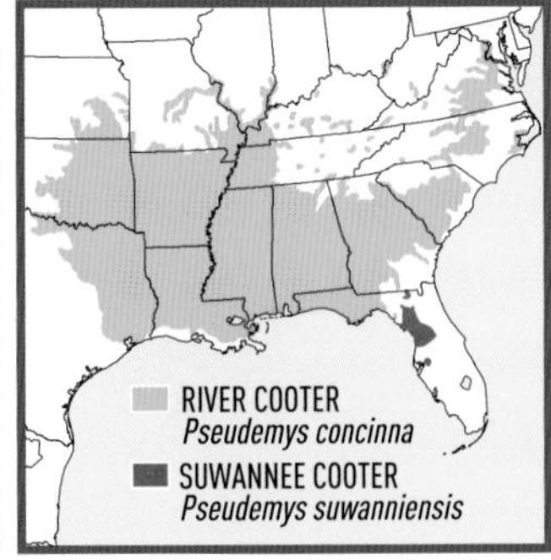

ward); concentric circles in conjunction with *C* and on other scutes (concentric circles on second pleurals absent in some western populations, which sometimes lack C-shaped marks). Five light stripes between eyes; small postorbital spots absent in western populations. Shell sometimes "pinched in" anterior to hindlimbs. Dark pattern on plastron tends to follow seams; X-shaped marks frequently present anteriorly. Undersides of all (or most) marginals have dark spots, usually doughnut-shaped; some may touch dark markings on bridge. Notch and cusps on upper jaw absent or faint. **SIMILAR SPECIES:** (1) Coastal Plain Cooter has wide vertical stripes on second pleurals and a plain or faintly patterned plastron with hollow circles on marginals. (2) Other cooters in range have a saw-toothed upper jaw with a notch flanked by cusps and lack a *C* on second pleurals. (3) Pond Slider has broad yellow or reddish stripes or patches behind eyes and a rounded lower jaw. **HABITAT:** Streams of varying sizes; less abundant in reservoirs. **RANGE:** MD south to the FL Panhandle and west to se. KS and e. TX.

COASTAL PLAIN COOTER *Pseudemys floridana* **PL. 17**

9–13 in. (23–33 cm); record $15^5/_8$ in. (39.7 cm). Unmarked plastron and *dark "doughnuts" or thick-edged, hollow ovals* on undersides of marginals; numerous stripes on head rarely join to form "hairpins" except in areas of interbreeding with the Peninsula Cooter. One or more vertical light stripes on second pleurals (as in Fig. 83, p. 180). Notch and cusps on upper jaw absent or faint. **SIMILAR SPECIES:** See River Cooter. **HABITAT:** Permanent ponds, lakes, swamps, marshes, rivers. **RANGE:** Coastal Plain from se. VA to extreme se. MS, but excluding peninsular FL. **REMARKS:** Sometimes considered a subspecies of the River Cooter, with which it interbreeds in some areas. Also interbreeds with Peninsula and Suwannee Cooters.

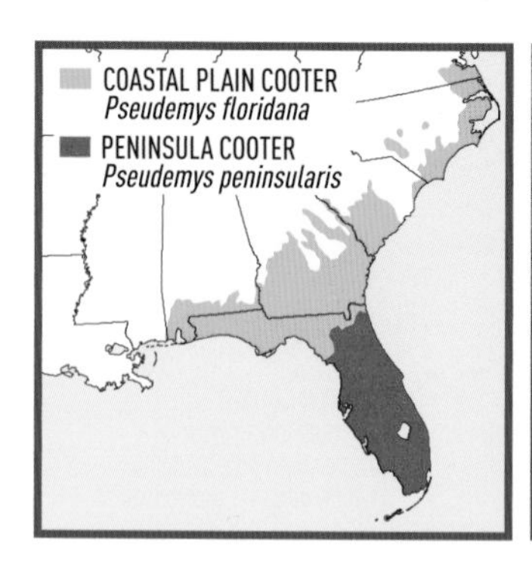

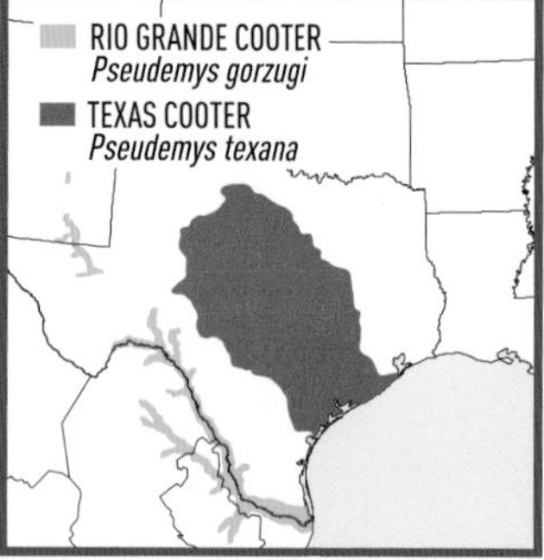

RIO GRANDE COOTER *Pseudemys gorzugi* **FIG. 95**

8–12 in. (20.3–30.6 cm); record 14⅝ in. (37.2 cm). *Four or 5 distinct concentric whorls on second pleurals*; centers of whorls light yellow, rarely with a C-shaped mark. Plastron with narrow dark lines along midline and anterior edges of seams; these fade in adult. Oval spots or blotches behind eyes; median line from snout onto neck; broad Y-shaped mark on chin. Large adults often melanistic. Notch and cusps on upper jaw absent or faint. **SIMILAR SPECIES:** See Texas Cooter. **HABITAT:** Large river pools. **RANGE:** Rio Grande drainage from the Gulf to the Pecos and Devil's R., TX; isolated populations in the Pecos R. system of s. NM and adjacent TX.

FLORIDA RED-BELLIED COOTER *Pseudemys nelsoni* **PL. 17, FIGS.83, 87**

8–12 in. (20.3–30.6 cm); record 14¾ in. (37.5 cm). Usually *reddish*, at least on plastron. Light vertical bands on second pleurals often wide but vary in width (Fig. 83, p. 180). Few light head stripes include a slender "arrow" with shaft between eyes usually separated from point at tip of snout. Notch in upper jaw flanked by strong cusps (Fig. 83, p. 180). Adult plastron almost always tinted with orange, red, or coral, at least along edges; juvenile plastron orange to scarlet-orange (rarely yellow); dark plastral markings usually solid semicircles with flat sides along seams (Fig. 87, p. 190). **SIMILAR SPECIES:** See Peninsula Cooter. **HABITAT:** Shallow water along shorelines of slow-moving streams, spring runs, ponds, lakes, ditches, sloughs, marshes, mangrove-bordered creeks; occasionally enters brackish water. **RANGE:** Peninsular FL and some Keys north to se. GA; isolated populations in the FL Panhandle. **REMARK:** Interbreeds with Peninsula and Suwannee Cooters.

FIG. 95. *The Rio Grande Cooter* (Pseudemys gorzugi) *is designated as Near Threatened on the IUCN Red List due to a relatively small range and habitat degradation attributable to pollution and water control and diversion projects along the Rio Grande and its major tributaries.*

PENINSULA COOTER *Pseudemys peninsularis* **PLS. 17, 22, FIG. 83**

9–13 in. (23–33 cm); record 15⅞ in. (40.3 cm). *Light "hairpins" on top of head* (Fig. 83, p. 180) occasionally broken or incomplete. Plastron usually unmarked, but marginal spots sometimes prominent. Juvenile with yellow plastron possibly tinged with orange, never strongly orange or reddish. Notch and cusps on upper jaw absent or faint. **SIMILAR SPECIES:** (1) Suwannee Cooter has light C-shaped marks on second pleurals and heavy dark ventral markings. (2) Coastal Plain Cooter has dark "doughnuts" on undersides of marginals. (3) Florida Red-bellied Cooter has reddish, orange, or coral plastron, almost always with some dark marks. (4) Pond Slider has a rounded lower jaw and broad stripes or blotches behind eyes. **HABITAT:** Lakes, sloughs, wet prairies, canals, springs, spring runs; also in the Everglades north of and along the Tamiami Trail. **RANGE:** Peninsular FL. **REMARKS:** Sometimes considered a subspecies of the Coastal Plain or River Cooter. Interbreeds with Suwannee, Coastal Plain, and Florida Red-bellied Cooters.

NORTHERN RED-BELLIED COOTER *Pseudemys rubriventris*

PL. 17, FIG. 87

10–12½ in. (25.4–32 cm); record 15¾ in. (40 cm). Female with vertical reddish lines on first 3 pleurals; old male mottled reddish brown. Color and pattern highly variable, melanism common. Reddish markings usually persist even in virtually black individuals, although often vague. Plastron yellow with large gray smudges, borders washed in pink or orange-red. Plastron of old male often mottled with pink and light charcoal gray. Juvenile has slight keels but carapace patterned (yellow or olive on green) as in adult. Plastron has complex dark pattern on coral red ground color (Fig. 87, p. 190). Notch in upper jaw flanked by cusps (as in Fig. 83, p. 180). **SIMILAR SPECIES:** (1) River and Coastal Plain Cooters lack a prominent notch and cusps in upper jaw. (2) Eastern Painted Turtle (a subspecies of the Painted Turtle) has 2 bright yellow spots on sides of head. **HABITAT:** Ponds, rivers, any relatively large body of fresh water. **RANGE:** Coastal Plain from cen. NJ to NC, inland to e. WV; isolated populations in MA. Introduced in NY (including Long Island). **CONSERVATION:** Populations in MA once were considered subspecifically distinct (USFWS: Endangered).

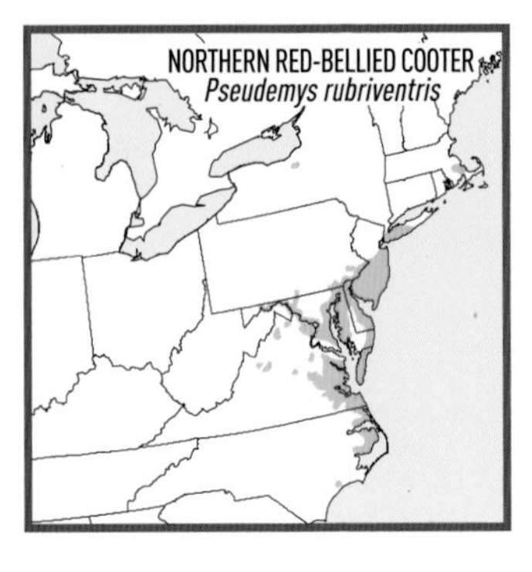

SUWANNEE COOTER *Pseudemys suwanniensis* **PL. 18, FIG. 83, P. 172**

9–13 in. (23–33 cm); record 17¼ in. (43.8 cm). Darkest and largest cooter. Carapace virtually plain black when dry; legs and head also dark, but with 5 whitish to greenish yellow head stripes between eyes (Fig. 83, p. 180). Plastron usually yellow, sometimes tinged with orange; dark spots on marginals in contact with dark markings on bridge. Notch and cusps on upper jaw absent or faint. Hatchling with dark gray blotches on pale gray ground color, but blotches quickly change to brownish green separated by a network of yellowish green lines. *Pale C-shaped marks on second pleurals* clearly defined (Fig. 83, p. 180); plastron lemon yellow with grayish brown along seams. **SIMILAR SPECIES:** See Peninsula Cooter. **HABITAT:** Clear spring runs, sometimes into springs, coastal wetlands, bays, lagoons, turtle-grass flats off mouths of streams, and occasionally far into the Gulf. **RANGE:** Vicinity of Tampa Bay, FL, north through the Suwannee R. drainage, possibly into extreme s. GA. **REMARKS:** Sometimes considered a subspecies of River Cooter. Interbreeds with Peninsula, Coastal Plain, and Florida Red-bellied Cooters.

TEXAS COOTER *Pseudemys texana* PL. 18

7–10 in. (18–25.5 cm); record 12½ in. (31.9 cm). Variable yellow head markings with many lateral stripes, vertical bars on sides of head near angle of jaw, and small, round postorbital spots. Lateral head stripes curve above bars near angles of the jaw. Second pleurals with *5–6 concentric dark-centered whorls*, marginals with narrow light vertical lines. Plastron with red-tinged rim and dark lines along both sides of seams. Upper jaw with a notch flanked by cusps (as in Fig. 83, p. 180); 2 swollen ridges extend down from nostrils to terminate in cusps. Many old males and occasional females with shell, head, and limbs uniformly mottled, obscuring any pattern. **SIMILAR SPECIES:** (1) River and Rio Grande Cooters lack a prominent notch and cusps on upper jaw. (2) Sliders (*Trachemys*) have a rounded lower jaw. **HABITAT:** Rivers, ditches, cattle tanks. **RANGE:** Colorado, Brazos, Guadalupe, and San Antonio R. drainages from cen. TX to the Gulf.

MEXICAN PLATEAU SLIDER *Trachemys gaigeae* FIG. 96

5–8 in. (12.5–20.3 cm); record 11¾ in. (29.8 cm). Sides of head with a *large black-bordered orange spot* and a second much smaller spot directly behind eyes (Fig. 96, below); larger spot usually apparent even in melanistic individuals. Carapace pale olive brown with numerous pale orange to red curved lines, including one on each marginal. Plastron pale orange and olive with a median series of elongated, concentric dark lines; eyelike spots on undersides of marginals. Melanism develops with age; pattern in large adults often obliterated. Nails on forelimbs of male not enlarged. **SIMILAR SPECIES:** See Pond Slider. **HABITAT:** Permanent water in the upper Rio Grande and some tributaries, including Elephant Butte Reservoir; also ponds, tanks, sloughs, canals. **RANGE:** Rio Grande Valley in the Big Bend Region of TX and s.-cen. NM; also the Río Conchos and Río Nazas drainages in Mex. **SUBSPECIES:** **Big Bend Slider** (*T. g. gaigeae*). **CONSERVATION:** Vulnerable.

BIG BEND SLIDER
A large light spot on side of head, small one back of eye

FIG. 96. *Head pattern of the Big Bend Slider* (Trachemys gaigeae gaigeae), *the subspecies of the Mexican Plateau Slider that occurs in our area.*

POND SLIDER *Trachemys scripta* PLS. 18, 22, FIG. 97

5–8 in. (12.5–20.3 cm); record 11⅞ in. (30.2 cm). *Yellow or reddish stripes or blotches behind eyes*, often obscure in large males; greenish

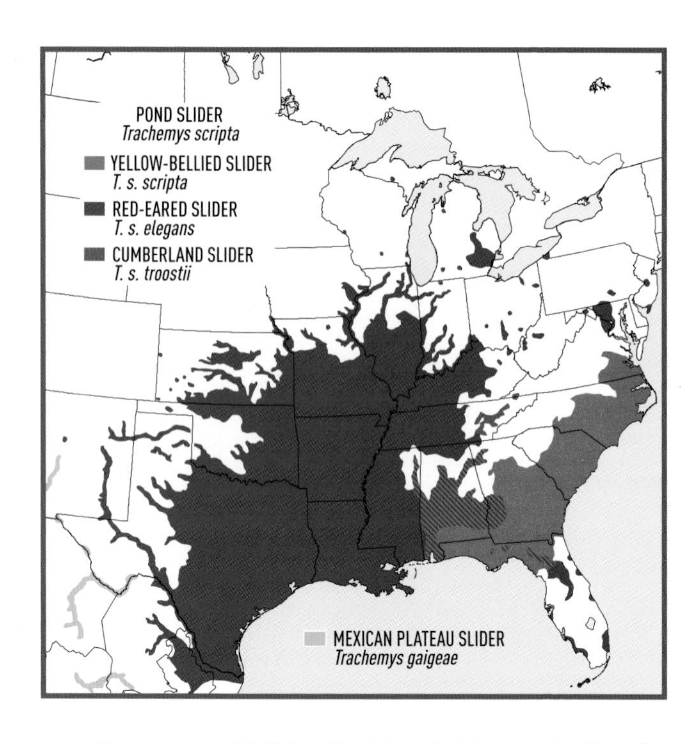

or olive carapace with light yellowish vertical bars; undersides of marginals and plastron yellow with dark blotches, spots, or ocelli. All markings often obscure in older turtles, which become dark and mottled. Carapace of young distinctly marked and often with a low mid-dorsal keel; undersides of marginals and plastron with many dark smudges, spots, or ocelli, often with marks more abundant and well defined anteriorly. **SIMILAR SPECIES:** (1) Big Bend Slider (a subspecies of Mexican Plateau Slider) has a large black-bordered orange spot and a much smaller spot behind each eye. (2) Cooters have a flattened lower jaw and narrow light stripes behind eyes (less than half the width of eyes). (3) Painted turtles have a smooth posterior shell margin (serrated in sliders). (4) Chicken Turtle has stripes on back of thighs (like Yellow-bellied Slider; see Subspecies), but yellow stripes on forelimbs are broad and carapace is long, narrow, and with a smooth posterior margin. **HABITAT:** Rivers, ditches, sloughs, lakes, ponds. **RANGE:** Extreme s.-cen. PA to n. FL, west to NM, and south into ne. Mex. Introduced widely in the U.S. and elsewhere (many U.S. introductions not mapped). **SUBSPECIES:** (1) **Yellow-bellied Slider** (*T. s. scripta*); yellow blotches behind eyes; distinct vertical yellow bands on pleurals; round dusky smudges on undersides of marginals and front

FIG. 97. *This young adult male Red-eared Slider* (Trachemys scripta elegans) *clearly demonstrates the broad red namesake stripe behind the eyes. Because of their popularity in the pet trade, Red-eared Sliders probably are the most widely introduced turtle in the world.*

of plastron; vertical stripes on back of hindlimbs and narrow yellow stripes along front of forelimbs (Fig. 91, p. 200). (2) **Red-eared Slider** (*T. s. elegans*); broad reddish stripes behind eyes in most individuals, red occasionally replaced by yellow; vertical bands on carapace often indistinct. (3) **Cumberland Slider** (*T. s. troostii*); stripes behind eyes narrower and yellow; fewer and much wider stripes on legs, neck, and head; vertical bands on carapace often indistinct; dark spots under marginals smaller in diameter than intervening light spaces. **REMARKS:** The Red-eared Slider is the most widely introduced turtle in the world and is included among 100 of the World's Worst Invasive Alien Species (http://www.issg.org/database/species/search.asp?st=100ss). Many introductions not mapped.

BOX TURTLES: *Terrapene*

Four species of these strictly North American turtles range widely in our area and also in the Southwest and Mexico. All are essentially terrestrial but can swim and will sometimes soak in mud or shallow water. The upper jaw ends in a down-turned beak; toes are not webbed; and movable lobes of the plastron are connected by a broad hinge (Fig. 98, p. 220) that allows complete closure of the shell. Hatchlings, which have a nonfunctional hinge, measure $1\frac{1}{8}$–$1\frac{3}{8}$ in. (2.8–3.5 cm). Young have median dorsal ridges, which sometimes persist in adults.

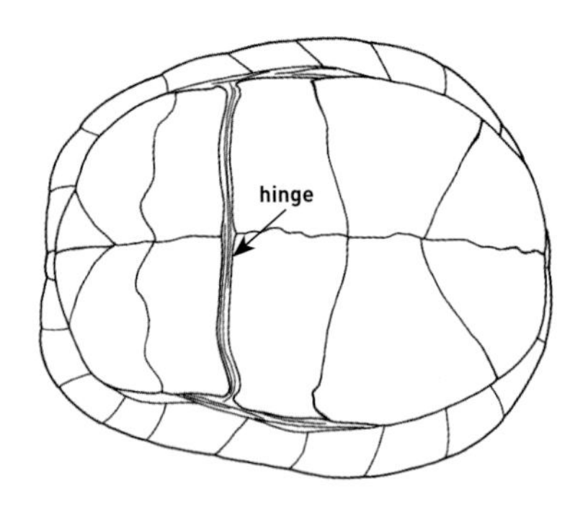

FIG. 98. *Box turtles* (Terrapene) *have a hinge across the plastron (lower shell). Blanding's Turtle* (Emydoidea blandingii) *has a similar hinge, although its shell cannot be completely closed.*

Populations of box turtles along the Gulf Coast, until recently recognized as a subspecies of the Eastern Box Turtle, are of uncertain status. Many are considered intergrades between the Eastern Box Turtle and a form known only from fossils, but at least some populations, especially from coastal areas in Louisiana, Mississippi, and probably Alabama, might comprise one or more distinct taxa.

FLORIDA BOX TURTLE *Terrapene bauri* **PLS. 18, 21**

5–6½ in. (12.5–16.5 cm); record 7½ in. (19 cm). High, domelike shell with broken or irregular *light radiating lines* on at least some scutes; posterior portion flares little or not at all. *Two lines on head* often interrupted or incomplete; usually 3 toes on hindlimbs. Plastron solid yellow to black or variously marked with lines or spots; male plastron deeply concave. Young has a yellowish middorsal stripe that involves a keel and can persist in adult; pattern mottled yellowish or greenish on dark brown. **SIMILAR SPECIES:** See Eastern Box Turtle. **HABITAT:** Moist woodlands with underbrush. **RANGE:** Se. GA and peninsular FL.

EASTERN BOX TURTLE *Terrapene carolina* **PLS. 18, 21**

4½–6 in. (11.5–15.2 cm); record 9¼ in. (23.5 cm). High, domelike shell, usually with a *middorsal keel* that can fade in adult. Extremely variable in color and pattern, but some pattern almost always present; both carapace and plastron may be yellow, orange, or olive on black or brown, either dark or light colors predominating. Carapace slopes downward posteriorly but sometimes flares outward and occasionally turns upward (especially in the FL Panhandle). Plastron *shorter* than carapace. Four toes on hindlimbs. Occasionally with white blotches on head. Rear plastral lobes of male usually with a central concave area; eyes of male often red. Plastron of female flat or slightly convex; eyes normally brown. Carapace of young much flatter, middorsal keel distinct and sometimes with a yellow middorsal stripe; usually brown to black with yellow spots on pleurals. **SIMILAR SPECIES:** (1) Ornate Box Turtle has a flattened carapace usually without a keel. (2) Florida

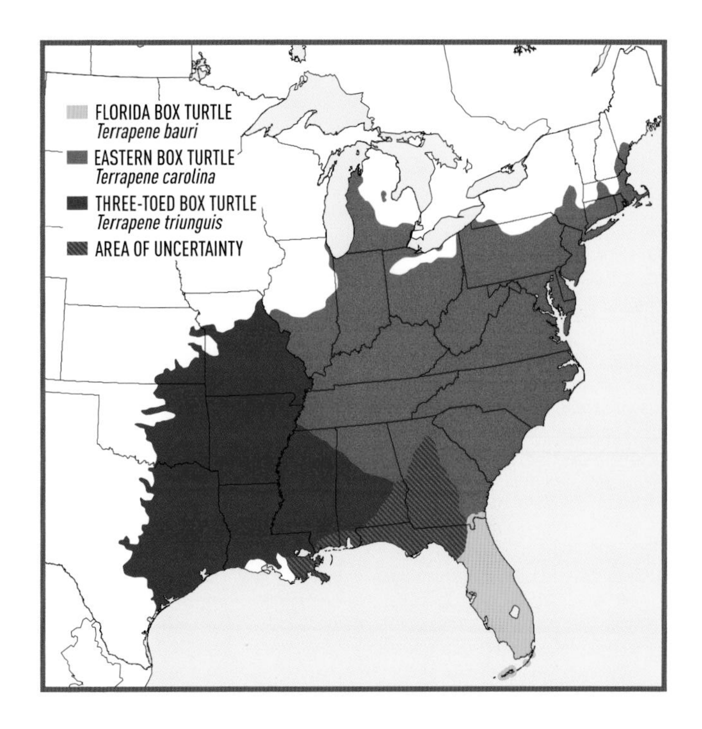

Box Turtle has a carapace that flares little or not at all posteriorly and usually has 3 toes on hindlimbs (Eastern Box Turtles in areas of contact almost always have a flared carapace and 4 toes on hindlimbs). (3) Three-toed Box Turtle has a plain olive or horn-colored carapace, plain yellow or horn-colored plastron, and usually has 3 toes on hindlimbs. (4) Gopher Tortoise lacks plastral hinges. (5) Blanding's Turtle has a flatter shell, profusion of light dots, and plastral lobes do not shut completely. **HABITAT:** Woodlands and thickets, occasionally coastal marshes. **RANGE:** S. ME to s. IL and south to AL and GA. Many populations declining or extirpated. **REMARK:** Florida Box Turtle, Three-toed Box Turtle, and populations in Mex. until recently were considered subspecies of the Eastern Box Turtle. **CONSERVATION:** Vulnerable.

ORNATE BOX TURTLE *Terrapene ornata* **PL. 18, FIG. 99**

4–5 in. (10–12.5 cm); record 6⅛ in. (15.4 cm). *Carapace flat on top*, sometimes even concave, and rarely with any trace of a keel in adult; plastron *large*, usually as long as or longer than carapace. Usually with light marks on dark ground color above and below (Fig. 99, p. 222), although pattern fades in desert populations, which frequently

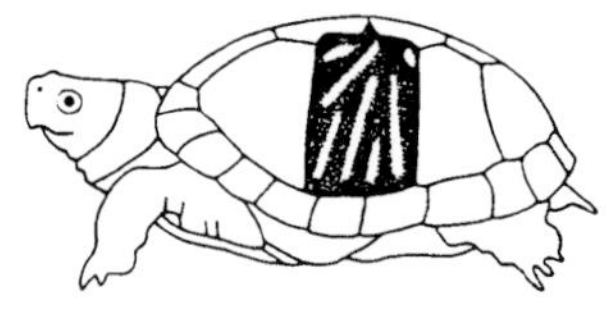

2 pattern variations

FIG. 99. *Radiating lines on the second pleural scutes of an Ornate Box Turtle* (Terrapene ornata).

are uniformly yellowish, straw, or horn colored. Male usually with red eyes and uniformly colored head; female usually has brown eyes and head with dull yellow blotches. **SIMILAR SPECIES:** Three-toed and Eastern Box Turtles have high-domed shells with a middorsal keel. **HABITAT:** Plains, prairies, desert grasslands, oak savannas, forest edges, often sandy areas. **RANGE:** IN to se. WY south to the Gulf and west to se. AZ and adjacent Mex. **REMARK:** Populations in portions of the Chihuahuan and Sonoran Deserts until recently were considered subspecifically distinct.

THREE-TOED BOX TURTLE *Terrapene triunguis* **PL. 18, FIG. 100**

4½–5 in. (11.5–12.5 cm); record 7 in. (17.9 cm). High, domelike shell usually with a *middorsal keel* that can fade in adult. Carapace pattern faint even in juvenile; adult often uniformly olive or horn colored, sometimes with scattered light spots. Plastron plain yellow or horn colored, distinctly *shorter* than carapace. Usually with 3 toes on hindlimbs. Orange or yellow spots frequently conspicuous on head and forelimbs. Plastron of male flat or faintly concave; eyes often red. Plastron of female flat or slightly convex; eyes normally brown. Carapace of young much flatter and with a distinct middorsal keel and yellow stripes; usually plain grayish brown, but with yellow spots on large scutes. **SIMILAR SPECIES:** See Eastern Box Turtle. **HABITAT:** Woodlands and thickets; sometimes far into prairies but usually retreats to wooded draws in fall and winter. **RANGE:** MO and e. KS south to the Gulf.

FIG. 100. *The Three-toed Box Turtle* (Terrapene triunguis) *typically has a rather plain carapace that is uniformly colored or has faint light spots or streaks. However, some individuals have conspicuous bright orange or yellow spots on the head and forelimbs.*

MUD and MUSK TURTLES: Kinosternidae

Two genera occur in our area, two others in Central and South America. Semiaquatic mud turtles (*Kinosternon*) range from New England to northern Argentina. They have a relatively large plastron with two transverse hinges (Fig. 101, p. 224). Pectoral scutes are triangular in most species. Strongly aquatic musk turtles (*Sternotherus*) range from southern Ontario south to the Gulf states. They have a relatively small plastron with an inconspicuous transverse hinge between the second and third pairs of scutes (Fig. 101, p. 224). Pectoral scutes are squarish. Males usually have broad areas of soft skin between the plastral scutes; females less so. Turtles in both genera have barbels (fleshy projections) on the chin and neck and almost always have 23 marginals, including the cervical (most turtles have 25). Males have a long, stout tail, usually with a clawlike tip; the tail of females is short and stubby.

STRIPED MUD TURTLE *Kinosternon baurii* **PL. 19**

3–4 in. (7.5–10 cm); record 5 7/16 in. (13.8 cm). *Three light stripes on smooth shell* sometimes obscure, particularly in older turtles and those in northern portions of the range. Sides of head with *2 light*

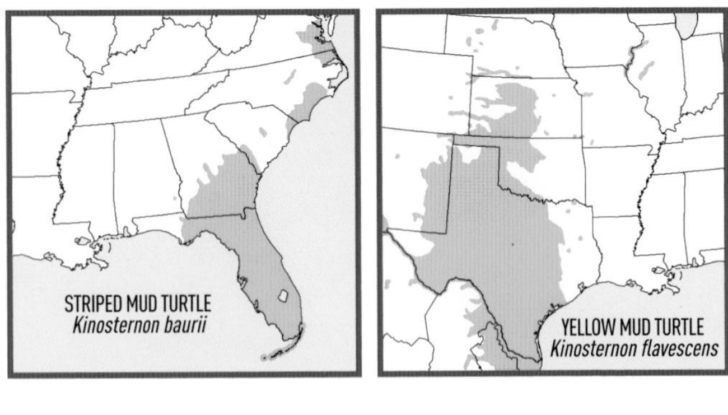

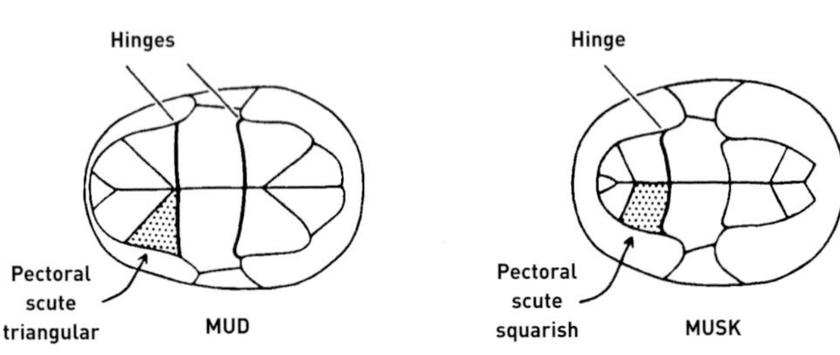

FIG. 101. *Plastrons (lower shells) of mud* (Kinosternon) *and musk* (Sternotherus) *turtles.*

stripes. Young with a narrow middorsal keel, rough carapace, and light spots on each marginal. Hatchling 9/16–1 in. (1.4–2.5 cm). **SIMILAR SPECIES:** See Eastern Mud Turtle. **HABITAT:** Drainage canals, blackwater rivers, sloughs, ponds, lakes in cypress swamps, wet meadows, ditches, other small, shallow, even temporary bodies of water; frequently on land. **RANGE:** Coastal Plain from e. VA through FL and the Keys.

YELLOW MUD TURTLE *Kinosternon flavescens* **PL. 19, FIGS. 102, 103**

4–5 in. (10–12.5 cm); record 6 5/8 in. (16.8 cm). *Yellowish chin and throat* (only barbels and front half of lower jaw yellow in populations in IL, IA, and ne. MO). Ninth marginals of adult distinctly higher than eighth marginals (tenth marginals also high). Carapace usually flat or even depressed. Pectorals pointed and only narrowly in contact (Fig. 102, facing page). Carapace olive brown to olive green, head and neck olive above (carapace dark brown, head and neck black or dark gray in isolated populations). Ninth and tenth marginals of hatchling as low as eighth, but ninth marginals always distinctly peaked, with peaks rising above upper edges of adjacent marginals. Young with bold black

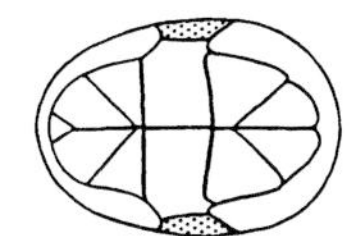

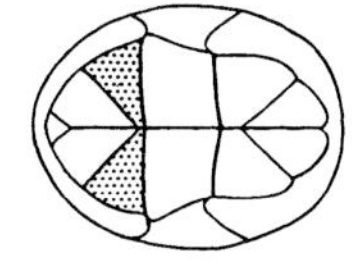

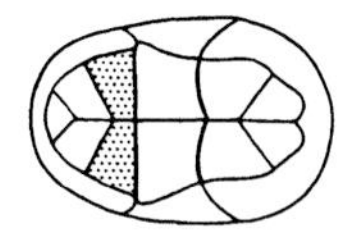

FIG. 102. *Plastrons (lower shells) of some mud turtles* (Kinosternon).

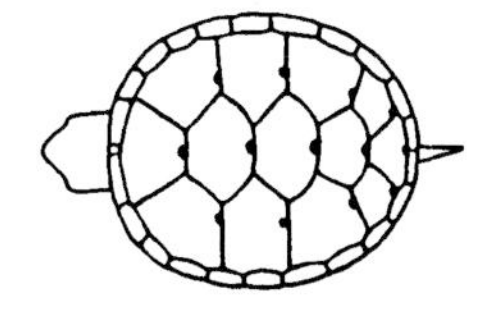

FIG. 103. *Carapace (upper shell) of a hatchling Yellow Mud Turtle* (Kinosternon flavescens).

dots at posterior borders of carapacial scutes (Fig. 103, above). Hatchling 13/16–1 3/16 in. (2.1–3 cm), with a bold yellow and black plastron. **SIMILAR SPECIES:** See Eastern Mud Turtle. **HABITAT:** Almost any body of water, including temporary pools, usually with a muddy bottom; artificial habitats such as cattle tanks, irrigation ditches, cisterns, sewer drains in western portions of range; on land during rains or while migrating when pools go dry. **RANGE:** NE and se. AZ to the Gulf. Many peripheral isolates are indicative of a wider historial distribution. Isolated populations in IL, IA, and n. MO once were considered subspecifically distinct.

ROUGH-FOOTED MUD TURTLE *Kinosternon hirtipes* **PL. 19, FIG. 102**

3 3/4–6 11/16 in. (9.5–17 cm); record 7 11/16 in. (19.5 cm). Similar to the Eastern Mud Turtle but larger and more elongate. *Tenth marginals distinctly higher than other marginals.* Faint middorsal keel present at all ages, lateral keels of juvenile lost in adult. Carapace olive brown to almost black, scutes narrowly bordered by black. Plastron yellowish;

seams bordered by black or dark brown; pectorals broadly in contact (Fig. 102, p. 225). Head brown or olive, profusely marked with small dark and light spots and streaks. **SIMILAR SPECIES:** Yellow Mud Turtle has a yellowish chin and throat. **HABITAT:** Permanent water; spring-fed cattle tanks in sw. TX; probably also in the Rio Grande R. **RANGE:** Sw. TX south into Mex. **SUBSPECIES: MEXICAN PLATEAU MUD TURTLE** (*K. hirtipes murrayi*).

FLORIDA MUD TURTLE *Kinosternon steindachneri* **FIG. 102**

3–4 in. (7.5–10 cm); record 4½ in. (12.1 cm). Adult carapace *smooth*, brownish, varying from olive horn to almost black. *Small*, double-hinged plastron plain yellowish brown or slightly to heavily marked with black or dark brown. *Posterior lobe of plastron short*, often shorter than anterior lobe. Bridges *narrow* (Fig. 102, p. 225). Head spotted, mottled, or irregularly streaked with yellow; enlarged in adult male. Often with patches of skin between plastral scutes. Hatchling 11/16–1 1/16 in. (1.7–2.6 cm); carapace rough, black or very dark brown, with a middorsal keel and additional imperfect lateral keels. Plastron usually black in center and along sutures (sometimes solid black); lighter parts yellow, orange, or reddish; bright spots on each marginal. **SIMILAR SPECIES:** See Eastern Mud Turtle. **HABITAT:** Highly aquatic; shallow water in ditches, wet meadows, marshes, sloughs, small ponds, marshes, bayous, lagoons, great swamps, brackish water at edges of tidal marshes and some islands. **RANGE:** Peninsular FL and Key Largo. **REMARK:** Until recently considered a subspecies of the Eastern Mud Turtle.

EASTERN MUD TURTLE *Kinosternon subrubrum* **PLS. 19, 21**

2¾–4¾ in. (7–12.1 cm); record 4⅞ in. (12.5 cm). Adult carapace *smooth*, brownish, varying from olive horn to almost black. Large double-hinged plastron plain yellowish brown or slightly to heavily marked with black or dark brown; *anterior lobe shorter than posterior*

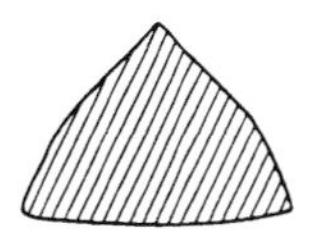

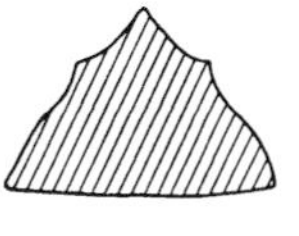

FIG. 104. *Transverse cross sections showing the shapes of shells in young musk turtles* (Sternotherus).

lobe. Bridges *broad* (Fig. 102, p. 225). Head spotted, mottled, or irregularly streaked with yellow (with 2 stripes in turtles from the Mississippi River Valley and farther west). Hatchling $\frac{11}{16}$–$1\frac{1}{16}$ in. (1.7–2.6 cm); carapace rough, black or very dark brown, with a middorsal keel and additional imperfect lateral keels; plastron usually black in center and along sutures (sometimes virtually solid black); lighter parts yellow, orange, or reddish; bright spots on each marginal. **SIMILAR SPECIES:** (1) Florida Mud Turtle has narrow bridges, posterior plastral lobe often shorter than anterior lobe, and male has an enlarged head. (2) Striped Mud Turtle has striped head and carapace. (3) Yellow Mud Turtle has ninth marginals higher than eighth. (4) Musk turtles have a small plastron with a single hinge. **HABITAT:** Shallow water in ditches, wet meadows, marshes, sloughs, small ponds, marshes, bayous, lagoons, great swamps, brackish water at edges of tidal marshes and many islands. **RANGE:** Long Island, NY, south through n. FL and west to e. TX and OK and north along the Mississippi River Valley into s. IL and IN. **SUBSPECIES:** (1) **Southeastern Mud Turtle** (*K. s. subrubrum*); no light stripes on head. (2) **Mississippi Mud Turtle** (*K. s. hippocrepis*); 2 light stripes on sides of head.

RAZOR-BACKED MUSK TURTLE *Sternotherus carinatus* **PL. 19, FIGS. 86, 104**

4–5 in. (10–12.5 cm); record $6\frac{15}{16}$ in. (17.6 cm). Prominent *keel* at all ages, sides of carapace slope down like a tent (Fig. 104, above). Head with dark spots, streaks, or blotches; carapace brown or horn colored with darker streaks; old adults may lose pattern and become almost plain horn colored. Scutes slightly overlapping. Plastron small, with a single hinge, usually *no gular*. Barbels on *chin only*. Margins of carapace often irregular. Hatchling with sharp middorsal keel and shell toothed along sides (Fig. 86, p. 188). Similar species: See Loggerhead Musk Turtle. **HABITAT:** Deep waters of streams, oxbows, great river swamps, usually with slow to moderate currents, often cobble, rock, or sandy bottoms with little vegetation. **RANGE:** S. MS to e. TX and north to cen. AR and se. OK. Introduced in s. FL (not mapped).

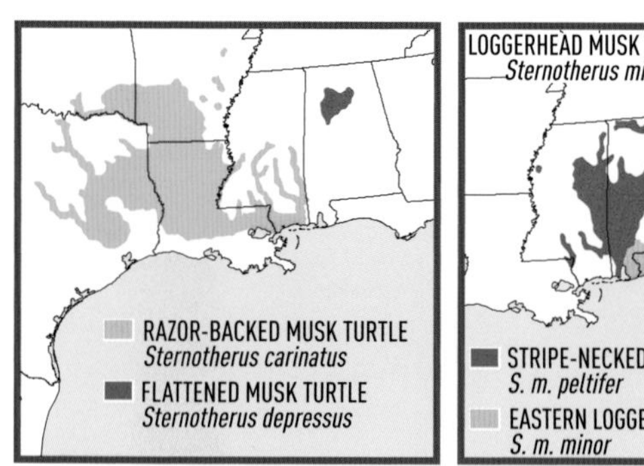

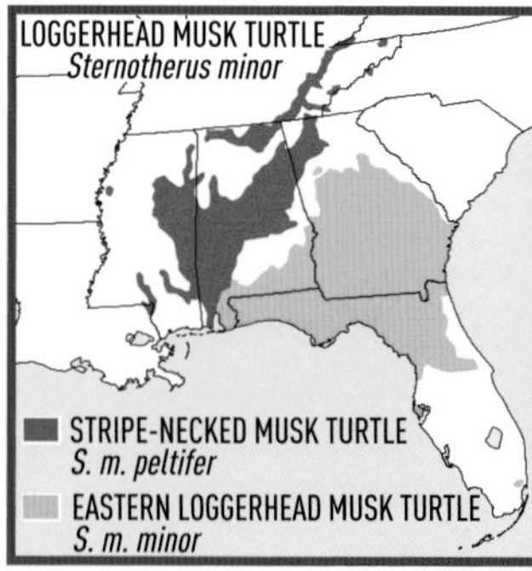

FLATTENED MUSK TURTLE *Sternotherus depressus* **PL. 19, FIG. 104**

3–4 in. (7.5–10 cm); record 4⁷⁄₈ in. (12.5 cm). Carapace *flattened* with round edges. Head and neck with a network of dark lines on light ground color. Carapacial scutes slightly overlapping. Plastron small, gular present. Barbels on *chin only*. Hatchling about 1 in. (2.5 cm), with a blunt dorsal keel and laterally flared carapace (Fig. 104, p. 227). **SIMILAR SPECIES:** See Loggerhead Musk Turtle. **HABITAT:** Nearly exclusively aquatic; medium-sized, clear-water streams and small rivers with sandy to rock bottoms and abundant crevices. **RANGE:** Black Warrior R. system of n.-cen. AL. **CONSERVATION:** Critically Endangered (USFWS: Threatened).

LOGGERHEAD MUSK TURTLE *Sternotherus minor*

PLS. 19, 21, FIG. 104, P. ix

3–4½ in. (7.5–11.5 cm); record 5⁵⁄₈ in. (14.5 cm). Carapace with *1 or 3 distinct keels* (Fig. 104, p. 227), often lost in old adults. Plastron small with a single hinge and one gular scute. Barbels on *chin only*; head normally with dark spots or stripes on lighter ground color; carapace streaked, spotted, or blotched (especially in female). Old males have an enormously enlarged head. Hatchling ⁷⁄₈–1¹⁄₈ in. (2.2–2.8 cm), with one or 3 prominent keels, pink plastron, and carapace strongly streaked with dark rays. **SIMILAR SPECIES:** (1) Eastern Musk Turtle has barbels on chin and neck and 2 light lines on sides of head. (2) Razor-backed Musk Turtle lacks a gular and has a tentlike carapace. (3) Flattened Musk Turtle has a wider, flatter carapace. (4) Mud turtles (*Kinosternon*) have a large plastron with 2 hinges. **HABITAT:** Springs, clear to muddy streams and rivers. **RANGE:** Extreme sw. VA and w. NC south to cen. FL and extreme e. LA. **SUBSPECIES:** (1) **EASTERN LOGGERHEAD MUSK TURTLE** (*S. m. minor*); 3 distinct keels, head with dark spots. (2) **STRIPE-NECKED MUSK TURTLE** (*S. m. peltifer*); one mid-dorsal keel (traces of additional keels apparent in hatchling); dark stripes on sides of head and neck; carapace gray or brown with dark streaks or spots.

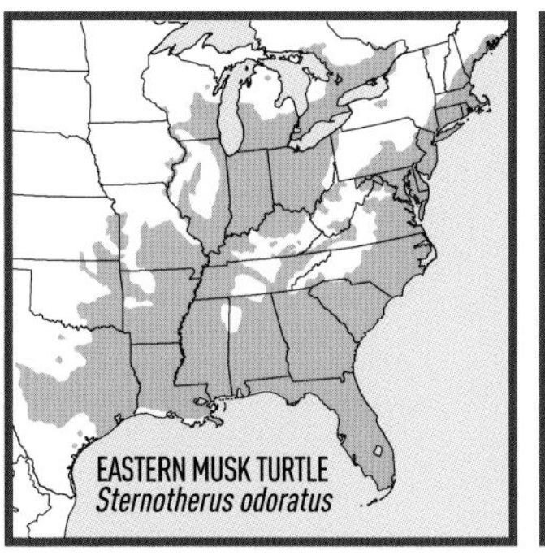

EASTERN MUSK TURTLE *Sternotherus odoratus* **PLS. 19, 21**

2–4½ in. (5.1–11.5 cm); record 5⅞ in. (15 cm). *Two light stripes* (rarely obscured) on dark head; *barbels on chin and throat*. Carapace smooth, light olive brown to almost black, sometimes irregularly streaked or spotted with dark pigment. Plastron small with a single hinge. Male with a thick tail terminating in a blunt, horny nail; tail very small and with or without sharp, horny nail in female. Hatchling ⅞–1 in. (2.2–2.5 cm); head stripes prominent; rough carapace dark gray to black, light spots on marginals, and a prominent middorsal keel; smaller dorsolateral keels vary from traces to prominent. **SIMILAR SPECIES:** (1) Other musk turtles have a head with dark spots, stripes, or streaks on lighter ground color and barbels restricted to chin. (2) Mud turtles (*Kinosternon*) have a large plastron with 2 hinges. **HABITAT:** Still bodies of water, sluggish streams. **RANGE:** New England, s. ON, and WI south to s. FL and TX. **REMARK:** Also called the Common Musk Turtle or Stinkpot in allusion to its malodorous secretions.

TORTOISES: Testudinidae

Tortoises occur on all continents except Australia and Antarctica and include the world's largest land turtles, with giants of the Galápagos and Aldabra Archipelagos exceeding 660 lb. (300 kg). Males are larger than females. All five native North American tortoises, two in our area, are in the genus *Gopherus*. Feet are stumpy, without webs, and hindlimbs are elephant-like. The carapace is high and rounded, the plastron rigid and distinctly indented in males. The gular often is elongated, forked, and curved upward, especially in adult males, which use it to overturn rivals during courtship.

Tortoises are common pets, and many populations are established outside their native ranges. At least four non-native species have been recorded, but none appear to have established breeding populations. However, the Red-footed Tortoise (*Chelonoidis carbonarius*), native to the Neotropics, and the African Spurred Tortoise (*Geochelone sulcata*) from Africa are popular pets and likely to become entrenched.

TEXAS TORTOISE *Gopherus berlandieri* **PL. 19**

5½–8 in. (14–20.3 cm); record 9 in. (22.8 cm). Carapace rounded, tan to dark brown, *nearly as broad as long and less than twice as long as high*; growth rings evident in younger animals. Head *wedge-shaped* when viewed from above. Large carapacial scutes of young with yellow centers, marginals with yellow edges. Hatchling 1½–2 in. (3.8–5.1 cm). **SIMILAR SPECIES:** Box turtles have a hinged plastron. **HABITAT:** Arid areas. **RANGE:** S. TX into Mex.

GOPHER TORTOISE *Gopherus polyphemus* **PL. 19, FIG. 205, P. 459**

6–9½ in. (15.2–24 cm); record 16⅜ in. (41.6 cm). Carapace brown or tan, *longer than wide and at least twice as long as high*; plastron dull yellowish, skin grayish brown. Front of head *rounded* when viewed from above. Shells of old adults virtually smooth, young with conspicuous growth rings, also with considerable orange or yellow on skin, plastron, and marginals; large carapacial scutes yellowish but bordered by brown. Hatchling 1–2 in. (2.5–5.1 cm). **SIMILAR SPECIES:** Box turtles have a hinged plastron. **HABITAT:** Sandy areas; extensive burrows used by many other animals, including the Eastern Diamond-backed Rattlesnake. **RANGE:** S. SC through FL to extreme e. LA. Introduced on Cumberland I., GA. **CONSERVATION:** Vulnerable (USFWS: Threatened).

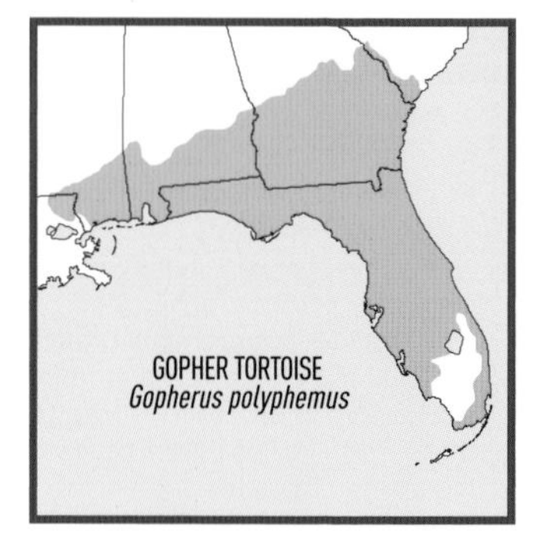

SOFTSHELLS: Trionychidae

Softshells also occur in Africa, Asia, and the East Indies. All North American softshells are in the genus *Apalone*. Soft, leathery shells lack scutes. Vague outlines of underlying bones often show through the plastral skin. Softshells are highly aquatic, frequently buried in mud or sand in shallow water with neck extended and only the eyes and snout breaking the surface. Young often distinctly patterned. Males usually retain the juvenile coloration, but patterns in larger females often are obliterated by mottling and blotches (although all Florida Softshells tend to become drab). Males have a longer, stouter tail than females. Hatchlings 1¼–2 in. (3.2–5.1 cm).

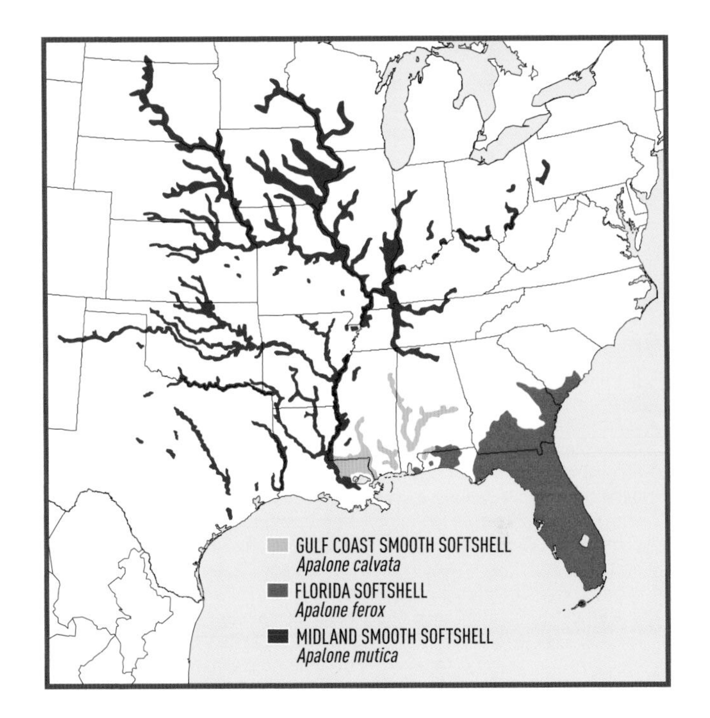

GULF COAST SMOOTH SOFTSHELL *Apalone calvata* **FIG. 85**

MIDLAND SMOOTH SOFTSHELL *Apalone mutica* **PLS. 20, 22, FIG. 85**

Female 6½–14 in. (16.5–35.6 cm); male 4½–10½ in. (11.5–26.6 cm); Gulf Coast Smooth Softshell smaller, female to 11¼ in. (28.7 cm). Shell smooth, *without spines, bumps, or sandpapery projections* along edges. *No ridges in nostrils* (Fig. 85, p. 186). Feet not strongly streaked or spotted. Midland Smooth Softshell has ill-defined pale stripes in front of eyes and pale postocular stripes with *narrow* dark borders. Carapace of male and young olive gray or brown, dots and dashes only slightly darker than ground color. Gulf Coast Smooth Softshell has no stripes on snout, pale postocular stripes with *thick* black borders, and carapace of young has large, circular, dusky spots that disappear in adult female but persist at least indistinctly in adult male. Female carapace of both species is mottled with gray, brown, or olive. Plastron of juvenile is paler than underside of carapace. **SIMILAR SPECIES:** See Spiny Softshell. **HABITAT:** Small creeks and large rivers, rarely in lakes; usually absent where Spiny Softshell is abundant. **RANGE:** Gulf Coast Smooth Softshell: streams draining into the Gulf from w. FL to e. LA. Midland Smooth Softshell: Ohio, Mississippi, and Missouri R. and tributaries, also streams draining into the Gulf in TX. **REMARK:** These species until recently were considered subspecifically related.

FLORIDA SOFTSHELL *Apalone ferox* **PLS. 20, 22, FIGS. 85, 203, P. 451**

Female 11–26½ in. (28–67.3 cm); male 6–12¾ in. (15.2–32.4 cm). Dark brown or dark brownish gray, nearly uniform in coloration or with vague dark spots. *Several rows of small flattened bumps* on front of carapace extend back as far as forelimbs. Ridges in nostrils (Fig. 85, p. 186). Carapace of young with large dark round spots, head dark with bright lines and spots. **SIMILAR SPECIES:** See Spiny Softshell. **HABITAT:** Lakes, ponds, springs, canals, roadside ditches, occasionally quiet portions of rivers. **RANGE:** S. SC through FL to Mobile Bay, AL. Possibly introduced on Big Pine Key, FL.

SPINY SOFTSHELL *Apalone spinifera* **PLS. 20, 22, FIGS. 85, 105, 106**

Female 7–21¼ in. (18–54 cm); male 5–12³⁄₁₆ in. (12.5–31 cm). Carapace sandpapery, especially in male, anterior edges with single row of *spines or conelike projections* (absent in some females from cen. TX). Ridges in nostrils (Fig. 85, p. 186). Dark brown or black spots or ocelli on olive gray to brown ground color in most populations, but white on darker ground color in southwestern subspecies. Spots variable in size. One or more dark (sometimes broken) lines parallel rear margin of carapace. Two light lines on sides of head, one extending back from eyes, the other from jaw (Fig. 106, facing page). Juvenile pattern fades in adult, especially female. **SIMILAR SPECIES:** (1) Florida Softshell has 2 or more rows of flattened bumps along front edges of carapace extending back to forelimbs. (2) Smooth softshells lack bumps or spines along front edges of carapace and have no ridges in nostrils. **HABITAT:** Rivers, especially in western parts of the range, lakes, oxbows, quiet bodies of water with sand and mud bars. **RANGE:** Disjunct; MA and extreme s. QC to MT south to the Gulf; Rio Grande drainage from NM, TX, and n. Mex. Introduced in Norfolk, VA, the Maurice R. system of s. NJ, and the Gila-Colorado R. system from sw. NM and extreme sw. UT to the Gulf of California. **SUBSPECIES:** (1) **Eastern Spiny Softshell** (*A. s. spinifera*); light lines on head usually fail to join; male olive gray to yellowish brown with dark ocelli (sometimes reduced to spots in western parts of the range), entire carapace sandpapery; adult female with brown or olive brown blotches producing a camouflaged effect, carapace smooth; young pale yellowish brown with small dark

FIG. 105. *Western populations of the Spiny Softshell* (Apalone spinifera) *once were thought to be subspecifically distinct because dark spots and ocelli on the carapace are greatly reduced and sometimes completely absent in large adults, especially in females such as this individual.*

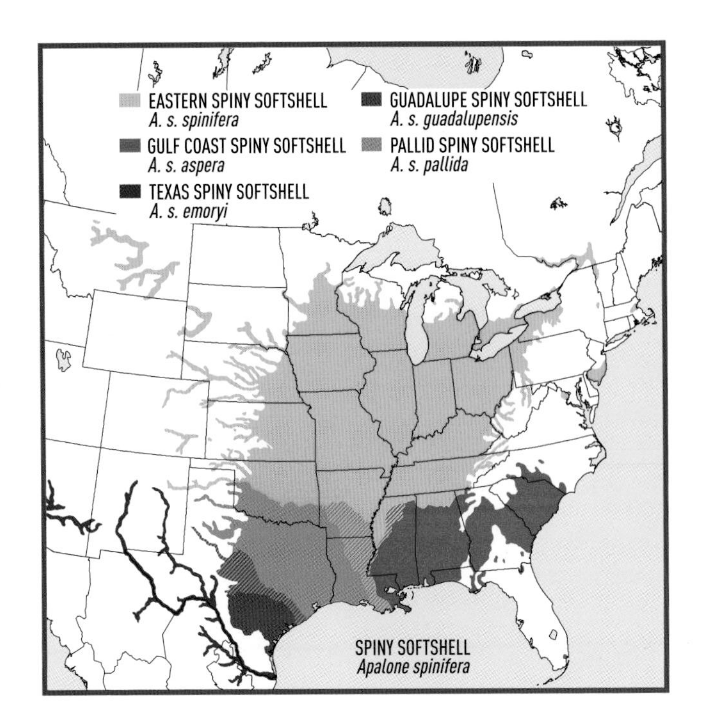

spots or ocelli. (2) **Gulf Coast Spiny Softshell** (*A. s. aspera*); 2 or more dark lines (sometimes broken) parallel rear margin of shell (single dark line in other subspecies); light lines on head usually unite. (3) **Texas Spiny Softshell** (*A. s. emoryi*); small white spots confined to rear third of brown or olive carapace, often lacking in large females; pale rim of carapace conspicuously widened along rear edge, 4 or 5 times wider than pale rim of lateral edges. (4) **Guadalupe Spiny Softshell** (*A. s. guadalupensis*); small white spots cover most of brown or olive carapace, each spot often surrounded by narrow black rings. (5) **Pallid Spiny Softshell** (*A. s. pallida*); white spots on brown or olive carapace largely confined to posterior half of carapace and not ringed with black. **REMARK:** Almost certainly a complex of species that might not correspond to currently recognized subspecies.

FIG. 106. *Head patterns of two subspecies of the Spiny Softshell* (Apalone spinifera).

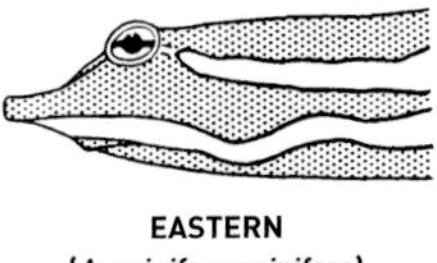

EASTERN
(*A. spinifera spinifera*)
Head lines separate

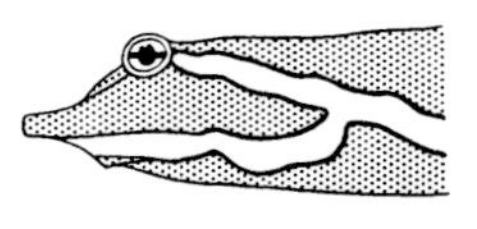

GULF COAST
(*A. spinifera aspera*)
Head lines meet

LIZARDS

PLATE 23

GLASS and ALLIGATOR LIZARDS (*Ophisaurus* and *Gerrhonotus*)

Lateral skinfolds lined with granular scales. See text and Fig. 107.

SLENDER GLASS LIZARD *O. attenuatus* **P. 255**
No legs; middorsal dark stripe; dark stripes on lower sides.

EASTERN GLASS LIZARD *O. ventralis* **P. 256**
No legs; greenish; no distinct middorsal stripe. Note regenerated tail.

TEXAS ALLIGATOR LIZARD *G. infernalis* **P. 254**
Large scales; irregular light crosslines.

FIG. 107. *Heads of glass lizards* (Ophisaurus).

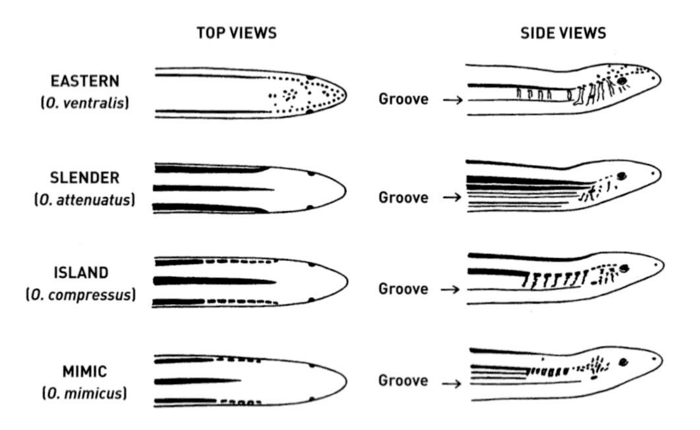

EYELID GECKOS (*Coleonyx*)

TEXAS BANDED GECKO *C. brevis* **P. 257**
Light crossbands (or mottling) on brown ground color; movable eyelids.

TRUE GECKOS (Gekkonidae)

MEDITERRANEAN GECKO *Hemidactylus turcicus* Non-native **P. 264**
Toepads broad (Fig. 108, p. 238); dorsum warty.

INDO-PACIFIC HOUSE GECKO *Hemidactylus garnotii* Non-native **P. 261**
Toepads broad (like Mediterranean Gecko; Fig. 108, p. 238); dorsum smooth; belly lemon yellow; frequently reddish under tail.

ROUGH-TAILED GECKO *Cyrtopodion scabrum* Non-native **P. 259**
Rough; enlarged, keeled scales on tail; sandy colored with dark brown spots.

TOKAY GECKO *Gekko gecko* Non-native **P. 260**
Pale bluish gray with numerous brick red to orange spots; prominent wart-like bumps on head, body, legs, and tail; enlarged toepads.

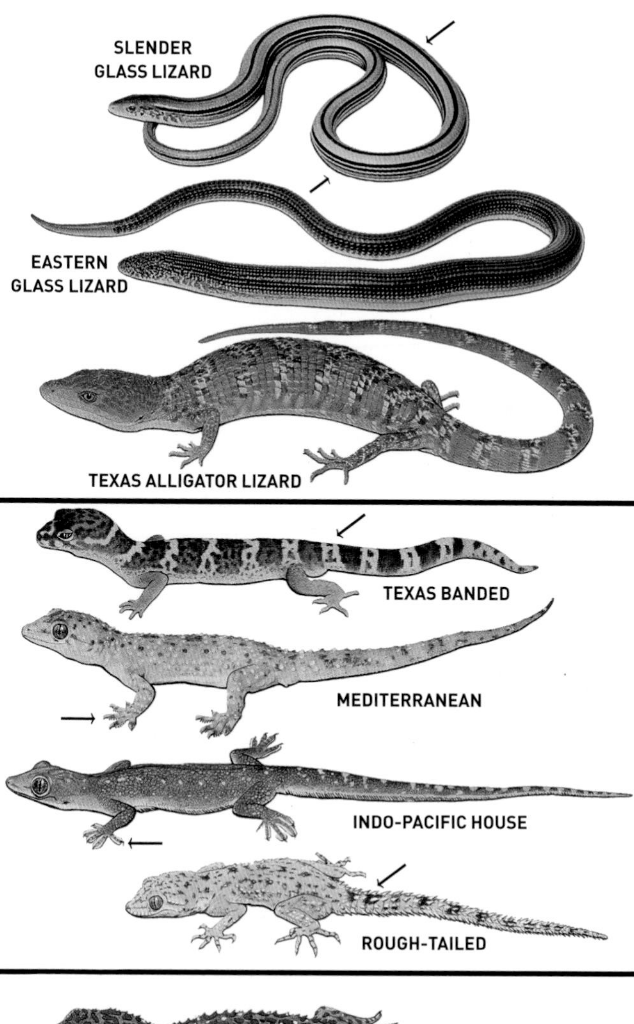
SLENDER
GLASS LIZARD
EASTERN
GLASS LIZARD
TEXAS ALLIGATOR LIZARD
TEXAS BANDED
MEDITERRANEAN
INDO-PACIFIC HOUSE
ROUGH-TAILED

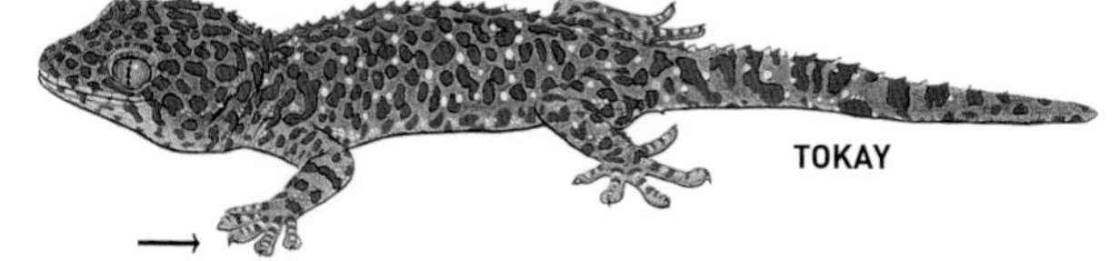
TOKAY

PLATE 24

SPHAERODACTYL GECKOS
(*Sphaerodactylus* and *Gonatodes*)

ASHY GECKO *S. elegans* **Non-native** **P. 269**
Adult with tiny light spots; young with dark crossbands, reddish tail.

FLORIDA REEF GECKO *S. notatus notatus* **P. 270**
Dark markings on lighter ground color. ♂ with numerous small dark spots; ♀ with dark head stripes, often two light spots on shoulders.

OCELLATED GECKO *S. argus* **Non-native** **P. 269**
Numerous white spots; tail light brown or reddish.

YELLOW-HEADED GECKO *G. albogularis* **Non-native** **P. 270**
Adult ♂ with yellowish head, bluish or black body; ♀ and young with light collar, body mottled.

FIG. 108. *Toes of geckos.*

SIDE VIEWS

BOTTOM VIEWS

TEXAS BANDED
(*Coleonyx brevis*)

OCELLATED, ASHY, and REEF
(*Sphaerodactylus* spp.)

YELLOW-HEADED
(*Gonatodes albogularis*)

MEDITERRANEAN
(*Hemidactylus turcicus*)

CASQUE-HEADED LIZARDS (*Basiliscus*)

BROWN BASILISK *B. vittatus* **Non-native** **P. 276**
♂ with triangular head crest; ♀ with hoodlike lobe on neck.

LEOPARD and COLLARED LIZARDS
(*Gambelia* and *Crotaphytus*)

LONG-NOSED LEOPARD LIZARD *G. wislizenii* **P. 278**
Many brown spots; whitish crosslines often prominent on young.

RETICULATE COLLARED LIZARD *C. reticulatus* **P. 278**
Conspicuous black dorsal spots; head large.

EASTERN COLLARED LIZARD *C. collaris* **P. 276**
Two black collars across neck; head large.

ASHY
adult
young
♂
♀
FLORIDA
REEF
OCELLATED
YELLOW-HEADED
♀
♂

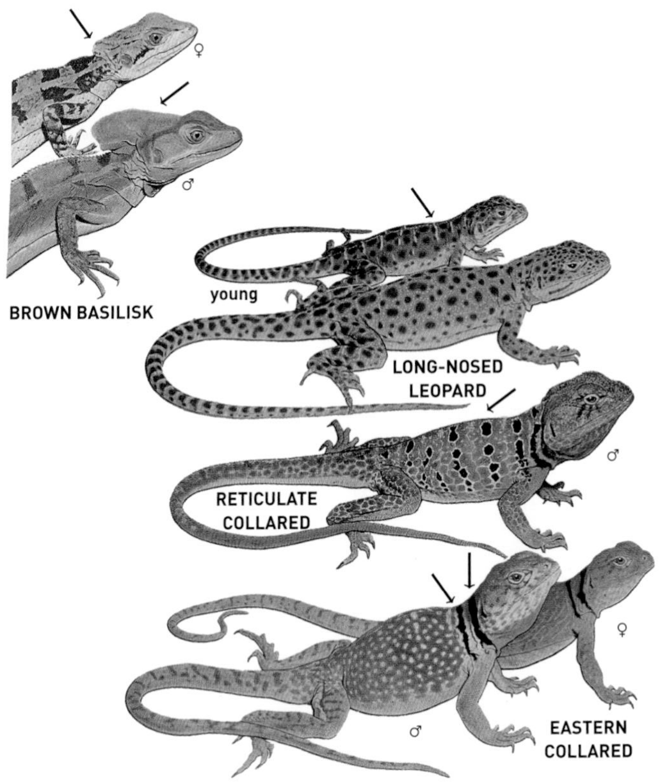
♀
♂
BROWN BASILISK
young
LONG-NOSED
LEOPARD
♂
RETICULATE
COLLARED
♀
♂
EASTERN
COLLARED

PLATE 25

ANOLES (*Anolis*)

BROWN ANOLE *A. sagrei* Non-native **P. 282**
Always brown or brownish gray. ♂ sometimes with light lateral lines or vertical rows of yellowish spots on sides. ♀ with a light middorsal stripe or row of diamond-shaped spots.

BARK ANOLE *A. distichus* Non-native **P. 281**
Crossbanded tail; dark line between eyes.

GREEN ANOLE *A. carolinensis* **P. 279**
Green or brown or mottled with both.

CRESTED ANOLE *A. cristatellus* Non-native **P. 280**
Brown to greenish gray. ♂ with crests on body and tail; sometimes with dark body bands. ♀ (not shown) with a dark-bordered light middorsal stripe.

HISPANIOLAN STOUT ANOLE *A. cybotes* Non-native **P. 280**
Pale to dark brown. ♂ with large head, often with light lateral stripes. ♀ (not shown) usually orange-brown with a light middorsal stripe.

JAMAICAN GIANT ANOLE *A. garmani* Non-native **P. 282**
Leaf green to brown. ♂ large; prominent crest on back of head and neck; nine or more body bands (sometimes faint). ♀ (not shown) small, with rows of spots instead of bands.

KNIGHT ANOLE *A. equestris* Non-native **P. 281**
Large; bony head casque; yellowish lines on shoulders and from beneath eyes to ears.

DEWLAPS OF MALE ANOLES

BROWN ANOLE *A. sagrei* **P. 282**
Orange-red with a light border (Fig. 109).

BARK ANOLE *A. distichus* **P. 281**
Yellow, sometimes with a pale orange blush.

GREEN ANOLE *A. carolinensis* **P. 279**
Usually pink.

CRESTED ANOLE *A. cristatellus* **P. 280**
Olive, mustard yellow, or orange-yellow, rarely with an orange border.

HISPANIOLAN STOUT ANOLE *A. cybotes* **P. 280**
Pale or grayish yellow, sometimes with a pale orange center.

JAMAICAN GIANT ANOLE *A. garmani* **P. 282**
Lemon yellow with an orange center.

KNIGHT ANOLE *A. equestris* **P. 281**
Pale pink.

BROWN
(*sagrei*)
Longitudinal light
streak on throat

FIG. 109. *The throat of a male Brown Anole* (Anolis sagrei) *has a longitudinal light streak formed by the lower edge of the dewlap.*

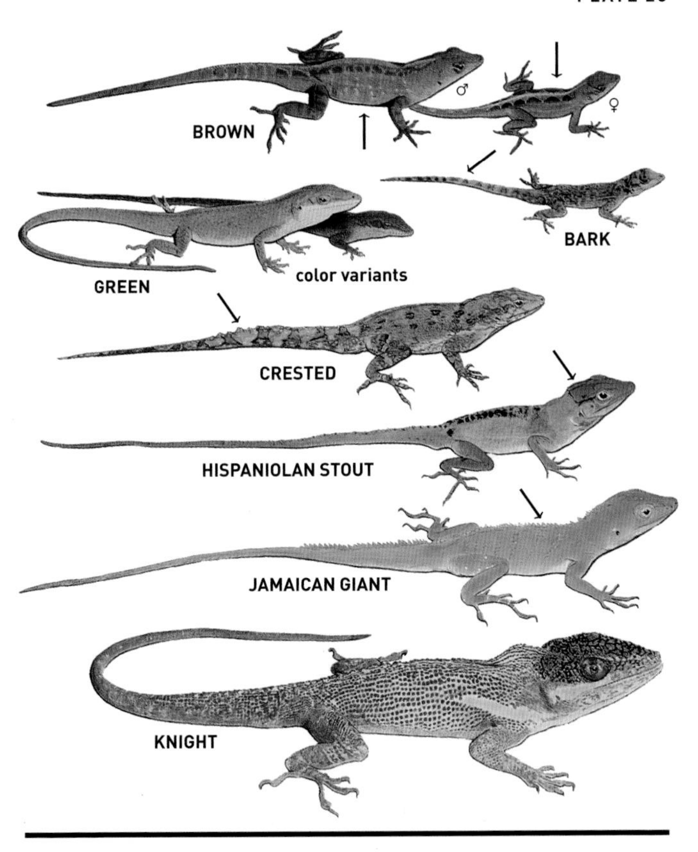
BROWN
♂
♀
BARK
GREEN
color variants
CRESTED
HISPANIOLAN STOUT
JAMAICAN GIANT
KNIGHT

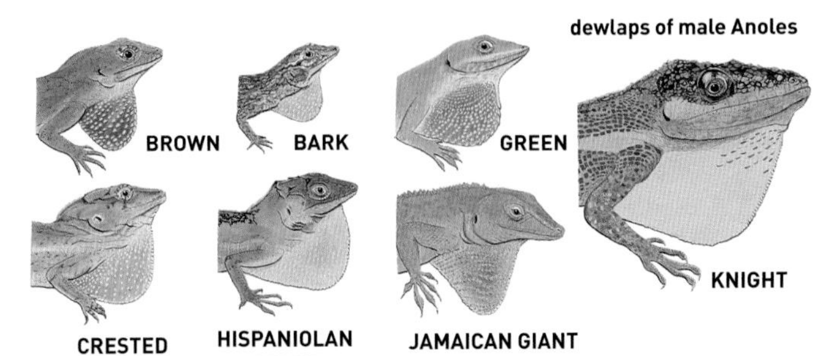
dewlaps of male Anoles
BROWN
BARK
GREEN
KNIGHT
CRESTED
HISPANIOLAN STOUT
JAMAICAN GIANT

IGUANAS (*Ctenosaura* and *Iguana*)

GREEN IGUANA *I. iguana* complex Non-native **P. 284**

Large scales below ears; middorsal spines strongly developed in ♂.

MEXICAN SPINY-TAILED IGUANA *C. pectinata* Non-native **P. 284**

No enlarged scales below ears; middorsal crest separated from tail crest by 5–10 or more scale rows; keeled spines ring tail.

CURLY-TAILED LIZARDS (*Leiocephalus*)

NORTHERN CURLY-TAILED LIZARD *L. carinatus* Non-native **P. 286**

Middorsal crest; dark bands on tail; tail curls over back.

RED-SIDED CURLY-TAILED LIZARD *L. schreibersii* Non-native **P. 286**

Low middorsal crest; ♂ flanks with reddish orange streaks.

EARLESS LIZARDS (*Cophosaurus* and *Holbrookia*)

GREATER EARLESS LIZARD *C. texanus* **P. 287**

TEXAS GREATER EARLESS LIZARD (*C. t. texanus*): Black crossbars under tail (Fig. 110); ♂ with two black lines on flanks near groin and invading blue field on belly.

CHIHUAHUAN GREATER EARLESS LIZARD (*C. t. scitulus*): Black ventrolateral bars broad; body color differs from that of hindquarters and tail.

SPOT-TAILED EARLESS LIZARD *H. lacerata* **P. 288**

Dark dorsal blotches with light borders; dark streaks at edges of belly; dark spots under tail (Fig. 110).

KEELED EARLESS LIZARD *H. propinqua* **P. 289**

Tail long; ♂ with two short black lines near armpits; ♀ markings absent or indistinct.

COMMON LESSER EARLESS LIZARD *H. maculata* **P. 288**

GREAT PLAINS EARLESS LIZARD (*H. m. maculata*): Tail short; ♂ with two short black bars near armpits; faint longitudinal stripes; pale speckling indistinct.

PRAIRIE EARLESS LIZARD (*H. m. perspicua*): ♀ with distinct dorsal blotches or spots; suggestion of black bars near armpits; faint longitudinal stripes.

CHIHUAHUAN LESSER EARLESS LIZARD (*H. m. flavilenta*): Tail short; light stripes largely restricted to neck; dorsum with pale speckling, especially in ♂.

FIG. 110. *Undersurface of the tails of earless lizards* (Cophosaurus *and* Holbrookia).

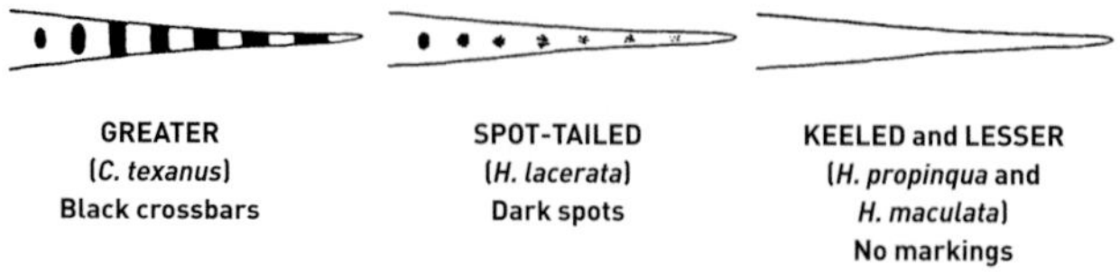

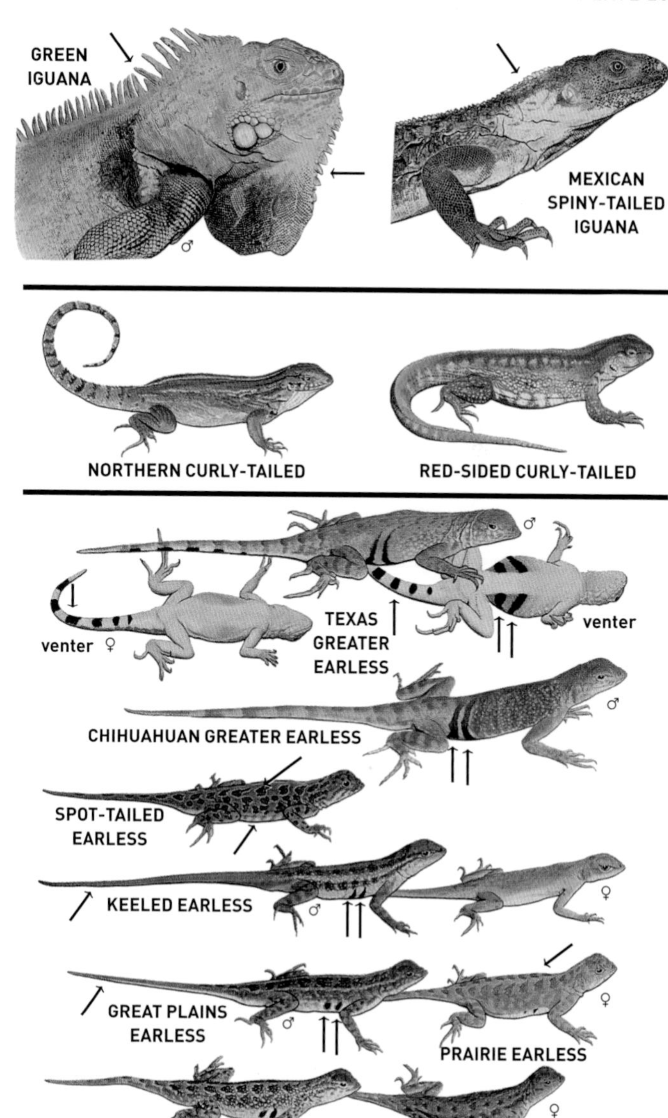
GREEN
IGUANA
♂
MEXICAN
SPINY-TAILED
IGUANA
NORTHERN CURLY-TAILED
RED-SIDED CURLY-TAILED
♂
venter
venter ♀
TEXAS
GREATER
EARLESS
♂
CHIHUAHUAN GREATER EARLESS
SPOT-TAILED
EARLESS
KEELED EARLESS
♂
♀
GREAT PLAINS
EARLESS
♂
♀
PRAIRIE EARLESS
♂
♀
CHIHUAHUAN LESSER
EARLESS

PLATE 27

HORNED LIZARDS (*Phrynosoma*)

GREATER SHORT-HORNED LIZARD *P. hernandesi* **P. 290**
Horns little more than nubs (Fig. 111).

TEXAS HORNED LIZARD *P. cornutum* **P. 290**
Two central horns greatly elongated (Fig. 111).

ROUND-TAILED HORNED LIZARD *P. modestum* **P. 290**
Crossbanded tail round in cross section; large horns about equal in size (Fig. 111).

FIG. 111. *Heads of horned lizards* (Phrynosoma*)*.

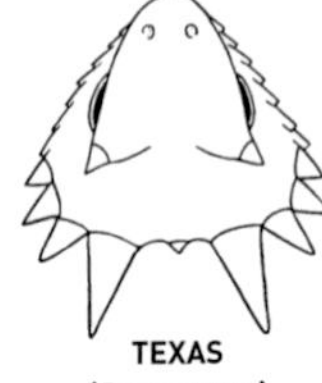

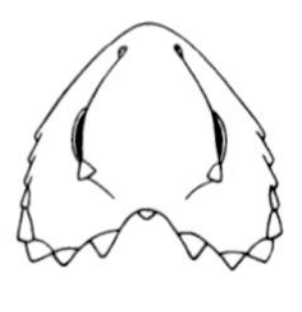

SPINY LIZARDS (*Sceloporus*)

Males of most species have blue patches on sides of belly. See also Pl. 28.

DUNES SAGEBRUSH LIZARD *S. arenicolus* **P. 292**
Pale, nearly unicolored; small granular scales on rear of thighs.

PRAIRIE LIZARD *S. consobrinus* **P. 293**
Bold light dorsolateral and lateral stripes. Populations from eastern portions of range like the Eastern Fence Lizard.

COMMON SAGEBRUSH LIZARD *S. graciosus* **P. 295**
Four rows of dark longitudinal spots may coalesce to form stripes; small granular scales on rear of thighs.

GRAPHIC SPINY LIZARD *S. grammicus* **P. 295**
Wavy dark crosslines in ♀; lateral neck scales much smaller than scales on nape (Fig. 140, p. 296).

CANYON LIZARD *S. merriami* **P. 296**
Vertical black bars in front of forelimbs; throat fold partially developed (Fig. 141, p. 297).

EASTERN FENCE LIZARD *S. undulatus* **P. 298**
♀ with wavy dark crosslines; ♂ nearly unicolored above, with a dark blue throat patch bordered by black.

ROSE-BELLIED LIZARD *S. variabilis* **P. 298**
Row of dark spots bordered below by light dorsolateral stripes; skin pockets behind thighs (Fig. 142, p. 298). ♂ with dark spots above armpits; large pink patches on sides of belly.

FLORIDA SCRUB LIZARD *S. woodi* **P. 299**
Prominent dark brown lateral stripes; ♀ with dark wavy lines across back.

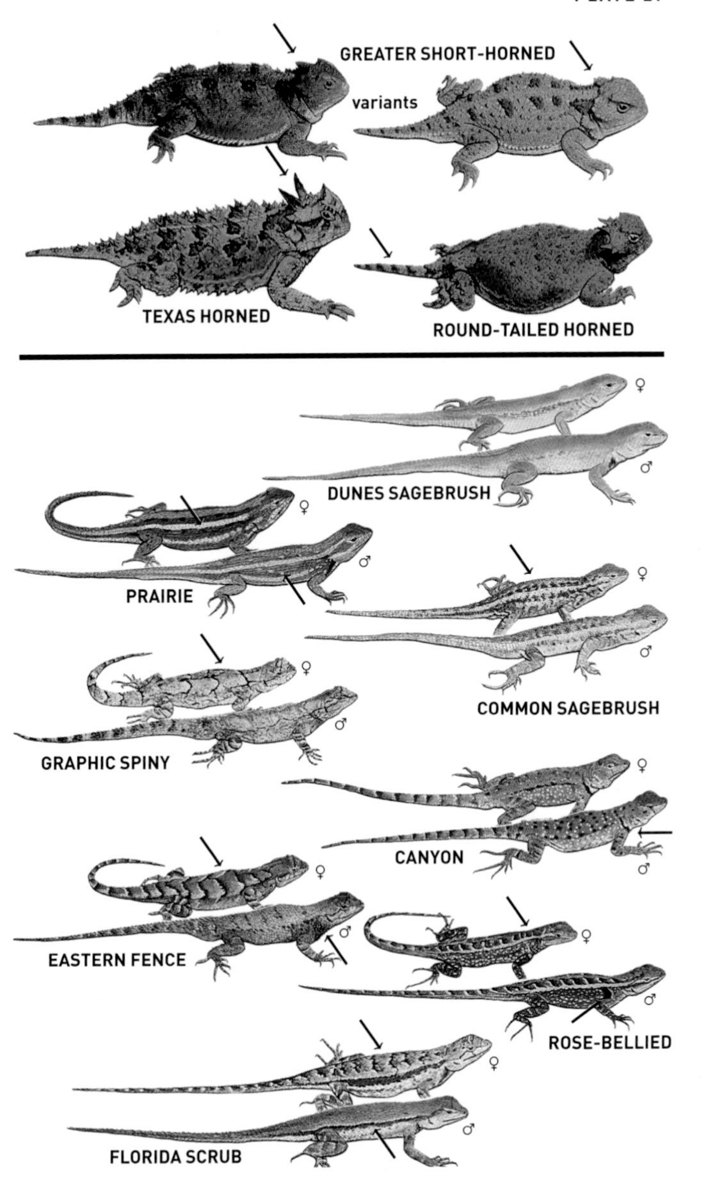
GREATER SHORT-HORNED
variants
TEXAS HORNED
ROUND-TAILED HORNED
♀
DUNES SAGEBRUSH
♂
♀
♂
PRAIRIE
♀
♂
COMMON SAGEBRUSH
♀
♂
GRAPHIC SPINY
♀
CANYON
♂
♀
♂
EASTERN FENCE
♀
♂
ROSE-BELLIED
♀
♂
FLORIDA SCRUB

SPINY LIZARDS (*Sceloporus*) (cont.)

Males have blue or blue-green patches on sides of belly. See also Pl. 27.

TEXAS SPINY LIZARD *S. olivaceus* **P. 297**
Vague light dorsolateral stripes.

TWIN-SPOTTED SPINY LIZARD *S. bimaculosus* **P. 292**
Black wedges or blotches on shoulders; twin spots on back not sharply defined.

CREVICE SPINY LIZARD *S. poinsettii* **P. 297**
White-bordered dark collar; tail strongly patterned near tip. ♀ and young with dark bands across back.

BLUE SPINY LIZARD *S. cyanogenys* **P. 295**
White-bordered dark collar; scattered light dorsal spots; tail markings not distinct near tip.

TREE LIZARDS (*Urosaurus*)

ORNATE TREE LIZARD *U. ornatus* **P. 299**
Irregular dark spots; dorsal scales variable, some large, some tiny; skinfold across throat (Fig. 112).

SIDE-BLOTCHED LIZARDS (*Uta*)

EASTERN SIDE-BLOTCHED LIZARD *U. stejnegeri* **P. 300**
Black spot behind armpits.

FIG. 112. *The single well-developed throat fold in tree lizards* (Urosaurus) *and side-blotched lizards* (Uta).

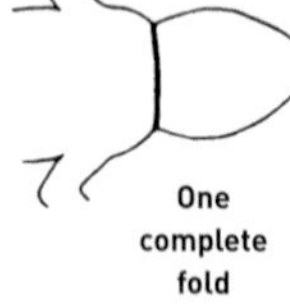

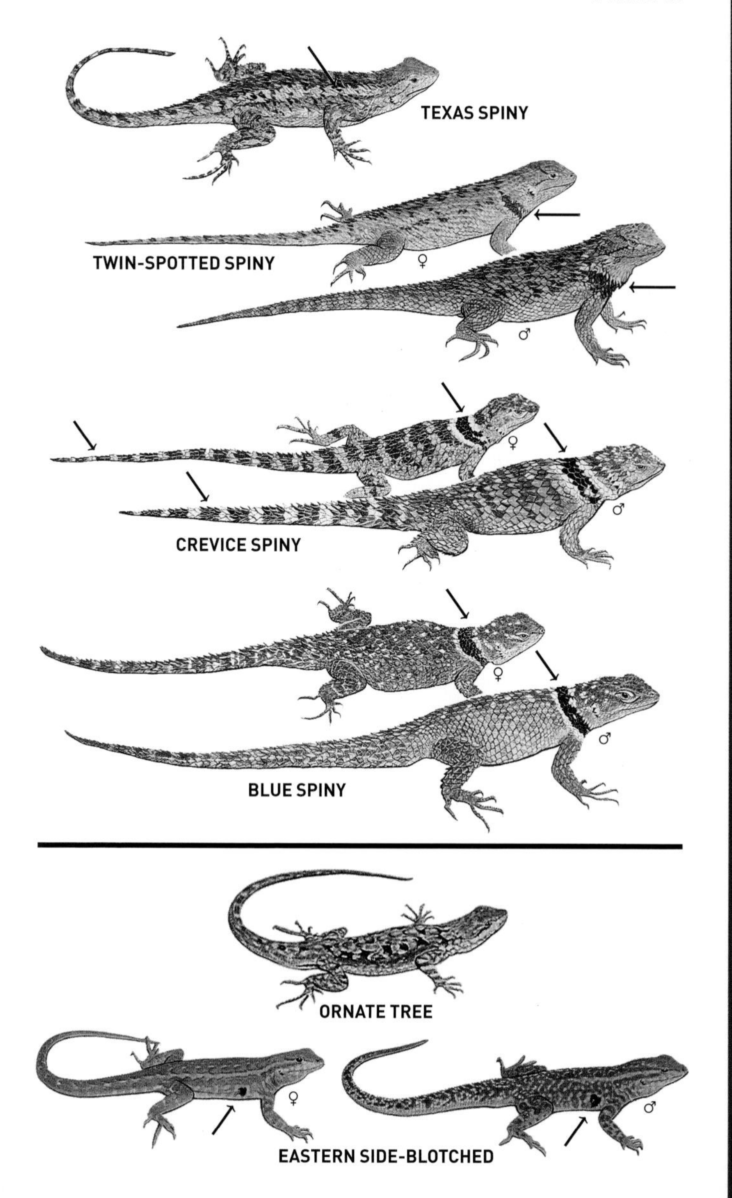
TEXAS SPINY
TWIN-SPOTTED SPINY
♀
♂
CREVICE SPINY
♀
♂
BLUE SPINY
♀
♂
ORNATE TREE
EASTERN SIDE-BLOTCHED
♀
♂

PLATE 29

WALL LIZARDS (*Podarcis*)

COMMON WALL LIZARD *P. muralis* **Non-native** **P. 301**
Often with dark markings on throat; blue spots on shoulders; six rows of belly scales.

ITALIAN WALL LIZARD *P. siculus* **Non-native** **P. 302**
Throat usually uniformly light; six rows of belly scales; dorsum often green.

TOOTHY SKINKS (*Plestiodon*)

See also Pl. 30.

GREAT PLAINS SKINK *P. obsoletus* **P. 310**
Dark-edged scales. Young black with bright head spots and a blue tail.

COAL SKINK *P. anthracinus* **P. 305**
Light stripes extend onto tail; broad dark lateral stripes 2½–4 scales wide.

MOLE SKINK *P. egregius* **P. 306**
Slender body; short legs.

PENINSULA MOLE SKINK (*P. e. onocrepis*): Tail variously colored.

FLORIDA KEYS MOLE SKINK (*P. e. egregius*): Tail red or brownish red.

BLUE-TAILED MOLE SKINK (*P. e. lividus*): Tail blue in young, may fade to salmon in some adults.

MANY-LINED SKINK *P. multivirgatus* **P. 310**
Tail swollen at base.

NORTHERN MANY-LINED SKINK (*P. m. multivirgatus*): Middorsal light stripe flanked by bold dark stripes.

VARIABLE SKINK (*P. m. epipleurotus*): Unicolored or striped.

SOUTHEASTERN FIVE-LINED SKINK *P. inexpectatus* **P. 308**
♀ with five light stripes, middle one narrow. ♂ and young (not shown) similar to Common Five-lined Skink.

COMMON FIVE-LINED SKINK *P. fasciatus* **P. 307**
♀ with five broad light stripes. ♂ with traces of stripes; reddish on head. Young with blue tail.

BROAD-HEADED SKINK *P. laticeps* **P. 309**
♂ olive-brown; head reddish; grows very large. ♀ and young (not shown) like Common Five-lined Skink.

FIG. 113. *Scales on the chins of some skinks in the genus* Plestiodon.

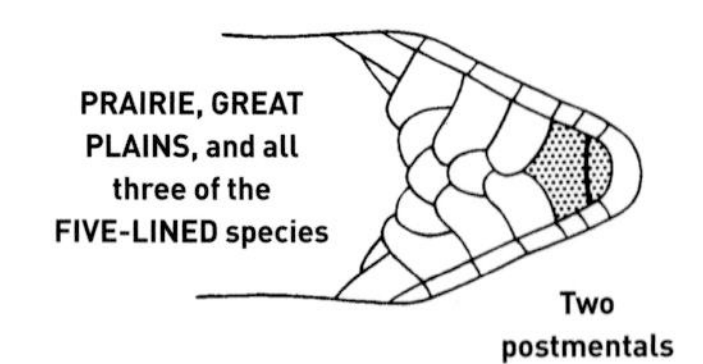

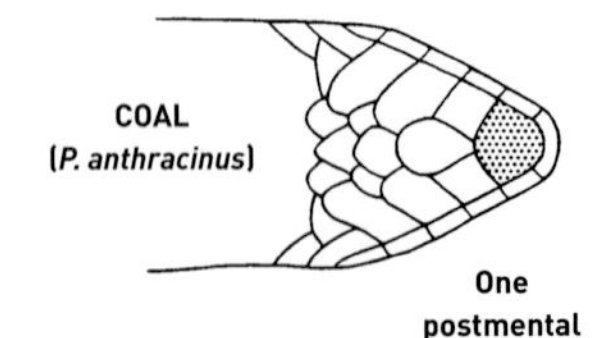

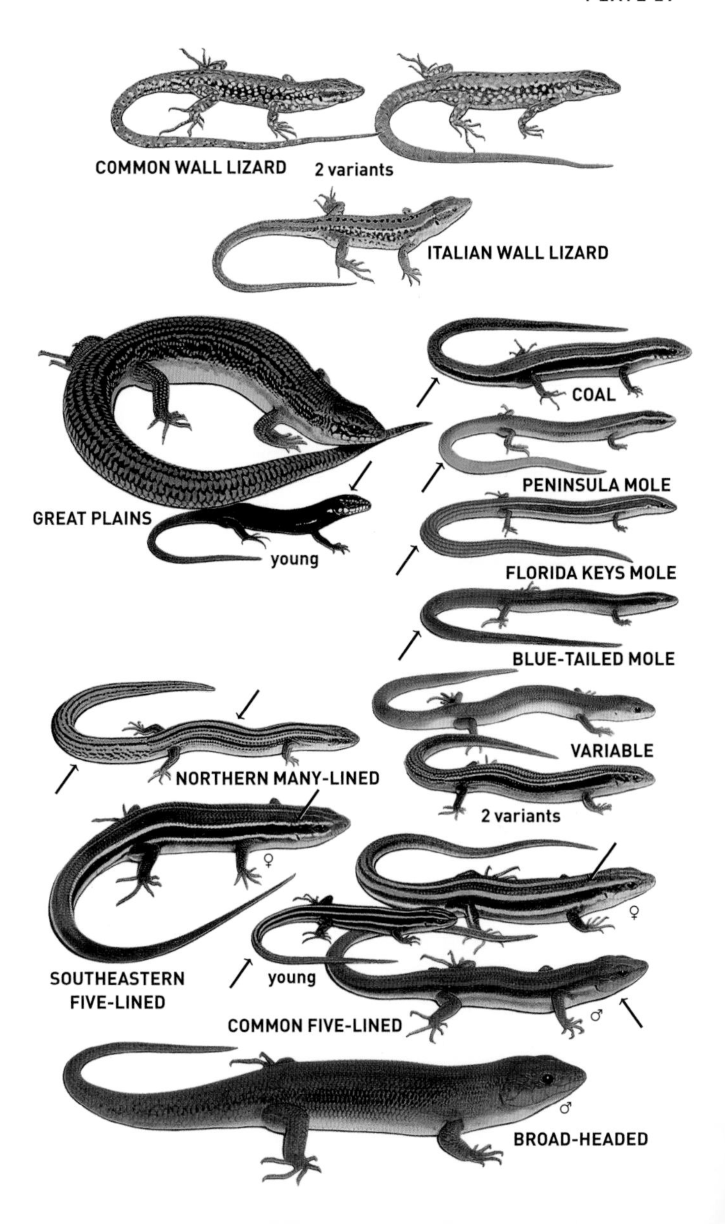
COMMON WALL LIZARD
2 variants
ITALIAN WALL LIZARD
GREAT PLAINS
young
COAL
PENINSULA MOLE
FLORIDA KEYS MOLE
BLUE-TAILED MOLE
VARIABLE
2 variants
NORTHERN MANY-LINED
♀
SOUTHEASTERN
FIVE-LINED
young
♀
♂
COMMON FIVE-LINED
♂
BROAD-HEADED

TOOTHY SKINKS (*Plestiodon*) (cont.)

See also Pl. 29.

FOUR-LINED SKINK *P. tetragrammus* **P. 312**
Young with blue tail.

SHORT-LINED SKINK (*P. t. brevilineatus*): Light stripes end at shoulders.

LONG-LINED SKINK (*P. t. tetragrammus*): Light stripes end near groin.

SOUTHERN PRAIRIE SKINK *P. obtusirostris* **P. 311**
Middorsal area plain or weakly patterned; broad dark lateral stripes not more than two scales wide.

NORTHERN PRAIRIE SKINK *P. septentrionalis* **P. 312**
Stripes in middorsal area; broadest dark stripes not more than two scales wide.

FLORIDA SAND SKINK *P. reynoldsi* **P. 311**
Tiny legs; only one or two toes.

GROUND SKINKS (*Scincella*)

LITTLE BROWN SKINK *S. lateralis* **P. 313**
Dark dorsolateral stripes; no light stripes.

WIDE-SNOUTED WORMLIZARDS (*Rhineura*)

FLORIDA WORMLIZARD *R. floridana* **P. 323**
No legs; rings of scales encircling body; eyes covered by scales; no external ear openings.

AMEIVAS and WHIPTAILS (*Ameiva* and *Cnemidophorus*)

BORRIGUERRO AMEIVA *A. praesignis* Non-native **P. 314**
Charcoal to bluish gray (never green); 10–12 rows of large rectangular ventral plates.

RAINBOW WHIPTAIL *C. lemniscatus* complex Non-native **P. 320**
Dark middorsal stripe; two narrow indistinct dorsolateral light stripes. Sides of head and tail blue, latter becoming greenish; flanks golden yellow to yellowish brown with light spots.

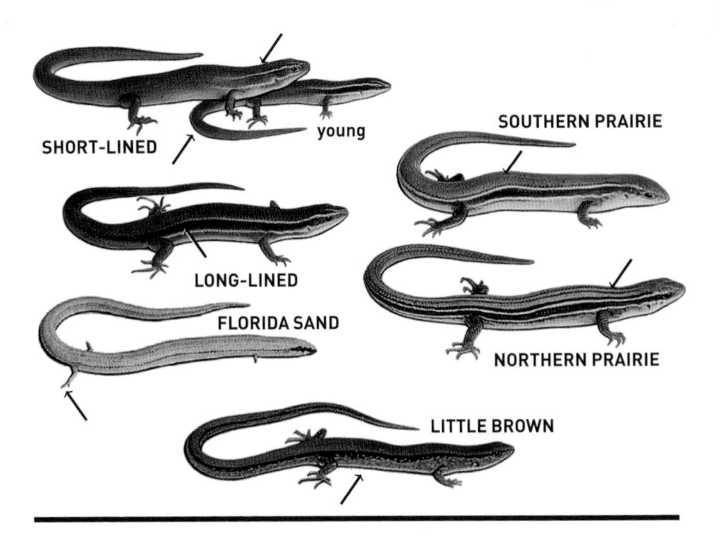
SHORT-LINED
young
SOUTHERN PRAIRIE
LONG-LINED
FLORIDA SAND
NORTHERN PRAIRIE
LITTLE BROWN

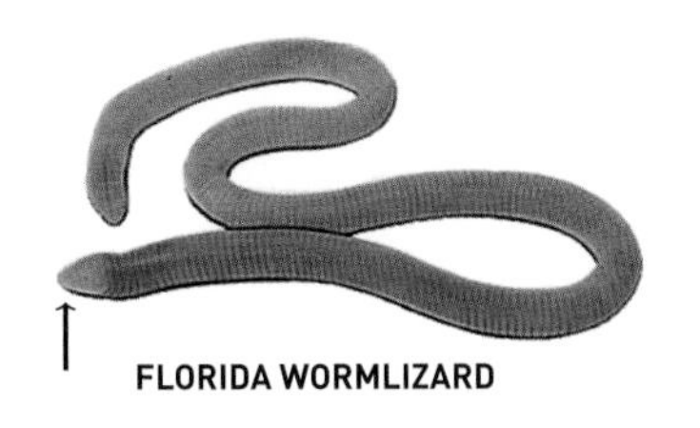
FLORIDA WORMLIZARD

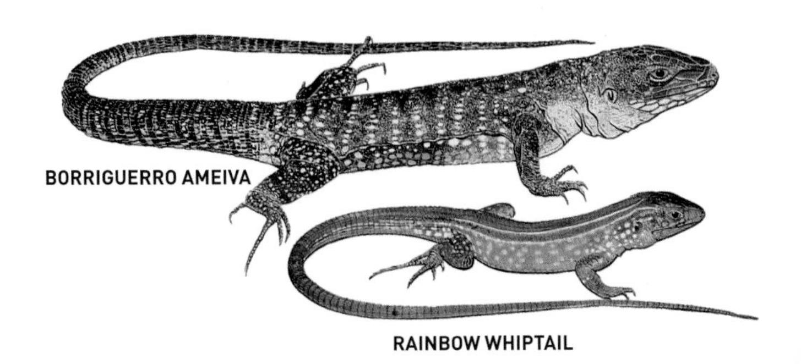
BORRIGUERRO AMEIVA
RAINBOW WHIPTAIL

PLATE 31

RACERUNNERS and WHIPTAILS (*Aspidoscelis*)

SIX-LINED RACERUNNER *A. sexlineata* **P. 319**

EASTERN SIX-LINED RACERUNNER (*A. s. sexlineata*): Six light stripes; dark fields solid.

PRAIRIE RACERUNNER (*A. s. viridis*): Bright green; seven light stripes.

PLATEAU SPOTTED WHIPTAIL *A. scalaris* **P. 318**

Rump and base of tail rust colored.

COMMON SPOTTED WHIPTAIL *A. gularis* **P. 315**

Prominent light spots in dark lateral fields; tail pink, pale orange-brown, or reddish.

DESERT GRASSLAND WHIPTAIL *A. uniparens* **P. 320**

Six light stripes (sometimes with suggestion of a middorsal stripe on neck); no light spots in dark fields.

NEW MEXICO WHIPTAIL *A. neomexicana* **P. 318**

Seven light stripes, middorsal one wavy; obscure light spots in dark fields.

LITTLE STRIPED WHIPTAIL *A. inornata* **P. 315**

Tail blue; ♂ with blue on belly and sides of head.

CHIHUAHUAN SPOTTED WHIPTAIL *A. exsanguis* **P. 314**

Six light stripes; pale spots on both dark fields and light stripes. Young with light stripes in strong contrast to dark fields.

MARBLED WHIPTAIL *A. marmorata* **P. 316**

Brownish to brownish gray but highly variable.

COMMON CHECKERED WHIPTAIL *A. tesselata* **P. 320**

Yellowish to cream, with black spots or squares.

FIG. 114. *Some characteristics of whiptails and racerunners* (Aspidoscelis *and* Cnemidophorus).

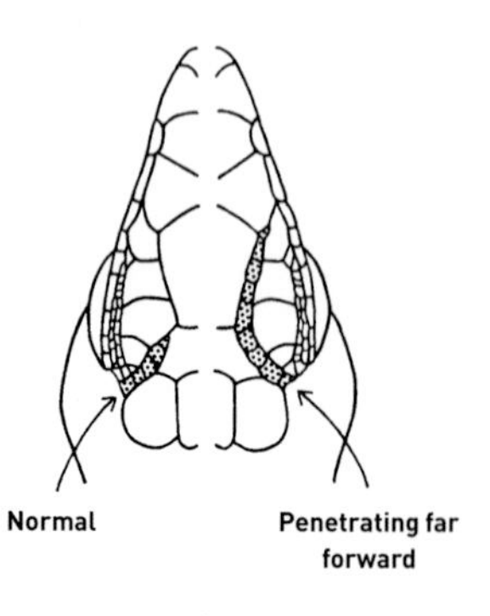

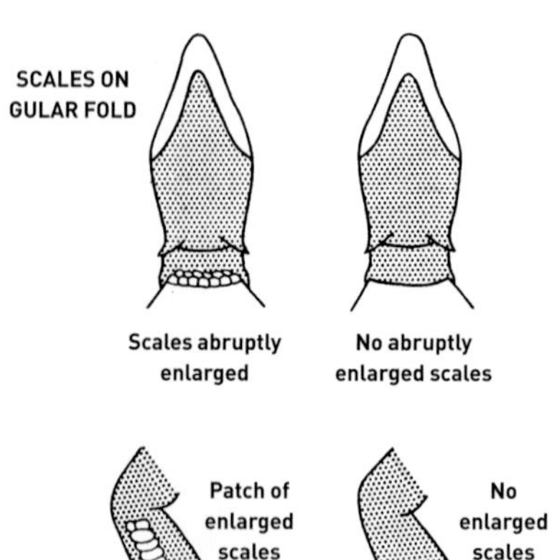

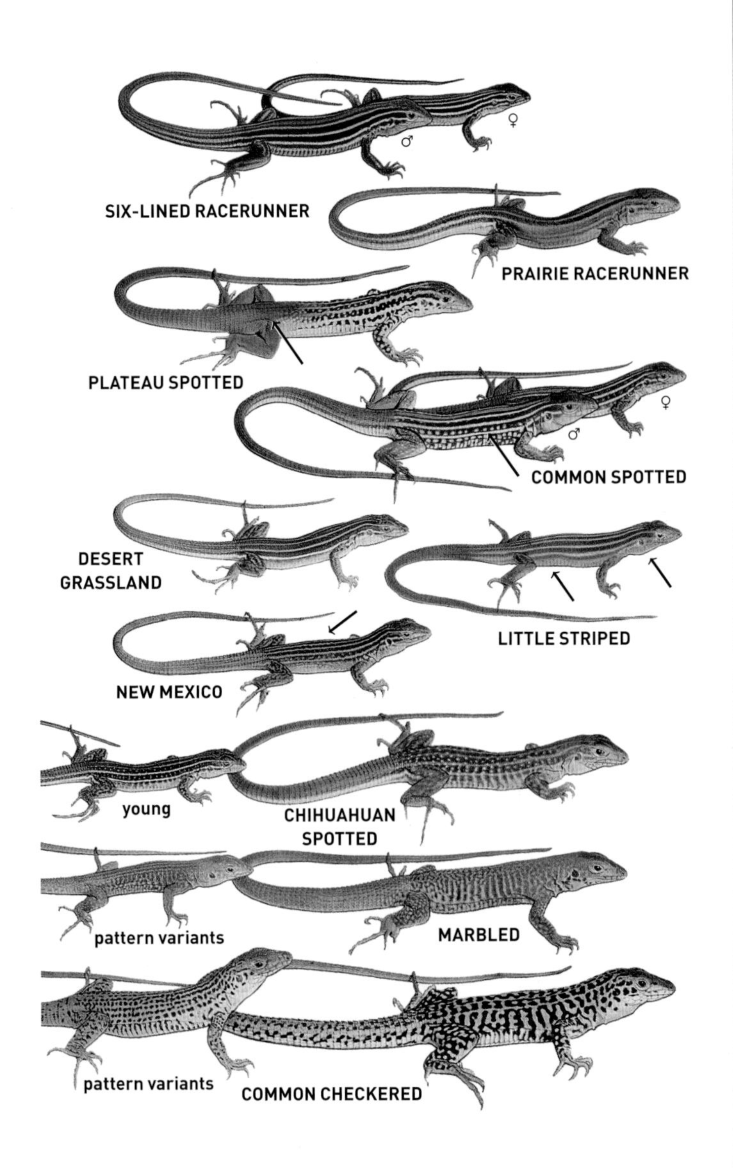
♀
♂
SIX-LINED RACERUNNER
PRAIRIE RACERUNNER
PLATEAU SPOTTED
♀
♂
COMMON SPOTTED
DESERT
GRASSLAND
LITTLE STRIPED
NEW MEXICO
young
CHIHUAHUAN
SPOTTED
pattern variants
MARBLED
pattern variants
COMMON CHECKERED

LIZARDS AND SNAKES (SQUAMATA)

Squamates comprise the vast majority of living reptiles. Lizards, snakes, and wormlizards (or amphisbaenians) once were considered formal taxonomic entities: Lacertilia or Sauria for lizards, Amphisbaenia for wormlizards, and Serpentes or Ophidia for snakes. However, phylogenetic studies clearly show that snakes and wormlizards are specialized legless lizards. We retain the distinction between lizards and snakes solely for convenience. In our area (but not necessarily elsewhere), what we call "lizards" always have at least one of the following: movable eyelids, limbs, external ears, or rings of scales that circle the body; and "snakes" (p. 324) lack limbs, movable eyelids, and external ears, and scales around the body never form distinct rings.

LIZARDS

Lizards range from above the Arctic Circle (in the Eastern Hemisphere) to the southern tips of Africa, Australia, and South America. More than 6,000 species have been described; only two (neither in our area) are venomous. The tails of many lizards break easily. This serves as an anti-predator strategy (predator occupied with writhing tail while the lizard escapes). Breakage usually occurs at preformed fracture planes that are surrounded by muscles that close the resulting wound.

DIPLOGLOSSAN LIZARDS

Only one of four families is represented in our area.

ALLIGATOR and GLASS LIZARDS: Anguidae

Alligator lizards range from British Columbia to Panama; legless glass lizards occur in eastern North America, Mexico, Europe, Africa, and Asia. Unlike snakes, glass lizards have movable eyelids and external ear openings. The tail is very long (as much as 2.75 times the head-body length) and often fragile. Hatchlings average 5–8 inches (12.7–20.3 cm).

Scales are reinforced with bony plates called osteoderms. Species from our area compensate for the resultant stiff body by having deep, flexible, lateral folds of skin. Folds lined with granular scales permit expansion of the body when distended with food or eggs.

TEXAS ALLIGATOR LIZARD *Gerrhonotus infernalis* **PL. 23**

10–16 in. (25.4–41 cm); record 20 in. (50.8 cm); head-body max. 8 in. (20.3 cm). Yellow to reddish brown; our only lizard with legs *and* lateral grooves. Scales large, platelike. Broken, irregular light lines cross back and tail. Tail somewhat prehensile. Dark brown to black juvenile vividly marked with narrow whitish crossbands; head tan. Hatchling 3½–4 in. (9–10 cm). **HABITAT:** Rocky hillsides, wooded can-

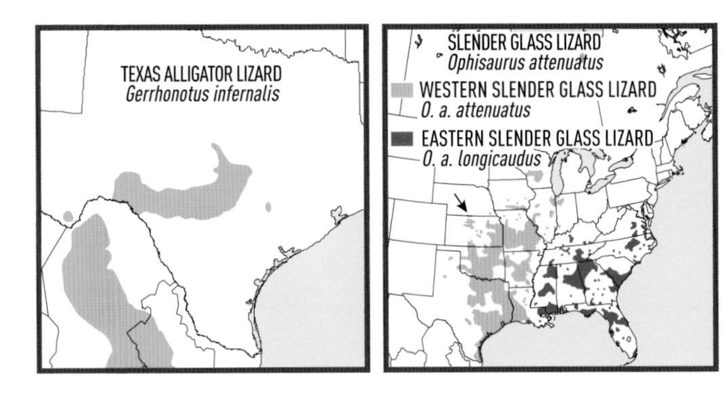

yons. **RANGE:** Edwards Plateau and the Big Bend Region of TX; disjunct in Mex.

SLENDER GLASS LIZARD *Ophisaurus attenuatus* **PL. 23, FIG. 107**

22–46½ in. (56–118.1 cm); head-body max. 14¼ in. (35.9 cm). Narrow dark longitudinal stripes *below lateral grooves and under tail* (Fig. 107, p. 236) black in juvenile, paler and less prominent in adult. Dark middorsal stripe or series of dashes evident in young and medium-sized individuals. Old adults brown with irregular dark-bordered light crossbands on back and tail. *Middle* of scales with white marks. Female in some areas strongly patterned; male flecked with whitish; old males often with a salt-and-pepper appearance. Scales along lateral grooves 98 or more. **SIMILAR SPECIES:** See Eastern Glass Lizard. **HABITAT:** Dry grasslands, open woodlands. **RANGE:** E. VA to s. FL west to cen. TX and north to s. WI. **SUBSPECIES:** (1) **Western Slender Glass Lizard** (*O. a. attenuatus*); unregenerated tail less than 2.4 times head-body length. (2) **Eastern Slender Glass Lizard** (*O. a. longicaudus*); larger and with a longer tail (2.4 or more times head-body length).

ISLAND GLASS LIZARD *Ophisaurus compressus* **FIG. 107**

15–24 in. (38–61 cm); head-body max. 7⅝ in. (19.5 cm). *Single* dark solid lateral stripe on scale rows 3 and 4 on each side above grooves. Middorsal dark stripe sometimes reduced to a series of dark dashes (Fig. 107, p. 236). *No dark stripes below grooves.* Unmarked undersurfaces pinkish buff or yellowish. Numerous more or less vertical light bars on neck, more numerous and usually more conspicuous than those of the Eastern Glass Lizard. Top and sides of neck mottled with bronze in older individuals. Scales along lateral grooves 97 or fewer. One or 2 upper labials extend to eyes. Fracture planes lacking in caudal vertebrae (present in other glass lizards); tail not fragile. **SIMILAR SPECIES:** See Eastern Glass Lizard. **HABITAT:** Coastal dune (including offshore islands) and sand pine scrub, coastal hammocks,

and inland pine flatwoods of peninsular FL. **RANGE:** Se. SC through peninsular FL.

MIMIC GLASS LIZARD *Ophisaurus mimicus* **FIG. 107**

15–25¾ in. (38–65.7 cm); head-body max. 7 3/16 in. (18.3 cm). Smallest of our glass lizards; dark brown or black middorsal stripe weak anteriorly, more distinct posteriorly and extending onto tail; *3 or 4 dark stripes or rows of spots above lateral grooves separated by pale stripes* (Fig. 107, p. 236). Scales along lateral grooves 97 or fewer. One or 2 upper labials extend to eyes, sometimes separated by very small scales. **SIMILAR SPECIES:** See Eastern Glass Lizard. **HABITAT:** Pine flatwoods, open woodlands. **RANGE:** Disjunct; coastal NC to n. FL and east to extreme e. LA.

EASTERN GLASS LIZARD *Ophisaurus ventralis* **PL. 23, FIG. 107**

18–42 5/8 in. (45.7–108.3 cm); head-body max. 12 in. (30.6 cm). *No* dark longitudinal stripes below lateral grooves or under tail; *no* distinct middorsal dark stripe (Fig. 107, p. 236). Largely vertical white marks on neck irregular. White marks on *posterior corners* of scales. Older individuals with numerous longitudinal dark dorsal lines or dashes; sometimes similar parallel lines occupy entire middorsal area. Old adults often greenish above and yellow below (making them the only glass lizards that may look *green*). Scales along lateral grooves 98 or more. Young khaki colored, usually with broad dark dorsolateral stripes. **SIMILAR SPECIES:** (1) Slender Glass Lizard usually has a dark middorsal stripe and series of dashes and dark stripes below lateral grooves. (2) Mimic Glass Lizard has a weak dark middorsal stripe and several lateral stripes above grooves. (3) Island Glass Lizard has a weak dark middorsal stripe and a single lateral stripe above grooves. **HABITAT:** Wet meadows, grasslands, pine flatwoods; tropical hardwood hammocks in s. FL. **RANGE:** Extreme se. VA through FL and some Keys and west to e. LA.

GECKOS

Four of seven families are represented in our area. Many geckos are vocal, and both common and scientific names refer to calls of an Asian species that sounds like *geck'-o*. Except for the aptly named day geckos, most species are nocturnal, have slitlike pupils, and often congregate to feed around night-lights. Many climb, and some are closely associated with buildings.

EYELID GECKOS: Eublepharidae

These geckos have functional eyelids and, lacking expanded toepads (Fig. 108, p. 238), are largely terrestrial. Thirty species in six genera occur in North and Central America, Asia, and Africa.

TEXAS BANDED GECKO *Coleonyx brevis* **PL. 23, FIGS. 108, 115**

4–4⅞ in. (10–12.4 cm); head-body max. 2 5/16 in. (5.9 cm). Scales tiny. Pattern changes with age. Juvenile has broad chocolate crossbands alternating with narrower cream or yellow bands; adult has dark pigment in light bands and light areas in chocolate bands, producing a mottled effect increasingly evident in larger individuals. Hatchling 1¾ in. (4.4 cm). **SIMILAR SPECIES:** Reticulate Banded Gecko is larger and

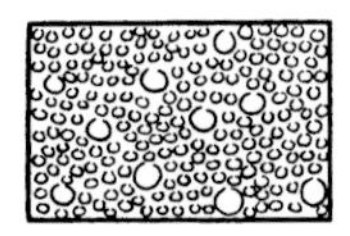

RETICULATE
(*C. reticulatus*)
Enlarged tubercles

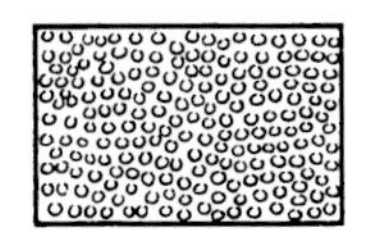

TEXAS BANDED
(*C. brevis*)
All scales small

FIG. 115. *Dorsal scales of eyelid geckos* (Coleonyx) *in our area.*

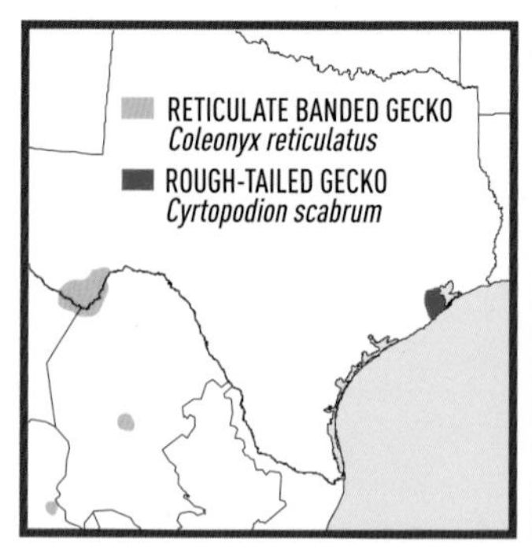

has enlarged tubercles scattered among tiny dorsal scales. **HABITAT:** Dry, rocky areas; shelter in crevices and under debris. **VOICE:** Faint squeaks, most frequently emitted when alarmed. **RANGE:** S. NM through s. TX and south into Mex.

RETICULATE BANDED GECKO *Coleonyx reticulatus* **FIG. 115**

5½–6¾ in. (14–17.2 cm); head-body max. 3⁹⁄₁₆ in. (9 cm). *Enlarged tubercles* scattered among tiny dorsal scales (Fig. 115, p. 257). Brown spots or streaks on light brown ground color often form a reticulate pattern. Hatchling 3⅛ in. (8 cm); body and tail with distinct chocolate brown bands on pinkish ground color. **SIMILAR SPECIES:** See Texas Banded Gecko. **HABITAT:** Rocky canyons. **VOICE:** Squeaks when captured. **RANGE:** Big Bend Region of TX; isolated records in Mex.

TRUE GECKOS: Gekkonidae

True geckos are widely distributed in tropical and subtropical regions in both the Eastern and Western Hemispheres. Many are human commensals, often called house geckos, that establish themselves around buildings and other human structures. All in our area are non-native. Most probably arrived by "hitchhiking" in cargoes of produce, lumber, or ornamental plants, but some undoubtedly are descendants of escaped pets. The four species described in the previous edition of this guide have grown to 13, and that number is likely to increase (individuals of several additional species have been found in southern states).

True geckos lack eyelids; transparent scales called spectacles cover the eyes. Many climb smooth walls and even ceilings. Brushlike pads (lamellae) bearing microscopic bristles (setae) cover the expanded toes. Setae (some geckos have a million of them) end in as many as 1,000 tiny plates that use attractive forces between molecules (van der Waals force) and friction to interact with surfaces. Claws are used when climbing rough surfaces.

FIG. 116. *Until recently, Bibron's Sand Gecko* (Chondrodactylus bibronii) *was assigned to the genus* Pachydactylus. *Many pets labeled* "Chondrodactylus bibronii" *actually are Turner's Thick-toed Geckos* (C. turneri). *Both species have been found in Florida.*

BIBRON'S SAND GECKO *Chondrodactylus bibronii* **Non-native** **FIG. 116**

6–8 in. (15.2–20.3 cm); head-body max. 5½ in. (14 cm). Stout, head large, tail thick. Brown, with darker crossbands (often broken in adult) and scattered white (more abundant in male) and dark spots. Belly white to light brown. Scales enlarged and keeled, forming *whorls on tail*; toepads thick, rounded. Hatchling 2⅛ in. (5.5 cm); crossbands distinct, rarely broken. **SIMILAR SPECIES:** No other FL geckos have enlarged, keeled scales on limbs. **HABITAT:** Buildings, power poles, trees. **RANGE:** Bradenton and Manatee Cos., FL (established breeding populations need verification). **NATURAL RANGE:** S. Africa. **REMARK:** Also called Bibron's Thick-toed Gecko.

ROUGH-TAILED GECKO *Cyrtopodion scabrum* **Non-native** **PL. 23**

3–4⅝ in. (7.5–11.7 cm); head-body max. 2 in. (5.1 cm). Sand colored with dark brown spots forming a fairly regular dorsal pattern. Belly white. Toetips barely wider than toes (no toepads). Tail with dark brown crossbands or spots, covered with *enlarged, keeled scales*. Hatchling 1⅝–2⅜ in. (4.1–6 cm). **SIMILAR SPECIES:** Mediterranean and Common House Geckos have wider toepads and lack rough, keeled scales on tail. **HABITAT:** Buildings. **RANGE:** Commercial shipping docks of Galveston, TX; also introduced in AZ. **NATURAL RANGE:** W. Mediterranean south to Sudan and east to nw. India.

FIG. 117. *Many references to the Golden Gecko* (Gekko badenii) *use the name* G. ulikovskii, *which was formerly applied to this species introduced into Florida from Vietnam.*

GOLDEN GECKO *Gekko badenii* Non-native **FIG. 117**

7–8⅝ in. (17.8–22 cm); female smaller; head-body max. 5½ in. (14 cm). Gray with a *golden wash*; forehead with light spots; faint pale, dark-edged, often yellowish dorsal crossbands usually evident. Belly light. Dorsum without enlarged tubercles. First digits lack claws. Juvenile lacks golden wash (head often yellowish) and dark gray or brown crossbands are distinct. **SIMILAR SPECIES:** (1) Tokay Gecko is larger and has dorsal tubercles and brick red dorsal spots. (2) Mourning Gecko is much smaller and has V- or W-shaped dorsal marks. (3) House geckos are smaller and have claws on all 5 digits. (4) Wall geckos have visible claws only on third and fourth toes (female has very small claws on other toes). **HABITAT:** Buildings, other human structures, debris. **RANGE:** Hollywood, FL. **NATURAL RANGE:** Vietnam. **REMARK:** Many references refer to this species as *G. ulikovskii.*

TOKAY GECKO *Gekko gecko* Non-native **PL. 23**

8–14 in. (20.3–36 cm); head-body max. 7 in. (18 cm). Largest gecko in our area; bluish gray with numerous white and *brick red to orange dorsal spots and prominent tubercles.* Enlarged toepads with up to 20 lamellae. First digits lack claws. Hatchling 3½–4 in. (9–10 cm), tail with wide dark bluish bands alternating with narrower white bands. **SIMILAR SPECIES:** See Golden Gecko. **HABITAT:** Buildings. **VOICE:** Cackles followed by *to-kay* repeated about 6 times, dropping off in loudness. **RANGE:** S. FL; scattered records farther north probably not established. **NATURAL RANGE:** Se. Asia and the Malay Archipelago.

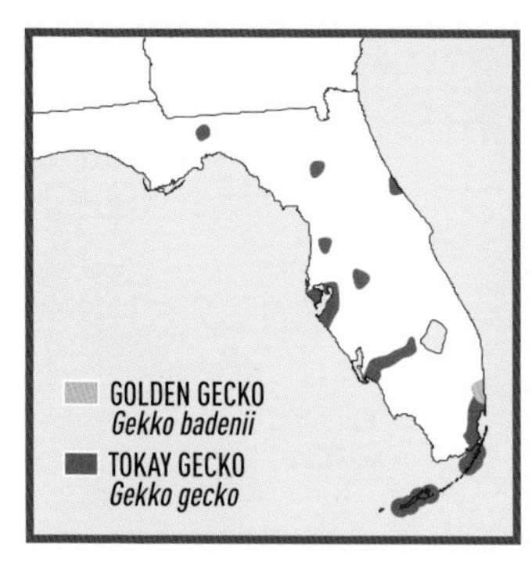

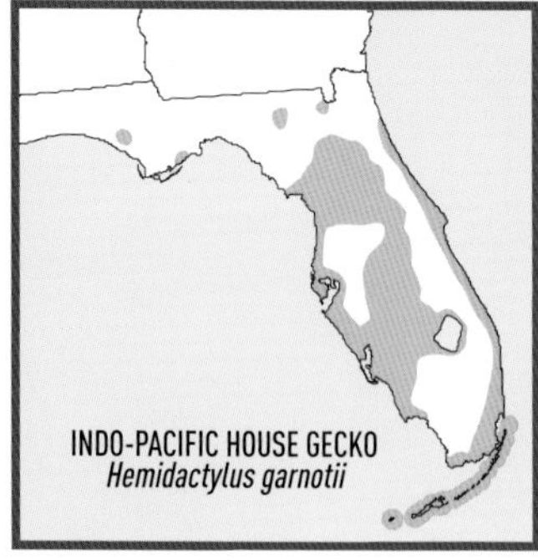

COMMON HOUSE GECKO *Hemidactylus frenatus* Non-native P. 448

4–5½ in. (10–14 cm); head-body max. 2¼ in. (5.7 cm). Broad toepads extending nearly full length of toes (as in Fig. 119, p. 263). Tan or gray, sometimes unicolored, but darker lateral stripes or small light or dark dorsal spots usually evident; darker individuals often well patterned; pale areas whitish; light lines through eyes. Body smooth; *scattered* dorsal tubercles small; rows of *enlarged spines circle some portions of unregenerated tail*. Femoral pores present. Hatchling head-body length ¾–13/16 in. (1.9–2.1 cm). **SIMILAR SPECIES:** (1) Other house geckos have a flattened tail (Asian Flat-tailed), toepads less than full length of toes (Tropical), dorsal tubercles small and in dorsolateral rows, spines only along sides of tail (Indo-Pacific), dorsal tubercles prominent, femoral pores absent, and pattern of irregular flecks and spots (Mediterranean). (2) Mourning Gecko has V- or W-shaped dorsal marks and lacks claws on first digits. (3) Tokay and Golden Geckos are larger and lack claws on first digits. (4) Wall geckos have small but distinctly visible claws on only third and fourth toes. **HABITAT:** Buildings. **VOICE:** Faint chirps often repeated 3 times; squeaks when captured. **RANGE:** Dallas and Galveston Cos., TX (not mapped) and s. FL. **NATURAL RANGE:** S. and se. Asia, many East Indian islands. Widely introduced in tropical and subtropical areas, including HI.

INDO-PACIFIC HOUSE GECKO *Hemidactylus garnotii* Non-native PL. 23

4–5½ in. (10–14 cm); head-body max. 2½ in. (6.4 cm). All-female species reproduces by parthenogenesis; consequently one individual can establish a new population. Small ovoid dorsal scales, *small tubercles in dorsolateral rows, spines restricted to sides of flattened tail*. Uniformly brownish gray or marbled with darker brown; small whitish dorsal spots vary in size and shape; sometimes faint crossbands on tail. Belly often lemon yellow; underside of tail frequently with a pale reddish cast. Hatchling head-body length ⅞–1 in. (2.3–2.5 cm); pattern distinct, with dark crossbands on tail. **SIMILAR SPECIES:** See Common House Gecko. **HABITAT:** Buildings. **VOICE:** Squeaks when

FIG. 118. *The Tropical House Gecko* (Hemidactylus mabouia) *appears to be displacing some other species of house geckos in southern Florida. These geckos often are called woodslaves on English-speaking West Indian islands, but that name also is used for many other species of geckos.*

fighting. **RANGE:** Peninsular FL through the Keys; isolated records in the FL Panhandle, also Dallas, TX (not mapped), and HI. Records in AL and GA (not mapped) probably not established. **NATURAL RANGE:** Se. Asia, East Indies, many South Sea islands.

TROPICAL HOUSE GECKO *Hemidactylus mabouia* **Non-native**

FIGS. 118, 119

4–5 in. (10–12.7 cm); head-body max. 2⅜ in. (6 cm). Toepads *do not extend full length of fourth toe* (Fig. 119, facing page). Moderately warty, *dorsal tubercles scattered*. Uniformly pale or gray to brown with dark chevrons; crossbands usually evident on tail. Undersides pale tan or gray to whitish. Hatchling head-body length ⅞–1 in. (2.2–2.5 cm). **SIMILAR SPECIES:** See Common House Gecko. **HABITAT:** Buildings, also mangroves, debris piles, under rocks or loose bark. **VOICE:** Chirps when disturbed or fighting (especially male). **RANGE:** Peninsular FL through the Keys. **NATURAL RANGE:** Tropical Africa. Also introduced in S. America and the W. Indies. **REMARK:** Often called the Woodslave.

SRI LANKAN HOUSE GECKO *Hemidactylus parvimaculatus* **Non-native**

FIG. 120

4–4½ in. (10–11.5 cm); head-body max. 2⅜ in. (6 cm). Brownish, 3 longitudinal *rows of irregular dark brown spots* smaller than eyes. Enlarged, *keeled tubercles* in 16–20 more or less regular longitudinal rows. Short, continuous rows of 4–8 precloacal pores. Juvenile and young adult have scattered small dark spots. **SIMILAR SPECIES:** (1) Mediterranean Gecko has larger, more prominent dorsal tubercles

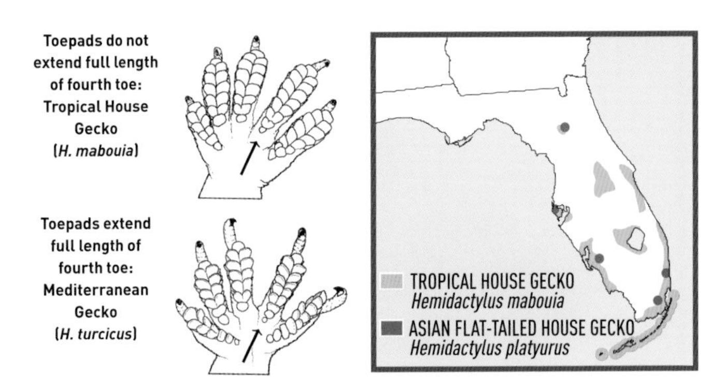

FIG. 119. *Toepads of the Tropical House Gecko* (Hemidactylus mabouia) *do not extend the full length of the fourth toes, whereas those of the Mediterranean Gecko* (H. turcicus) *do. The Common House Gecko* (H. frenatus) *has toepads like those of the Mediterranean Gecko.*

FIG. 120. *A Mediterranean Gecko* (Hemidactylus turcicus; *above) and a Sri Lankan House Gecko* (H. parvimaculatus; *below) from building walls in the Audubon Zoo, New Orleans. The Sri Lankan House Gecko is in its dark phase and its dorsal pattern is obscure. The Mediterranean Gecko has a more boldly banded tail with smaller, more recumbent tubercles but larger, more prominent tubercles on the back and sides.*

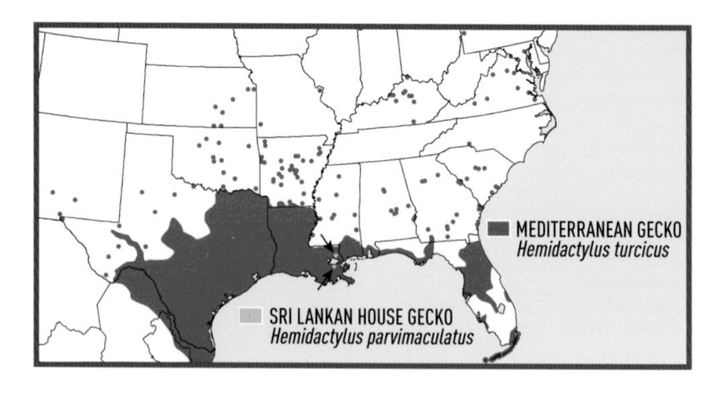

arranged in 14–16 rows and a proportionately shorter, more distinctly banded tail with smaller tubercles. (2) Other house geckos established in the U.S. lack regular rows of enlarged dorsal tubercles. **HABITAT:** Buildings. **VOICE:** *Chuk-chuk-chuk*. **RANGE:** Audubon Park, New Orleans, LA, and possibly north of Lake Pontchartrain. **NATURAL RANGE:** S. India, Sri Lanka, Indian Ocean islands.

ASIAN FLAT-TAILED HOUSE GECKO *Hemidactylus platyurus* Non-native

FIG. 121

4–5½ in. (10–14 cm); head-body max. 2⁷⁄₁₆ in. (6.1 cm). *Toes webbed, skinfolds along sides, tail distinctly flattened*. Skin smooth. Brownish gray with faint to distinct marbled pattern; dark streaks from eyes to shoulders; undersides lighter, almost white. Hatchling head-body length ¾–1 in. (2–2.5 cm). **SIMILAR SPECIES:** See Common House Gecko. **HABITAT:** Buildings. **VOICE:** Male clicks; distressed geckos emit high-pitched squeaks. **RANGE:** Scattered in peninsular FL; some populations have disappeared. **NATURAL RANGE:** Nepal and e. India throughout se. Asia. **REMARK:** Previously assigned to the genus *Cosymbotus*.

FIG. 121. *The webbed toes, lateral skinfolds, and flattened tail of the Asian Flat-tailed House Gecko* (Hemidactylus platyurus) *provide a low profile that blends with the substrate without leaving a shadow.*

MEDITERRANEAN GECKO *Hemidactylus turcicus* Non-native

PL. 23, FIGS. 108, 119, 120

4–5 in. (10–12.7 cm); head-body max. 2³/₈ in. (6 cm). Broad toepads extend nearly full length of toes (Fig. 108, p. 238; Fig. 119, p. 263); *tubercles prominent*; light and dark brown or gray spots on pale pinkish to whitish ground color. Hatchling head-body length ¾–1¼ in. (2–3 cm). **SIMILAR SPECIES:** See Common and Sri Lankan House Geckos. **HABITAT:** Buildings. **VOICE:** Faint mouselike squeaks repeated at more or less regular intervals. Dominant males may emit clicking noises in rapid succession. **RANGE:** FL west to TX and adjacent Mex., scattered localities in other states. Widely introduced (not all introductions mapped). **NATURAL RANGE:** Mediterranean Basin into w. Asia.

MOURNING GECKO *Lepidodactylus lugubris* Non-native **FIG. 122**

2¾–3¾ in. (7–9.5 cm); head-body max. 1⁵/₈ in. (4.1 cm). All-female species reproduces by parthenogenesis; one individual can establish a new population. Almost white to brown or gray with paired dark spots or V- and W-shaped dorsal markings. *First toe reduced, lacks claws.* Tail slightly flattened. Hatchling head-body length ⁹/₁₆–¾ in. (1.5–1.9 cm). **SIMILAR SPECIES:** Sphaerodactyl geckos lack toepads or pads are limited to a single round scale near

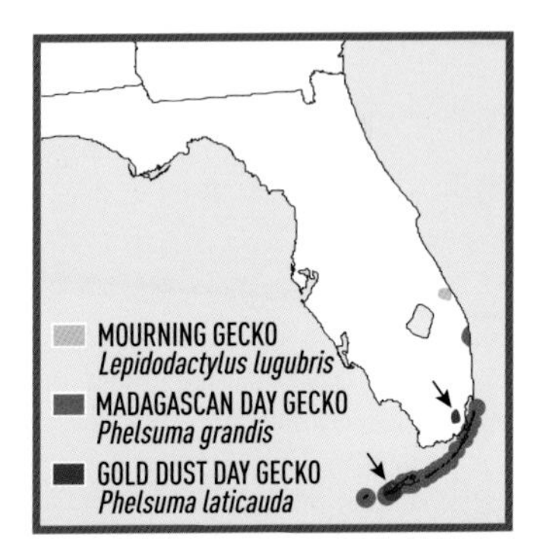

FIG. 122. *The Mourning Gecko* (Lepidodactylus lugubris) *is an all-female species that reproduces by parthenogenesis (development of unfertilized eggs). Consequently, a single individual can establish a new population.*

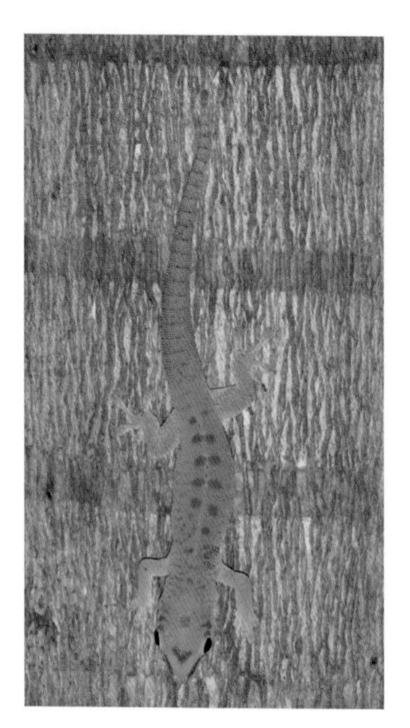

FIG. 123. *Until recently, the Madagascan Day Gecko* (Phelsuma grandis) *was considered a subspecies of* P. madagascariensis, *another giant day gecko from Madagascar. As the common name implies, these geckos are active by day.*

FIG. 124. *The Gold Dust Day Gecko* (Phelsuma laticauda) *in Florida (and Hawaii) often functions as a human commensal, exploiting buildings, fences, and other artificial perches. It appears to have a lifestyle very similar to that of anoles, the consequences of which are unknown.*

tips of digits. See also Common House Gecko. **HABITAT:** Buildings, trees; shelter beneath rocks and loose bark. **VOICE:** Chirps repeated 5 or more times. **RANGE:** S. FL; some populations have disappeared. **NATURAL RANGE:** Many Pacific islands. Widely introduced, including HI.

MADAGASCAN DAY GECKO *Phelsuma grandis* Non-native **FIG. 123**

10–11¾ in. (25.4–30 cm), male larger than female; head-body max. 4 in. (10.1 cm). Bright green with orange or red blotches or spots and orange stripes from nostrils to eyes. Pupils round. Hatchling 2¾ in. (7 cm); dull olive. **SIMILAR SPECIES:** Gold Dust Day Gecko is smaller and has extensive yellow speckling on neck. **HABITAT:** Trees, especially palms, buildings. **RANGE:** S. FL through the Keys. **NATURAL RANGE:** Madagascar.

GOLD DUST DAY GECKO *Phelsuma laticauda* Non-native **FIG. 124**

4–5⅛ in. (10–13 cm), male larger than female; head-body max. 2⁵⁄₁₆ in. (5.8 cm). Bright green with light blue around upper portions of eyes, toes, and tailtip; *brilliant yellow specks from back of head onto back*; 3 elongated red dorsal markings continue as scattered spots onto tail; additional red marks on head and snout. Pupils round. Hatchling 1⅝ in. (4 cm). **SIMILAR SPECIES:** See Madagascan Day Gecko. **HABITAT:** Buildings, fences, walls, gardens, tall plants such as palms. **RANGE:** S. FL and some Keys, also HI. **NATURAL RANGE:** Madagascar.

LEAF-TOED GECKOS: Phyllodactylidae

More than 100 species in nine genera are distributed throughout much of the Western Hemisphere, northern Africa, southern Europe, and the Middle East. Two introduced species, both in the genus *Tarentola* (wall geckos), have toepad lamellae extending undivided across the entire pad and only the third and fourth toes have visible claws (females have small claws on other digits).

RINGED WALL GECKO *Tarentola annularis* Non-native **FIG. 125**

7–8 in. (17.8–20.3 cm), male larger than female; head-body max. 5½ in. (14 cm). Dark brown-gray to light sandy gray; back and tail with regular bands of low, smooth tubercles separated by 6–7 small scales; sides of head with large tubercles widely separated by small

FIG. 125. *This Ringed Wall Gecko* (Tarentola annularis) *has lost and regenerated part of its tail. Note, however, that neither the scales nor the pattern of the regenerated portion match the original.*

FIG. 126. *The Moorish Gecko* (Tarentola mauritanica) *has much more prominent tubercles than the Ringed Wall Gecko* (T. annularis). *As in many introduced populations of geckos throughout the world, the Moorish Gecko is closely associated with human structures and benefits by eating insects attracted to lights.*

scales; 4–5 dark and light dorsal crossbars, usually 4 distinctive dark-bordered white spots on shoulders (often lacking in juvenile); dark lines through eyes. Tail banded. Belly off-white. **SIMILAR SPECIES:** Moorish Gecko is smaller, has more prominent tubercles, and lacks 4 dark-bordered white spots on shoulders. **HABITAT:** Buildings. **RANGE:** S. FL. **NATURAL RANGE:** N. Africa.

MOORISH GECKO *Tarentola mauritanica* **Non-native** **FIG. 126**

4½–6 in. (11.4–15 cm); head-body max. 3⁵⁄₁₆ in. (8.4 cm). Robust, spiny, with a whorled tail. Prominent dorsal tubercles separated by 4–5 small scales; sides of head with large tubercles widely separated by small scales. Light yellowish gray, with 4–5 crossbars that extend onto tail. Belly white. Hatchling head-body length ¾–1 in. (2–2.5 cm). **SIMILAR SPECIES:** See Ringed Wall Gecko. **HABITAT:** Buildings, other human structures. **RANGE:** S. FL. **NATURAL RANGE:** Atlantic and Mediterranean coastal regions. Widely introduced, including CA.

SPHAERODACTYL GECKOS: Sphaerodactylidae

Approximately 200 species and 12 genera range throughout Central and South America, the Caribbean, northern Africa, southern Europe, the Middle East, and central Asia. One species occurs naturally in our area. Geckos in the genus *Sphaerodactylus* have small round pads consisting of a single enlarged scale on the tips of their toes; those in the genus *Gonatodes* lack toepads (Fig. 108, p. 238).

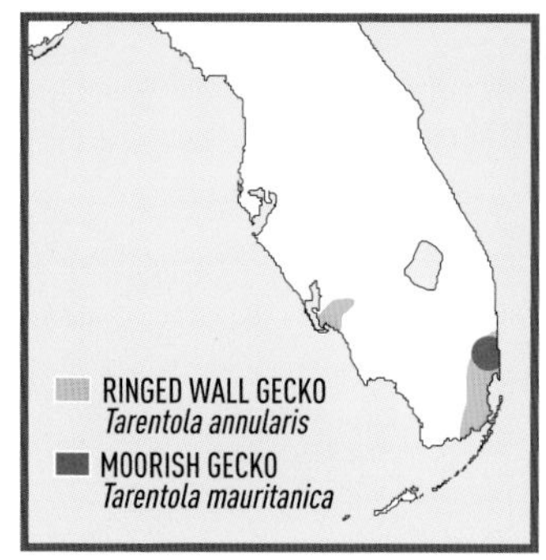

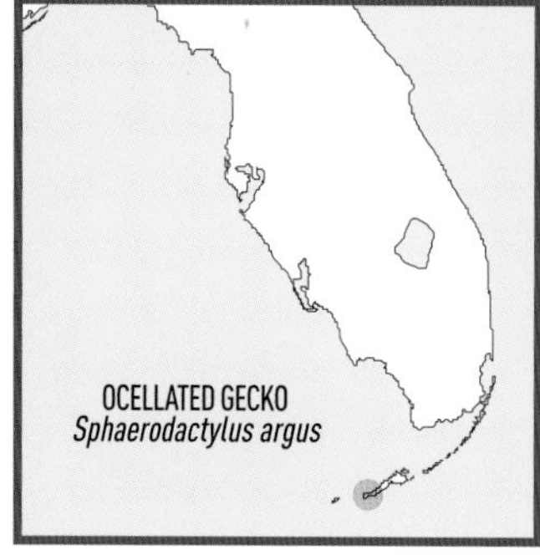

OCELLATED GECKO *Sphaerodactylus argus* Non-native **PL. 24**

1⁷⁄₈–2³⁄₈ in. (4.8–6 cm); head-body max. 1⁵⁄₁₆ in. (3.3 cm). Brown to olive brown, tail brown or reddish; tiny *white eyelike spots* (ocelli) on nape often continue onto back, sometimes fuse with one another or with light lines on head, rarely patternless. Dorsal scales keeled, 57–73 around midbody. Young similar to adult, but tends to be lineate, tail often reddish. Hatchling 1¹⁄₁₆ in. (2.6 cm). **SIMILAR SPECIES:** (1) Reef Gecko has 2 light spots on nape or none, head stripes (if present) are dark, dorsal scales around midbody 48 or fewer. (2) Ashy Gecko has granular dorsal scales. (3) Yellow-headed Gecko lacks toepads. **HABITAT:** Under debris or treebark. **RANGE:** Key West and Stock I., FL. **NATURAL RANGE:** Jamaica, Cuba, some islands in the Bahamas, Corn I. (Nicaragua). Widely introduced. **SUBSPECIES:** **Ocellated Gecko** (*S. a. argus*). **REMARK:** Also called the Jamaican Stippled Sphaero.

ASHY GECKO *Sphaerodactylus elegans* Non-native **PL. 24**

2¾–2⁷⁄₈ in. (7–7.3 cm); head-body max. 1⁷⁄₁₆ in. (3.7 cm). Reddish to pale grayish brown with many tiny white or yellow spots that often fuse to form light lines on head. Snout pointed, slightly flattened. Dorsal scales granular. Young with striking dark crossbands, reddish tail. **SIMILAR SPECIES:** See Ocellated Gecko. **HABITAT:** Buildings and under debris. **RANGE:** S. FL, including the Keys. **NATURAL RANGE:** Cuba. **SUBSPECIES:** **Ashy Gecko** (*S. e. elegans*).

REEF GECKO *Sphaerodactylus notatus* **PL. 24**

2–2¼ in. (5.1–5.7 cm); head-body max. 1³⁄₁₆ in. (3 cm). Male brown with small dark dorsal spots; old males often uniformly brown; female and young with 3 broad, dark, light-centered, longitudinal stripes on head, often with pairs of light spots on shoulders. Pointed, slightly flattened snout. Strongly keeled dorsal scales, 41–48 around midbody. **SIMILAR SPECIES:** See Ocellated Gecko. **HABITAT:** Pinelands,

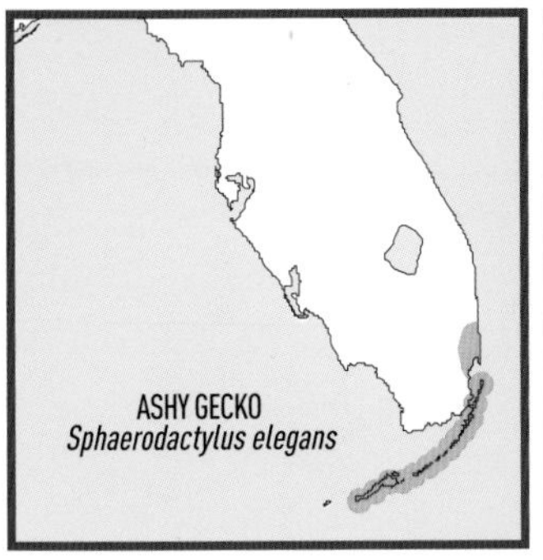

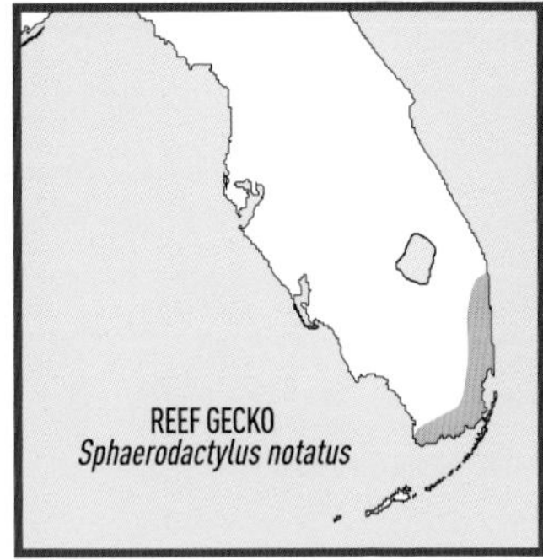

hammocks, vacant lots, around buildings; often under debris. **RANGE:** S. FL, including the Keys, Bahamas, Cuba. **SUBSPECIES: FLORIDA REEF GECKO** (*S. n. notatus*). **REMARK:** Also called the Brown-speckled Sphaero.

YELLOW-HEADED GECKO *Gonatodes albogularis* Non-native **PL. 24**

2½–3½ in. (6.4–8.9 cm); head-body max. 1 9/16 in. (4 cm). Male with *yellow head* and uniformly dark body, legs, and tail; female and young mottled with brown, gray, and yellow; narrow light collar usually present in both sexes. Unregenerated tailtip white. Pupils round. **SIMILAR SPECIES:** See Ocellated Gecko. **HABITAT:** Buildings, docks, usually under debris. **RANGE:** S. FL, including some Keys. **NATURAL RANGE:** Cuba, Jamaica, Hispaniola, mainland Cen. and S. America. **SUBSPECIES: YELLOW-HEADED GECKO** (*G. a. albogularis*). **REMARK:** Also called the Neotropical Clawed Gecko.

IGUANIAN LIZARDS

Iguanian lizards are represented by three major groups, two of which (Agamidae and Chamaeleonidae) occur naturally only in the Eastern Hemisphere, although species in both are established in our area. The third, divided into as many as 12 families, is largely confined to the Western Hemisphere. Especially the males of many species employ display patterns that include head-bobbing and push-ups to declare territories and attract mates.

AGAMID LIZARDS: Agamidae

More than 440 species occur naturally in Africa, Asia, Australia, and southern Europe. Introduced populations are undoubtedly descendants of escaped or released pets. No agamids were included in previous editions of this guide. Many are popular pets. Individuals of at least three additional species have been captured in Florida, and the number of established populations will undoubtedly increase.

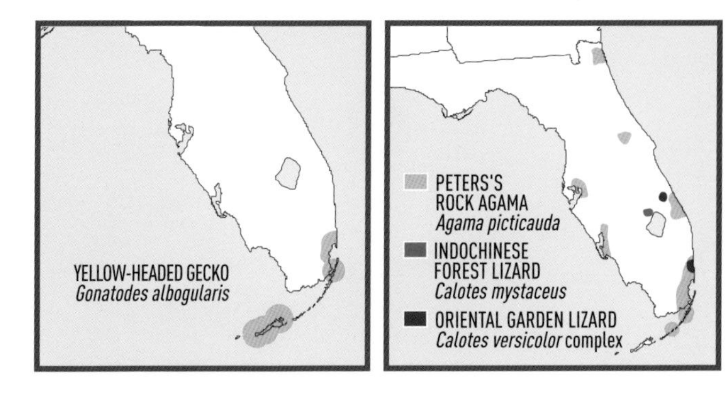

PETERS'S ROCK AGAMA *Agama picticauda* **Non-native** **FIG. 127**

7⅞–11⅞ in. (20–30 cm); max. total length 12 in. (30.5 cm). Body robust, slightly flattened. Male black, bluish gray, or charcoal with a *yellow, orange, or red head*, color often extending onto nape; light middorsal line often present; tail base bluish white, midsection orange, tip black (cooler or stressed individuals less intensely colored). Nonbreeding female brown, grayish, or yellowish brown, often with a vague light middorsal line and scattered darker and lighter brown

FIG. 127. *The brightly colored head of a male Peters's Rock Agama* (Agama picticauda) *presumably functions in communication, simultaneously warning other males to stay away while also attracting the attention of females.*

FIG. 128. *The Oriental Garden Lizard* (Calotes versicolor *complex*) *is part of a widely distributed and exceedingly variable species complex that has yet to be resolved in many parts of the natural range. These lizards sometimes are called bloodsuckers, alluding to the red color of breeding males.*

spots or flecks; breeding female often with an orangish or bluish head. Belly generally lighter than back. **HABITAT:** Disturbed areas around buildings, often associated with rock piles or trees. **RANGE:** Scattered urban areas in peninsular FL. **NATURAL RANGE:** W. Africa. **REMARK:** Populations in FL until recently were assigned to the African Rainbow Lizard (*A. agama*), the name used in many publications.

INDOCHINESE FOREST LIZARD *Calotes mystaceus* **Non-native** **NOT ILLUS.**

10–15 in. (25.4–38.1 cm); head-body max. 5½ in. (14 cm). Body somewhat triangular in cross section; *distinct middorsal crest* most prominent on neck, more developed in male. Gray, olive, or light brown, often with a bluish cast; pale white or yellowish stripes extend from upper lip onto body and are followed by 1–3 (occasionally 5) orange-red spots, irregularly edged in white. Breeding male becomes bright blue or turquoise, most intense on head, and flank spots become brighter. Belly grayish cream. Juvenile lacks bright colors. **SIMILAR SPECIES:** Oriental Garden Lizard is similar, but breeding male is red (not blue) anteriorly. **HABITAT:** Trees, fenceposts in urban and suburban areas. **RANGE:** S. FL. **NATURAL RANGE:** S. Asia.

ORIENTAL GARDEN LIZARD *Calotes versicolor* complex **Non-native** **FIG. 128**

9–14½ in. (23–37 cm); head-body max. 4¼ in. (10.8 cm). Body somewhat triangular in cross section; *distinct middorsal crest* higher in male, most prominent on neck but extending beyond base of tail. Adult gray to light brown, darker middorsally than on lower sides. Breeding male becomes bright rusty red anteriorly and yellowish

FIG. 129. *The Common Butterfly Lizard* (Leiolepis belliana) *in Florida is associated with lawns and gardens and rarely ventures far from its burrow.*

posteriorly; throat, jowls, parts of forelimbs jet-black. Breeding female yellowish. Juvenile often with dark, irregular transverse bands on body and tail, white dorsolateral stripes from neck to halfway down tail. Hatchling head-body length 1⅛ in. (3 cm). **SIMILAR SPECIES:** See Indochinese Forest Lizard. **HABITAT:** Trees in urban and suburban areas. **RANGE:** S. FL. **NATURAL RANGE:** S. Asia. **REMARK:** A complex of several species; the origin and specific identity of FL populations is unknown.

COMMON BUTTERFLY LIZARD *Leiolepis belliana* Non-native **FIG. 129**

9–15¾ in. (22.9–40 cm); head-body max. 611/16 in. (17 cm). Body robust, somewhat flattened, scales small, loose skinfolds along sides. Back blackish brown to grayish olive with black-edged yellow eyelike spots and 3 longitudinal stripes sometimes broken into series of spots, especially anteriorly. Flanks bluish black with *7–9 broad vertical orange bars.* Belly cream. Female less intensely colored, often with yellowish spots on top of limbs. Juvenile has longitudinal dorsal stripes. **SIMILAR SPECIES:** Red-banded Butterfly Lizard has an irregular dorsal network of oval black-edged bluish gray spots and lacks lateral orange bars. **HABITAT:** Lawns and gardens. **RANGE:** S. FL. **NATURAL RANGE:** Indochina and the Malay Peninsula.

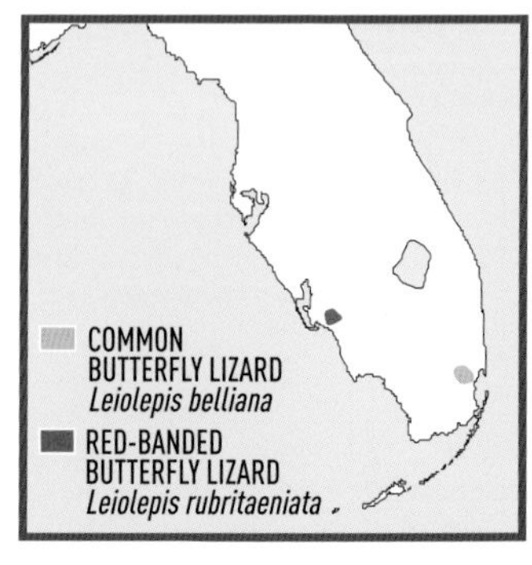

RED-BANDED BUTTERFLY LIZARD *Leiolepis rubritaeniata* Non-native **NOT ILLUS.**

7⅞–13¾ in. (20–35 cm); head-body max. 5¼ in. (13.4 cm). Similar to Common Butterfly Lizard, but back has a network of oval dark-edged bluish gray spots and orange lateral stripes with variable black vertical bars are most prominent anteriorly. Belly cream. **SIMILAR SPECIES:** See Common Butterfly Lizard. **HABITAT:** Lawns and gardens. **RANGE:** S. FL. **NATURAL RANGE:** Indochina.

CHAMELEONS: Chamaeleonidae

Nearly 200 species range across Africa, Madagascar, southern Europe, and western Asia. Chameleons are popular pets. Two species are established in Florida; at least two others, the Panther Chameleon (*Furcifer pardalis*) and Jackson's Chameleon (*Triceros jacksonii*), are encountered in our area, although neither is known to have established breeding populations.

Chameleons have a laterally compressed body, prehensile tail, and digits fused into sets of two and three, forming opposable, mittenlike hands and feet. The eyes work independently, allowing one to look ahead while the other looks behind. Many species have a protrusion called a "casque" on the head; these vary in size but are more prominent in males. Tongues, sometimes longer than the head and body, can be extended rapidly to catch prey. Chameleons move slowly, often rocking from side to side.

VEILED CHAMELEON *Chamaeleo calyptratus* Non-native **FIG. 130**

12–18 in. (30–45 cm); max. total length 24 in. (61 cm), male much larger than female. *Triangular*, laterally compressed bony casque; closely packed spikelike scales form dorsal, gular (throat), and ventral crests; *gular and ventral crests continuous*. Male has small "spurs" on heels. Male primarily green (lime to reddish olive) with stripes and spots of yellow, brown, blue, and black; colors most intense when animal is stressed. Nonbreeding female uniformly green with some white markings. Breeding and gravid females very dark green with blue and yellow spots. Juvenile like nonbreeding female, casque very small. Hatchling 2–3 in. (5–7.5 cm). **SIMILAR SPECIES:** Oustalet's Chameleon has a low, rounded casque, gular crest not continuous with ventral crest. **HABITAT:** Trees in degraded habitats, including vacant lots and residential areas. **RANGE:** Scattered in peninsular FL, also HI. **NATURAL RANGE:** Arabian Peninsula.

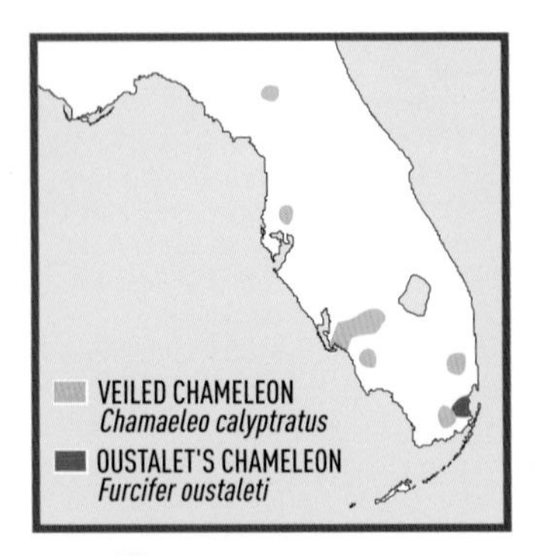

FIG. 130. *Like most chameleons, the Veiled Chameleon* (Chamaeleo calyptratus) *is renowned for changing colors. However, individuals of most species are somewhat cryptic when resting and develop brighter colors when alarmed or breeding.*

FIG. 131. *Oustalet's Chameleon* (Furcifer oustaleti) *is the world's longest chameleon; males, such as this individual, are much larger than females.*

OUSTALET'S CHAMELEON *Furcifer oustaleti* **Non-native** **FIG. 131**

13¾–23½ in. (35–60 cm); max. total length 27½ in. (70 cm). *Rounded, laterally compressed casque on back of head;* closely packed spike-like scales form dorsal, gular (throat), and ventral crests; *gular and ventral crests not continuous.* Male with vague gray or brown camouflage pattern. Female yellowish, greenish, or reddish, with vague vertical bars often present on sides. Juvenile like female, casque very small. **SIMILAR SPECIES:** See Veiled Chameleon. **HABITAT:** Ecotonal and disturbed areas, including orchards. **RANGE:** S. FL. **NATURAL RANGE:** Madagascar.

CASQUE-HEADED LIZARDS: Corytophanidae

None of the nine Neotropical species in three genera are native to our area.

BROWN BASILISK *Basiliscus vittatus* **Non-native** **PL. 24**

11–27 in. (28–68.8 cm); head-body max. 6⅞ in. (17.5 cm), male larger than female. Body slender, legs and tail long. Male with *triangular head crest*, female with hoodlike lobe on neck. Brown to olive brown, often with 6–7 dark brown or black crossbands or blotches; young with 2 dorsolateral light-colored stripes extending from head onto body, upper stripe more distinct and sometimes reaching groin; older adults more uniformly patterned. Belly cream to light gray with keeled scales. Chin and upper lip lighter than top of head. Tail lighter than body, with dark bands. **HABITAT:** Open areas near water. **RANGE:** S. FL and some Keys. **NATURAL RANGE:** Cen. Mex. to Colombia. **REMARKS:** Fringelike scales on toes of hindlimbs allow smaller individuals to run across water. Sometimes called the Jesus Lizard because it "walks on water."

COLLARED and LEOPARD LIZARDS: Crotaphytidae

Seven currently recognized species range from eastern Missouri to the Pacific and south into Mexico.

EASTERN COLLARED LIZARD *Crotaphytus collaris* **PL. 24, FIG. 132**

8–14 in. (20.3–35.6 cm); head-body max. 4⅝ in. (11.6 cm). Large-headed, small-scaled, long-tailed, with *2 black collars* (often broken at nape). Male yellowish, greenish, brownish, or bluish with a profusion of light dorsal spots. Female tan; when gravid, develops red spots or bars on sides and sometimes on sides of neck. Hatchling 3–3½ in. (7.5–9 cm), with broad dark crossbands consisting of rows of dark spots alternating with yellowish crossbands; young of both sexes may develop red bars like gravid female, but red disappears at head-body length of about 3 in. (7.5 cm). **SIMILAR SPECIES:** Reticulate Collared Lizard has a network of light lines and light-bordered black spots on back and sides. **HABITAT:** Hilly, rocky, often arid or semiarid areas, often associated with limestone ledges or rock piles. **RANGE:** MO west to AZ and south into Mex.

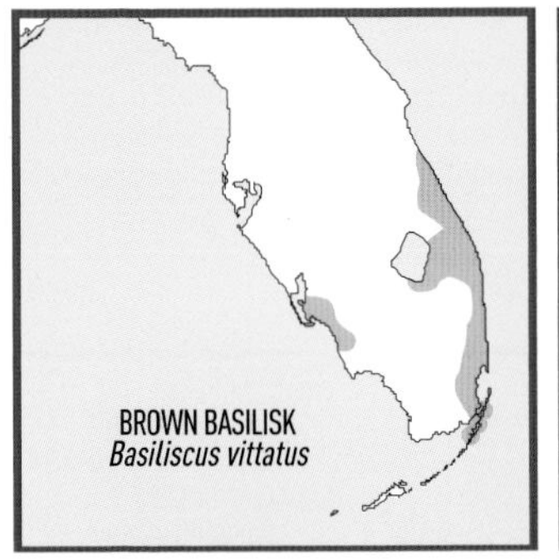

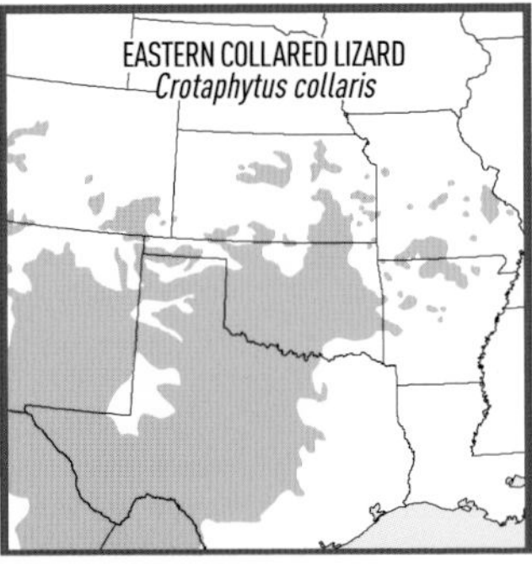

FIG. 132. *The male Eastern Collared Lizard* (Crotaphytus collaris) *often is brightly colored, especially during the breeding season. These lizards are intensely territorial and quite pugnacious; one account describes an individual threatening an oncoming train.*

RETICULATE COLLARED LIZARD *Crotaphytus reticulatus* **PL. 24**

8–16¾ in. (20.3–42.5 cm); head-body max. 5⅜ in. (13.7 cm). Large-headed, small-scaled, long-tailed, with *transverse rows of light-bordered black spots*; light markings on brownish back form an open but often broken network. Male with vertical black bars on neck, breeding male with a bright yellow suffusion on chest. Female lacks vertical black bars on neck, breeding female has vertical brick red bars between lateral rows of black spots and a pinkish suffusion on throat.

Hatchling 3½–4 in. (9–10 cm), light gray with 4–6 yellow or yellow-orange crossbands interspersed with transverse rows of black spots. **SIMILAR SPECIES:** See Eastern Collared Lizard. **HABITAT:** Thornbrush deserts, usually rocky areas. **RANGE:** Rio Grande Valley in s. TX and adjacent Mex. **CONSERVATION:** Vulnerable.

LONG-NOSED LEOPARD LIZARD *Gambelia wislizenii* **PL. 24**

8½–15⅛ in. (21.5–38.4 cm); head-body max. 5¾ in. (14.6 cm). Gray or brown with numerous brown to black spots; sometimes with narrow whitish vertical lines on sides meeting to form narrow crossbands (especially prominent in juvenile), throat streaked with gray. Flanks and underside of tail in gravid female with salmon red spots and streaks. Young often with reddish spots and curved yellow lines above eyes. Hatchling 4–5⅛ in. (10–13 cm). **HABITAT:** Arid or semiarid flats with clumped vegetation, usually sandy or gravelly soils. **RANGE:** W. TX to e. CA north to sw. OR and s. ID and south into Mex.

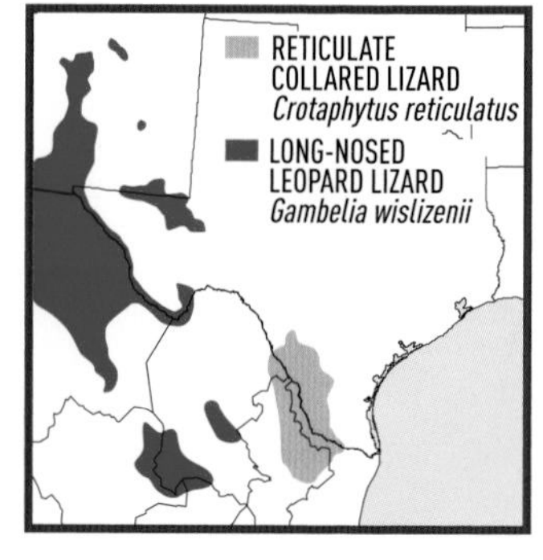

ANOLES: Dactyloidae

This strictly American group of nearly 400 species sometimes assigned to as many as eight genera ranges from the southern United States to Bolivia and Paraguay. Only one species is native in our area, but nine West Indian species are established in Florida and another (the Marie-Galante Anole, *Anolis ferreus*) presumably has been extirpated. Due to their popularity as pets and the proximity of the West Indies, additional species probably will become established.

Anoles have dewlaps (see Pl. 25), throat fans extended downward by flexible rods of cartilage to expose colors used to communicate during courtship or territorial defense. Especially the males of many species also have an erectile nuchal and dorsal crest.

Anoles change color in response to temperature, humidity, emotion, and activity. However, they are not chameleons, although they are sometimes called American chameleons. Anoles have expanded toepads that aid in climbing; all of our species are primarily arboreal. Males are larger than females (listed measurements for adults are for males).

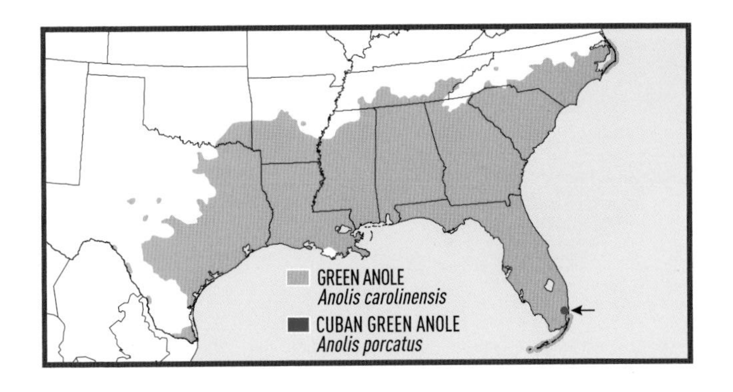

GREEN ANOLE *Anolis carolinensis* **PL. 25, P. 234**

CUBAN GREEN ANOLE *Anolis porcatus* **Non-native** **NOT ILLUS.**

5–9 in. (12.5–23.1 cm); head-body max. 3⁵⁄₁₆ in. (8.4 cm). These species cannot be reliably distinguished without employing molecular methods. Elongated head, pointed snout, keeled ventral scales, tail round in cross section. Uniformly green to mottled green and brown to uniformly brown, lips and belly white; *dewlap usually pink* but can vary from virtually white or light gray and greenish through pinks and magentas to blues and purples. Variable dark streaks or spots or even light-bordered ocelli (eyelike spots) sometimes present on shoulders. Female and juvenile often with vague light middorsal stripes. Hatchling 2¹⁄₁₆–2⁵⁄₈ in. (5.2–6.7 cm). **SIMILAR SPECIES:** (1) Brown Anole is never green, has a less elongated head, laterally compressed tail, and rust-colored dewlap with light borders. (2) Other introduced anoles have smooth ventral scales. (3) Day geckos lack functional eyelids and have extensive yellow dorsal speckling or distinct red or orange bars or spots. **HABITAT:** Trees, shrubs, vines, buildings, fences. **RANGE:** Green Anole: NC through FL and west to se. OK and cen. TX; established in the Lower Rio Grande Valley; widely introduced, including CA and HI; Cuban Green Anole: apparently restricted to s. FL. **NATURAL RANGE** (Cuban Green Anole): Cuba. **REMARK:** These species interbreed in s. FL.

HISPANIOLAN GREEN ANOLE *Anolis chlorocyanus* **Non-native** **FIG. 133**

5–8½ in. (12.5–22 cm); head-body max. 3⁵⁄₁₆ in. (8.4 cm). Elongated head, pointed snout, smooth ventral scales, tail round in cross section. Uniformly green to mottled green and brown to uniformly brown. Lips and belly whitish or yellowish. Dewlap with white or bluish scales on a *bicolored ground color*, whitish, yellowish, or greenish to pale gray or tan anteriorly, dark purple to black posteriorly. Male often with tan spots on shoulders and sides of neck, sometimes with tan to light purplish spots behind eyes. Female and juvenile with 2–4 longitudinal tan stripes and a tan middorsal zone; areas between

FIG. 133. *The Hispaniolan Green Anole* (Anolis chlorocyanus) *is the only anole in Florida with white or bluish scales on a bicolored dewlap.*

stripes often darker than ground color. Hatchling 2 1/16–2 5/8 in. (5.2–6.7 cm). **SIMILAR SPECIES:** (1) St. Vincent Bush Anole has dark coloration around eyes and a yellow dewlap. (2) Other introduced anoles are stockier, have a less elongated head, are never green, or are much larger and have a prominent casque on head. See also Green Anole. **HABITAT:** Trees, shrubs, vines, buildings. **RANGE:** S. FL. **NATURAL RANGE:** Hispaniola. **REMARK:** Sometimes assigned to the genus *Deiroptyx*.

CRESTED ANOLE *Anolis cristatellus* **Non-native** **PL. 25**

4–7 in. (10–18 cm); head-body max. 3 in. (7.7 cm). Stout body, smooth ventral scales, laterally compressed tail. Dark to light brown to greenish gray; darker bands sometimes evident on body and tail. Male with *distinct crests* on body and tail (like a little dragon). Dewlap olive to mustard, rarely with orange borders. Female and juvenile with a *light middorsal stripe* bordered by thinner dark stripes. **SIMILAR SPECIES:** See Brown Anole. **HABITAT:** Low on trunks of trees, human-made structures. **RANGE:** S. FL. **NATURAL RANGE:** PR and Virgin Is. Also introduced in Costa Rica and elsewhere in the W. Indies. **SUBSPECIES: PUERTO RICAN CRESTED ANOLE** (*A. c. cristatellus*). **REMARK:** Sometimes assigned to the genus *Ctenonotus*.

HISPANIOLAN STOUT ANOLE *Anolis cybotes* **Non-native** **PL. 25**

7–8 in. (18–20.3 cm); head-body max. 3 3/16 in. (8.1 cm). *Large-headed*, stocky, brown to tan, smooth ventral scales, tail round in cross section. Male with light *longitudinal stripes* on sides and dark brown crossbands on body, legs, and at least base of tail. Dewlap pale yellowish, sometimes with a pale orange center. Female with a middorsal light stripe, often with pairs of small yellowish spots on back and shoulders. Belly pale tan to white. **SIMILAR SPECIES:** See Brown Anole. **HABITAT:** Low on trunks of trees, human-made structures. **RANGE:** S. FL. **NATURAL RANGE:** Hispaniola. **SUBSPECIES: HISPANIOLAN STOUT ANOLE** (*A. c. cybotes*). **REMARKS:** Introduced populations sometimes called the Large-headed Anole. Sometimes assigned to the genus *Audantia*.

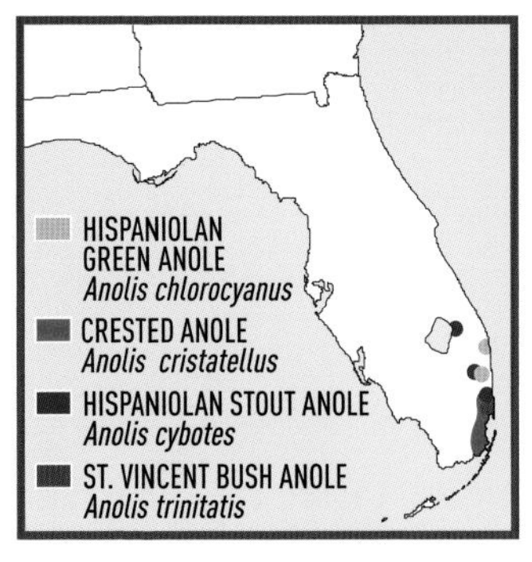

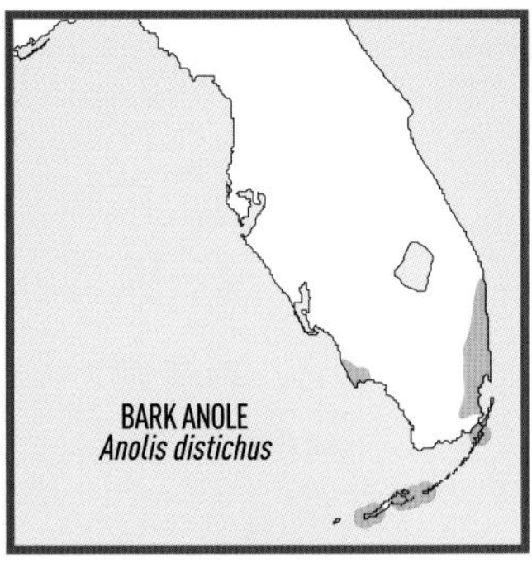

FIG. 134. *Dorsal view of a Bark Anole* (Anolis distichus) *showing faint chevronlike markings.*

BARK ANOLE *Anolis distichus* **Non-native** **PL. 25, FIG. 134**

3½–5 in. (9–12.7 cm); head-body max. 2¼ in. (5.8 cm). Gray, brown, or green with vague dorsal chevrons or irregular spots and streaks resembling lichen-covered bark (Fig. 134, above). Dark line between eyes, *crossbanded tail* (most pronounced near tip); sometimes with 2 ocelli on nape. Dewlap yellow, sometimes tinged with orange. *Scales on snout in paired rows*, top of tail a sawlike ridge. **SIMILAR SPECIES:** See Brown Anole. **HABITAT:** Treetrunks. **RANGE:** S. FL and some Keys. **NATURAL RANGE:** Bahamas and Hispaniola. **REMARKS:** Subspecies from different areas freely interbreed in FL. Sometimes assigned to the genus *Ctenonotus*.

KNIGHT ANOLE *Anolis equestris* **Non-native** **PL. 25**

13–19⅜ in. (33–49.2 cm); head-body max. 7 in. (18 cm). *Prominent bony casque*, especially in large males. Bright apple green, but sometimes dull grayish brown, often with yellow areas that may disappear, especially on tail. *White or yellowish longitudinal lines* on shoulders and extending from below eyes to ears. Dewlap very large, typically pink. **SIMILAR SPECIES:** Jamaican Giant Anole lacks light lines on shoulders and below eyes, has a less prominent casque, larger middorsal spines, and a yellowish orange dewlap. **HABITAT:** Trees, rarely descending to ground. **RANGE:** S. FL and some Keys; numerous records from elsewhere in FL (most probably not established). **NATURAL RANGE:** Cuba. **SUBSPECIES:** **WESTERN KNIGHT ANOLE** (*A. e. equestris*). **REMARKS:** Also called the Cuban Giant Anole. Sometimes assigned to the genus *Deiroptyx*.

JAMAICAN GIANT ANOLE *Anolis garmani* **Non-native** **PL. 25**

5⅞–10⅝ in. (15–27 cm); head-body max. 5¼ in. (13.3 cm). Large head with bony casque. Leaf green to brown, sometimes with 9 or more lighter crossbands. Prominent *spiny crest* from neck onto body. Dewlap lemon yellow with an orangish center. Female with rows of dorsal spots instead of bands; dewlap small and dusky colored. **SIMILAR SPECIES:** See Knight Anole. **HABITAT:** Crowns of trees, rarely descending to ground. **RANGE:** S. FL. **NATURAL RANGE:** Jamaica. **REMARK:** Sometimes assigned to the genus *Norops*.

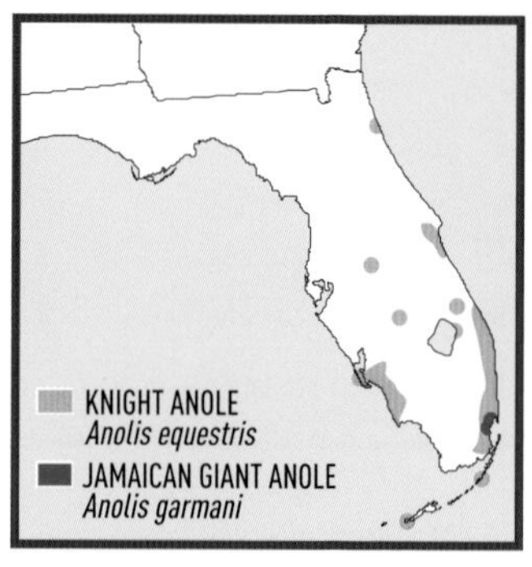

BROWN ANOLE *Anolis sagrei* **Non-native** **PL. 25, FIG. 109**

5–8⅜ in. (12.5–21.3 cm); head-body max. 2¾ in. (7 cm). Stocky, brown or gray, keeled ventral scales, laterally compressed tail, male sometimes with a tail crest (less pronounced than in the Crested Anole). Dewlap usually orange-red with light borders, evident as a white streak on throat of male (Fig. 109, p. 240). Male essentially unicolored to variously marked; small yellow spots in vertical rows on sides and light lateral lines sometimes present. Female and juvenile with a *light middorsal stripe or series of diamond-shaped marks*. **SIMILAR SPECIES:** (1) Hispaniolan Stout Anole lacks a tail crest, has a yellow dewlap, tail round in cross section, and smooth ventral scales. (2) Crested Anole has a prominent tail crest (male only), mustard or olive dewlap, and smooth ventral scales. (3) Green, Cuban Green, and Hispaniolan Green Anoles can be bright green, are slender, have an elongated snout, and tail is round in cross section. (4) Bark Anole is smaller, less chunky, and has crossbands on tail. (5) Knight and Jamaican Giant Anoles are much larger and often green. **HABITAT:** Bushes, rocks, treetrunks, rarely high above ground, often in open areas. **RANGE:** FL and the Keys, s. GA, and along the Gulf Coast to s. TX. Numerous records in other states, including HI. **NATURAL RANGE:** Cuba, Bahamas, Little Cayman I. Widely introduced, including elsewhere in the West Indies. **REMARKS:** At least 2 different subspecies freely interbreed in FL. Sometimes assigned to the genus *Norops*.

ST. VINCENT BUSH ANOLE *Anolis trinitatis* **Non-native** **FIG. 135**

3½–5 in. (9–12.7 cm); head-body max. 2⅞ in. (7.4 cm). Slender, green, often bluish, especially on head and shoulders, gray or brown when

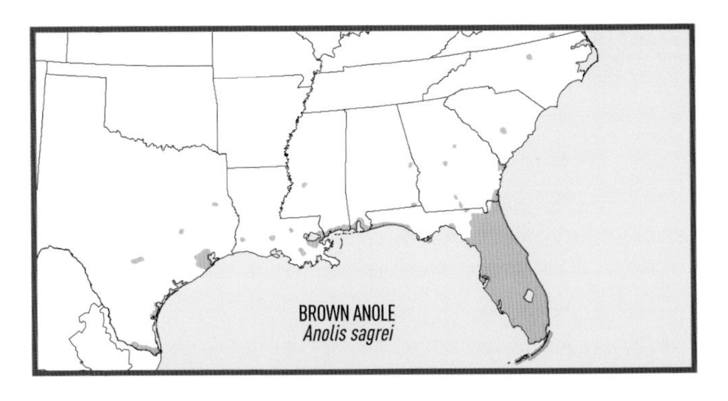

FIG. 135. *The St. Vincent Bush Anole* (Anolis trinitatis) *has dark areas surrounding its eyes. These are much more vivid in males, such as this individual.*

cool or stressed. *Area around eyes blue, gray, or black*; lower jaw, chin, and venter yellow. Female less vivid, with light stripes along lower sides and a mottled middorsal line. Dewlap lemon yellow with scattered blue scales, sometimes greenish. Two rows of middorsal scales enlarged; ventrals smooth. **SIMILAR SPECIES:** See Green and Hispaniolan Green Anoles. **HABITAT:** Trees, shrubs, vines, buildings, rarely on ground. **RANGE:** S. FL (possibly extirpated). Also introduced to Trinidad. **NATURAL RANGE:** St. Vincent. **REMARK:** Sometimes assigned to the genus *Dactyloa*.

IGUANAS: Iguanidae

Large, primarily herbivorous lizards with more than 40 species in eight or nine genera, none native in our area. Three established Neotropical species probably are descendants of animals imported as pets. All have become garden pests, eating mostly non-native ornamental plants.

MEXICAN SPINY-TAILED IGUANA *Ctenosaura pectinata* Non-native **PL. 26**

12–48 in. (30.6–122 cm); head-body max. 13⅝ in. (34.8 cm). Large, robust. *Raised middorsal crest* usually separated from tail crest by 5–10 or more dorsal scale rows. Small, smooth body scales contrast sharply with large, rough, *keeled spines ringing tail*, rings separated by *3 rows* of smaller scales. Adult almost unicolored black, gray, brown, or yellowish brown or with ill-defined darker crossbands. Upper parts of head and neck usually brown, legs darker, belly somewhat lighter, tail with wide yellow and brown bands. Young green with brown, black, and gold markings, tail with prominent dark and light rings. **SIMILAR SPECIES:** (1) Gray's Spiny-tailed Iguana has a continuous crest of scales extending onto tail, rings of spiny scales on tail separated by 2 rows of smaller scales. (2) Green Iguana has greatly enlarged scales beneath ears and lacks whorls of keeled scales on tail. **HABITAT:** Open, sandy, or rocky areas; hides in rock walls, piles of wood, rock, trash, tree hollows, abandoned buildings. **RANGE:** Extreme s. TX (not mapped) and s. FL. **NATURAL RANGE:** W. Mex. **REMARK:** Also called the Western Spiny-tailed or Black Iguana.

GRAY'S SPINY-TAILED IGUANA *Ctenosaura similis* Non-native **FIG. 136**

12–48 in. (30.6–122 cm); head-body max. 19¼ in. (49 cm). Large, robust. *Raised middorsal crest* usually continuing onto tail. Small, smooth body scales contrast sharply with large, rough, *keeled spines ringing tail*, rings separated by *2 rows* of smaller scales. Back and head of adult nearly black to gray, brown, or yellowish brown with broad, dark crossbands; legs densely patterned, appear darker; belly usually without any pattern and lighter; tail with wide yellow and brown bands increasingly indistinct in very large individuals. Young bright green with brown, black, and gold markings. **SIMILAR SPECIES** and **HABITAT:** See Mexican Spiny-tailed Iguana. **RANGE:** S. FL. **NATURAL RANGE:** S. Mex. to Colombia. **REMARK:** Also called the Black Iguana.

GREEN IGUANA *Iguana iguana* complex Non-native **PL. 26**

30–79 in. (76–201 cm); head-body max. 21⅝ in. (55 cm). Large, robust. *Enlarged scales* below ears, prominent *black rings on tail*, middorsal spines extending onto tail. Green, brown, tan, or gray, sometimes with dark bands on shoulders; often darkens and becomes more uniformly colored with age. Male sometimes bright reddish brown or orange, with massive head and jaws, well-developed spines, and a pendulous skinfold beneath throat. Female usually dull green with less-

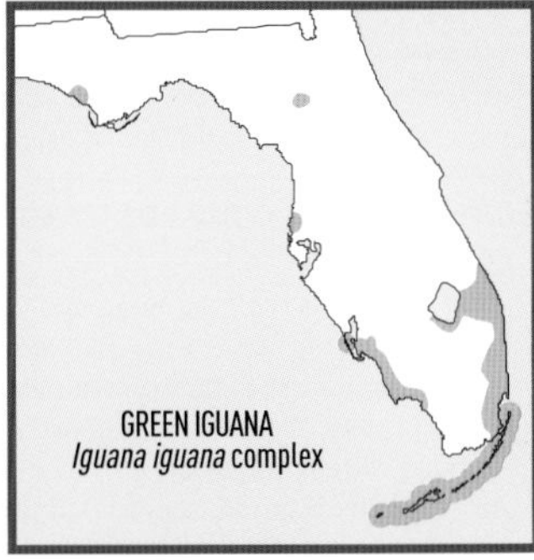

FIG. 136. *Iguanas, such as this Gray's Spiny-tailed Iguana* (Ctenosaura similis), *have become pests in many parts of their introduced ranges, where they burrow under sidewalks and walls and eat ornamental vegetation.*

developed spines and a smaller dewlap. Young bright green. Hatchling head-body length 2½–3⅜ in. (6.4–8.6 cm). **HABITAT:** Trees, especially near water, also open areas. **RANGE:** Peninsular FL and the Keys. Scattered records in other states (not mapped). **NATURAL RANGE:** Mex. to S. America and some W. Indian islands. Widely introduced, including PR and HI. **REMARKS:** Also called the Common Iguana. Genetically unique populations occur in different mainland regions and on some islands; introduced populations are of mixed descent.

CURLY-TAILED LIZARDS: Leiocephalidae

Twenty-nine currently recognized species are native to the West Indies.

NORTHERN CURLY-TAILED LIZARD *Leiocephalus carinatus* **Non-native**
PL. 26, P. vi

7–10½ in. (18–26 cm); head-body max. 5¼ in. (13.3 cm). Gray to dark brown; often with irregular dorsolateral rows of dark spots or blotches, usually most prominent anteriorly; sometimes with small dark dorsal spots and lighter lateral spots. *Large, keeled* dorsal scales. Especially males have a low middorsal crest extending onto indistinctly banded, slightly compressed, *frequently curled tail.* **SIMILAR SPECIES:** Red-sided Curly-tailed Lizard is less "spiny," usually much paler, has prominent skinfolds between fore- and hindlimbs, and male has light blue and reddish orange lateral marks. **HABITAT:** Rocky areas or artificial equivalents (sidewalks, parking lots, stonewalls). **RANGE:** Scattered in peninsular FL, including many Keys. **NATURAL RANGE:** Bahamas, Cuba, Cayman and Swan Is. **SUBSPECIES:** **LITTLE BAHAMA CURLY-TAILED LIZARD** (*L. c. armouri*). **REMARK:** Called the Lion Lizard on English-speaking islands.

RED-SIDED CURLY-TAILED LIZARD *Leiocephalus schreibersii* **Non-native**
PL. 26

7–10½ in. (18–26 cm); head-body max. 4¼ in. (10.7 cm). *Low middorsal crest* becomes more pronounced on the laterally compressed tail; *skinfolds*, composed of very small scales, along sides. Pale tan to sandy; belly grayish or pale blue. Male sprinkled with buffy, yellow, or golden dots, washed laterally with light blue, *reddish orange streaks* along sides and onto belly; area around eyes often reddish orange; throat grayish to purplish with scattered pale blue to green scales; underside of tail orange, mixed with pale blue or brick red scales. Female less brightly colored, with series of grayish transverse dorsal bars, often with small gray to black spots in armpits, and throat streaked or clouded with dark gray. **SIMILAR SPECIES:** See Northern Curly-tailed Lizard. **HABITAT:** Open areas along railroads or fencerows, occasionally residential areas. **RANGE:** S. FL. **NATURAL RANGE:** Hispaniola. **SUBSPECIES:** **RED-SIDED CURLY-TAILED LIZARD** (*L. s. schreibersii*). **REMARK:** Also called the Hispaniolan Khaki Curlytail.

EARLESS, SPINY, TREE, SIDE-BLOTCHED, and HORNED LIZARDS: Phrynosomatidae

Ten genera with more than 135 species range from southern Canada through Central America.

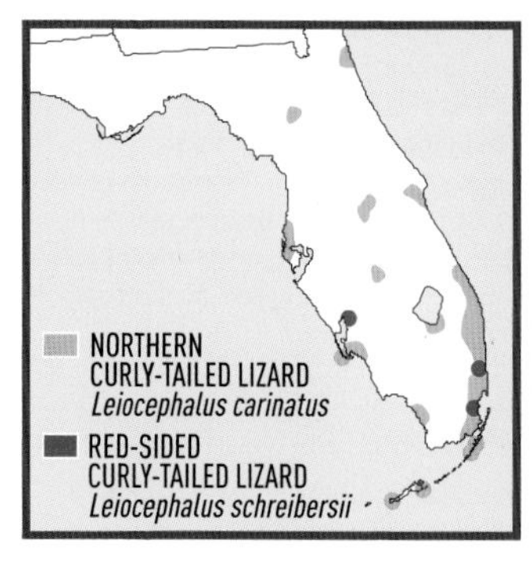

EARLESS LIZARDS: *Cophosaurus* and *Holbrookia*

Earless lizards have no visible ear openings and two folds across the throat, one strongly indicated, the other weak, irregular, or both (Fig. 137, below). They are well adapted for dry, sandy, or loamy soil. The long legs and toes are useful for running, and the head is shaped for diving headfirst into sand. Males often have pairs of bold black lateral bars, which are less distinct or lacking entirely in females. Gravid females exhibit a bright overwash of orange, reddish orange, pink, or yellow. Hatchling Greater Earless Lizard about 2 in. (5.1 cm), hatchlings of other species about 1½ in. (3.8 cm).

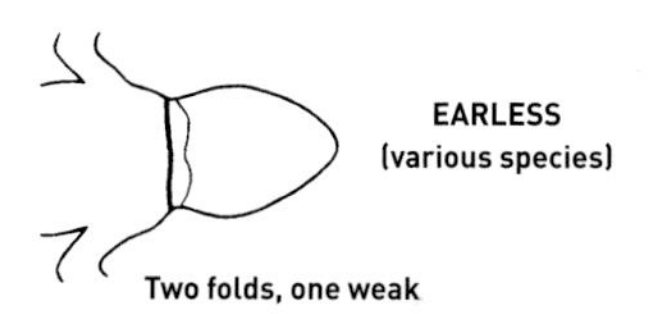

FIG. 137. *Throat folds of earless lizards* (Cophosaurus *and* Holbrookia).

GREATER EARLESS LIZARD *Cophosaurus texanus* **PL. 26, FIG. 110**

Male 3¼–7¼ in. (8.3–18.4 cm); head-body max. 3¼ in. (8.3 cm). *Black bars under tail* (Fig. 110, p. 242) evident when tail is curled over back. General coloration resembles habitats; lizards living on gray soil often grayish, those on reddish soil reddish, etc. *Two bold, black, crescent-shaped lines near groin* in male surrounded by blue or green fields; lines and fields faint or absent in female. Light-bordered *dark stripes on rear of thighs* conspicuous in female and young, less so in large males. **SIMILAR SPECIES:** See Spot-tailed Earless Lizard. **HABITAT:** Rocky desert flats, rocky streambeds, sandstone or limestone outcrops, open sandy washes. **RANGE:** N.-cen. TX to w.-cen. AZ and south into Mex. **SUBSPECIES:** (1) **TEXAS GREATER EARLESS LIZARD** (*C. t.*

texanus); fields surrounding crescent-shaped lines on flanks usually blue. (2) **Chihuahuan Greater Earless Lizard** (*C. t. scitulus*); brightly colored; fields surrounding crescent-shaped lines on flanks blue or green; male with orange or pink markings on anterior body; hindquarters, legs, and tail green or yellowish; gravid female with a rosy tint on flanks. **REMARKS:** Currently recognized subspecies warrant further study. Populations near Eagle Pass, TX, appear to be more closely related to populations in Mex. currently assigned to another subspecies than to the Texas Greater Earless Lizard.

SPOT-TAILED EARLESS LIZARD *Holbrookia lacerata* PL. 26, FIG. 110

4½–6 in. (11.5–15.2 cm); head-body max. 2$^{13}/_{16}$ in. (7.1 cm). Tan to brown, with *dark light-bordered dorsal spots, rounded dark spots under tail* (Fig. 110, p. 242), usually dusky to black oval streaks on sides of abdomen. **SIMILAR SPECIES:** (1) Greater Earless Lizard has black bars under tail. (2) Common Lesser and Keeled Earless Lizards lack dark spots under tail. **HABITAT:** Arid, dark-soil flats, mesquite-prickly pear associations. **RANGE:** Cen. and s. TX and adjacent Mex. **SUBSPECIES:** (1) **Northern Spot-tailed Earless Lizard** (*H. l. lacerata*); dark dorsolateral blotches usually fused to form dorsolateral rows; average 13 femoral pores under each hindlimb. (2) **Southern Spot-tailed Earless Lizard** (*H. l. subcaudalis*); 2 distinct rows of dark dorsolateral blotches; average 16 femoral pores under each hindlimb.

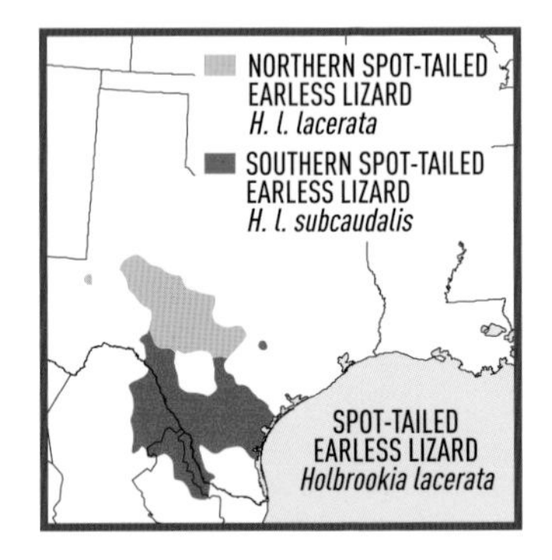

COMMON LESSER EARLESS LIZARD *Holbrookia maculata* PL. 26, FIG. 110

4–5⅛ in. (10–13 cm); head-body max. 2¾ in. (7 cm). Tan to brown, with *pale longitudinal stripes* extending from eyes to base of tail and from armpits to groin, lower stripes less pronounced, often reduced to the neck in western populations; an obscure middorsal stripe sometimes present; 5–15 (average 11–12) femoral pores. Male with pairs of black bars behind armpits, often in small blue fields; dark dorsal blotches sometimes fused to form dorsolateral rows of broad blotches; back usually sprinkled with small white specks, absent in eastern populations. Female and juvenile have reduced black bars without blue field; light-bordered dorsal spots well defined; gravid female strongly tinted with orange or orange-red, brightest on lateral stripes. **SIMILAR SPECIES:** (1) Greater and Spot-tailed Earless Lizards have black bars or spots under tail. (2) Keeled Earless Lizard lacks blue around black bars near armpits of male, has a longer tail and

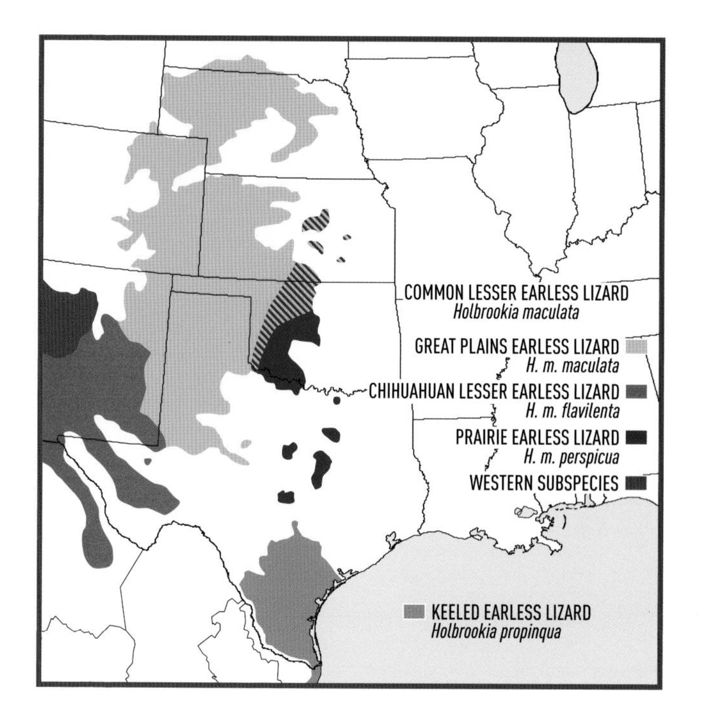

more femoral pores. **HABITAT:** Tallgrass prairie with sandy or loamy soils in northeastern parts of the range, open or sparsely vegetated sandy grasslands to desert grasslands and slopes of intermontane basins farther west; often on dry sandbars in streambeds and fields under cultivation. **RANGE:** S. SD to cen. TX west to AZ and south into Mex. **SUBSPECIES:** (1) **Great Plains Earless Lizard** (*H. m. maculata*); light stripes extending to tail and groin, blotches usually not fused into dorsolateral rows of blotches. (2) **Chihuahuan Lesser Earless Lizard** (*H. m. flavilenta*); back in male strongly speckled, light stripes often restricted to neck. (3) **Prairie Earless Lizard** (*H. m. perspicua*); dark dorsal blotches usually fused to form dorsolateral rows of broad blotches, no dorsal speckling. **REMARK:** Prairie Earless Lizard might not be distinct from the Great Plains Earless Lizard.

KEELED EARLESS LIZARD *Holbrookia propinqua* **PL. 26, FIG. 110**

4½–5⁹⁄₁₆ in. (11.5–14.1 cm); head-body max. 2⅜ in. (6 cm). Tan to brown, *tail long* (original tail noticeably longer than head-body length in male and as long or longer in female). Dorsal scales keeled (keels very small). Two black lines on sides of male (behind armpits) *not* surrounded by blue; dorsal pattern usually a combination of dark blotches and dark longitudinal stripes covered by many small light dots. Females from coastal areas lack prominent markings, inland females darker, with dorsal blotches. Ten to 21 (usually 14–15) femoral pores. Young with a defined pattern of paired blotches. **SIMILAR SPECIES:** See Common Lesser Earless Lizard. **HABITAT:** Sand dunes and barrier island beaches. **RANGE:** S. TX south into Mex. **SUBSPECIES: NORTHERN KEELED EARLESS LIZARD** (*H. p. propinqua*). **REMARK:** Northwestern populations represent a lineage that might be worthy of taxonomic distinction.

HORNED LIZARDS: *Phrynosoma*

Seventeen currently recognized species range from southern Canada to Guatemala. When stressed, most can squirt blood from the eyes. Apparently, chemicals in blood acquired from eating ants cause lizards to be distasteful to at least some predators (e.g., coyotes). Hatchlings 1⅛–1½ in. (2.8–3.8 cm). Horned lizards are protected by law in many states.

TEXAS HORNED LIZARD *Phrynosoma cornutum* **PL. 27, FIGS. 111, 138**

2½–4 in. (6.4–10 cm); record 7⅛ in. (18.1 cm); head-body max. 5⅛ in. (13 cm). Yellowish, reddish, grayish brown or tan, sometimes gray; dark dorsal spots have light posterior borders. Two central horns *much longer* than others (Fig. 111, p. 244). *Two rows of fringe scales* at sides of abdomen. **SIMILAR SPECIES:** (1) Round-tailed Horned Lizard has a slender, rounded tail that broadens abruptly near base, larger horns about equal in length, and no fringe scales along edges of abdomen. (2) Greater Short-horned Lizard has greatly reduced horns and one row of fringe scales. **HABITAT:** Flat, open terrain with sparse plant cover. **RANGE:** KS to se. AZ south into Mex.; many peripheral isolates. Widely introduced in the Southeast (not mapped).

GREATER SHORT-HORNED LIZARD *Phrynosoma hernandesi* **PL. 27, FIG. 111**

2½–5 in. (6.4–12.5 cm); record 6 5/16 in. (15.9 cm); head-body max. 4⅜ in. (11.2 cm). Brown or gray, often with orange to reddish brown on back and belly, especially in southern populations. *Horns short and stubby* (Fig. 111, p. 244), *single row of fringe scales* at sides of abdomen. **SIMILAR SPECIES:** See Texas Horned Lizard. **HABITAT:** Semiarid, shortgrass prairies to forested slopes and uplands high in the Rocky Mts. **RANGE:** S. AB and SK south to Mex. **REMARK:** A very recent study suggested that this species is part of a complex of as many as 7 species, 4 of which might occur in our area.

ROUND-TAILED HORNED LIZARD *Phrynosoma modestum* **PL. 27, FIG. 111**

3–4⅛ in. (7.5–10.5 cm); head-body max. 2 13/16 in. (7.1 cm). Yellowish gray or ashy white to light brown; adult often matches predominant soil color, less evident in young. Intensity of dark blotches can change

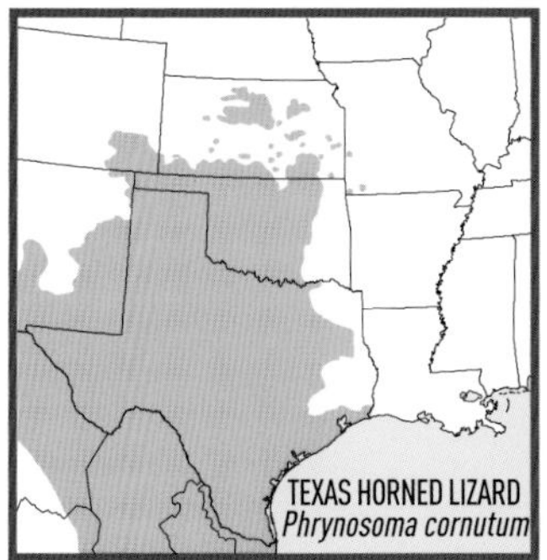

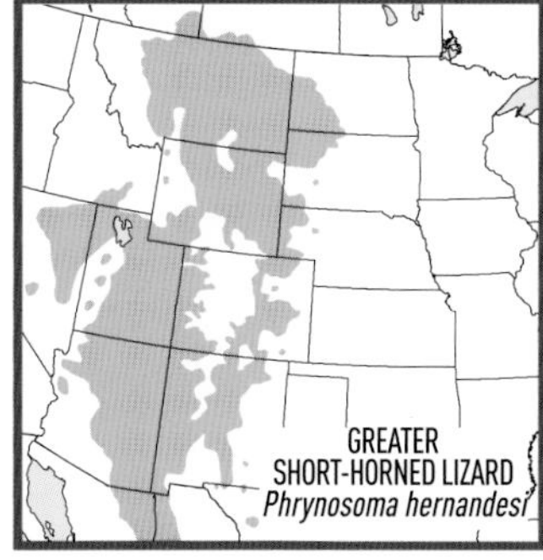

and varies from black to slightly darker than ground color. *Tail round and slender*, broadens abruptly near base. *Four horns about equal in length* (Fig. 111, p. 244). *No fringe scales* along sides of abdomen. **SIMILAR SPECIES:** See Texas Horned Lizard. **HABITAT:** Desert flats and washes, arid and semiarid plains with shrubby vegetation. **RANGE:** W. TX to se. AZ south into Mex.; peripheral isolates as far north as CO.

FIG. 138. *The pattern of a Texas Horned Lizard* (Phrynosoma cornutum) *is an effective camouflage and the rows of lateral fringe scales break up the lizard's outline when viewed from above. The combined effect renders this harmless little lizard quite cryptic.*

SPINY LIZARDS: *Sceloporus*

Nearly 100 species range from the Pacific Northwest and southern New York to Panama. Most in our area are arboreal to some extent, although they readily exploit elevated perches on rock outcrops or boulders, large rotting stumps or logs, and fences or walls. Males of most species have blue to black patches on the belly and some blue on the throat. These are exhibited when individuals defend territories or seek to attract mates. Blue areas in females are absent or only slightly developed. Middorsal dark markings, which serve as camouflage, often are faint or reduced in males but prominent in females.

DUNES SAGEBRUSH LIZARD *Sceloporus arenicolus* **PL. 27**

4½–6 in. (11.5–15.2 cm); head-body max. 2¾ in. (7 cm). *Very pale* with faint grayish brown dorsolateral stripes extending from ears onto tail. Dorsum light yellowish brown, belly white or yellowish. *Scales on rear of thighs small and granular*, not keeled. Male with large bright blue patches on sides of belly, bordered midventrally by bands of deeper blue or black. No blue on throat. Hatchling 1¾ in. (4.4 cm). **SIMILAR SPECIES:** (1) Other spiny lizards in range have large scales on rear of thighs. (2) Eastern Side-blotched Lizard has dark spots posterior to armpits. **HABITAT:** Sparsely vegetated sandhills. **RANGE:** W. TX and se. NM. **CONSERVATION:** Vulnerable.

TWIN-SPOTTED SPINY LIZARD *Sceloporus bimaculosus* **PL. 28**

7½–13 in. (19–33 cm); head-body max. 5½ in. (14 cm). Stocky, pale gray or brown to straw color, some scales on sides may be yellow. Large pointed scales, narrow dark lines running back from lower corners of eyes, *black wedges or blotches on sides of neck*; twin spots on back not sharply defined. Legs finely striped longitudinally with pale and dark gray. Male has a black groin, blue-green throat patches, and 2 elongated blue-green belly patches edged with black and often joined. Mature male may have a dorsal wash of pale blue about 5 scales wide on anterior back. Female essentially lacks blue pigment. Markings in young prominent; black wedges bordered by yellow and 4 rows of black spots on back and sides. Hatchling 2⅞–3⅜ in. (7.3–8.6 cm). **SIMILAR SPECIES:** Blue and Crevice Spiny Lizards have a white-bordered dark band across the neck. **HABITAT:** Arid and semiarid areas, seldom far from thickets, rock piles, old buildings. **RANGE:** W. TX to se. AZ and south into Mex.

PRAIRIE LIZARD *Sceloporus consobrinus* **PL. 27, FIG. 139**

3½–7½ in. (9–19.1 cm); head-body max. 3 in. (7.6 cm). Populations from the Mississippi R. west to e. KS, OK, and TX are essentially indistinguishable from the Eastern Fence Lizard. Populations farther west are much lighter, light to reddish brown, have a light brown middorsal

FIG. 139. *The Prairie Lizard* (Sceloporus consobrinus) *typically occupies open areas with sufficient brush or mammal burrows in which to seek refuge. However, eastern populations occur primarily along edges or in open patches of wooded areas, where they often exploit elevated perches on logs, stumps, and fences. In this, they are much like the closely related Eastern Fence Lizard* (S. undulatus), *which they strongly resemble.*

stripe and *bold, light dorsolateral and lateral longitudinal stripes*, with dorsolateral stripes often brighter than lateral stripes. Dark dorsal markings reduced to spots bordering light dorsolateral stripes. Populations in northwestern parts of the range have orange or red coloration on chin and lips. Male has long narrow light blue patches on sides of belly bordered medially with black and often in contact.

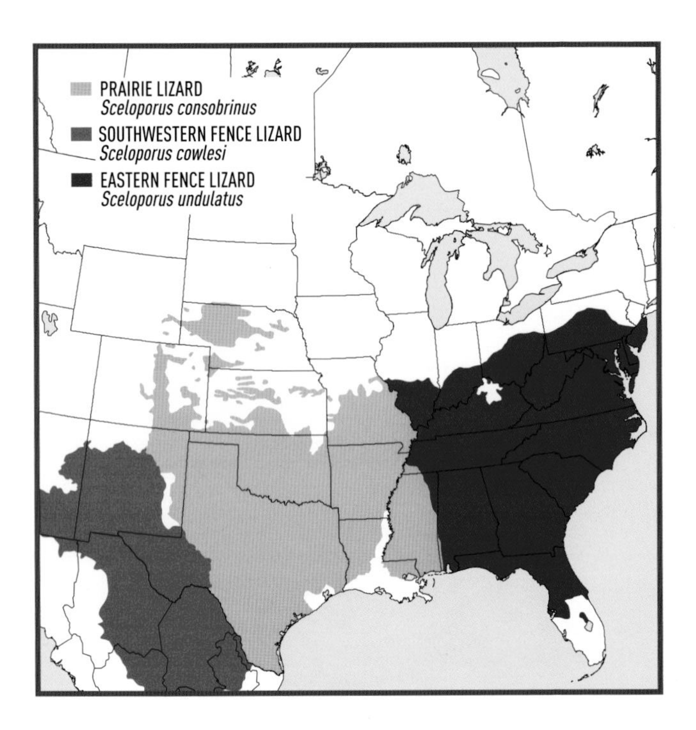

Throat markings absent or reduced to 2 small, widely separated blue patches. Female uniformly white below or with black flecks. Hatchling 1¾–2 in. (4.4–5.1 cm). **SIMILAR SPECIES:** (1) Eastern and Southwestern Fence Lizards can be impossible to distinguish from eastern and western populations of the Prairie Lizard and are best identified by range. (2) Rose-bellied Lizard has skin pockets behind thighs and male has pink patches on sides of belly. (3) Texas Spiny Lizard is larger and has very large, spiny dorsal scales (33 or fewer from back of head to base of tail vs. 35 or more in the Prairie Lizard). **HABITAT:** Eastern populations similar to that of the Eastern Fence Lizard; western populations in dunes, other sandy areas, open prairies (with weeds, brush, or mammal burrows for cover), brushy flatlands, rocky mountain slopes, cliffs, lava flows, gully banks. **RANGE:** Extreme s. SD south to the Gulf in TX and as far east as extreme sw. AL. **REMARKS:** Until recently considered conspecific with Eastern and Southwestern Fence Lizards. The correct name for this species might be *S. thayerii*.

SOUTHWESTERN FENCE LIZARD *Sceloporus cowlesi* **NOT ILLUS.**

3½–7 in. (9–18 cm); head-body max. 3 in. (7.5 cm). Essentially indistinguishable from the Prairie Lizard and best identified by range. Populations in the White Sands of NM are very pale gray to almost white and virtually patternless. **SIMILAR SPECIES:** See Prairie Lizard. **HABITAT:** Dunes, other sandy areas, open prairies, brushy flatlands, rocky mountain slopes. **RANGE:** W. TX to e. AZ and south into Mex. **REMARK:** Until recently considered conspecific with Eastern Fence and Prairie Lizards.

BLUE SPINY LIZARD *Sceloporus cyanogenys* **PL. 28**

5–14¼ in. (12.5–36.2 cm); head-body max. 5 13/16 in. (14.8 cm). *Dark, white-bordered band across neck* and *light spots on nape and back*; tail markings *not* conspicuous at tip. Adult male metallic greenish blue over brown ground color; upper sides of legs bronzy; throat bright blue; large light blue patches on sides of belly bordered medially by black bands. Female and young gray or brown. Young 2½–2¾ in. (6.4–7 cm) at birth. **SIMILAR SPECIES:** (1) Crevice Spiny Lizard lacks light spots on nape and back and has conspicuous black bands near tailtip. (2) Desert Spiny Lizard has black wedges or blotches on sides of neck but no dark band across neck. **HABITAT:** Boulders, rocky or earthen cliffs, stone bridges, abandoned houses. **RANGE:** S. TX into Mex. **REMARK:** Formerly considered a subspecies of *S. serrifer*, a name now restricted to populations in Cen. America.

COMMON SAGEBRUSH LIZARD *Sceloporus graciosus* **PL. 27**

4½–5⅞ in. (11.5–14.9 cm); head-body max. 2¾ in. (7 cm). Pale brown, with 4 longitudinal rows of darker brown spots that sometimes run together to suggest dark longitudinal stripes; usually 2 pale dorsolateral stripes, one extending back from eyes and terminating on tail; often with irregular black spots on shoulders. *Small granular scales on rear of thighs.* Male with large elongated blue patches on sides of belly; no throat patches, although throat sometimes mottled with blue. Hatchling 2½ in. (6.4 cm). **SIMILAR SPECIES:** Other spiny lizards in range have large keeled scales on rear of thighs. **HABITAT:** Sagebrush flats, rocky areas, open forests, canyon bottoms. **RANGE:** W. ND to nw. NM and west to the Pacific Coast.

GRAPHIC SPINY LIZARD *Sceloporus grammicus* **PL. 27, FIG. 140**

4–6⅞ in. (10–17.5 cm); head-body max. 2⅞ in. (7.3 cm). Gray or olive gray, dorsum in male sometimes with a metallic greenish luster; markings obscure; dark vertical lines in front of forelimbs; sides of belly pale blue bordered medially by black; throat mottled with black except for pinkish or pale blue center. Female with 4–5 dark wavy lines across back; forelimbs distinctly barred. Lateral neck scales *much smaller* than those on nape (Fig. 140, p. 296) Young 1⅝–1 15/16 in.

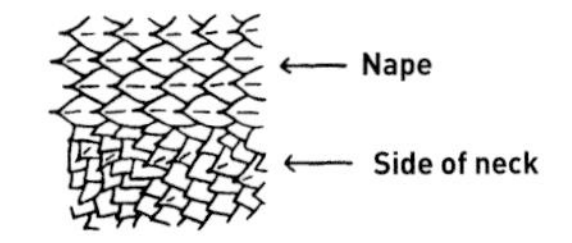

FIG. 140. *Lateral neck scales in the Graphic Spiny Lizard* (Sceloporus grammicus).

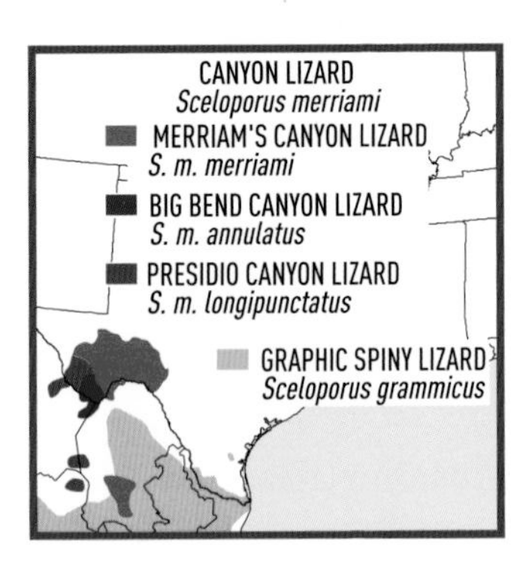

(4.1–4.9 cm) at birth. **SIMILAR SPECIES:** (1) Rose-bellied and Southwestern Fence Lizards usually have light dorsolateral stripes. (2) Texas Spiny Lizard is very large and has 33 or fewer dorsal scales from back of head to base of tail (50 or more in the Graphic Spiny Lizard). (3) Ornate Tree Lizard has skinfolds across throat. **HABITAT:** Mesquite, also other scrubby trees. **RANGE:** S. TX south into Mex. **REMARKS:** Populations currently assigned to the Graphic Spiny Lizard almost certainly represent a complex of species. The status of U.S. populations likely will change after further study.

CANYON LIZARD *Sceloporus merriami* **PL. 27, FIG. 141**

4½–6¼ in. (11.5–15.9 cm); head-body max. 2 7/16 in. (6.2 cm). Gray, tan, reddish brown, often matching general coloration of cliffs and boulders. Four rows of dark dorsal spots most conspicuous in western populations. *Vertical black bars* on shoulders sometimes obscured by skinfolds. *Scales on sides granular*; keeled dorsal scales small; scales on sides of neck abruptly smaller than those on nape. Throat folds partially developed (Fig. 141, facing page). Male has 2 large blue belly patches with black margins midventrally and posteriorly; a light midventral stripe might separate blue patches or black margins can merge. Blue and black lines on throat. Transverse lines often present under tail. Male with a small but conspicuous dewlap; female with less extensive ventral markings and a considerably smaller dewlap. Hatchling about 2 in. (5.1 cm). **SIMILAR SPECIES:** (1) Other spiny lizards in range have large keeled scales on sides. (2) Ornate Tree and Eastern Side-blotched Lizards have complete skinfolds across throat. **HABITAT:** Boulders, rocky outcrops, rocky walls of canyons. **RANGE:** W. TX and adjacent Mex.; disjunct in n. Mex. **SUBSPECIES:** (1) **MERRIAM'S CANYON LIZARD** (*S. m. merriami*); pale ground color, dorsal spots faint or absent, usually 58 or more dorsal scales from back of head to base of tail, narrow light midventral stripes separate blue belly patches in male, dark ventral markings reduced. (2) **BIG BEND CANYON LIZARD** (*S. m. annulatus*); dark dorsal spots well defined, usually 52 or fewer dorsal scales from back of head to base of tail, ventral markings bold and

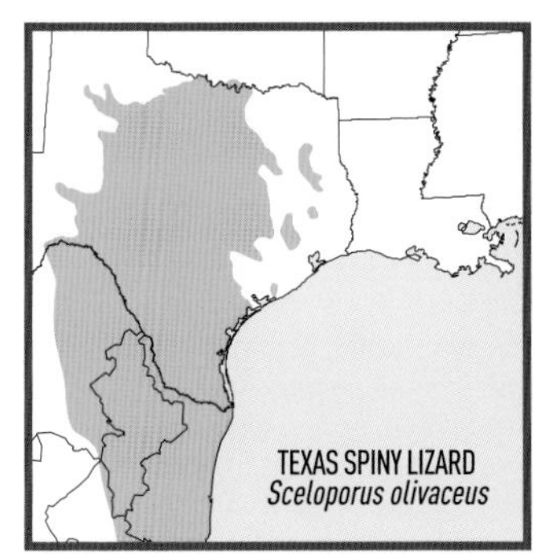

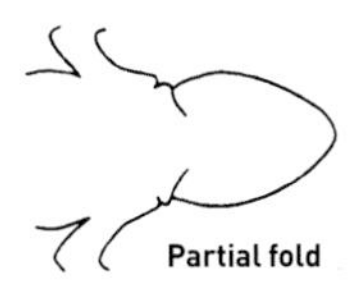

FIG. 141. *Partially developed throat fold of the Canyon Lizard* (Sceloporus merriami).

extensive in male. (3) **Presidio Canyon Lizard** (*S. m. longipunctatus*); dorsal spots comma-shaped, with tail of commas pointing outward, ventral markings paler, less extensive than in the Big Bend Canyon Lizard.

TEXAS SPINY LIZARD *Sceloporus olivaceus* **PL. 28**

7½–11¾ in. (19–29.9 cm); head-body max. 4¾ in. (12.1 cm). Gray brown to rusty brown; vague dorsolateral light stripes more pronounced in male, wavy dark lines across back more conspicuous in female. *Very large dorsal scales*, 33 or fewer (average 30) from back of head to base of tail. Male with narrow light blue areas *without* black borders along sides of belly. Hatchling 2½–2⅞ in. (6.4–6.7 cm). **SIMILAR SPECIES:** See Prairie Lizard. **HABITAT:** Usually in trees, also on fences, old bridges, abandoned houses, patches of prickly pear, other elevated perches. **RANGE:** Extreme s. OK south into Mex.

CREVICE SPINY LIZARD *Sceloporus poinsettii* **PL. 28**

5–11½ in. (12.5–29.2 cm); head-body max. 5⅛ in. (13 cm). Gray or greenish gray to orange or reddish, with *white-bordered dark collar across neck* and a *strongly barred tail* (especially *near tip*). Female and young with bold black crossbands, retained in adult male only across neck and on tail. Male has bright blue throat and sides of belly, latter bordered medially by broad black bands. Young sometimes with a narrow dark middorsal stripe connecting crossbands; 2½–3 in. (6.4–7.5 cm) at birth. **SIMILAR SPECIES:** See Blue Spiny Lizard. **HABITAT:** Boulders, rocky outcrops. **RANGE:** Cen. TX and s. NM south into Mex. **REMARK:** Subspecies (not shown) warrant further study.

EASTERN FENCE LIZARD *Sceloporus undulatus* **PL. 27**

4–7¼ in. (10–18.4 cm); head-body max. 3⅜ in. (8.6 cm). Gray or brown, with more or less complete *dark lines along rear of thighs*. Male brown, sides of belly purplish to greenish blue, bright colors medially bordered by black; broad bluish area at base of throat, blue surrounded by black and often split in two; dorsal crosslines indistinct or absent. Female gray or grayish brown, with a series of dark wavy (undulating) lines across back; frequently some yellow, orange, or red at base of tail; belly whitish with scattered black flecks; small amounts of pale blue at sides of belly and throat. Young like female but darker and duller. Hatchling 1⅝–2¼ in. (4.1–5.7 cm). **SIMILAR SPECIES:** (1) Florida Scrub Lizard has distinct dark brown dorsolateral stripes. (2) E. populations of the Prairie Lizard are impossible to distinguish and best identified by range. **HABITAT:** Elevated perches such as fences, logs, stumps. **RANGE:** S. NJ west to the Mississippi R. and south to cen. FL and the Gulf Coast east to AL. **REMARK:** Prairie Lizard, Southwestern Fence Lizard, and populations farther west were assigned to this species before it was shown to be a complex of species.

ROSE-BELLIED LIZARD *Sceloporus variabilis* **PL. 27, FIG. 142**

3¾–5½ in. (9.5–14 cm); head-body max. 2¼ in. (5.7 cm). Buffy to olive brown; *light dorsolateral stripes* along 2 rows of brown spots; *skin pockets behind thighs* (Fig. 142, below). Male with large *pink areas* on sides of belly bordered by dark blue; dark color extending onto sides to form *prominent dark spots in armpits* and smaller spots in groin. Young 2 in. (5.1 cm) at birth. **SIMILAR SPECIES:** See Prairie Lizard. **HABITAT:** Arid areas, often on fenceposts, clumps of cactus, occasionally rocks, mesquite, or other scrubby trees. **RANGE:** S. TX south into Mex. **SUBSPECIES:** **TEXAS ROSE-BELLIED LIZARD** (*S. v. marmoratus*). **REMARK:** Populations in our area might be distinct at the species level from subspecies in Mex.

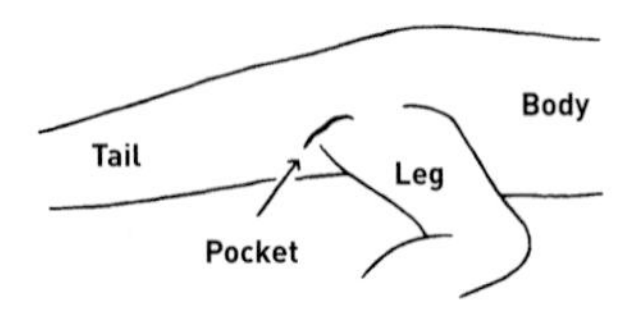

FIG. 142. *Skin pocket behind the thigh of a Rose-bellied Lizard* (Sceloporus variabilis).

FLORIDA SCRUB LIZARD *Sceloporus woodi* **PL. 27**

3 13/16–5 11/16 in. (9.7–14.4 cm); head-body max. 2 9/16 in. (6.5 cm). Brown or gray brown, with *dark brown* dorsolateral stripes; male with largely unmarked middorsal area, medially black-bordered blue patches on sides of belly, pairs of blue spots at base of throat, throat otherwise black except for median white stripes. Female and juvenile with 7–10 dark brown wavy lines across back, rarely fused or otherwise altered to form dark longitudinal lines; dark spots often present on chest and underside of head; some blue on throat and sides of belly not bordered by black; young usually paler than female. Belly whitish. Hatchling 1¾ in. (4.4 cm). **SIMILAR SPECIES:** See Eastern Fence Lizard. **HABITAT:** Sand pine scrub, adjacent beach dune scrub, longleaf pine–turkey oak woodlands, citrus groves with areas of open sandy ground. **RANGE:** Disjunct in cen. and s. FL.

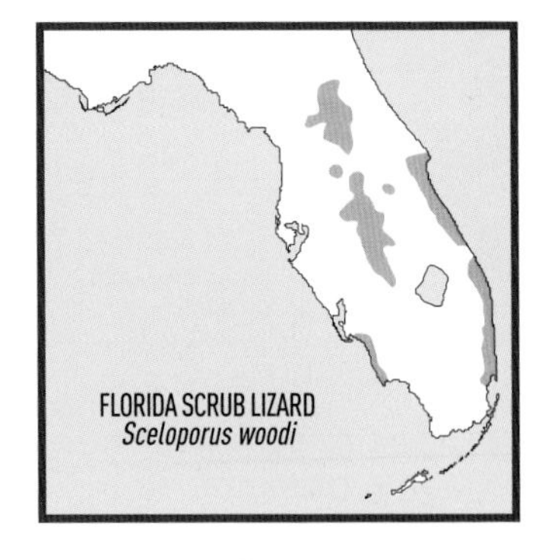

TREE and SIDE-BLOTCHED LIZARDS: *Urosaurus* and *Uta*

Nine currently recognized species in the genus *Urosaurus* and eight in the genus *Uta* range from the southwestern United States through Mexico.

ORNATE TREE LIZARD *Urosaurus ornatus* **PL. 28**

4–5⅜ in. (10–13.7 cm); head-body max. 2 in. (5.1 cm). Gray or grayish brown, *skinfolds across throat* (Fig. 112, p. 246), *dorsal scale rows variable in size, 2 long skinfolds* along sides, tail long. Back with irregular dark spots or crossbands, throat yellow, pale orange, whitish or blue in adult. Male has blue patches on throat and sides of belly, latter sometimes joined to form entirely blue belly. Female lacks belly patches and throat is rarely bluish. Hatchling 1¼–1¾ in. (3.2–4.4 cm). **SIMILAR SPECIES:** (1) Eastern Side-blotched Lizard has distinct spots behind armpits. (2) Spiny lizards lack complete throat folds. **HABITAT:** Elevated perches such as trees or rocks. **RANGE:** Cen. TX and the Rio Grande Valley and adjacent Mex. west to se. CA, north to sw. WY and south into w. Mex. **SUBSPECIES:** (1) **Texas Tree Lizard** (*U. o. ornatus*); enlarged scales of inner dorsolateral series about twice as large as those of outer series. (2) **Big Bend Tree Lizard** (*U. o. schmidti*); inner series of enlarged dorsal scales not twice as large as those of outer series.

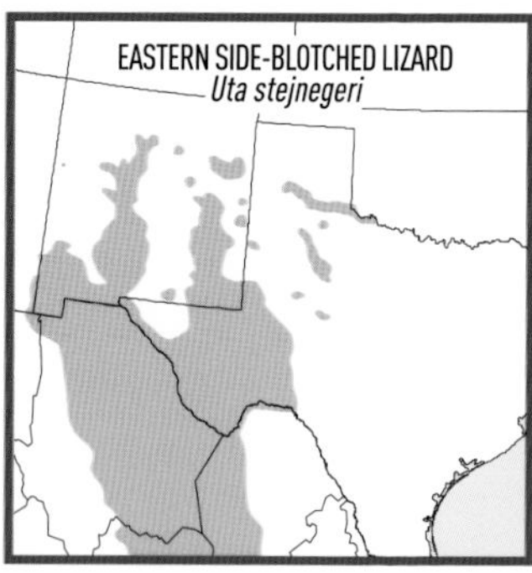

EASTERN SIDE-BLOTCHED LIZARD *Uta stejnegeri* **PL. 28**

4–5³/₈ in. (10–13.7 cm); head-body max. 2¹/₈ in. (5.4 cm). Brown, with *single bluish black spot behind armpit* visible in all but very smallest individuals. *Single skinfold across throat* (Fig. 112, p. 246); middorsal scales keeled, larger than small, smooth lateral scales. Female and juvenile with pale, dark-bordered stripes from eyes to base of tail, stripes or series of pale spots from upper lip to groin. Stripes in adult male reduced or absent; *back with profusion of blue flecks* and yellow to pale orange-brown spots on sides; throat and lower sides washed with pale blue or pale bluish gray. Hatchling 2⁵/₁₆ in. (5.9 cm). **SIMILAR SPECIES:** See Ornate Tree Lizard. **HABITAT:** Sandy regions, desert flats and foothills, rocky areas. **RANGE:** Extreme sw. OK and w. TX to se. AZ and south into Mex. **REMARK:** Until recently considered a subspecies of the Common Side-blotched Lizard (*U. stansburiana*), which occurs west of our area.

SCINCOMORPH and VARANOID LIZARDS

Five of the 15 families of scincomorph lizards occur in our area. Three (Scincidae, Sphenomorphidae, Teiidae) include species native to our area; two (Lacertidae, Mabuyidae) are represented only by non-native species. The families Mabuyidae, Scincidae, and Sphenomorphidae, along with six other families not represented in our area, are sometimes considered subfamilies of a more broadly defined family Scincidae. Regardless, all are called skinks. Lizards in the families Lacertidae and Teiidae might be distinct from other scincomorph taxa. The one varanoid family (Varanidae) with a presence in eastern North America is represented by a single introduced species.

WALL LIZARDS and LACERTAS: Lacertidae

More than 300 species in more than 40 genera occur throughout much of Europe, Africa, and Asia. Granular dorsal scales contrast sharply with rows of enlarged rectangular ventral scales. Superficially similar to native whiptails and racerunners (Teiidae), wall lizards have a single enlarged scale

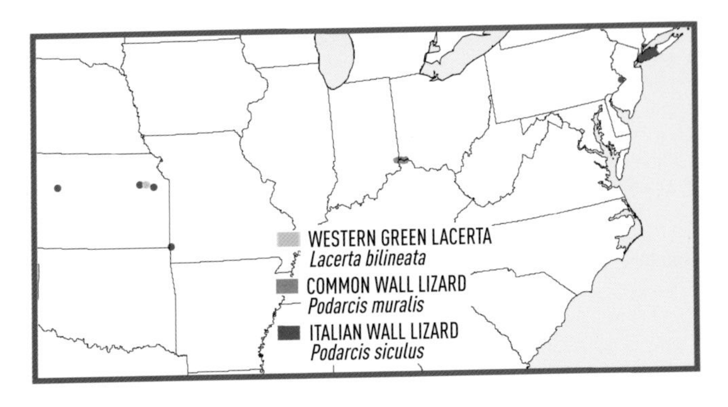

immediately in front of the vent (racerunners and whiptails have several enlarged scales). At least one other species has been reported from our area but is not established. Many subspecies of all three established species are currently recognized; the subspecific identity of the introduced populations is uncertain.

WESTERN GREEN LACERTA *Lacerta bilineata* Non-native **FIG. 143**

6–12⅝ in. (15.2–32 cm); head-body max. 4⅛ in. (10.5 cm) (larger in native range). Back green (brighter in male) variously speckled with dark spots, head usually darker with fewer speckles; throat white,

FIG. 143. *Introduced populations of the Western Green Lacerta* (Lacerta bilineata) *are associated with yards and gardens near buildings, often in and around piles of rubble and other debris.*

chest and belly yellowish or cream, sometimes with dark spots; tail green or olive. Male often with blue on lips extending back to below ears. Female brownish or more subdued green. Young brown on back, sides pale green to yellowish. Belly with 4 rows of enlarged rectangular scales. **SIMILAR SPECIES:** See Common Wall Lizard. **HABITAT:** Yards and gardens, often around buildings. **RANGE:** Topeka, KS. **NATURAL RANGE:** W. Europe.

COMMON WALL LIZARD *Podarcis muralis* **Non-native** **PL. 29**

5½–8⅛ in. (14–20.5 cm); head-body max. 3 in. (7.5 cm). Back usually uniformly brown to gray (occasionally tinged with green), often with a row of dark middorsal spots sometimes fused into a stripe; some individuals have a reticulate pattern with dark markings on sides and many scattered white spots that may be blue in the shoulder region. Tail brown, gray, or rusty red, sometimes with light and dark bars on sides. Belly with 6 rows of large, smooth, uniformly white to reddish, pink, or orange rectangular scales; dark markings sometimes present on throat. **SIMILAR SPECIES:** (1) Italian Wall Lizard is similar and best distinguished by range; almost never has spots on throat; light markings, if present on sides of tail, usually not as spots or bars. (2) Western Green Lacerta has 4 rows of enlarged belly scales. (3) Six-lined Racerunner has enlarged belly scales in 8 rows, a distinctly striped back, and several enlarged scales immediately in front of vent. **HABITAT:** Crumbling rock walls, debris piles. **RANGE:** Sw. OH, adjacent IN and KY. Also introduced in s. England and Vancouver I., BC. **NATURAL RANGE:** S. Europe into Turkey.

ITALIAN WALL LIZARD *Podarcis siculus* **Non-native** **PL. 29**

5½–9½ in. (14–24 cm); head-body max. 3 9/16 in. (9.1 cm). Head, neck, and upper body often green with a brown middorsal stripe or row of spots; some individuals have an irregular reticulate pattern of dark markings on light green, yellow, olive, or light brown ground color. Blue spot or spots sometimes present on shoulders. Tail generally brown or gray, sometimes with light and dark markings on sides. Belly and throat uniformly white or gray. Six rows of large, smooth, rectangular ventral scales. **SIMILAR SPECIES:** See Common Wall Lizard. **HABITAT:** Urban areas, vegetation or debris piles. **RANGE:** New York City and Long Island, NY; Greenwich, CT; Topeka, Hays, and Lawrence, KS; Joplin, MO. Presumably extirpated in Philadelphia, PA. Also introduced in CA. **NATURAL RANGE:** S. Europe.

Skinks

Skinks occur on all habitable continents, but mostly in the Eastern Hemisphere. Three of nine families occur in our area. All 15 species have smooth, shiny scales underlaid with bony plates called osteoderms. Dorsal and ventral scales are similar in size and shape.

FIG. 144. *Until recently, the Common Sun Skink* (Eutropis multifasciata) *was assigned to the genus* Mabuya. *This species also is called the Brown Mabuya, Many-lined or Many-striped Skink, or Golden Skink.*

SUN SKINKS: Mabuyidae

Twenty-two genera and about 190 species range throughout much of tropical America, Africa, and southern Asia.

COMMON SUN SKINK *Eutropis multifasciata* **Non-native** **FIG. 144**

7–9 in. (17.5–22.5 cm); head-body max. 5⅜ in. (13.7 cm). Back bronze brown; black-edged scales appear as narrow, black longitudinal lines; pale dorsolateral lines often evident but sometimes essentially unicolored; usually with yellow or red stripes and series of white spots, ocelli, or streaks along flanks; yellow or red stripes in breeding male turn bright orange or reddish orange. Belly cream to greenish or bluish white. Dorsal scales with 3–5 keels, less prominent laterally. Young head-body length 1¼–1¾ in. (3.3–4.3 cm) at birth. **SIMILAR SPECIES:** Ocellated Skink has regular transverse rows of ocelli and "windows" in lower eyelids. **HABITAT:** Open sunny patches in landscaped areas. **RANGE:** S. FL. **NATURAL RANGE:** S. Asia, Indonesian and

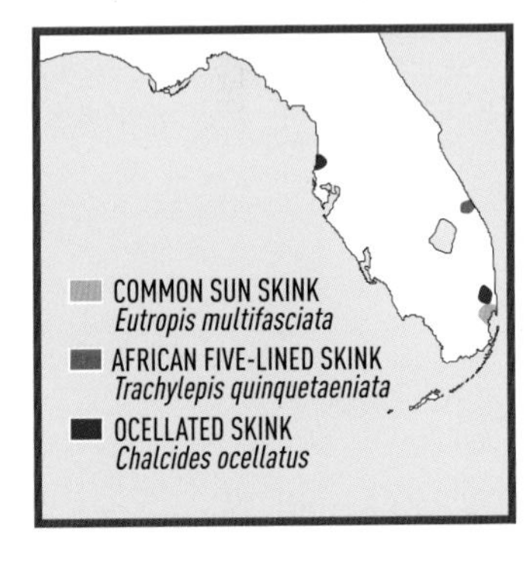

FIG. 145. *The male (left) and female African Five-lined Skink* (Trachylepis quinquetaeniata) *differ primarily in the intensity of the pattern and color. Like the Common Sun Skink* (Eutropis multifasciata), *this species until recently was assigned to the genus* Mabuya. *It often is called the Rainbow Skink.*

Philippine Archipelagos, probably New Guinea. Also introduced in Australia. **REMARK:** Also called the Brown Mabuya.

AFRICAN FIVE-LINED SKINK *Trachylepis quinquetaeniata* Non-native

FIG. 145

3½–5¼ in. (8.9–13.3 cm); head-body max. 4 in. (10.1 cm). Stocky with a relatively long tail (1½ times head-body length). Back brown to olive with 5 cream to yellowish longitudinal stripes, *most lateral stripes extending forward onto upper lip.* Three dorsal stripes fade in adult male, but yellowish orange lateral stripes usually obvious at least from lips to forelimbs; male often with a black throat and sides of neck, sometimes with white specks. Tail yellowish brown, belly grayish white. Juvenile dark brown to black with prominent stripes and a blue tail. Lower eyelids with clear "windows"; dorsal scales with 3–5 keels. **HABITAT:** Concrete slabs and curbs, along buildings, on trees. **RANGE:** S. FL. **NATURAL RANGE:** Sub-Saharan Africa and possibly the Nile Valley.

TYPICAL SKINKS: Scincidae

Thirty-four genera and more than 270 species range throughout much of North and Central America, Africa, Madagascar, and southern Asia. All native species are Toothy Skinks in the genus *Plestiodon*. Until recently, all were

FIG. 146. *The Ocellated Skink* (Chalcides ocellatus) *has a broad natural range and might represent a complex of species.*

assigned to the genus *Eumeces*, which now is restricted to species in the Eastern Hemisphere.

Red or orange coloration appears on the head of males in some species only during the breeding season, whereas males in other species retain bright colors throughout the year. The position of longitudinal stripes is important for distinguishing some species. Count downward from the dorsal midline. Thus, "stripes on third row," for example, indicates that stripes occupy the third row of scales on either side of the midline.

OCELLATED SKINK *Chalcides ocellatus* Non-native **FIG. 146**

6–12 in. (15.2–30.5 cm); head-body max. 6 11/16 in. (17 cm). Almost cylindrical body with short limbs; a dark brown middorsal stripe bordered by tan dorsolateral stripes, all extending to base of tail. Darker brown sides fade into white to light tan belly. *Dark-bordered light ocelli* in regular transverse rows on body and entire unregenerated tail; often more prominent in male, occasionally lacking. Young with only faint traces of adult pattern and a greenish yellow tail. Lower eyelids with clear "windows." **SIMILAR SPECIES:** Common Sun Skink sometimes has ocelli-like spots on flanks, but never in regular transverse rows, and lacks "windows" in lower eyelids. **HABITAT:** Debris in residential areas. **RANGE:** S. FL; also introduced in AZ. **NATURAL RANGE:** S. Europe, n. Africa, w. Asia, many Mediterranean islands.

COAL SKINK *Plestiodon anthracinus* **PL. 29, FIG. 113**

5–7 in. (12.5–17.8 cm); head-body max. 2¾ in. (7 cm). Four light stripes extending *onto tail*. Broad dark lateral stripes 2½–4 scales wide. *No light lines on top of head*. Dorsolateral light stripes on edges of third and fourth scale rows. *One postmental* (Fig. 113, p. 248). Male with reddish sides of head during spring breeding season in some parts of the range. Juvenile plain black or patterned like adult. Hatchling 1⅞–2 in. (4.8–5.1 cm). **SIMILAR SPECIES:** (1) Northern and Southern Prairie Skinks have dark lateral stripes not more than 2 scales wide, dorsolateral light stripes on fourth (or fourth and fifth) scale rows. (2) Four-lined Skink has light stripes that do not extend onto the tail. (3) Common Five-lined, Southeastern Five-lined, and Broad-headed Skinks have 2 light lines on the head in all but old adults and 2 postmentals. (4) Great Plains Skink is black when young but has bold white and orange spots on the head and 2 postmentals. **HABITAT:** Moist wooded hillsides, vicinity of springs or rocky bluffs. **RANGE:** Discontinuous from NY to the FL Panhandle and west to e. KS, OK, and TX. **SUBSPECIES:** (1) **NORTHERN COAL SKINK** (*P. a. anthracinus*); continuous light stripes extend through posterior supralabials, usually 25 or fewer rows of scales around midbody; young with blue tail, otherwise like adult. (2) **SOUTHERN COAL SKINK** (*Eumeces a. pluvialis*); posterior supralabials with light centers and dark edges produce a spotted appearance; 26 or more rows of scales around midbody; young black and unpatterned, but often with a suggestion of light stripes or whitish labial spots or both; snout and lips may be reddish, tail is blue.

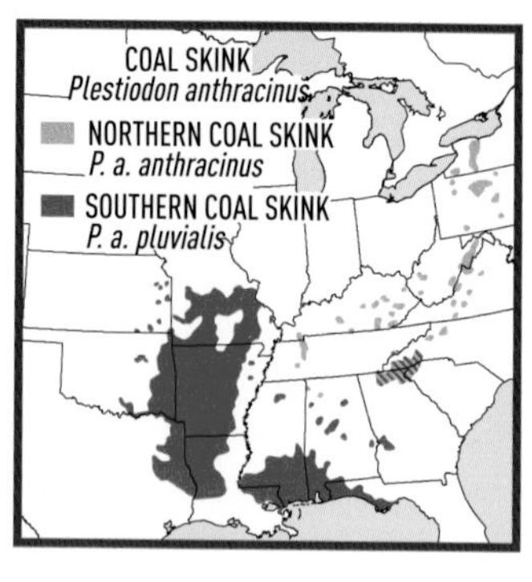

MOLE SKINK *Plestiodon egregius* **PL. 29**

3½–6⅜ in. (9–16.2 cm); head-body max. 2½ in. (6.4 cm). Slender, *very short legs*, light gray brown to dark chocolate brown; dorsolateral and lateral *light stripes* vary in length, terminating near shoulders or extending full length of body; stripes sometimes widen posteriorly and diverge to other scale rows. Tail blue, red, reddish brown, orange, yellow, pinkish, brown, or lavender (see Subspecies). Hatchling 1⅞–2⅜ in. (4.8–6 cm). **SIMILAR SPECIES:** (1) Florida Sand Skink has even smaller limbs, only one or 2 digits, and lacks external ear openings. (2) Little Brown Skink lacks light stripes and has transparent "windows" in lower eyelids. **HABITAT:** Rocky areas (especially in northern parts of the range), sandy soils in sandhill scrub or dry hammock vegetation; often in piles of debris, driftwood, tidal wrack along shores. **RANGE:** Cen. AL and GA south through FL to the Dry Tortugas. **SUBSPECIES:** (1) **FLORIDA KEYS MOLE SKINK** (*P. e. egregius*); red or brownish red tail color persists through life, light stripes neither widen nor diverge to other scale rows, lateral stripes usually continue to groin but dorsolateral stripes often terminate much farther forward, usually 22 or more scales around midbody; male with reddish

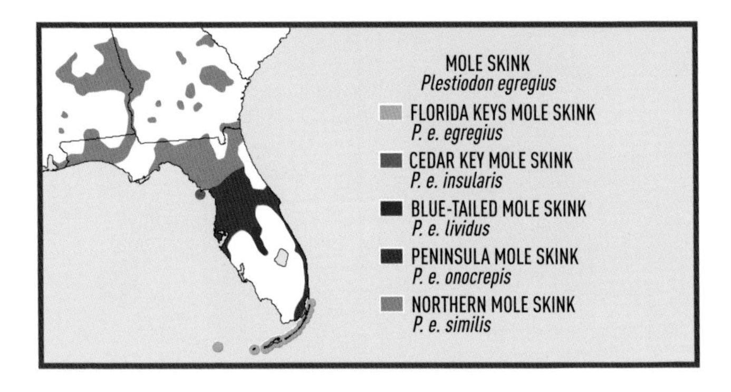

or orange suffusion extending onto belly during the mating season. (2) **Cedar Key Mole Skink** (*P. e. insularis*); dorsolateral light stripes inconspicuous; 21 or fewer scale rows around midbody; hatchling almost uniformly black. (3) **Blue-tailed Mole Skink** (*P. e. lividus*); blue tail of juvenile retained in some adults but fades to salmon in others; dorsolateral light stripes widen posteriorly, diverge to involve another scale row, or both. (4) **Peninsula Mole Skink** (*P. e. onocrepis*); tail red, orange, yellow, pinkish, brown, or lavender; dorsolateral light stripes widen posteriorly, diverge to involve another scale row, or both. (5) **Northern Mole Skink** (*P. e. similis*); 6 upper labials (usually 7 in other subspecies), 21 or fewer scales around midbody, tail red, orange, or reddish brown, length of stripes highly variable. **REMARK:** At least some subspecies might not warrant recognition. **CONSERVATION:** Blue-tailed Mole Skink (USFWS: Threatened).

COMMON FIVE-LINED SKINK *Plestiodon fasciatus*

PL. 29, FIGS. 113, 147, 148, P. 493

5–8¾ in. (12.5–22.2 cm); head-body max. 3⅜ in. (8.6 cm). Hatchling black with 5 white or yellowish stripes and blue tail. Pattern becomes less conspicuous with age, as stripes darken, ground color lightens, and tail turns gray. Female retains indications of stripes, with broad dark bands extending from eyes along sides. Adult male usually shows traces of stripes, but often becomes nearly uniform brown or olive; orange-red on jaws during spring breeding season. Usually *4* labials anterior to large labial reaching the eye, *2 enlarged postlabials* (Fig. 147, p. 308), dorsolateral light stripes on *third and fourth* (or fourth only) scale rows; *26–30 longitudinal rows of scales* around midbody; middle row of scales under tail *wider* than adjacent scale rows (Fig. 148, p. 308). Hatchling 2–2½ in. (5.1–6.4 cm). **SIMILAR SPECIES:** (1) Broad-headed Skink usually has 30–32 rows of scales at midbody, 5 labials anterior to large labial, and no enlarged postlabials. (2) Southeastern Five-lined Skink has a middle row of scales under tail about same width as adjacent rows. **HABITAT:** Wooded areas, especially near rock outcrops, cutover woodlots with rotting stumps and

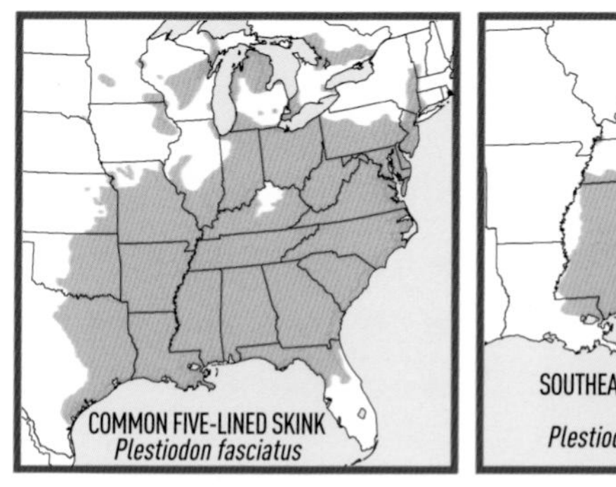

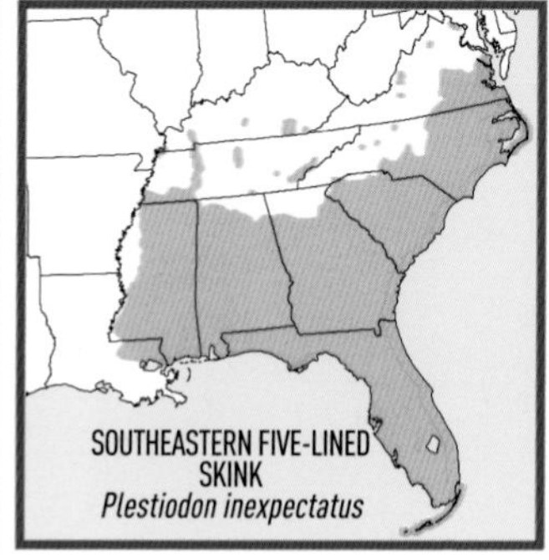

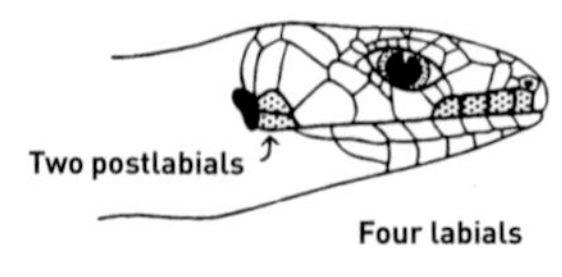

No large postlabials

Five labials

COMMON FIVE-LINED
(*P. fasciatus*)

BROAD-HEADED
(*P. laticeps*)

FIG. 147. *Heads of two species of skinks in the genus* Plestiodon.

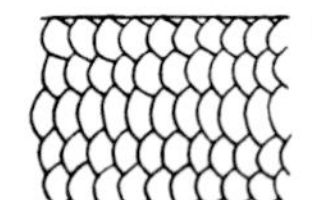

SOUTHEASTERN FIVE-LINED (*P. inexpectatus*) All scales about same size

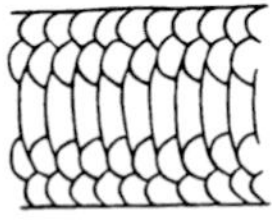

COMMON FIVE-LINED and BROAD-HEADED (*P. fasciatus* and *P. laticeps*) Middle row enlarged

FIG. 148. *Undersurface of the tails of some skinks in the genus* Plestiodon.

logs, abandoned board or sawdust piles, rock piles, decaying debris in or near woods. **RANGE:** New England to n. FL west to MN and TX; many northeastern populations extirpated.

SOUTHEASTERN FIVE-LINED SKINK *Plestiodon inexpectatus*
PL. 29, FIGS. 113, 148

5½–8½ in. (14–21.6 cm); head-body max. 3½ in. (8.9 cm). Hatchling black with 5 narrow yellowish or orange stripes often brighter (even reddish orange) on head; sometimes additional faint light stripes on sides of belly; middorsal stripe narrow; tail bright blue or purple. Pattern becomes less conspicuous with age, although juvenile coloration may persist in adult, orange head stripes remaining prominent, giving head overall orange-brown appearance; purplish hues on tail even in large individuals. Orange-red on head and jaws of adult male during spring breeding season. Rows of scales under original (not regenerated) tail all *about same width* (Fig. 148, facing page); dorsolateral stripes on *fifth* (or fourth and fifth) scale rows. Hatchling 2–2½ in. (5.1–6.4 cm). **SIMILAR SPECIES:** See Common Five-lined Skink. **HABITAT:** Relatively dry areas in open woodlands, ridgetops with well-drained soils, islands with little vegetation. **RANGE:** VA through FL to the Dry Tortugas and west to the Mississippi R.

BROAD-HEADED SKINK *Plestiodon laticeps* **PL. 29, FIGS. 113, 147, 148, 149**

6½–12¾ in. (16.5–32.4 cm); head-body max. 5⅝ in. (14.3 cm). Hatchling black with 5 (sometimes 7 in eastern parts of range) white or yellowish stripes and a blue tail. Pattern becomes less conspicuous with age as stripes darken, ground color lightens, and tail turns gray. Female usually retains indications of stripes. Adult male may show traces of stripes, but many become uniform brown or olive; orange-red on jaws during spring breeding season. Usually *30–32 longitudinal rows of scales* around midbody, usually *5* (occasionally 4) labials anterior to large labial reaching the eye, *no enlarged postlabials* (Fig. 147, facing page), dorsolateral light stripes on *third and fourth* (or fourth only) scale rows; middle row of scales under tail *wider* than adjacent scale rows (Fig. 148, facing page). Hatchling 2¼–3⅜ in. (5.7–8.6 cm). **SIMILAR SPECIES:** See Common Five-lined Skink. **HABITAT:** Woodlands, swamp forests, urban lots strewn with debris. **RANGE:** Se. PA to cen. FL west to e. KS and TX.

FIG. 149. *This adult female Broad-headed Skink* (Plestiodon laticeps) *attends a clutch of eggs. Female skinks of many species remain with their eggs until they hatch. Note the vague stripes, a vestige of the juvenile pattern often retained by adult females.*

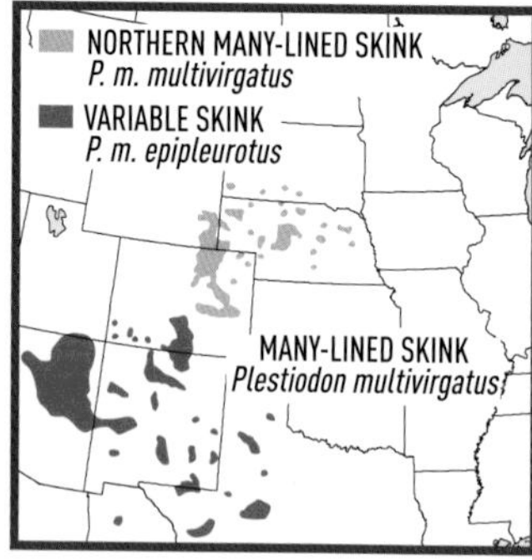

MANY-LINED SKINK *Plestiodon multivirgatus* **PL. 29**

5–7⅝ in. (12.5–19.4 cm); head-body max. 2⅞ in. (7.3 cm). Entirely plain to distinctly striped; striped individuals change with age. *Prominent light stripes restricted to third scale rows.* Dark stripes well defined, weak, mere rows of dark dots, or absent. Tail swollen at base, especially in large adults. Juvenile darker than adult, often only prominent lines evident, and these dim; fainter secondary stripes develop later in striped individuals; tail blue. Hatchling 2–2½ in. (5.1–6.4 cm). **SIMILAR SPECIES:** Northern Prairie Skink has dorsolateral light stripes on fourth (or fourth and fifth) scale rows. **HABITAT:** Mountains, plateaus, creosote bush deserts, open plains and sandhills, often under debris in towns. **RANGE:** S. SD to w. TX and west to AZ. **SUBSPECIES:** (1) **Northern Many-lined Skink** (*P. m. multivirgatus*); well-defined *light* middorsal stripe flanked by *dark* stripes (rarely stripeless or virtually stripeless). (2) **Variable Skink** (*P. m. epipleurotus*); plain-colored and striped variants, striped variant more common in upland areas, plain-colored more abundant in open, arid landscapes at lower elevations. Juvenile of striped morph dark, with well-defined light middorsal and dorsolateral stripes. As lizards grow, ground color becomes paler, middorsal stripes disappear, dark dorsal stripes reduced to narrow zigzag lines or absent, broad dark lateral stripes invaded by pale ground color and replaced by 2 or 3 dark lines. **REMARK:** Subspecies sometimes are considered separate species.

GREAT PLAINS SKINK *Plestiodon obsoletus* **PL. 29, FIGS. 113, 150**

6½–13¾ in. (16.5–34.9 cm); head-body max. 5⅝ in. (14.3 cm). Scales on sides arranged in *oblique* instead of horizontal rows (Fig. 150, facing page). Light tan to gray, scales edged with black or dark brown sometimes forming longitudinal stripes. Two postmentals (Fig. 113, p. 248). Juvenile jet-black with white and orange spots on head; tail blue. Hatchling 2½ in. (6.4 cm). **SIMILAR SPECIES:** See Coal Skink. **HABITAT:** Grasslands, often along watercourses in drier areas. **RANGE:** S. NE and se. IA south into Mex. and west to AZ.

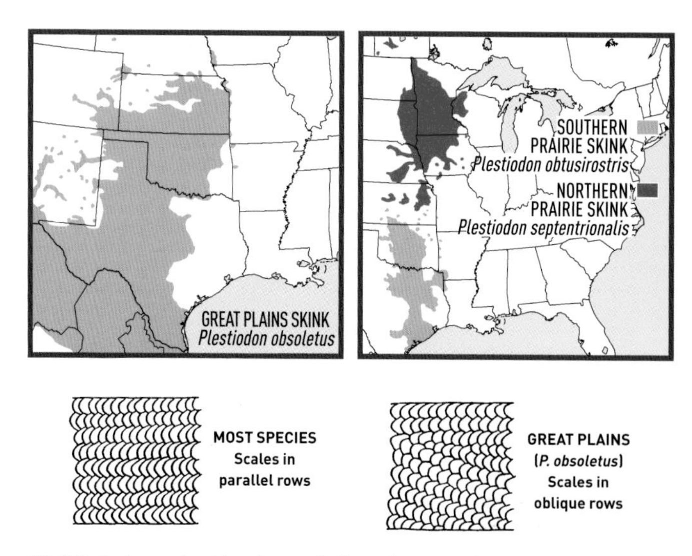

FIG. 150. *Scales on the sides of some skinks in the genus* Plestiodon.

SOUTHERN PRAIRIE SKINK *Plestiodon obtusirostris* **PL. 30, FIG. 104**

5–7$\frac{15}{16}$ in. (12.5–20.1 cm); head-body max. 2$\frac{15}{16}$ in. (7.5 cm). Dorsolateral *light stripes strongly bordered below and usually above by dark stripes* extending onto tail; light stripes on fourth (or fourth and fifth) scale rows. Middorsal markings *reduced or absent*, sometimes even without dark lines above light dorsolateral stripes, rarely without light stripes along lower sides. Broad dark lateral stripes not more than *2 scales wide*. Two postmentals (Fig. 113, p. 248). **SIMILAR SPECIES:** See Coal Skink. **HABITAT:** Streambeds or washes, often near clumps of prickly pear or other vegetation. **RANGE:** S.-cen. KS south to the Gulf.

FLORIDA SAND SKINK *Plestiodon reynoldsi* **PL. 30**

4–5$\frac{1}{8}$ in. (10–13 cm); head-body max. 2$\frac{9}{16}$ in. (6.5 cm). Whitish to deep tan; limbs *greatly reduced*, forelimbs fit into grooves along sides and bear only single toes; hindlimbs slightly larger, with 2 digits. Snout wedge-shaped, lower jaw partially countersunk into upper jaw. Belly flat or slightly concave and meets sides at an angle. Tiny eyes with "windows" in lower lids. *No external ear openings*. Hatchling 2$\frac{5}{16}$ in. (5.9 cm). **SIMILAR SPECIES:** See Mole Skink. **HABITAT:** A burrower that "swims" through sand largely where moist sand underlies dry surface sand. **RANGE:** Cen. FL. **REMARK:** Until recently assigned to the genus *Neoseps*. **CONSERVATION:** Vulnerable (USFWS: Threatened).

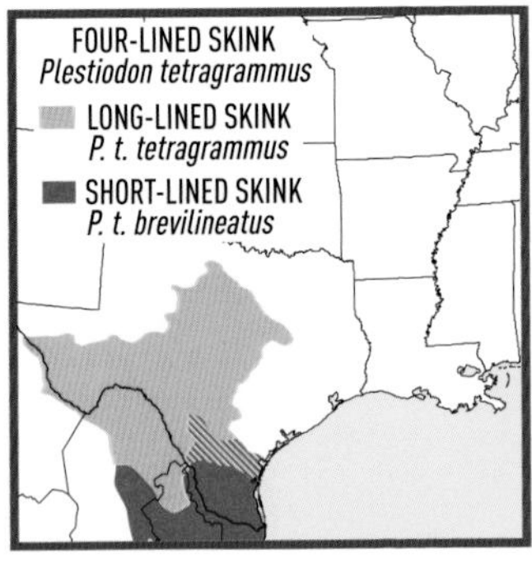

NORTHERN PRAIRIE SKINK *Plestiodon septentrionalis* **PL. 30, FIG. 113**

$5\frac{1}{4}$–$8\frac{13}{16}$ in. (13.3–22.4 cm); head-body max. $3\frac{1}{2}$ in. (9 cm). Olive to olive brown with multiple stripes; dorsolateral *light stripes strongly bordered above and below by dark stripes* extending onto tail; light stripes on fourth (or fourth and fifth) scale rows. Pairs of faint dark middorsal lines or a single pale middorsal stripe or both. Broadest dark stripes not more than *2 scales wide*. Two postmentals (Fig. 113, p. 248). Male with sides of head reddish orange during breeding season. Young with blue tail. Hatchling 2 in. (5.1 cm). **SIMILAR SPECIES:** See Many-lined and Coal Skinks. **HABITAT:** Open plains, usually along streambanks. **RANGE:** S. MB and w. WI south to KS.

FOUR-LINED SKINK *Plestiodon tetragrammus* **PL. 30**

5–$7\frac{7}{8}$ in. (12.5–20 cm); head-body max. 3 in. (7.5 cm). *Four light stripes terminate on body*, either at shoulders or in groin; light lateral stripes *cross* ear openings onto head. Pattern less distinct in old individuals, which fade to lighter brown. Adult male with orange on sides of throat, more prominent in southern populations. Juvenile black with orange head and neckline or with dark bands extending backward from eyes; tail blue. Hatchling $1\frac{1}{2}$–2 in. (3.8–5.1 cm). **SIMILAR SPECIES:** See Coal Skink. **HABITAT:** Brush- or grasslands, especially with sandy soil; rough, hilly country in northern parts of the range, riparian woodlands and gallery forests along rivers farther south; usually under debris. **RANGE:** N.-cen. and w. TX south into Mex.; isolated populations farther west. **SUBSPECIES:** (1) **LONG-LINED SKINK** (*P. t. tetragrammus*); dark gray or gray brown, light stripes terminate in groin, broad black bands along sides. (2) **SHORT-LINED SKINK** (*P. t. brevilineatus*); gray or brown or olive to grayish green, light stripes terminate on shoulders.

FOREST SKINKS: Sphenomorphidae

Thirty-five genera and nearly 550 species occur in North and Central America, southern Asia, and Australasia. Ground skinks (genus *Scincella*) are represented by a single species in eastern North America.

LITTLE BROWN SKINK *Scincella lateralis* **PL. 30**

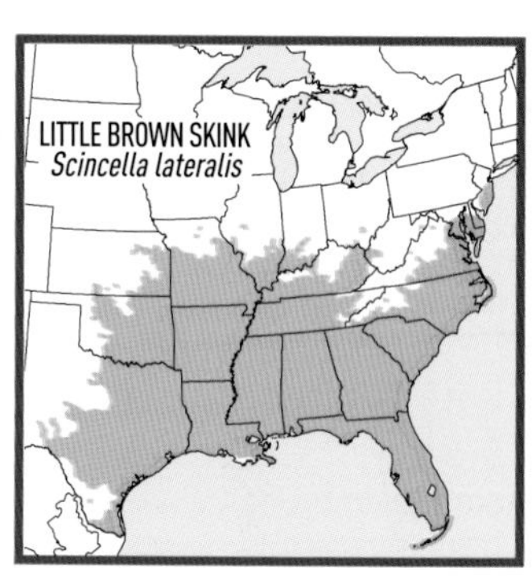

3–5¾ in. (7.5–14.6 cm); head-body max. 2¼ in. (5.7 cm). Golden brown to blackish brown with *dark dorsolateral stripes*; dark stripes almost blend with ground color in very dark individuals. Belly white or yellowish. Lower eyelids with "windows." Hatchling 1⅞ in. (4.8 cm). **SIMILAR SPECIES:** See Mole Skink. **HABITAT:** Leaf litter in woodlands, under debris in towns and gardens. **RANGE:** S. NJ through FL and the Keys west to e. KS and TX. **REMARK:** Often called the Ground Skink.

AMEIVAS, WHIPTAILS, RACERUNNERS, and TEGUS: Teiidae

About 17 genera and more than 140 species, all in the Western Hemisphere. They have a long tail, tiny granular dorsal scales, and eight or more rows of large rectangular ventral scales. "Unisexual" species are products of hybridization events and reproduce through parthenogenesis. Five of 10 native species are unisexual. The recently described Neaves' Whiptail (*Aspidoscelis neavesi*), currently known only from New Mexico, might also range into Texas, where both parental species (Chihuahuan Spotted and Little Striped Whiptail) occur. Native species in the genus *Aspidoscelis* until recently were assigned to the genus *Cnemidophorus*, a name now restricted to species in the Neotropics.

GIANT AMEIVA *Ameiva ameiva* Non-native **FIG. 151**

15–25 in. (38–63.5 cm); head-body max. 6⅝ in. (17 cm). Adult *brown or tan anteriorly and bright green posteriorly* (most evident in large males), dark-bordered light ocelli on sides rarely extend across back, spots on lower sides and outermost rows of belly scales bluish. No middorsal stripe. Female often retains 4 prominent tan to greenish stripes bordering dark brown sides of juvenile; stripes frequently fade posteriorly. Belly with blue in adult male, light gray to white in juvenile and female; *10–12 rows of large rectangular ventral plates*. **SIMILAR**

FIG. 151. *Like whiptails and racerunners, the Giant Ameiva* (Ameiva ameiva) *is an active forager, moving frequently and rooting in leaves and other debris in search of prey.*

SPECIES: Borriguerro Ameiva is never green posteriorly, large adults are charcoal or bluish gray. **HABITAT:** Fields, often along railroad tracks and canal banks. **RANGE:** S. FL and some Keys. **NATURAL RANGE:** Amazonia.

BORRIGUERRO AMEIVA *Ameiva praesignis* **Non-native** **PL. 30**

15–25 in. (38–63.5 cm); head-body max. 9 7/16 in. (24 cm). Dark *charcoal to bluish gray*, male darker than female. Vertical lateral rows of yellowish circular spots occasionally extend onto back, bluish spots on outermost rows of ventrals and anterior surface of hindlimbs; brown or yellowish middorsal stripe (sometimes broken) usually present. Juvenile and smaller females have a middorsal stripe, 4 stripes bordering dark brown sides, and ocelli often extending across back. Belly with blue in adult male, light gray to white in juvenile and female; *10–12 rows of large rectangular ventrals*. **SIMILAR SPECIES:** See Giant Ameiva. **HABITAT:** Open grassy areas and parks. **RANGE:** Key Biscayne, FL. **NATURAL RANGE:** S. Costa Rica to n. S. America. **REMARK:** Until recently considered a subspecies of the Giant Ameiva.

CHIHUAHUAN SPOTTED WHIPTAIL *Aspidoscelis exsanguis* **PL. 31**

9½–12 3/8 in. (24–31.4 cm); head-body max. 3 7/8 in. (9.8 cm). Unisexual. Brown or reddish brown, *6 pale brown or gray stripes* most conspicuous and often yellowish on sides. Pale yellowish or whitish spots in dark fields *and on light stripes*. Tail usually brown, often with a greenish tinge. Unmarked belly whitish or faintly bluish. Enlarged scales along anterior edge of gular fold and on back of forearms; supraorbital semicircles "normal" (Fig. 114, p. 252); 62–86 granules around midbody. Juvenile with yellow or whitish stripes that contrast sharply with dark brown or blackish fields; paravertebral stripes wavy. *Pale reddish spots present in hatchling*. Light middorsal stripe may be present on nape, absent, or vaguely indicated in adult. Tail blue or greenish. Hatchling head-body length 1 5/16–1 9/16 in. (3.3–4 cm). **SIMILAR SPECIES:** See Common Spotted Whiptail. **HABITAT:** Open areas in desert grasslands and forested mountains, rocky hillsides, canyon bottoms,

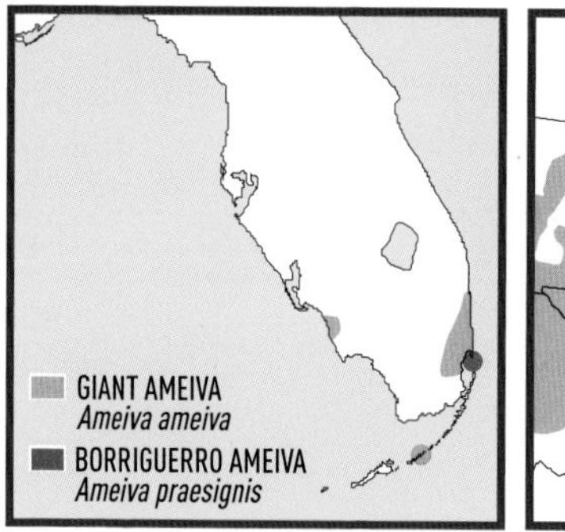

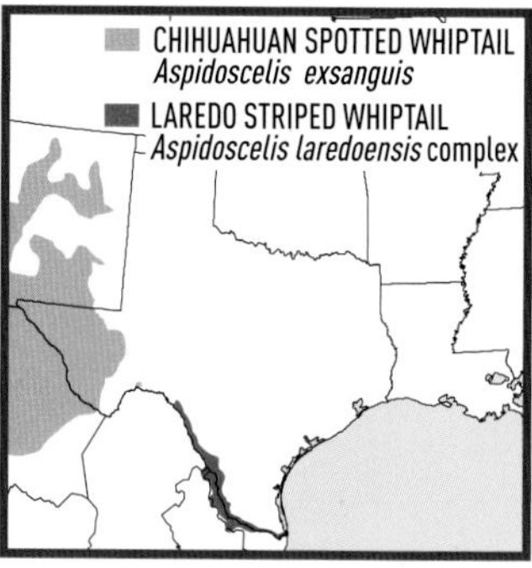

dry washes, frequently disturbed areas. **RANGE:** NM through Trans-Pecos TX and south into Mex.

COMMON SPOTTED WHIPTAIL *Aspidoscelis gularis* **PL. 31**

6½–11¹³⁄₁₆ in. (16.5–29.6 cm); head-body max. 3⅞ in. (10.5 cm). Liberally spotted, 7–8 stripes (broad middorsal stripe often split) whitish or yellowish, but often greenish anteriorly and brownish posteriorly. *Pale spots prominent in dark lateral fields* but faint or absent in dark dorsal fields; spots white or pale yellow, often brownish posteriorly. *Tail pink, pale orange-brown, or reddish*. Enlarged scales along anterior edge of gular fold and back of forearms; supraorbital semicircles "normal" (Fig. 114, p. 252); 76–97 granules around midbody. Male with pink, red, or orange chin; blue chest and belly, frequently with a large black patch. Female with unmarked whitish or cream belly. Juvenile striped; spots faint, when present, confined to dark lateral fields; tail pink or reddish; rump reddish. Hatchling head-body length 1⅜ in. (3.5 cm). **SIMILAR SPECIES:** (1) Six-lined Racerunner in range is green anteriorly, lacks spots in dark fields, and has light stripes extending backward from hindlimbs. (2) Chihuahuan Spotted Whiptail has 6 light lines, an unmarked belly, and pale spots in paravertebral and lateral dark fields. (3) Plateau Spotted Whiptail has dorsal stripes and dark fields that terminate on body. (4) Laredo Striped Whiptail is green with a narrow middorsal stripe. **HABITAT:** Prairie grasslands, river floodplains and washes, grasslands reverting to brush, rocky hillsides. **RANGE:** S. OK to se. NM and south into Mex. **SUBSPECIES:** **TEXAS SPOTTED WHIPTAIL** (*A. g. gularis*).

LITTLE STRIPED WHIPTAIL *Aspidoscelis inornata* **PL. 31**

6½–9⅜ in. (16.5–23.8 cm); head-body max. 2¹³⁄₁₆ in. (7.1 cm). Tail *bright to purplish blue*; juvenile pattern of 7 light stripes separated by dark, *unspotted fields* retained through life. Stripes yellow and fields black in young, whitish or cream and grayish or brownish in adult. Scales anterior to gular fold slightly enlarged; usually 2 rows of

slightly enlarged scales on back of forearms; supraorbital semicircles "normal" (Fig. 114, p. 252); 55–71 granules around midbody. Male with *blue belly and sides of head* (bluish to bluish white in female). Juvenile with less blue below than adult. Hatchling head-body length 1⅛ in. (2.8 cm). **SIMILAR SPECIES:** Desert Grassland and New Mexico Whiptails have a greenish tail. **HABITAT:** Rocky, grassy slopes, deteriorated grasslands, sometimes on alluvial flats. **RANGE:** NM through w. TX south into Mex. **SUBSPECIES:** **Trans-Pecos Striped Whiptail** (*A. i. heptagramma*).

LAREDO STRIPED WHIPTAIL *Aspidoscelis laredoensis* complex **FIG. 152**

7⅜–11⅜ in. (18.7–28.9 cm); head-body max. 3½ in. (9 cm); record 15⅝ in. (39.8 cm), head-body 3¾ in. (9.6 cm), presumably a hybrid. Unisexual. *Dark green to greenish brown*, with *7 cream or whitish stripes*; middorsal stripe very narrow; small pale or indistinct spots sometimes present posteriorly. Upper surface of hindlimbs with a reticulate pattern of irregular cream-colored lines. Belly white. Tail greenish brown above, light tan below, anterior third with dorsolateral stripes. Slightly enlarged scales on back of forearms; supraorbital semicircles "normal" (Fig. 114, p. 252); 84–98 granules around midbody. Juvenile distinctly striped with spots confined to dark fields. **SIMILAR SPECIES:** See Common Spotted Whiptail. **HABITAT:** Semiarid regions near streambeds, hills with little vegetation. **RANGE:** Lower Rio Grande Valley of TX and adjacent Mex. **REMARKS:** A complex of species. Individuals in some populations interbreed with the Common Spotted Whiptail.

MARBLED WHIPTAIL *Aspidoscelis marmorata* **PL. 31**

8–12 in. (20.3–30.5 cm); head-body max. 4⅛ in. (10.5 cm). Extremely variable, almost uniformly brownish to brownish gray or sometimes with 4–8 yellowish white stripes (often vague, broken, reticulate); dark lateral fields often broken into spots by light transverse bars. Belly chiefly white or pale yellow with black flecks anteriorly and

FIG. 152. *Variants of the Laredo Striped Whiptail* (Aspidoscelis laredoensis *complex*) *are probably derived from separate hybridization events and could represent two distinct species.*

laterally covering less than half of chest. Washed with peach on throat, also on chest and sides of abdomen in larger individuals. Chin and throat with round or irregular black flecks or spots in adult. Scales anterior to gular fold slightly enlarged, separated from fold by one or more rows of granules; no enlarged scales on forearms; *supraorbital semicircles extending well forward* (Fig. 114, p. 252), reaching or almost reaching frontals; 74–114 granules around midbody. Juvenile dark brown to black with numerous small, pale, often elongated spots in longitudinal rows. Hatchling head-body length 1⅜–1¾ in. (3.5–4.4 cm). **SIMILAR SPECIES:** Common Checkered Whiptail has abruptly enlarged scales anterior to gular fold, chin is never peach, and rarely has black spots. Other whiptails have at least 2 pairs of light stripes throughout life. **HABITAT:** Desert flats or sandy areas. **RANGE:** NM and se. AZ through w. TX and south into Mex. **SUBSPECIES:** (1) **Western Marbled Whiptail** (*A. m. marmorata*); upper body of adult more reticulate than striped. (2) **Eastern Marbled Whiptail** (*A. m. reticuloriens*); pattern of adult more striped than reticulate, but patternless variant occurs in some areas. **REMARK:** Sometimes considered a subspecies of the Tiger Whiptail (*A. tigris*), which occurs west of our area.

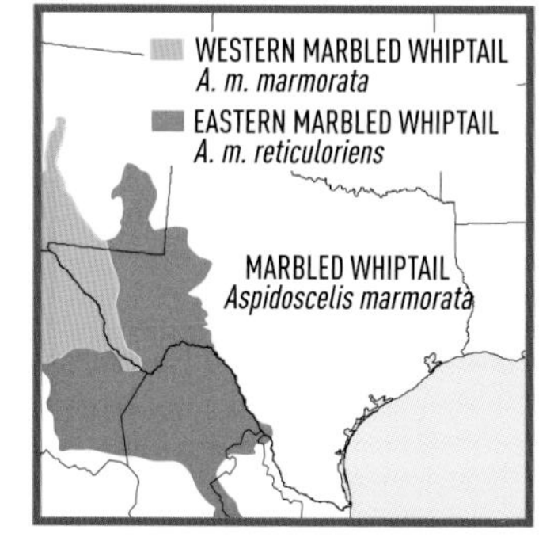

GIANT WHIPTAIL *Aspidoscelis motaguae* Non-native **FIG. 153**

6½–14 in. (16.5–35.5 cm); head-body max. 5¾ in. (14.5 cm). Back brown, head slightly darker, lower sides fading to grayish brown or gray. *Back, sides, and base of tail with yellowish spots*, those on sides of belly bluish gray to bright blue. Tail brownish gray grading to reddish

FIG. 153. *Like most whiptails, the introduced Giant Whiptail* (Aspidoscelis motaguae) *favors open areas such as fields, canal banks, grassy parking lot edges, and road shoulders.*

brown near tip. **SIMILAR SPECIES:** (1) Six-lined Racerunner is green, at least anteriorly, and has 6–7 stripes. (2) Rainbow Whiptail has a broad brown middorsal stripe bordered laterally by distinct dark stripes. **HABITAT:** Open, frequently disturbed areas. **RANGE:** S. FL. **NATURAL RANGE:** S. Mex. into Cen. America.

NEW MEXICO WHIPTAIL *Aspidoscelis neomexicana* **PL. 31**

8–11⅞ in. (20.3–30.2 cm); head-body max. 3⅜ in. (8.6 cm). Unisexual. Seven pale yellow stripes, all distinct, but *middorsal stripe wavy.* Fields brown to black, obscurely spotted throughout life. Tail *greenish gray to bluish green.* Unmarked belly white to pale blue. Usually no enlarged scales anterior to gular fold or on back of forearms; *supraorbital semicircles extending far forward* (Fig. 114, p. 252); 71–85 granules around midbody. Juvenile black with yellow stripes and spots in dark fields. Hatchling with blue tail; head-body length 1⅜–1⅝ in. (3.5–4.1 cm). **SIMILAR SPECIES:** See Little Striped Whiptail. **HABITAT:** Disturbed areas subject to periodic flooding or altered by humans, perimeters of playas, open sandy sites. **RANGE:** Upper Rio Grande Valley of w. TX and NM; isolated populations in NM and AZ.

PLATEAU SPOTTED WHIPTAIL *Aspidoscelis scalaris* **PL. 31**

8–12½ in. (20.3–31.8 cm); head-body max. 4⅛ in. (10.5 cm). Brownish green, 6–7 dorsal stripes terminate on body (middorsal stripe narrow), dark fields spotted, *rusty red or orange rump and base of tail* (sometimes also hindlimbs). Tail brown or gray. Belly white or pale blue; chin and chest sometimes with black spots. Enlarged scales along anterior edges of gular fold and on back of forearms; supraorbital semicircles "normal" (Fig. 114, p. 252); 77–98 granules around midbody. Female has orange throat. Light stripes in juvenile often wavy and middorsal stripe intermittent; fields very dark with pale spots chiefly in lowermost fields; tail blue or greenish. Hatchling

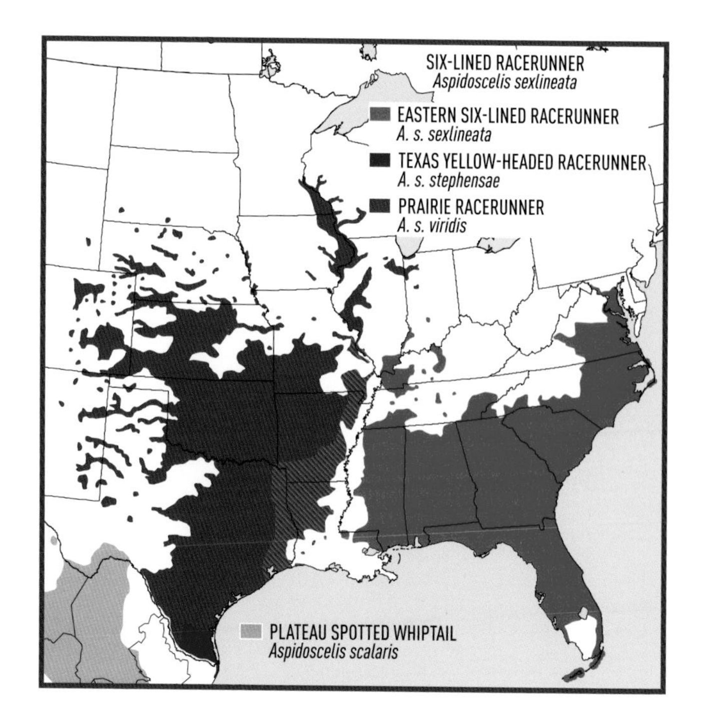

head-body length 1½ in. (3.8 cm). **HABITAT:** Mountains, desert foothills, lava flows, canyons; most frequently rocky situations with sparse vegetation. **RANGE:** Big Bend Region of TX south into Mex. **SUBSPECIES: Big Bend Spotted Whiptail** (*A. s. septemvittata*); sometimes considered a distinct species. **REMARK:** Also called the Rusty-rumped Whiptail.

SIX-LINED RACERUNNER *Aspidoscelis sexlineata* **PL. 31**

6–10½ in. (15.2–26.7 cm); head-body max. 3½ in. (9 cm). Brown, greenish brown, sometimes almost black, western populations often bright green anteriorly; 6–7 yellow, white, pale gray, or pale blue stripes; *no spots in dark fields. Short dark stripes on sides of tail* extend backward from legs and are bordered above by light stripes. Scales anterior to gular fold conspicuously enlarged; scales on back of forearms slightly enlarged or not at all; supraorbital semicircles "normal" (Fig. 114, p. 252); 62–110 granules around midbody. Belly of male washed with blue. Juvenile with light blue tail; yellow stripes contrast sharply with dark fields. Hatchling head-body length 1⅛–1¼ in. (2.8–3.2 cm). **SIMILAR SPECIES:** Other whiptails in range have pale spots in dark fields. **HABITAT:** Open areas with sand or loose soil, fields, grasslands, open woods, thicket margins, rocky outcrops, river floodplains. **RANGE:** MD south through FL into the Keys, west to CO and NM and north to WI, MN, and SD. **SUBSPECIES:** (1) **Eastern Six-lined Race-runner** (*A. s. sexlineata*); 6 light stripes, 89–110 granules around midbody. (2) **Texas Yellow-headed Racerunner** (*A. s. stephensae*); 6 light stripes (vestiges of a middorsal stripe might be present anteriorly), 69–99 granules around midbody, yellow face and sides of neck (brightest in male). (3) **Prairie Racerunner** (*A. s. viridis*); 7 light stripes, dark fields bright green anteriorly, 62–91 granules around midbody.

COMMON CHECKERED WHIPTAIL *Aspidoscelis tesselata* **PL. 31**

11–15½ in. (28–39.4 cm); head-body max. 4³⁄₁₆ in. (10.6 cm). Unisexual. Yellowish to cream, sometimes checkered or with rows of black spots; rump occasionally orange-brown; tail brown to yellow; belly plain whitish or with some black spots; chin usually unmarked. *Scales anterior to gular fold abruptly enlarged*; no enlarged scales on back of forearms; *supraorbital semicircles extend far forward* (Fig. 114, p. 252) in most individuals in TX; 81–112 granules around midbody. Juvenile with variable number (6–14) of light stripes, middorsal light stripe (if present) often a series of pale dots or dashes that appear wavy; small pale spots in dark fields; stripes become less conspicuous with age, often disappearing altogether on sides when invaded by black checks or spots. Hatchling head-body length 1½–1¾ in. (3.8–4.4 cm). **SIMILAR SPECIES:** See Marbled Whiptail. **HABITAT:** Plains, canyons, foothill uplands, river floodplains, almost always associated with rocky areas. **RANGE:** Se. CO through w. TX into Mex.

DESERT GRASSLAND WHIPTAIL *Aspidoscelis uniparens* **PL. 31**

6½–9⅜ in. (16.5–23.8 cm); head-body max. 2¹⁵⁄₁₆ in. (7.4 cm). Unisexual. Reddish brown to black, with 6 yellowish (whitish laterally) stripes, often with a suggestion of a seventh (middorsal) light stripe at least on neck; *no pale spots in dark fields. Tail bluish green or olive green*; virtually unmarked belly whitish; adult often with bluish wash on chin and sides of neck. Enlarged scales along anterior edge of gular fold and on back of forearms; supraorbital semicircles "normal" (Fig. 114, p. 252); 59–78 granules around midbody. Hatchling head-body length 1⁵⁄₁₆–1⁹⁄₁₆ in. (3.3–4 cm). **SIMILAR SPECIES:** See Little Striped Whiptail. **HABITAT:** Desert and mesquite grasslands, ascending river valleys into lower montane areas. **RANGE:** Extreme w. TX to AZ and south into Mex.

RAINBOW WHIPTAIL *Cnemidophorus lemniscatus* complex Non-native **PL. 30**

6½–13 in. (16.5–33 cm); head-body max. 4⅛ in. (10.5 cm). Brightly colored; male with 2 narrow light stripes flanking 2 broader dorsolateral green stripes anteriorly; narrow light stripes fade posteriorly, green stripes become dark green and border a broad, middorsal

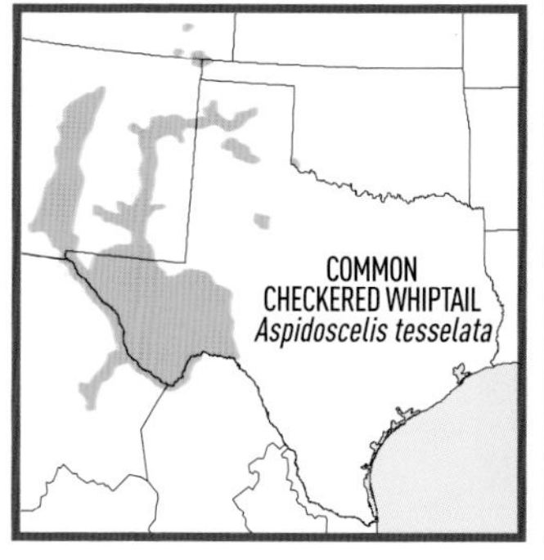

brown stripe. Flanks yellowish brown with light spots; tail and underside of limbs blue; hindlimbs brownish above with light spots; sides of head, chin, throat, and chest blue; belly light gray. Female usually with numerous light dorsal stripes; flanks with light stripes; sides of head orangish yellow; chin, throat, lower surface of forelimbs, and anterior belly white to yellow; posterior belly, lower surface of hindlimbs, and tail greenish yellow; top of hindlimbs dark with light spots. Scales anterior to gular fold enlarged, supraorbital semicircles "normal" (Fig. 114, p. 252); 103–128 granules around midbody. Hatchling head-body length 1³⁄₁₆ in. (3 cm). **SIMILAR SPECIES:** (1) Six-lined Racerunner has no spots on flanks and fewer (89–110) granules around midbody. (2) Giant Whiptail has spots but lacks a broad brown middorsal stripe bordered laterally by distinct dark stripes. **HABITAT:** Open areas, vacant lots, railroad rights of way. **RANGE:** S. FL. **NATURAL RANGE:** Guatemala south into n. S. America. **REMARK:** This complex consists of at least 4 forms, some of which are unisexual.

ARGENTINE GIANT TEGU *Salvator merianae* Non-native **FIG. 154**

25–35 in. (63.5–89 cm); head-body max. 19¾ in. (50 cm). *Very large, heavy-bodied.* Adult *black and white banded* with small black spots in white bands; limbs black with white spots. Light bands in juvenile green anteriorly and light brown posteriorly, fading to yellowish and then to white in large adults. Head scales not equal in size, belly scales rectangular and in transverse rows. **SIMILAR SPECIES:** Nile Monitor has a head covered with small scales about equal in size, small belly scales not in rows. **HABITAT:** Open areas in dry upland scrub and disturbed areas. **RANGE:** Cen. and s. FL, scattered records elsewhere in FL. **NATURAL RANGE:** Ne. Brazil south to n. Argentina. **REMARKS:** Until recently assigned to the genus *Tupinambis*. Also called the Argentine Black and White Tegu.

FIG. 154. *Also called the Argentine Black-and-White Tegu, the Argentine Giant Tegu* (Salvator merianae) *was until recently assigned to the genus* Tupinambis *(to which it might revert pending further studies).*

MONITORS: Varanidae

This family includes the largest known lizard (*Varanus priscus*, which is extinct) and the largest living species, the Komodo Dragon (*V. komodoensis*). More than 70 extant species occur in Africa, Asia, and Australasia.

NILE MONITOR *Varanus niloticus* **Non-native** **FIG. 155**

4–5½ ft. (122–168 cm); record 7 ft. 11 in. (243 cm). *Very large*, snout pointed, *neck longer than head*. Scales on body and top of head small and buttonlike. Black, brown, dark green, or grayish green variously spotted with yellow. Back with *6–11 crossbars of yellow spots or ocelli*, flanks with vertical yellow bars or blotches, tail with 10–18 vertical yellow bars on each side. Belly dirty yellow or cream with dark bluish gray crossbars, vermiculate marks, or blotches. Juvenile brighter than adult. Hatchling 7–12 in. (18–30.5 cm). **SIMILAR SPECIES:** See Argentine Giant Tegu. **HABITAT:** Near water along canals and coastal mangroves, often in urban areas. **RANGE:** Cen. and s. FL and some Keys, scattered records elsewhere in FL. **NATURAL RANGE:** Sub-Saharan Africa.

FIG. 155. *The Nile Monitor* (Varanus niloticus) *grows very large, will eat about anything it can catch, and might pose a serious threat to native wildlife in areas where it has become established.*

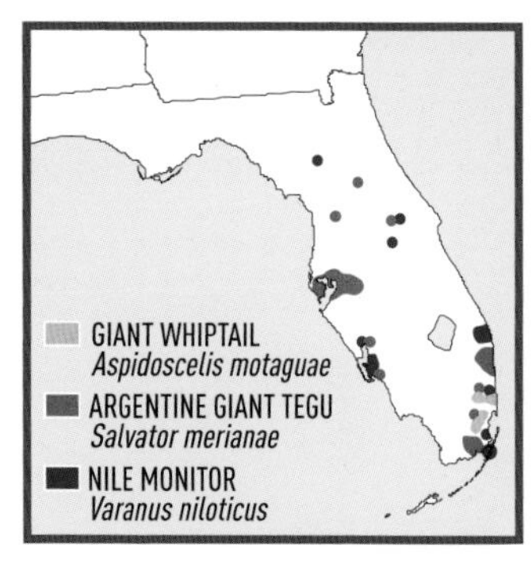

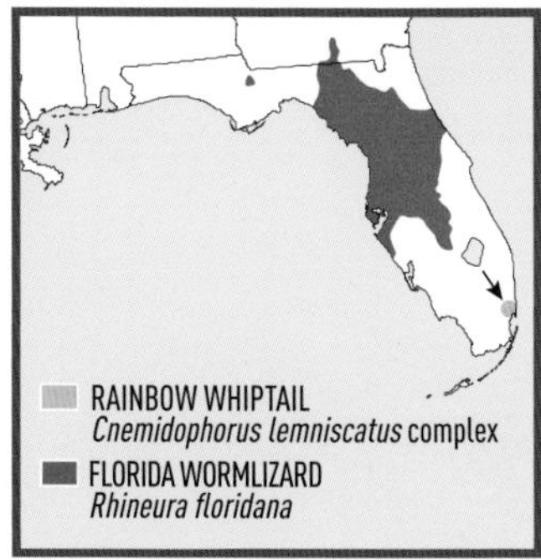

WORMLIZARDS

Once placed in a separate suborder, wormlizards lack limbs, have rings of scales encircling the body and tail, vestigial eyes covered by scales, and no external ear opening. Six families occur in the Americas, Africa, and the Middle East.

WIDE-SNOUTED WORMLIZARDS: Rhineuridae

Only one living species is endemic to eastern North America.

FLORIDA WORMLIZARD *Rhineura floridana* **PL. 30, FIG. 156**

7–11 in. (18–28 cm); record 16 in. (40.6 cm). Lizardlike head with a countersunk lower jaw to facilitate burrowing. Upper surface of *very short, blunt tail* flattened and covered with numerous small bumps (tubercles). Hatchling about 4 in. (10 cm). **HABITAT:** Underground in dry, sandy areas. **RANGE:** Cen. and n. FL into extreme s. GA.

FIG. 156. *The Florida Wormlizard* (Rhineura floridana) *is the only representative of its family, but five other families of wormlizards have representatives in many parts of the world. All are burrowers with vestigial eyes covered by scales and other adaptations for a fossorial lifestyle. The rings of scales encircling the body and tail cause them to superficially resemble earthworms and account for the common name.*

SNAKES

THREADSNAKES (*Rena*)

TEXAS THREADSNAKE *R. dulcis* **P. 362**
Wormlike; tail blunt; belly scales not enlarged.

GLOSSY SNAKES and SCARLETSNAKES (*Arizona* and *Cemophora*)

Scales smooth, cloacal undivided.

TEXAS GLOSSY SNAKE *A. elegans arenicola* **P. 366**
Cream or buff with brown blotches.

SCARLETSNAKE *C. coccinea* **P. 367**
Red saddles bordered by black; head largely reddish; belly whitish, unpatterned (see also Pl. 44).

FIG. 157. *Prefrontal scales in the Glossy Snake* (Arizona elegans).

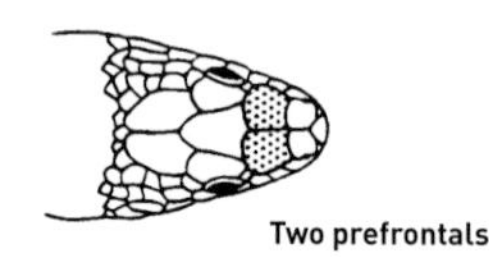

NORTH AMERICAN RACERS and COACHWHIPS (*Coluber*)

Scales smooth, cloacal divided. See also Pl. 33.

NORTH AMERICAN RACER *C. constrictor* **P. 369**
Young with dark middorsal blotches.

NORTHERN BLACK RACER (*C. c. constrictor*): Black above and below; some white on chin.

BLUE RACER (*C. c. foxii*): Blue above; belly paler; dark areas on sides of head.

EASTERN YELLOW-BELLIED RACER (*C. c. flaviventris*): Belly yellow; dorsum variable.

BUTTERMILK RACER (*C. c. anthicus*): Irregular white, buff, or blue spots.

COACHWHIP *C. flagellum* **P. 370**
Tail like braided whip; young with dark crosslines.

WESTERN COACHWHIP (*C. f. testaceus*): Plain above, reddish in some areas.

EASTERN COACHWHIP (*C. f. flagellum*): Black or dark brown, light brown posteriorly.

TEXAS THREADSNAKE

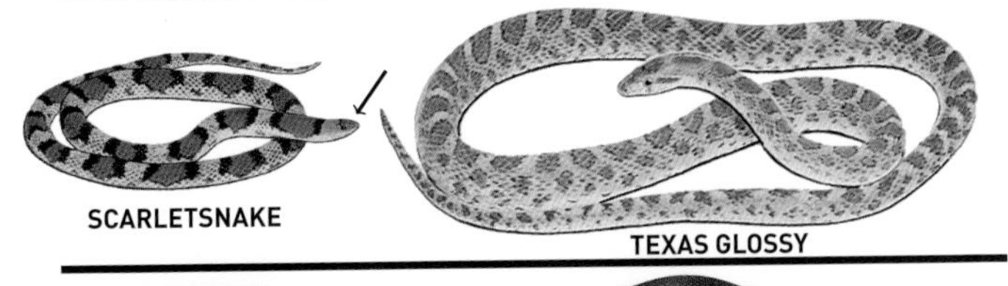

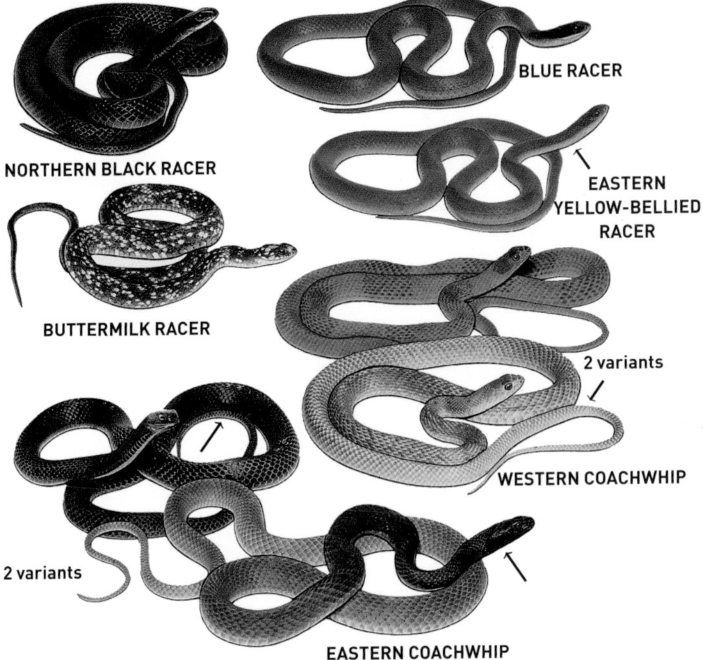

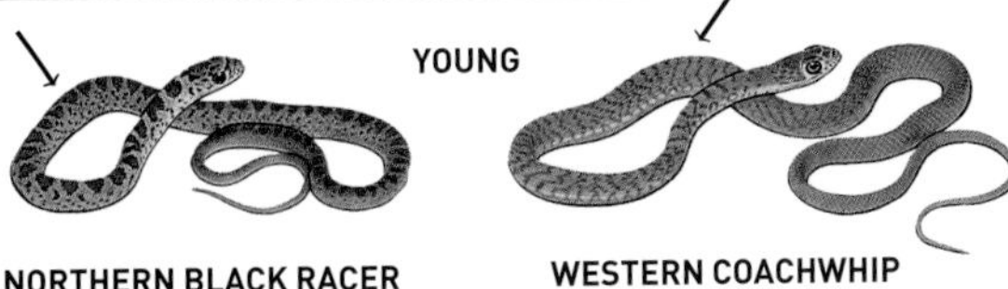

PLATE 33

WHIPSNAKES (*Coluber*) (cont.)

Scales smooth, cloacal divided. See also Pl. 32.

CENTRAL TEXAS WHIPSNAKE *C. taeniatus girardi* **P. 372**
Longitudinal white patches on sides.

SCHOTT'S WHIPSNAKE *C. schotti* **P. 371**
Dorsal scales with light edges (Fig. 176, p. 368).

SCHOTT'S STRIPED WHIPSNAKE (*C. s. schotti*): Striped on sides; reddish on neck.

RUTHVEN'S WHIPSNAKE (*C. s. ruthveni*): Suggestion of stripes on sides.

INDIGO SNAKES (*Drymarchon*)

Cloacal undivided, scales mostly smooth.

TEXAS INDIGO SNAKE *D. melanurus erebennus* **P. 373**
Black lines on upper lip; traces of pattern on forepart of body.

EASTERN INDIGO SNAKE *D. couperi* **P. 373**
Shiny bluish black; chin and sides of head often reddish.

FIG. 158. *Heads of indigo snakes* (Drymarchon).

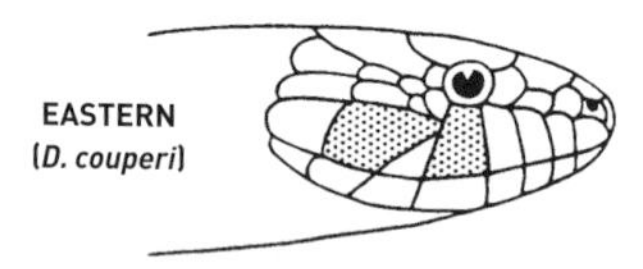

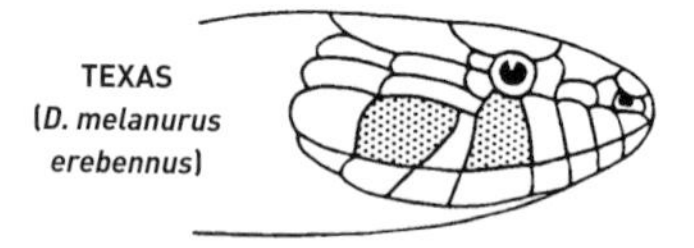

NEOTROPICAL RACERS (*Drymobius*)

Middorsal scale rows weakly keeled, outer rows smooth, cloacal divided.

SPECKLED RACER *D. margaritiferus* **P. 374**
Scales with light centers; dark stripes behind eyes.

HOOK-NOSED SNAKES (*Gyalopion* and *Ficimia*)

Snout upturned, scales smooth, cloacal divided

CHIHUAHUAN HOOK-NOSED SNAKE *G. canum* **P. 375**
Head strongly crossbanded.

TAMAULIPAN HOOK-NOSED SNAKE *F. streckeri* **P. 374**
Head virtually unpatterned.

CENTRAL TEXAS
WHIPSNAKE
SCHOTT'S
STRIPED
WHIPSNAKE
RUTHVEN'S
WHIPSNAKE

TEXAS INDIGO
EASTERN INDIGO

SPECKLED RACER

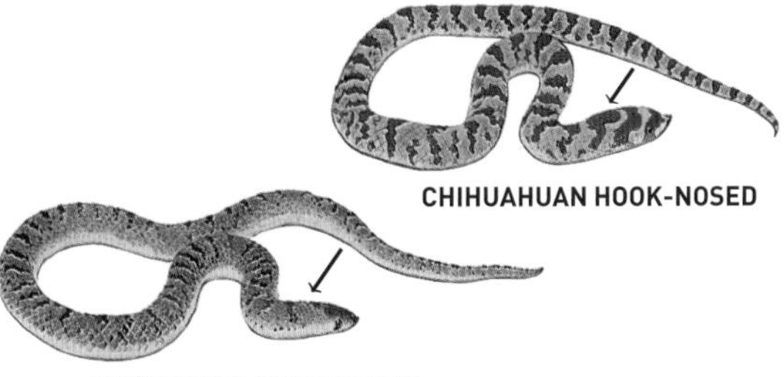
CHIHUAHUAN HOOK-NOSED
TAMAULIPAN HOOK-NOSED

PLATE 34

KINGSNAKES (*Lampropeltis*)

Scales smooth, cloacal undivided. See also Pl. 35.

GRAY-BANDED KINGSNAKE *L. alterna* **P. 375**

Gray crossbands with varying amounts of red or orange (sometimes absent).

YELLOW-BELLIED KINGSNAKE *L. calligaster* **P. 376**

PRAIRIE KINGSNAKE (*L. c. calligaster*): Brown or reddish brown blotches in middorsal and lateral rows; pattern sometimes obscure with suggestion of dark longitudinal stripes.

MOLE KINGSNAKE (*L. c. rhombomaculata*): Variable; uniformly brown or with well-separated, dark, often reddish spots.

EASTERN KINGSNAKE *L. getula* **P. 379**

Highly variable; bold light chainlike pattern, pale with lines faintly indicated, occasionally with suggestion of stripes; light bands very wide, dark blotches broad and few in number (n. FL and extreme s. GA); pale, crosslined pattern only faintly indicated.

SPECKLED KINGSNAKE *L. holbrooki* **P. 380**

Salt-and-pepper effect, occasionally nearly black.

EASTERN BLACK KINGSNAKE *L. nigra* **P. 380**

Black with a pattern faintly or incompletely indicated by white or yellow dots.

DESERT KINGSNAKE *L. splendida* **P. 381**

Black or dark brown middorsal blotches; sides speckled.

FIG. 159. *Diagrammatic belly patterns of snakes boldly ringed or blotched with red, yellow, and black.*

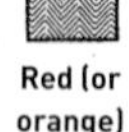

Red (or orange)

Yellow (or white)

Black

AMERICAN CORALSNAKES (*Micrurus*) Danger! Red and yellow touch

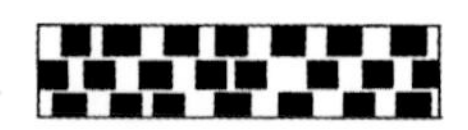

EASTERN MILKSNAKE (*L. triangulum*) Checkerboard effect

SCARLET KINGSNAKE (*L. elapsoides*) Red and yellow separated by black

WESTERN MILKSNAKE (*L. gentilis*; w. populations) H-shaped effect

WESTERN MILKSNAKE (*L. gentilis*; s. populations) Pattern may not cross belly

TAMAULIPAN MILKSNAKE (*L. annulata*; also some Western Milksnakes) Black predominates

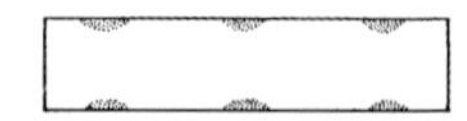

SCARLETSNAKE (*Cemophora coccinea*) Belly plain whitish

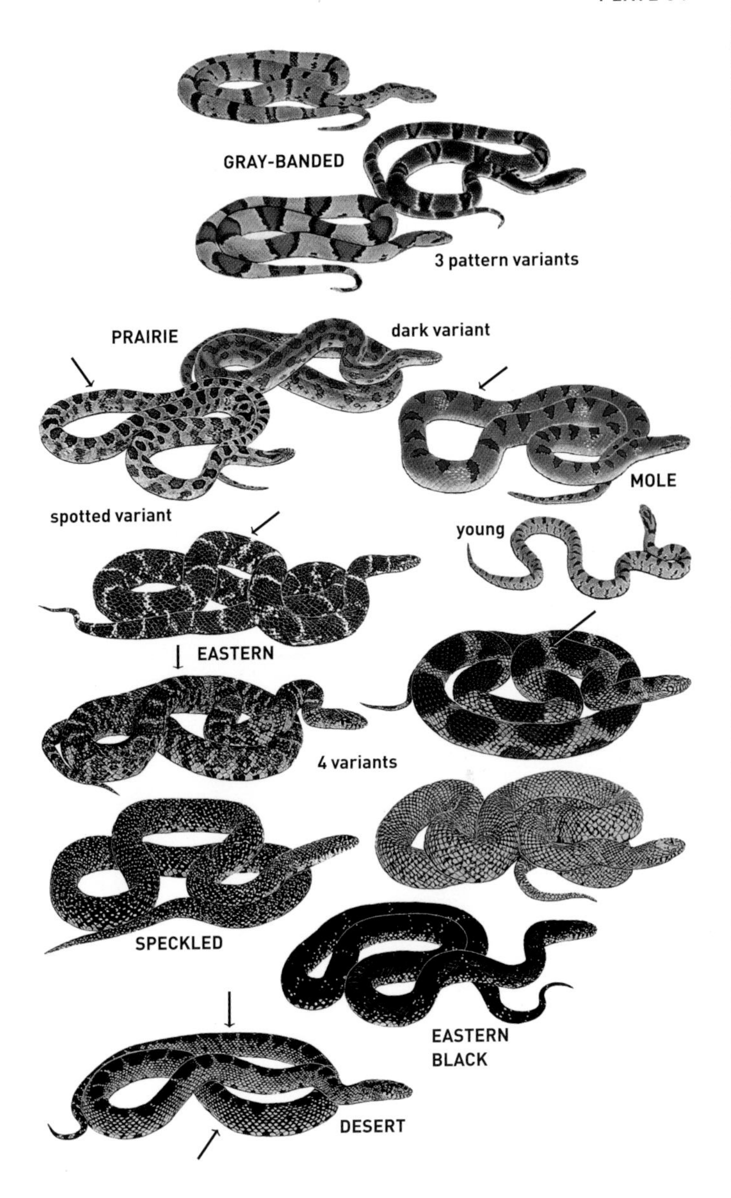
GRAY-BANDED
3 pattern variants
PRAIRIE
dark variant
MOLE
spotted variant
young
EASTERN
4 variants
SPECKLED
EASTERN
BLACK
DESERT

KING- and MILKSNAKES (*Lampropeltis*) (cont.)

Scales smooth, cloacal undivided. See also Pl. 34 and Fig. 159.

SHORT-TAILED KINGSNAKE *L. extenuata* **P. 378**
Dorsal blotches separated by orange, red, or yellow.

EASTERN MILKSNAKE *L. triangulum* **P. 381**
Variable; light Y or V at back of head and dorsal blotches alternating with smaller lateral blotches; light collar with reduced lateral blotches or dorsal blotches reaching ventral scales at least anteriorly. Western variant with light collar; lateral blotches reduced or absent.

SCARLET KINGSNAKE *L. elapsoides* **P. 378**
Rings enter or cross belly; snout red.

TAMAULIPAN MILKSNAKE *L. annulata* **P. 376**
Red rings broad (14–26 in number); snout black; belly chiefly black.

WESTERN MILKSNAKE *L. gentilis* **P. 378**
Red rings variable in width; head black; snout normally light.

GREENSNAKES (*Opheodrys*)

Plain green.

SMOOTH GREENSNAKE *O. vernalis* **P. 382**
Smooth scales.

ROUGH GREENSNAKE *O. aestivus* **P. 382**
Keeled scales.

DESERT RATSNAKES (*Bogertophis*)

Scales weakly keeled, cloacal divided.

TRANS-PECOS RATSNAKE *B. subocularis* **P. 384**
Head unicolored; black stripes on neck and dark blotches posteriorly sometimes poorly defined.

FIG. 160. *Diagrammatic cross section of ratsnakes* (Pantherophis *and* Bogertophis) *compared with that of most other snakes.*

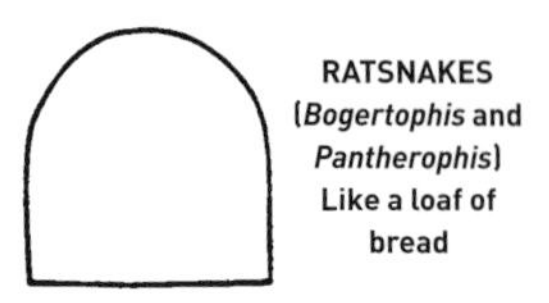

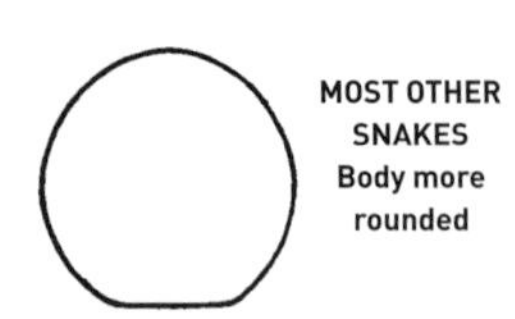

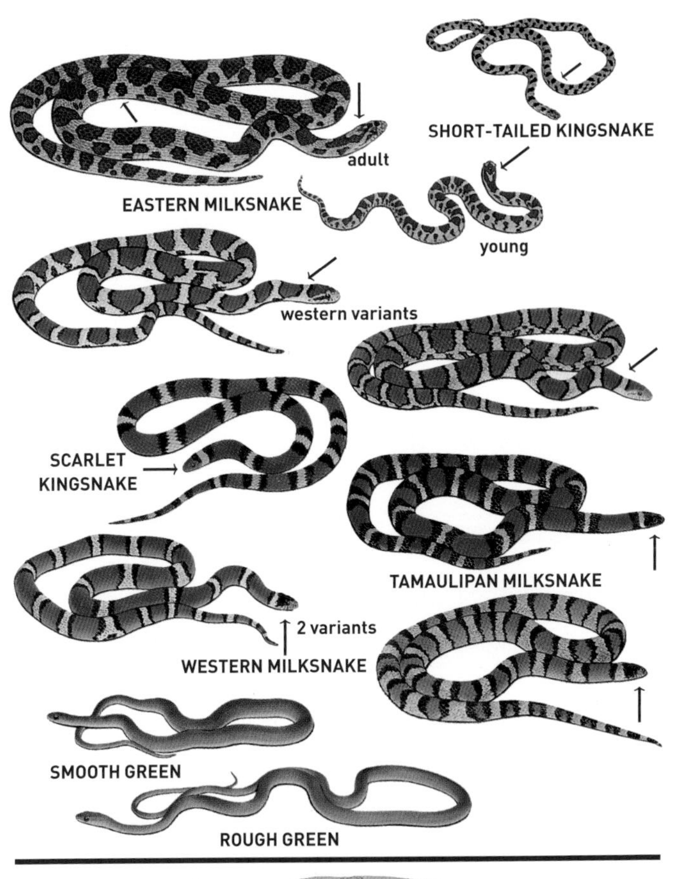
SHORT-TAILED KINGSNAKE
adult
EASTERN MILKSNAKE
young
western variants
SCARLET
KINGSNAKE
TAMAULIPAN MILKSNAKE
2 variants
WESTERN MILKSNAKE
SMOOTH GREEN
ROUGH GREEN

2 variants
TRANS-PECOS RATSNAKE

NORTH AMERICAN RATSNAKES (*Pantherophis*)

Scales weakly keeled, cloacal divided.

RED CORNSNAKE *P. guttatus* **P. 387**
Black-bordered reddish blotches; spearpoint on head (Fig. 161).

GREAT PLAINS RATSNAKE *P. emoryi* **P. 387**
Brown blotches often with dark outlines; spearpoint on head (Fig. 161).

WESTERN FOXSNAKE *P. ramspotti* **P. 389**
Head brown or reddish with no conspicuous markings (Fig. 161).

EASTERN RATSNAKE *P. alleghaniensis* **P. 384**
Variable; uniformly black, yellow or orange with four dark stripes (especially in FL), or with faint traces of blotched or striped patterns; throat light. Young like Gray Ratsnake (blotches not distinct in snakes from s. FL).

GRAY RATSNAKE *P. spiloides* **P. 389**
Variable; plain black (as in black Eastern Ratsnake) to blotched.

WESTERN RATSNAKE *P. obsoletus* **P. 388**
Variable; plain black (as in black Eastern Ratsnake) to blotched.

BAIRD'S RATSNAKE *P. bairdi* **P. 385**
Four poorly defined dark stripes.

FIG. 161. *Heads of some ratsnakes* (Pantherophis). *Patterns are most prominent in young snakes.*

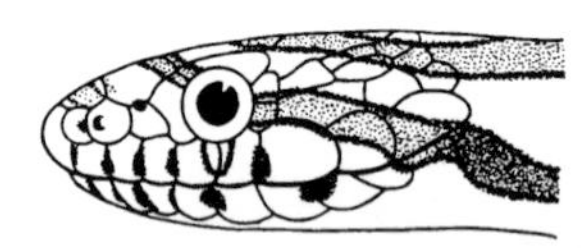

GREAT PLAINS RATSNAKE and CORNSNAKES
(*P. emoryi, P. guttatus, P. slowinskii*)
Postocular stripe has dark border, extends onto neck

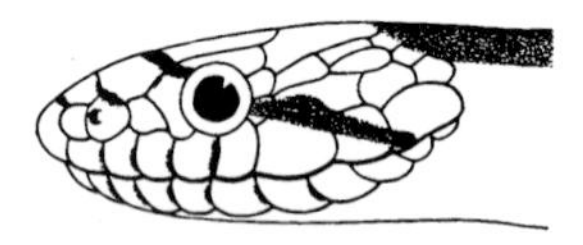

EASTERN, BAIRD'S, WESTERN, and GRAY RATSNAKES
(*P. alleghaniensis, P. bairdi, P. obsoletus, P. spiloides*)
Postocular stripe entirely dark, stops at mouth line

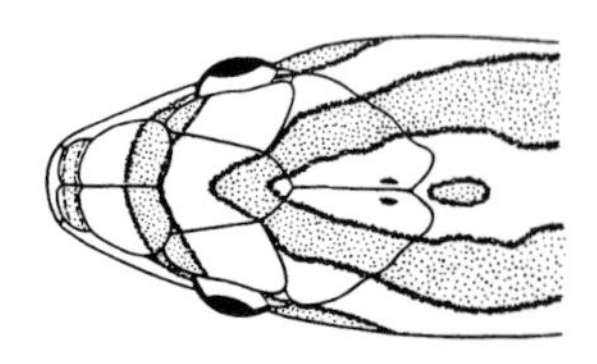

GREAT PLAINS RATSNAKE and CORNSNAKES
(*P. emoryi, P. guttatus, P. slowinskii*)
Dark neck lines unite to form a spearpoint between the eyes

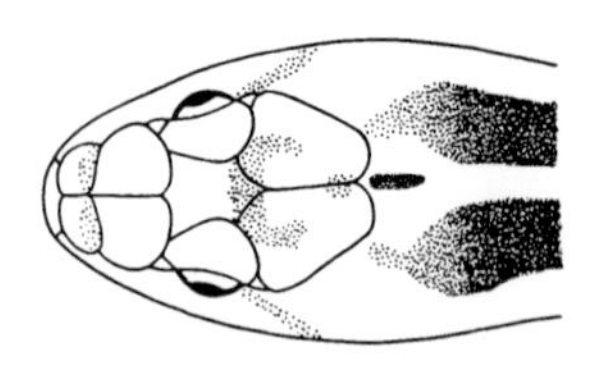

FOXSNAKES
(*P. ramspotti, P. vulpinus*)
No spearpoint; head brown or reddish

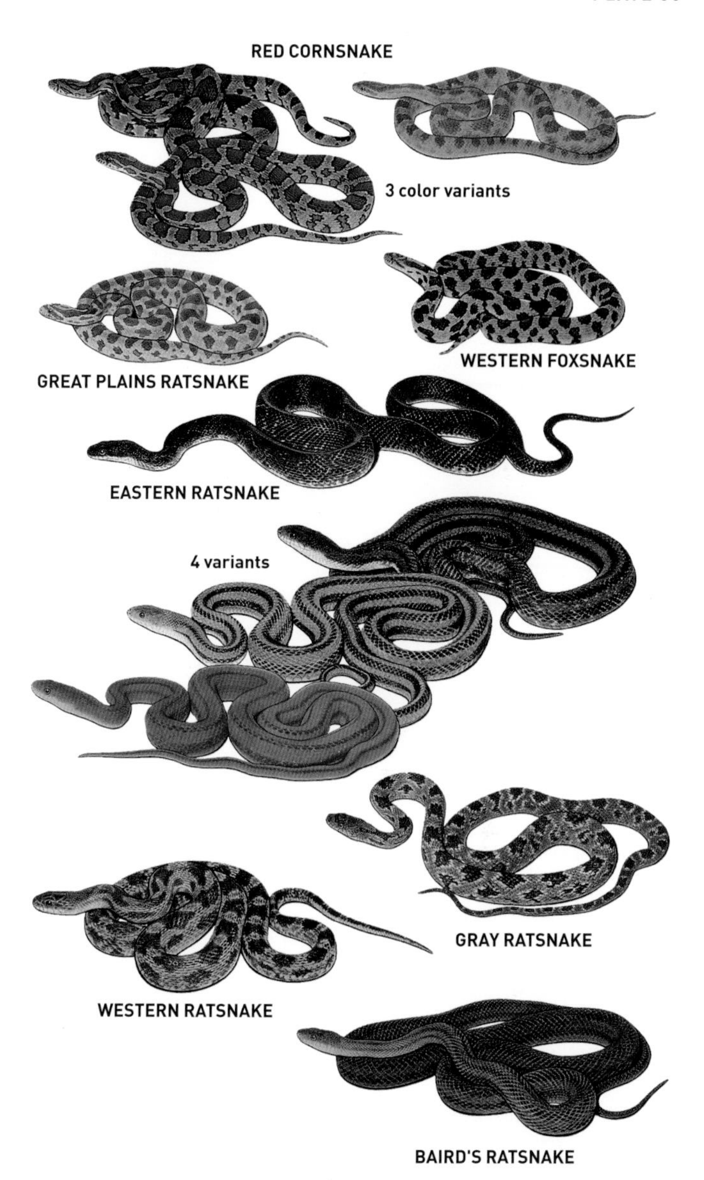
RED CORNSNAKE
3 color variants
WESTERN FOXSNAKE
GREAT PLAINS RATSNAKE
EASTERN RATSNAKE
4 variants
GRAY RATSNAKE
WESTERN RATSNAKE
BAIRD'S RATSNAKE

PLATE 37

PINE-, BULL-, and GOPHERSNAKES (*Pituophis*)

Cloacal undivided, scales keeled.

BULLSNAKE *P. catenifer sayi* **P. 391**

Dark lines from eyes to angles of jaw; 41 or more black or brown body blotches.

PINESNAKE *P. melanoleucus* **P. 392**

NORTHERN PINESNAKE (*P. m. melanoleucus*): White, yellowish, or pale gray with dark blotches.

BLACK PINESNAKE (*P. m. lodingi*): Nearly uniform black or dark brown.

FLORIDA PINESNAKE (*P. m. mugitus*): Rusty brown or brownish gray, blotches obscure anteriorly.

LOUISIANA PINESNAKE *P. ruthveni* **P. 392**

Forty-two or fewer dark body blotches, dark and obscure anteriorly, clear-cut on and near tail; no conspicuous head markings.

FIG. 162. *Prefrontal scales in pine-, bull-, and gophersnakes* (Pituophis).

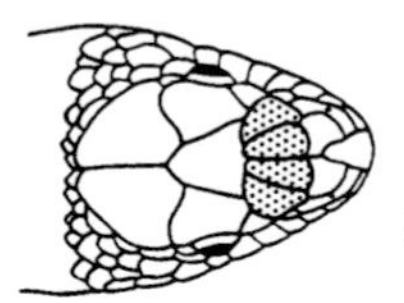

LONG-NOSED SNAKES (*Rhinocheilus*)

Scales smooth, cloacal undivided.

LONG-NOSED SNAKE *R. lecontei* **P. 392**

Crossbanded; red areas speckled with black; black areas speckled with yellow; head largely black; subcaudals in a single row.

PATCH-NOSED SNAKES (*Salvadora*)

Projecting flaps on sides of rostral, scales smooth, cloacal divided.

BIG BEND PATCH-NOSED SNAKE *S. deserticola* **P. 392**

Narrow dark lateral lines on fourth scale rows anteriorly; two or three scales between posterior chin shields (Fig. 182, p. 394).

EASTERN PATCH-NOSED SNAKE *S. grahamiae* **P. 394**

Posterior chin shields touch or separated by width of one scale (Fig. 182, p. 394).

MOUNTAIN PATCH-NOSED SNAKE (*S. g. grahamiae*): No narrow dark lateral lines (sometimes with faint lines on third scale rows).

TEXAS PATCH-NOSED SNAKE (*S. g. lineata*): Both broad and narrow dark lateral lines, latter on third scale rows anteriorly.

PATCH-NOSED SNAKES

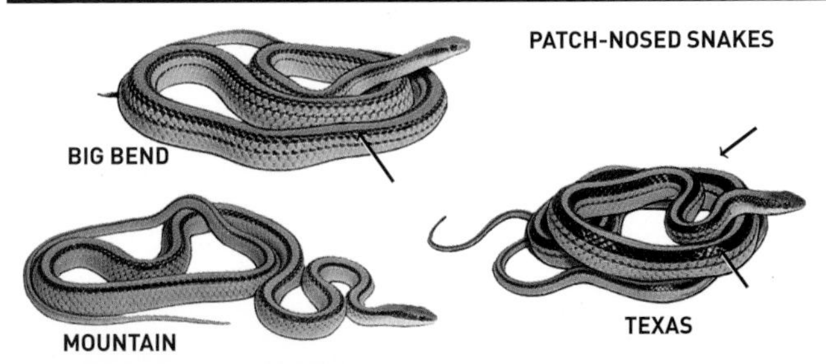

NORTH AMERICAN GROUNDSNAKES (*Sonora*)

Scales smooth, cloacal divided.

WESTERN GROUNDSNAKE *S. semiannulata* P. 394
Small; shiny; variable, uniformly colored or with numerous crossbands.

BLACK-HEADED, CROWNED, and FLAT-HEADED SNAKES (*Tantilla*)

Scales smooth in 15 rows, cloacal divided. See also Fig. 184 (p. 397).

MEXICAN BLACK-HEADED SNAKE *T. atriceps* P. 395
Short black cap straight or slightly convex posteriorly.

FLAT-HEADED SNAKE *T. gracilis* P. 398
Head only slightly darker than body; usually six upper labials.

SOUTHEASTERN CROWNED SNAKE *T. coronata* P. 396
Light band across rear of head.

CENTRAL FLORIDA CROWNED SNAKE *T. relicta neilli* P. 400
No light band; black cap extends far beyond head scales.

PLAINS BLACK-HEADED SNAKE *T. nigriceps* P. 399
Black cap rounded or pointed posteriorly; usually seven upper labials.

TRANS-PECOS BLACK-HEADED SNAKE *T. cucullata* P. 396
Entire head black or with light collar sometimes interrupted middorsally.

LYRESNAKES (*Trimorphodon*)

Vertically elliptical pupils; scales smooth, cloacal divided.

TEXAS LYRESNAKE *T. vilkinsonii* P. 400
Widely spaced dark brown saddles.

NORTH AMERICAN WORMSNAKES (*Carphophis*)

Scales smooth, cloacal divided.

EASTERN WORMSNAKE *C. amoenus amoenus* P. 401
Brown back; light pink belly

WESTERN WORMSNAKE *C. vermis* P. 402
Black back; bright pink belly.

BLACK-STRIPED SNAKES (*Coniophanes*)

Scales smooth, cloacal divided.

REGAL BLACK-STRIPED SNAKE *C. imperialis* P. 402
Three black (or dark brown) stripes; belly bright red or orange.

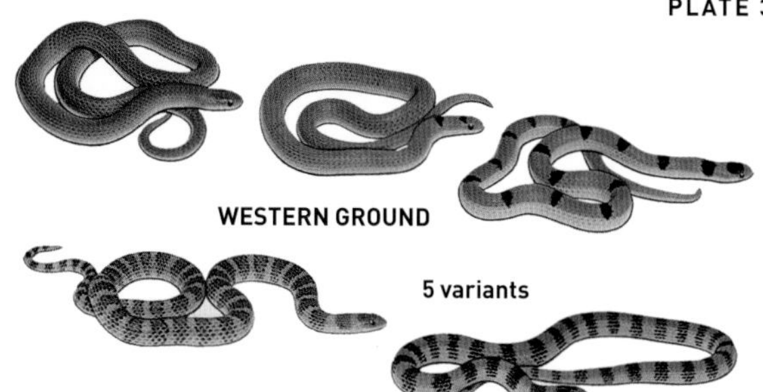

WESTERN GROUND

5 variants

MEXICAN BLACK-HEADED

FLAT-HEADED

SOUTHEASTERN CROWNED

CENTRAL FLORIDA CROWNED

PLAINS BLACK-HEADED

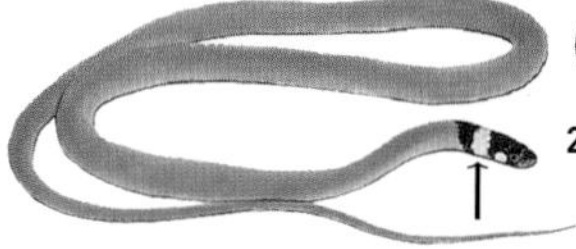

2 variants

TRANS-PECOS BLACK-HEADED

TEXAS LYRESNAKE

WESTERN WORM

EASTERN WORM

REGAL BLACK-STRIPED

RING-NECKED SNAKES (*Diadophis*)

Scales smooth, cloacal divided.

RING-NECKED SNAKE *D. punctatus* **P. 403**

SOUTHERN RING-NECKED SNAKE (*D. p. punctatus*): Ring interrupted; belly with row of black spots.

NORTHERN RING-NECKED SNAKE (*D. p. edwardsii*): Ring complete; belly plain.

MUD- and RAINBOW SNAKES (*Farancia*)

Scales smooth, cloacal usually divided.

RED-BELLIED MUDSNAKE *F. abacura* **P. 405**
Shiny black or dark gray; red or pink of belly encroaches on sides; black belly spots continuous with black of dorsum in neck region.

RAINBOW SNAKE *F. erytrogramma* **P. 406**
Red and black dorsal stripes; venter with double row of rounded black spots.

NORTH AMERICAN HOG-NOSED SNAKES (*Heterodon*)

Snout upturned, scales keeled, cloacal divided. See also Fig. 189 (p. 408).

SOUTHERN HOG-NOSED SNAKE *H. simus* **P. 409**
Belly unpatterned or mottled with grayish brown.

PLAINS HOG-NOSED SNAKE *H. nasicus* **P. 408**
Belly chiefly black.

EASTERN HOG-NOSED SNAKE *H. platirhinos* **P. 408**
Highly variable; underside of tail lighter than belly.

NIGHTSNAKES and CAT-EYED SNAKES (*Hypsiglena* and *Leptodeira*)

Vertically elliptical pupils, scales smooth, cloacal divided.

CHIHUAHUAN NIGHTSNAKE *H. jani* **P. 409**
Dark neck spots; middorsal spots alternate with smaller lateral spots.

CAT-EYED SNAKE *L. septentrionalis* **P. 409**
Cream to yellow or reddish tan with bold black or dark brown saddles.

LITTERSNAKES (*Rhadinaea*)

PINE WOODS LITTERSNAKE *R. flavilata* **P. 410**
Dark lines through eyes; scales smooth; cloacal divided.

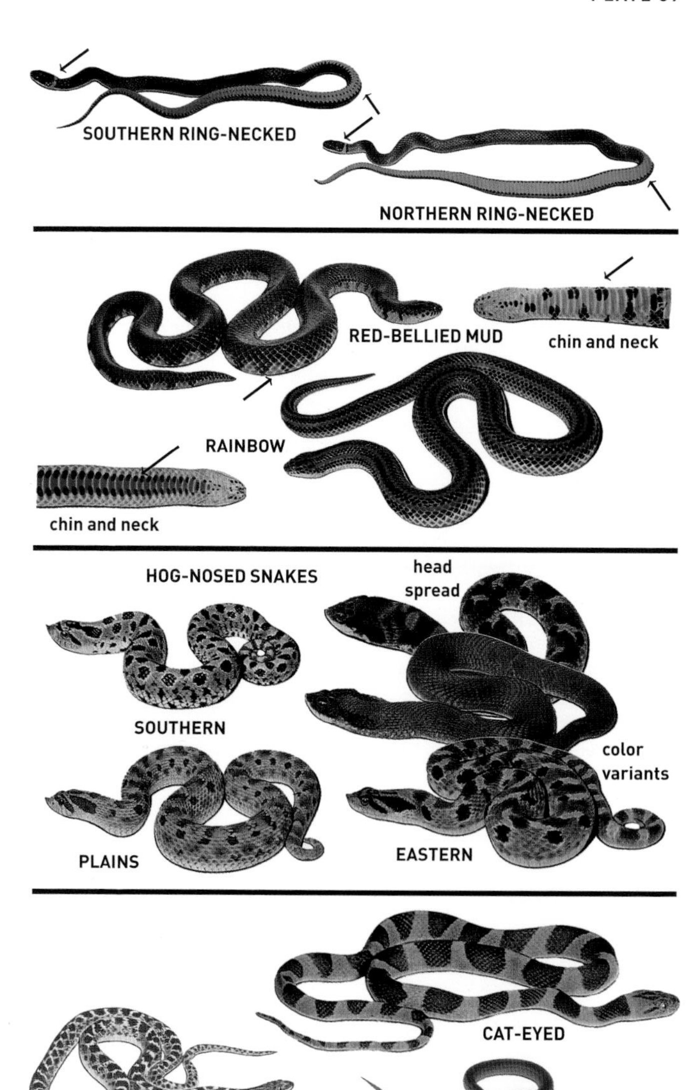
SOUTHERN RING-NECKED
NORTHERN RING-NECKED
RED-BELLIED MUD
chin and neck
RAINBOW
chin and neck
HOG-NOSED SNAKES
head spread
SOUTHERN
color variants
PLAINS
EASTERN
CAT-EYED
CHIHUAHUAN NIGHT
PINE WOODS LITTER

KIRTLAND'S SNAKES (*Clonophis*)

Scales keeled, cloacal divided.

KIRTLAND'S SNAKE *C. kirtlandii* **P. 410**

Two lateral rows of large dark spots; belly brick red with lateral rows of black spots.

NORTH AMERICAN WATERSNAKES (*Nerodia*)

Scales keeled, cloacal divided. See also Pl. 41.

SALTMARSH WATERSNAKE *N. clarkii* **P. 414**

GULF SALTMARSH WATERSNAKE (*N. c. clarkii*): Two dark stripes on sides; belly with row of light spots (sometimes three rows).

MANGROVE SALTMARSH WATERSNAKE (*N. c. compressicauda*): Irregular dark markings above and below; highly variable. Red variant with red or orange-red on back and belly.

MISSISSIPPI GREEN WATERSNAKE *N. cyclopion* **P. 415**

No distinctive pattern; subocular scales (Fig. 163); belly dark with light half-moons.

PLAIN-BELLIED WATERSNAKE *N. erythrogaster* **P. 416**

Variable; plain brown, reddish brown, gray, or greenish above, with traces of dark-bordered light transverse bands, or retaining juvenile pattern of mid-dorsal blotches alternating with lateral blotches (see Pl. 41); red or orange above; belly red, orange, or yellow, sometimes with faint suggestion of spots.

SOUTHERN WATERSNAKE *N. fasciata* **P. 417**

Eye stripes distinct (Fig. 163).

FLORIDA WATERSNAKE (*N. f. pictiventris*): Black, brown, or red crossbands; often secondary dark spots on sides; belly with wavy vermiculate crosslines.

BANDED WATERSNAKE (*N. f. fasciata*): Black, brown, or red crossbands full length of body; usually some red on sides even in darkest individuals; largest belly markings squarish.

BROAD-BANDED WATERSNAKE (*N. f. confluens*): Dark crossbands much wider than light interspaces; belly with large red to black squarish markings.

FIG. 163. *Heads of three species of watersnakes* (Nerodia).

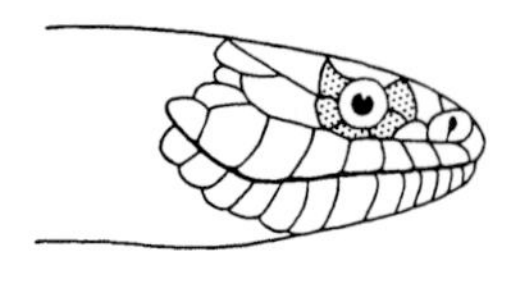

GREEN WATERSNAKES
(*Nerodia cyclopion* and *N. floridana*)
Scales between eye and lip plates

SOUTHERN WATERSNAKES
(*Nerodia fasciata*)
Dark stripe from eye to angle of jaw

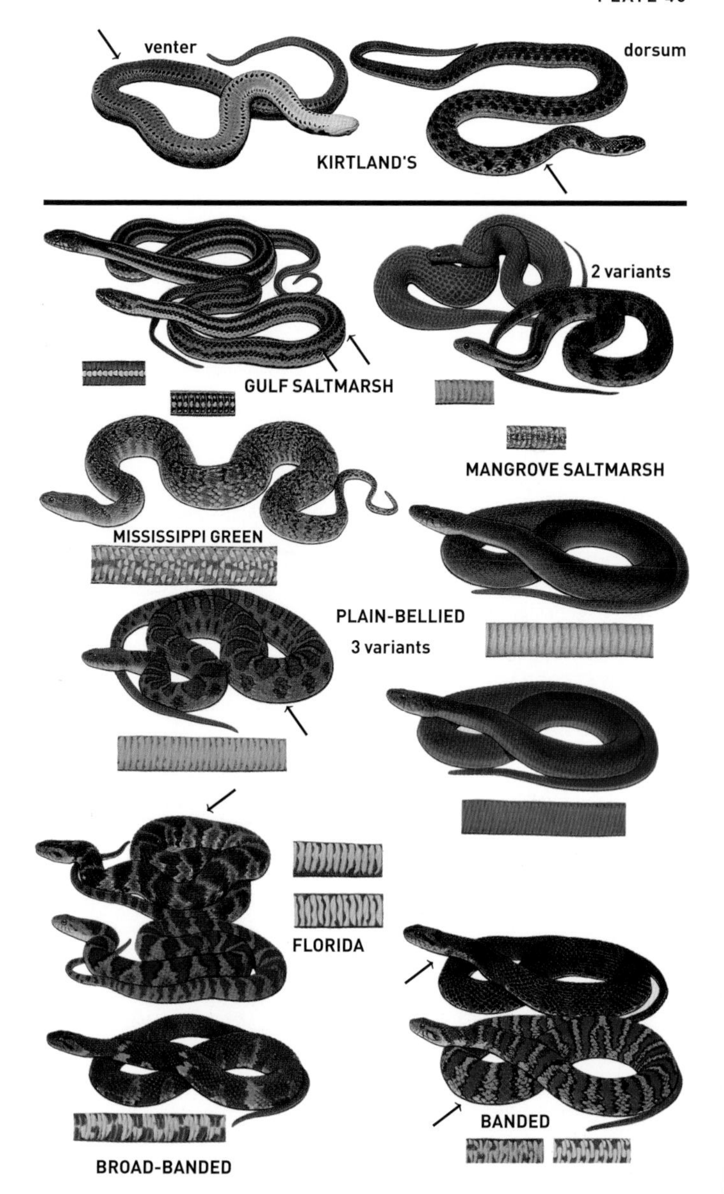
venter
dorsum
KIRTLAND'S
GULF SALTMARSH
2 variants
MANGROVE SALTMARSH
MISSISSIPPI GREEN
PLAIN-BELLIED
3 variants
FLORIDA
BROAD-BANDED
BANDED

NORTH AMERICAN WATERSNAKES (*Nerodia*) (cont.)

Scales keeled, cloacal divided. See also Pl. 40.

FLORIDA GREEN WATERSNAKE *N. floridana* **P. 417**
No distinctive pattern; subocular scales (Fig. 163, p. 342); belly light, virtually unicolored.

BRAZOS RIVER WATERSNAKE *N. harteri* **P. 418**
Two rows of spots on sides; belly with pink center and lateral rows of dark dots.

DIAMOND-BACKED WATERSNAKE *N. rhombifer* **P. 419**
Dark chainlike dorsal pattern (Fig. 191, p. 420); belly yellow with largest dark spots chiefly along sides.

COMMON WATERSNAKE *N. sipedon* **P. 420**
Crossbands anteriorly, alternating middorsal and lateral blotches posteriorly.

NORTHERN WATERSNAKE (*N. s. sipedon*): Dark markings wider than lighter intervening spaces, latter usually less than 2½ scales wide; belly variable with virtually no markings or paired half-moons.

MIDLAND WATERSNAKE (*N. s. pleuralis*): Dark markings narrower than lighter intervening spaces, latter usually more than 2½ scales wide; belly with double row of half-moons or crescents.

BROWN WATERSNAKE *N. taxispilota* **P. 421**
Dark middorsal blotches alternate with and are separated from lateral blotches (Fig. 191, p. 420); belly yellow with heavy dark markings.

PLAIN-BELLIED WATERSNAKE *N. erythrogaster* (young)
Dorsal blotches on neck alternate with lateral blotches.

NORTHERN WATERSNAKE *N. s. sipedon* (young)
Dorsal blotches on neck join lateral blotches to form crossbands.

CRAYFISH SNAKES and SWAMPSNAKES (*Regina* and *Liodytes*)

Cloacal divided.

GRAHAM'S CRAYFISH SNAKE *R. grahamii* **P. 422**
Broad yellowish lateral stripes; belly plain or with central row of dark spots.

QUEENSNAKE *R. septemvittata* **P. 422**
Yellowish lateral stripes; belly with four brown stripes.

GLOSSY SWAMPSNAKE *L. rigida* **P. 414**
Shiny; often traces of stripes; belly with double row of black half-moons.

STRIPED SWAMPSNAKE *L. alleni* **P. 412**
Broad yellowish lateral stripes; dark middorsal and dorsolateral stripes; scales smooth; belly plain yellow or orange or with midventral row of dark spots.

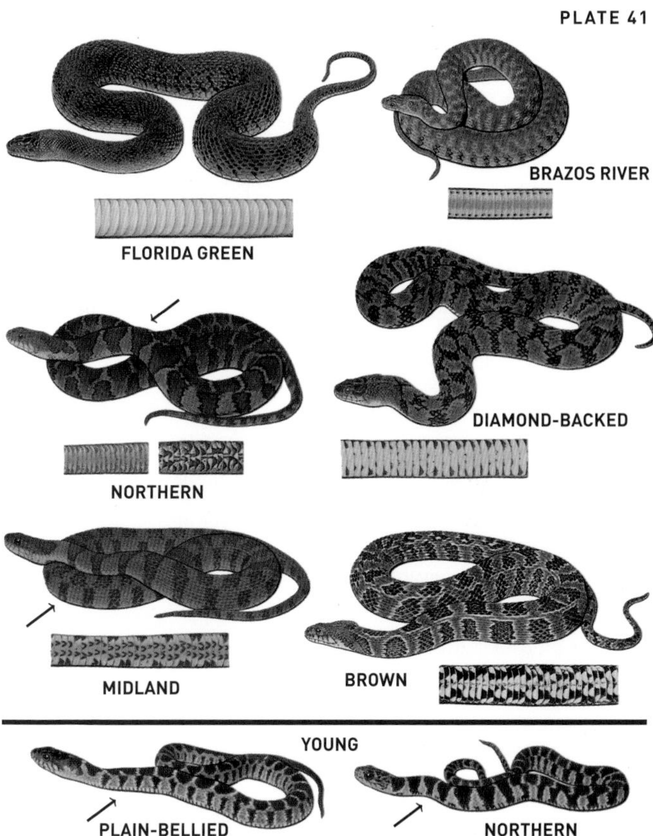
BRAZOS RIVER
FLORIDA GREEN
NORTHERN
DIAMOND-BACKED
MIDLAND
BROWN

YOUNG
PLAIN-BELLIED
NORTHERN

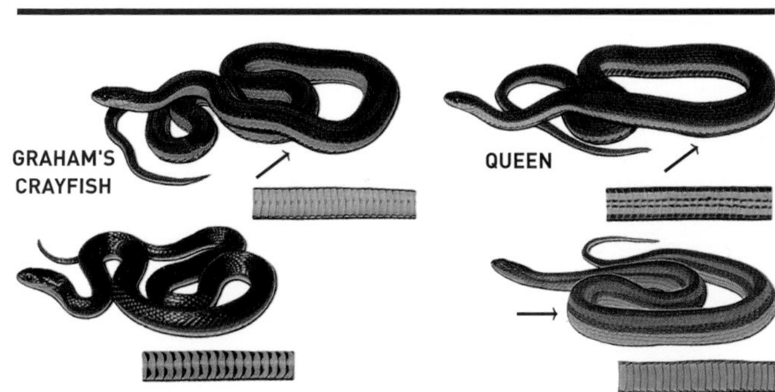
GRAHAM'S
CRAYFISH
QUEEN
GLOSSY SWAMP
STRIPED SWAMP

PLATE 42

SWAMPSNAKES (*Liodytes*) (cont.)

BLACK SWAMPSNAKE *L. pygaea* **P. 413**

Shiny black; scales smooth but with pale streaks that look like keels; belly red with black encroaching on edges of ventrals.

NORTH AMERICAN BROWNSNAKES (*Storeria*)

Scales keeled, cloacal divided.

DEKAY'S BROWNSNAKE *S. dekayi* **P. 423**

Dark vertical streaks on sides of head (Fig. 193, p. 424); no loreals (Fig. 192, p. 423); young with light band across neck.

FLORIDA BROWNSNAKE *S. victa* **P. 426**

Light band across head; no loreals (Fig. 192, p. 423).

RED-BELLIED SNAKE *S. occipitomaculata* **P. 424**

Light spots on nape; coloration variable; belly usually bright red or orange-red, sometimes blue-black.

GARTER- and RIBBONSNAKES (*Thamnophis*)

Scales keeled, cloacal almost always undivided. See also Pl. 43.

BUTLER'S GARTERSNAKE *T. butleri* **P. 426**

Small head; lateral stripes on rows 2, 3, 4; 19 scale rows.

SHORT-HEADED GARTERSNAKE *T. brachystoma* **P. 426**

Very small head; 17 scale rows.

BLACK-NECKED GARTERSNAKE *T. cyrtopsis* **P. 427**

Black blotches on sides of neck.

WESTERN BLACK-NECKED GARTERSNAKE (*T. c. cyrtopsis*): Lateral stripes on rows 2 and 3.

EASTERN BLACK-NECKED GARTERSNAKE (*T. c. ocellatus*): Lateral stripes wavy.

WANDERING GARTERSNAKE *T. elegans vagrans* **P. 428**

Eight upper labials; 21 scale rows at midbody; striped or spotted.

CHECKERED GARTERSNAKE *T. marcianus* **P. 428**

Checkerboard of black spots; light curved bands behind mouth followed by broad black blotches.

WESTERN RIBBONSNAKE *T. proximus* **P. 428**

Slender; tail long; lateral stripes on rows 3 and 4; parietal spots in contact.

ORANGE-STRIPED RIBBONSNAKE (*T. p. proximus*): Like Eastern Ribbonsnake but middorsal stripe often bright orange.

RED-STRIPED RIBBONSNAKE (*T. p. rubrilineatus*): Middorsal stripe red.

venter
dorsum
BLACK SWAMP
color variants
young
FLORIDA BROWN
DEKAY'S BROWN
venter
RED-BELLIED
color variants
BUTLER'S GARTER
SHORT-HEADED GARTER
WESTERN BLACK-NECKED GARTER
EASTERN BLACK-NECKED GARTER
CHECKERED GARTER
2 pattern variants
WANDERING GARTER
ORANGE-STRIPED RIBBON
RED-STRIPED RIBBON

PLATE 43

GARTER- AND RIBBONSNAKES (*Thamnophis*) (cont.)

Scales keeled, cloacal almost always undivided. See also Pl. 42.

PLAINS GARTERSNAKE *T. radix* **P. 429**

Black bars on lips; lateral stripes on rows 3 and 4.

EASTERN RIBBONSNAKE *T. sauritus* **P. 430**

Slender; tail long; lateral stripes on rows 3 and 4; parietal spots, if present, faint, not in contact.

COMMON RIBBONSNAKE (*T. s. sauritus*): Brown ventrolateral bands below lateral stripes.

PENINSULA RIBBONSNAKE (*T. s. sackenii*): Middorsal stripe fainter than lateral stripes or lacking.

BLUE-STRIPED RIBBONSNAKE (*T. s. nitae*): Lateral stripes blue; obscure middorsal stripe sometimes present.

COMMON GARTERSNAKE *T. sirtalis* **P. 431**

Lateral stripes on rows 2 and 3.

EASTERN GARTERSNAKE (*T. s. sirtalis*): Stripes or spots may predominate (rarely stripeless).

CHICAGO GARTERSNAKE (*T. s. semifasciatus*): Black bars cross lateral stripes in neck region.

BLUE-STRIPED GARTERSNAKE (*T. s. similis*): Lateral stripes blue.

RED-SIDED GARTERSNAKE (*T. s. parietalis*): Red or orange lateral bars.

TEXAS GARTERSNAKE (*T. s. annectens*): Broad orange middorsal stripe; lateral stripes on rows 2, 3, 4.

NEW MEXICO GARTERSNAKE (*T. s. dorsalis*): Red subdued, largely confined to skin between scales.

FIG. 164. *Positions of lateral stripes in gartersnakes* (Thamnophis). *Numbers refer to scale rows.*

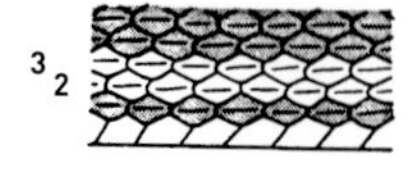

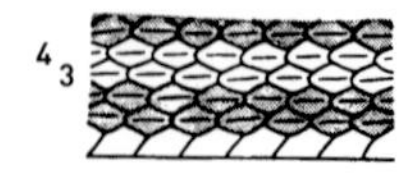

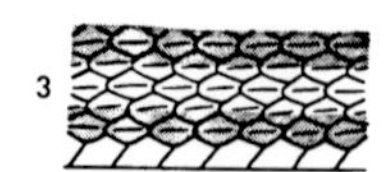

PLAINS GARTER
COMMON RIBBON
PENINSULA RIBBON
BLUE-STRIPED RIBBON
EASTERN GARTER
3 pattern variants
CHICAGO GARTER
BLUE-STRIPED GARTER
RED-SIDED GARTER
TEXAS GARTER
NEW MEXICO GARTER

LINED SNAKES (*Tropidoclonion*)

Scales keeled, cloacal undivided.

LINED SNAKE *T. lineatum* **P. 433**

Double row of black half-moons on belly; small head; lateral stripes on rows 2 and 3.

EARTHSNAKES (*Virginia* and *Haldea*)

Cloacal usually divided; loreals horizontal and touching eyes (Fig. 165).

SMOOTH EARTHSNAKE *V. valeriae* **P. 412**

Gray or brown with tiny black dots; scales smooth or weakly keeled.

ROUGH EARTHSNAKE *H. striatula* **P. 411**

Head pointed; scales keeled.

FIG. 165. *Head of an earthsnake* (Haldea *or* Virginia).

EARTHSNAKES (*Virginia* and *Haldea*)
Loreal scale horizontal and touching eye

AMERICAN CORALSNAKES (*Micrurus*) VENOMOUS

Red and yellow rings touch; tip of snout black. See also Fig. 159 (p. 330).

HARLEQUIN CORALSNAKE *M. fulvius* **P. 434**

Black neck ring does not touch parietals.

TEXAS CORALSNAKE *M. tener* **P. 434**

Black neck ring extends onto parietals.

Note: Nonvenomous red-and-yellow banded snakes in our area have *red and yellow separated by black* (see also milksnakes and Scarlet Kingsnake on Pl. 35).

SCARLETSNAKE *Cemophora coccinea* **P. 367**

Snout red; belly whitish, unpatterned (see also Pls. 32 and 34).

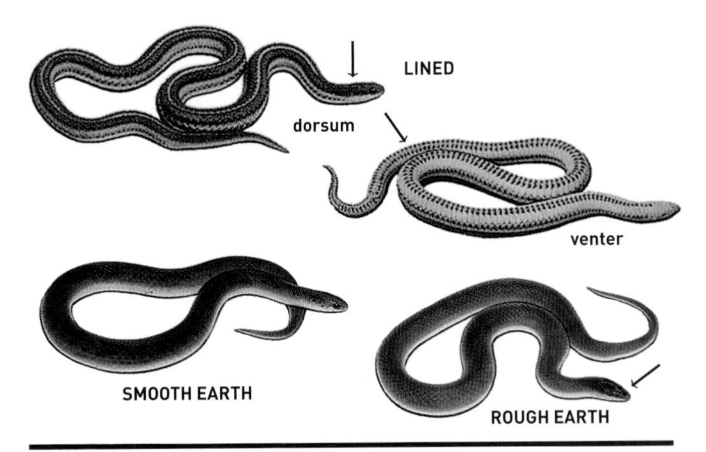
LINED
dorsum
venter
SMOOTH EARTH
ROUGH EARTH

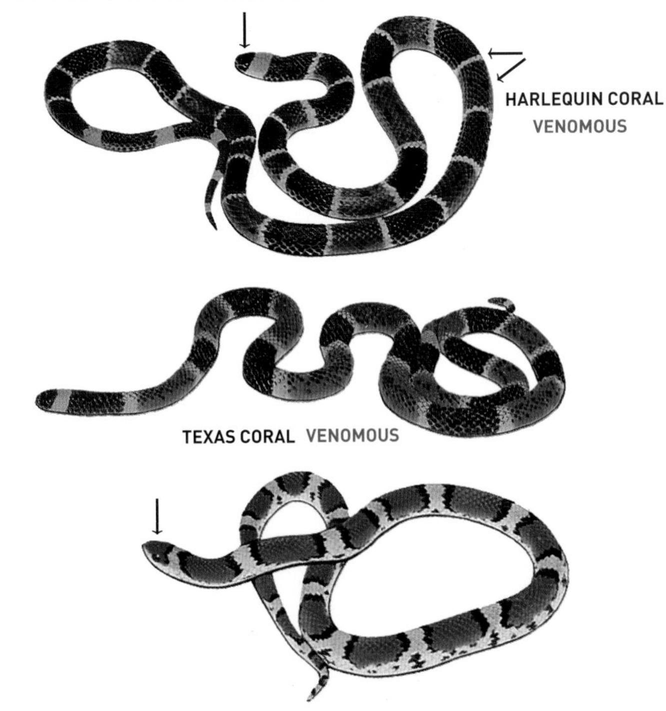
HARLEQUIN CORAL
VENOMOUS
TEXAS CORAL VENOMOUS
SCARLETSNAKE

PLATE 45

AMERICAN MOCCASINS (*Agkistrodon*) VENOMOUS

Scales weakly keeled, cloacal undivided. Young with yellow tailtip, narrow lines through eyes in copperheads, broad dark bands through eyes in cottonmouths.

EASTERN COPPERHEAD *A. contortrix* **P. 436**
Bands narrow middorsally, often fail to meet in southern populations, edged in white in the Midwest.

BROAD-BANDED COPPERHEAD *A. laticinctus* **P. 436**
Bands broad, nearly as wide dorsally as laterally.

NORTHERN COTTONMOUTH *A. piscivorus* **P. 437**
Head markings obscure or absent; body pattern variable.

FLORIDA COTTONMOUTH *A. conanti* **P. 437**
Head markings well defined.

FIG. 166. *Heads and undersides of tails of American moccasins* (Agkistrodon) *and watersnakes (*Nerodia*). Note also that moccasins, like the Northern Cottonmouth (*A. piscivorus*) shown below, tend to swim with most of the body at the surface, whereas in watersnakes only the head and a small part of the body break the surface.*

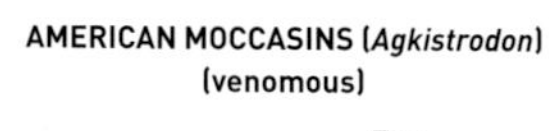

AMERICAN MOCCASINS (*Agkistrodon*)
(venomous)

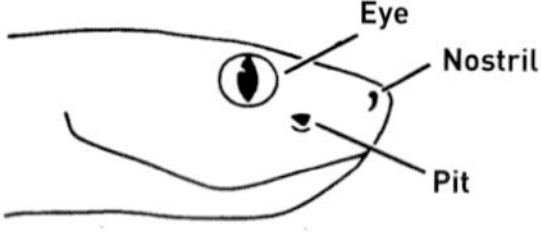

Facial pit; vertical pupil

WATERSNAKES (*Nerodia*)
(nonvenomous)

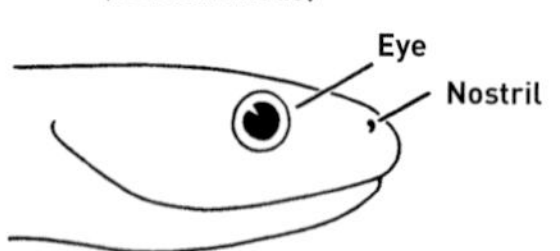

No pit; round pupil

Single row of scales;
single anal plate

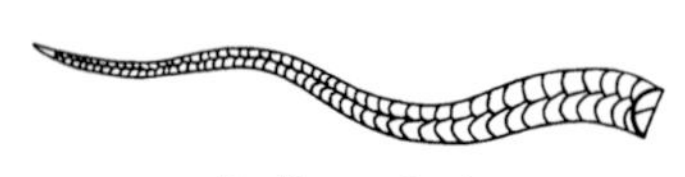

Double row of scales;
divided anal plate

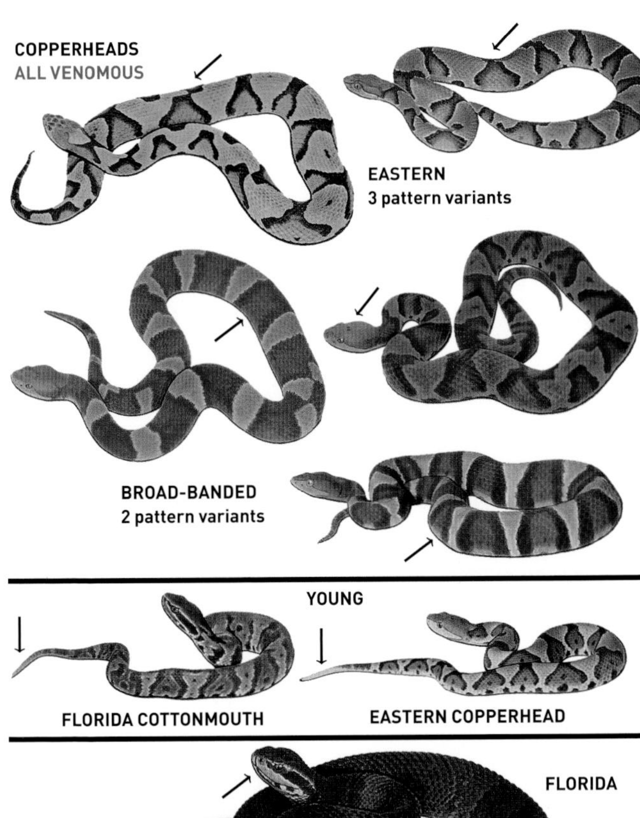
COPPERHEADS
ALL VENOMOUS
EASTERN
3 pattern variants
BROAD-BANDED
2 pattern variants
YOUNG
FLORIDA COTTONMOUTH
EASTERN COPPERHEAD

FLORIDA
COTTONMOUTHS
ALL VENOMOUS
NORTHERN
2 pattern variants

PLATE 46

RATTLESNAKES (*Sistrurus* and *Crotalus*) VENOMOUS

See also Pl. 47.

PYGMY RATTLESNAKE *S. miliarius* **P. 443**

Rattle tiny; tail slender.

WESTERN PYGMY RATTLESNAKE (*S. m. streckeri*): Dark dorsal bars.

DUSKY PYGMY RATTLESNAKE (*S. m. barbouri*): Rounded spots sometimes obscured by dark stippling.

CAROLINA PYGMY RATTLESNAKE (*S. m. miliarius*): Gray or reddish; markings clear-cut.

WESTERN MASSASAUGA *S. tergeminus* **P. 445**

Generally pale with large rounded spots; belly light, sometimes with dark markings (Fig. 201, p. 444).

EASTERN MASSASAUGA *S. catenatus* **P. 443**

Generally dark with large rounded spots; belly mostly black (Fig. 201, p. 444).

EASTERN DIAMOND-BACKED RATTLESNAKE *C. adamanteus* **P. 439**

Large, clear-cut, light-bordered diamonds.

TIMBER RATTLESNAKE *C. horridus* **P. 440**

Variable, but always with dark crossbands (except when all black). Western variant generally gray with orange-red middorsal stripe, dark markings often edged in white, and reddish brown postocular lines. Southern variant generally gray to pinkish with reddish brown middorsal stripe and dark postocular lines. Northeastern variant all black or yellow with no head markings (Fig. 168, p. 356).

FIG. 167. *Scales on top of the heads of rattlesnakes* (Crotalus *and* Sistrurus).

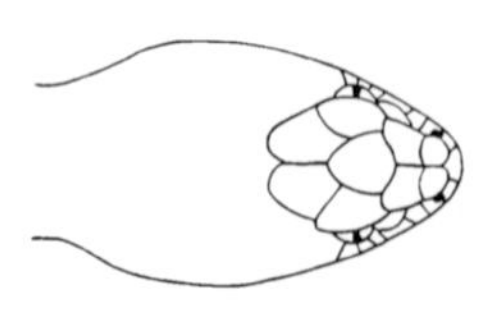

PYGMY RATTLESNAKES and MASSASAUGAS (*Sistrurus*)
A group of nine large scales (plates) on crown of head

RATTLESNAKES (*Crotalus*)
Crown of head with a mixture of large and small scales

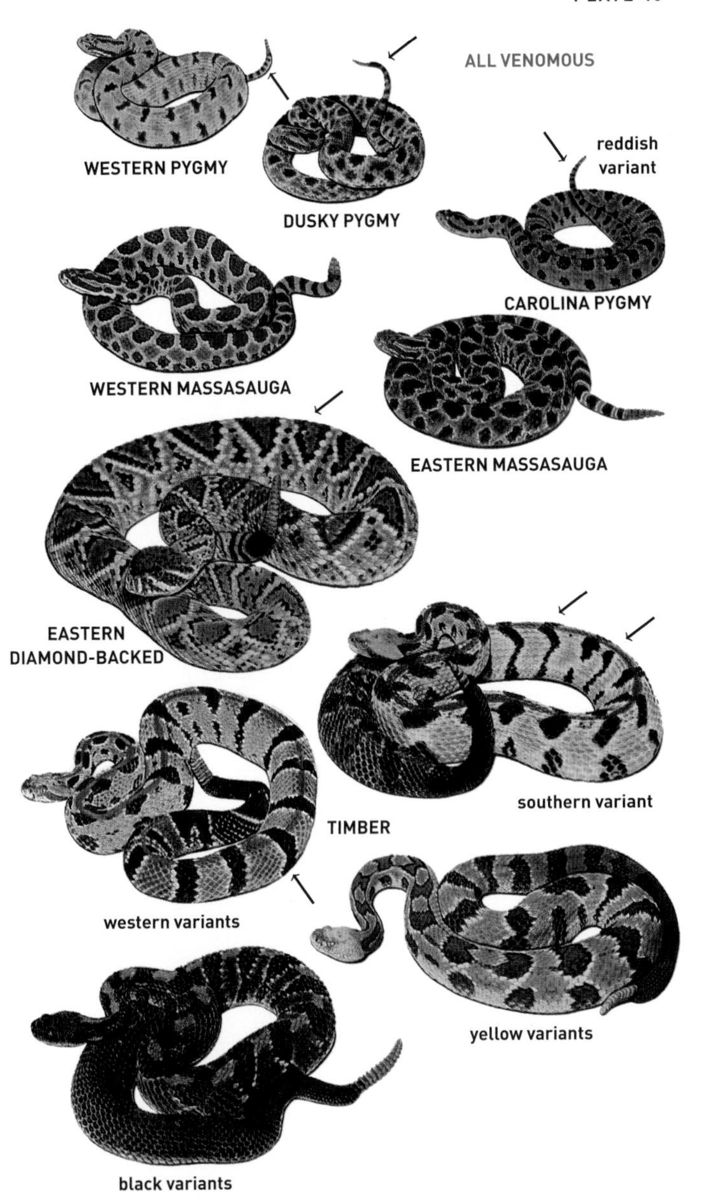
ALL VENOMOUS
WESTERN PYGMY
DUSKY PYGMY
reddish variant
CAROLINA PYGMY
WESTERN MASSASAUGA
EASTERN MASSASAUGA
EASTERN DIAMOND-BACKED
southern variant
TIMBER
western variants
yellow variants
black variants

PLATE 47

RATTLESNAKES (*Crotalus*) VENOMOUS

See also Pl. 46.

ROCK RATTLESNAKE *C. lepidus* **P. 441**

BANDED ROCK RATTLESNAKE (*C. l. klauberi*): Conspicuous dark crossbands throughout length of body; no dark stripes from eyes to angles of mouth.

MOTTLED ROCK RATTLESNAKE (*C. l. lepidus*): Dark crossbands conspicuous only posteriorly; dark stripes from eyes to angles of mouth.

EASTERN BLACK-TAILED RATTLESNAKE *C. ornatus* **P. 441**

Tail solid black.

PRAIRIE RATTLESNAKE *C. viridis* **P. 442**

Middorsal blotches anteriorly, joining with lateral blotches to form crossbands posteriorly; postocular light lines pass above corners of mouth (Fig. 168).

MOHAVE RATTLESNAKE *C. scutulatus* **P. 442**

Black rings on tail much narrower than white rings; postocular light lines pass above corners of mouth (Fig. 168).

WESTERN DIAMOND-BACKED RATTLESNAKE *C. atrox* **P. 439**

Black tail rings relatively broad; diamonds not clear-cut; postocular light lines end in front of corners of mouth (Fig. 168).

FIG. 168. *Sides of the heads of some rattlesnakes in the genus* Crotalus.

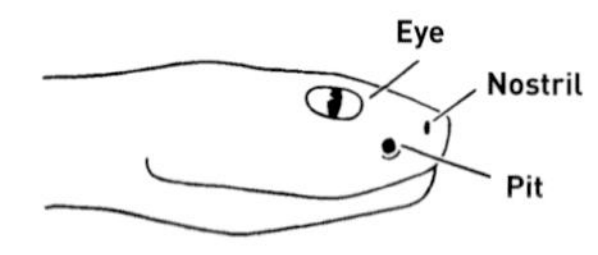

TIMBER
(*C. horridus*)
Northern variations. Head markings are normally lacking.

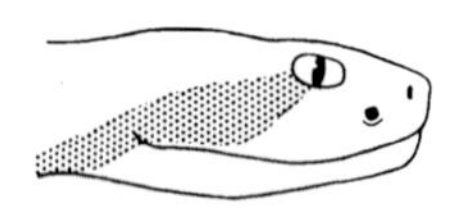

TIMBER
(*C. horridus*)
Southern and western variations. Dark stripe from eye to angle of jaw.

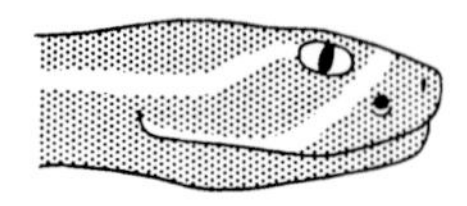

PRAIRIE (*C. viridis*) and MOHAVE (*C. scutulatus*)
Rear light stripe passes above mouthline

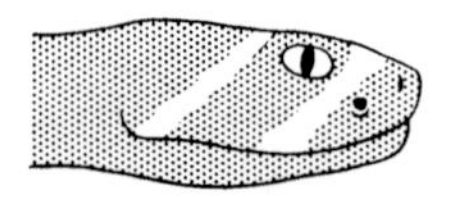

WESTERN DIAMOND-BACKED
(*C. atrox*)
Rear light strip meets mouthline in front of angle of jaw

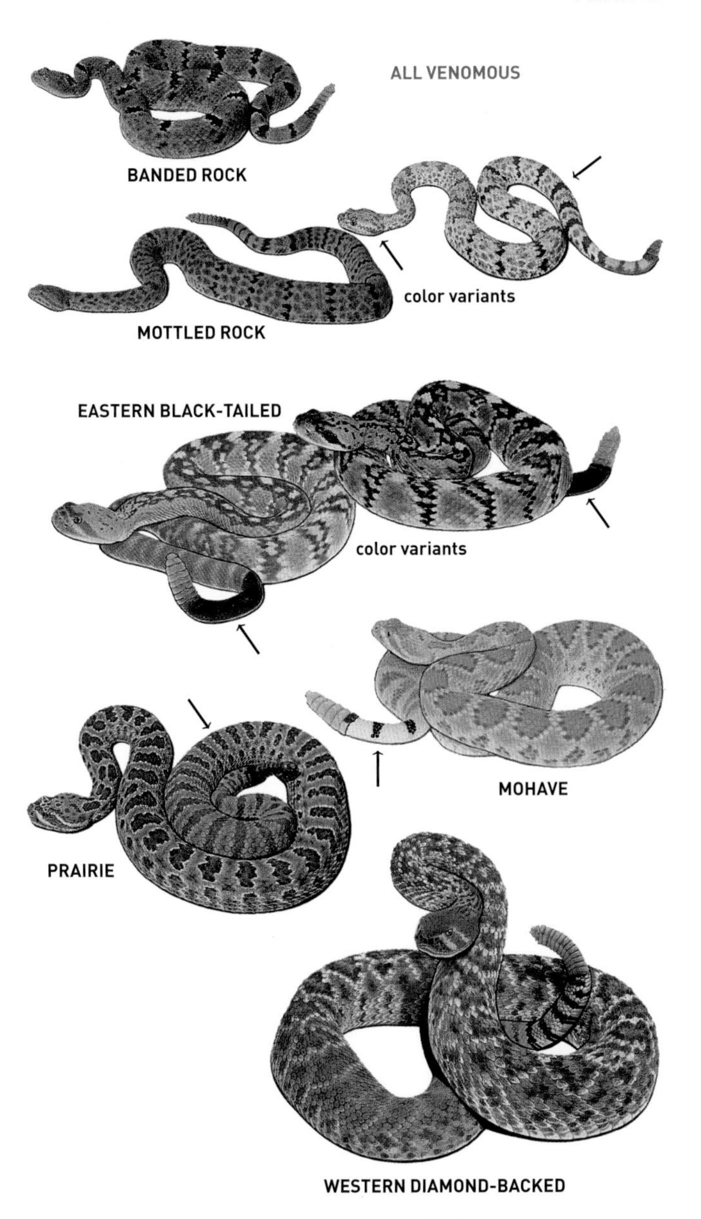
ALL VENOMOUS
BANDED ROCK
color variants
MOTTLED ROCK
EASTERN BLACK-TAILED
color variants
MOHAVE
PRAIRIE
WESTERN DIAMOND-BACKED

FIG. 169. *Characteristics of snakes.*

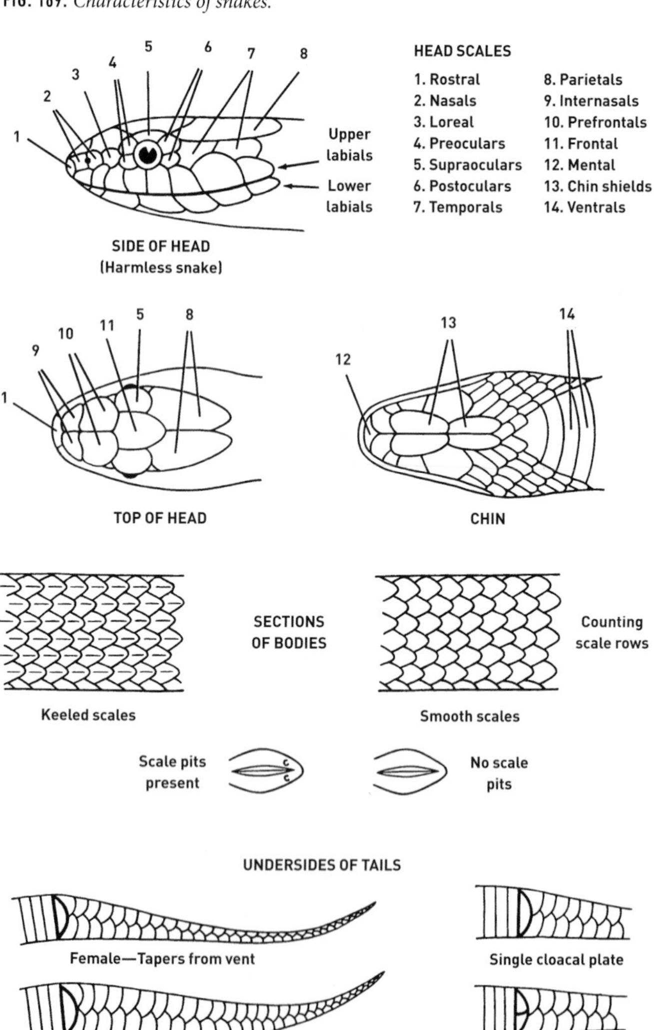

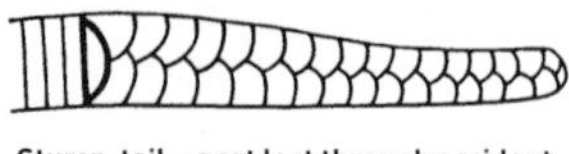

SNAKES

Snakes lack limbs, movable eyelids, and external ears, and scales around the body never form distinct rings (Fig. 169, facing page). Nearly 3,500 species range from above the Arctic Circle in Scandinavia to Tasmania, the Cape Region of Africa, and very nearly to the tip of South America. Few species are venomous. However, be extremely careful when approaching dangerous species; do not attempt to catch them. Learn to recognize venomous snakes on sight (study Pls. 44–47, pp. 350–357). If you are bitten, go immediately to the nearest physician or medical facility for treatment.

FILESNAKES: Acrochordidae

At least three species in one genus, also known as elephant-trunk snakes or wartsnakes, range from Indochina to northern Australia. All are aquatic and rarely leave water. Rough scales are used to hold fish, the primary prey. Filesnakes are harvested intensely for leather.

JAVANESE FILESNAKE *Acrochordus javanicus* Non-native **FIG. 170**

36–60 in. (90–152 cm); to ca. 83 in. (210 cm) in FL; larger females to ca. 114 in. (290 cm) in the native range. Brown above, flanks and belly lighter, often yellowish. *Skin wrinkled and baggy, scales around body small, rough, not overlapping* (ventrals not enlarged); broad head flattened, *eyes and nostrils on top of head*. Semiterrestrial young blotched, skin less baggy. **HABITAT:** Ponds and canals; into brackish and coastal

FIG. 170. *The Javanese Filesnake* (Acrochordus javanicus) *is aquatic, with eyes and nostrils on top of the flattened head.*

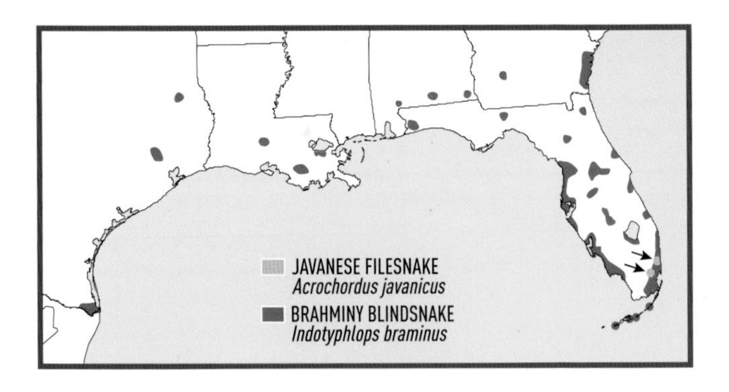

marine habitats in the native range. **RANGE:** S. FL. Natural range: S. Asia. **REMARK:** Possibly a complex of 2 species.

THREADSNAKES and BLINDSNAKES: Leptotyphlopidae and Typhlopidae

Two large families of small, burrowing, wormlike snakes contain more than 370 species in more than 30 genera. Threadsnakes range from the south-central United States to Argentina, and also occur in the West Indies, Africa, and southwestern Asia. Blindsnakes occur in the Neotropics, Mediterranean Region, southern Asia, Australia, Africa, Madagascar, the Philippines, and many South Pacific islands. The families are indistinguishable on the basis of external characters. Species in our area have vestigial eyes covered by scales, no neck constriction, an extremely short tail, and ventral and dorsal scales of essentially the same size.

BRAHMINY BLINDSNAKE *Indotyphlops braminus* Non-native **FIG. 171**

4⅜–6½ in. (11.2–16.5 cm); record 6⅞ in. (17.3 cm). Smallest snake in our area. Very slender, dark gray or brown or black above, belly lighter; snout, lower lip, chin, throat, tailtip, and area near the vent white to buffy yellow; *20 scale rows around midbody*. Hatchling 2½ in. (6.4 cm). **SIMILAR SPECIES:** Threadsnakes tend to be lighter in color, have only 14 scale rows around midbody. **HABITAT:** Areas with loose, moist soil, often heavily disturbed by human activities. Frequently among roots of potted plants, with which these snakes are transported and widely introduced. **RANGE:** Scattered throughout FL. Additional records in VA and MA (not mapped); also CA and HI. Natural range: S. Asia. Widely introduced. **REMARKS:** Unisexual, reproduces via parthenogenesis, so only one individual is needed to establish a new population. Until recently placed in the genus *Ramphotyphlops*.

FIG. 171. *Like other blindsnakes, the Brahminy Blindsnake* (Indotyphlops braminus) *has scales covering the vestigial eyes. Blindsnakes and threadsnakes have a blunt, rounded tail that superficially resembles a head but ends in a tiny spine that is used as an anchor when burrowing. The lower jaw is countersunk into the upper jaw to avoid ingesting grit. The limited gape restricts prey to small items such as termites and ants.*

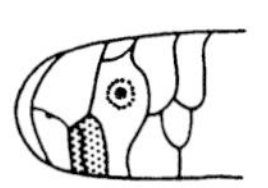

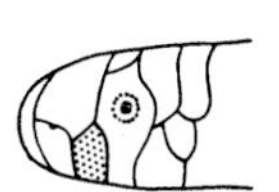

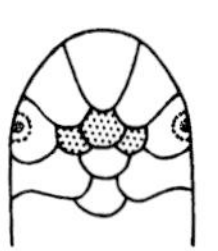

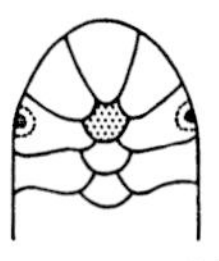

FIG. 172. *Heads of threadsnakes in the genus* Rena.

NEW MEXICO THREADSNAKE *Rena dissecta* **FIG. 172**

5–8 in. (12.5–20.3 cm); record 11½ in. (29.3 cm). Pale shiny brown or reddish brown above; whitish or pinkish below; *14 scale rows around midbody, 3* small scales between oculars, *divided* supralabial between nasal and ocular (Fig. 172, above). Hatchling 2½–2¾ in. (6.4–7 cm). **SIMILAR SPECIES:** (1) Texas Threadsnake has an undivided supralabial between nasal and ocular. (2) Western Threadsnake has one scale between oculars. (3) Brahminy Blindsnake is darker and has 20 scale rows around midbody. **HABITAT:** Plains and semiarid regions,

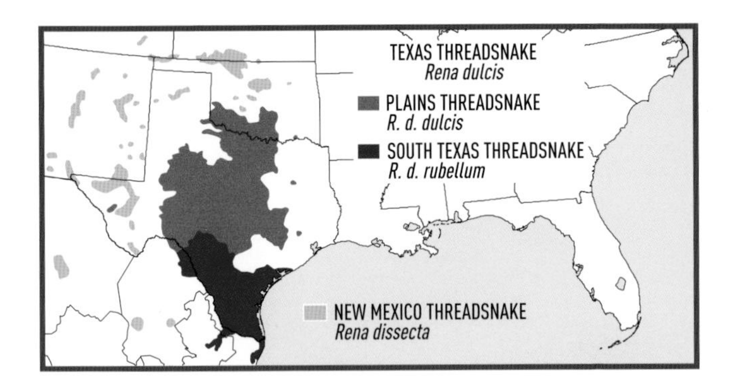

stony hillsides, prairies, sandy or rocky deserts, but almost always where some moisture is available. **RANGE:** S. CO and KS south into Mex. **REMARK:** Until recently considered a subspecies of the Texas Threadsnake.

TEXAS THREADSNAKE *Rena dulcis* **PL. 32, FIG. 172**

5–8 in. (12.5–20.3 cm); record 10¾ in. (27.3 cm). Pale shiny brown or reddish brown above with a pinkish or light to medium brown middorsal scale row; whitish or pinkish below; *14 scale rows around midbody*, *3* small scales between oculars (Fig. 172, p. 361), *undivided* supralabial between nasal and ocular. **SIMILAR SPECIES** and **HABITAT:** See New Mexico Threadsnake. **RANGE:** S. OK south into Mex. **SUBSPECIES:** (1) **PLAINS THREADSNAKE** (*R. d. dulcis*); 210–246 dorsal scales, middorsal scale row pinkish. (2) **SOUTH TEXAS THREADSNAKE** (*R. d. rubellum*); 222–257 dorsal scales, middorsal scale row light to medium brown. **REMARK:** Possible unnamed subspecies with 202–228 dorsal scales near Seminole, OK.

WESTERN THREADSNAKE *Rena humilis* **FIG. 172**

7–10 in. (18–25.4 cm); record 13 5/16 in. (33.9 cm). Brown or pale dull purplish above; venter paler, but usually pink or purplish; *one* scale between oculars (Fig. 172, p. 361); *14 scale rows around midbody*. Hatchling 3½ in. (8.9 cm). **SIMILAR SPECIES:** See New Mexico Threadsnake. **HABITAT:** Sandy and stony deserts, grassland-desert transition areas, foothill canyons, usually under cover, often near water. **RANGE:** W. TX to s.

CA and south into Mex. **SUBSPECIES: TRANS-PECOS THREADSNAKE** (*R. h. segrega*); sometimes considered a distinct species.

BOAS and PYTHONS: Boidae and Pythonidae

Boas occur in the Americas, southern Europe, northern Africa, the Middle East, Madagascar, central and southeastern Asia, and some Pacific islands. Pythons occur in Africa, Asia, and Australia. Boas and pythons include the world's largest snakes. The South American Green Anaconda (*Eunectes murinus*) can exceed 26 ft. (8 m) and weigh 440 lb. (200+ kg). The Asian Reticulated Python (*Malayopython reticulatus*) can reach 33 ft. (10 m) but is relatively slender and reaches a maximum weight of "only" 330 lb. (150 kg). Individuals of both species and several others have been captured in Florida but are not known to have established breeding populations.

BOA CONSTRICTOR *Boa constrictor* Non-native **FIG. 173**

6½–10 ft. (2–3 m); record 14½ ft. (4.45 m) and ca. 77 lb. (35 kg), although reports of snakes weighing 100 lb. (45 kg) are credible. Tan, light brown, cream, or grayish brown with brown or reddish brown crossbands or "saddles," more pronounced posteriorly, alternating with a series of large pale dorsal blotches. Top of head with a dark median line usually extending to snout (sometimes with a perpendicular line between eyes); usually dark marks in front of, below, and

FIG. 173. *Like the introduced pythons, the Boa Constrictor* (Boa constrictor) *in Florida is an unwelcome result of the trade in live animals. Apparently present since the 1970s, this species represents a potential threat to a wide range of native vertebrates.*

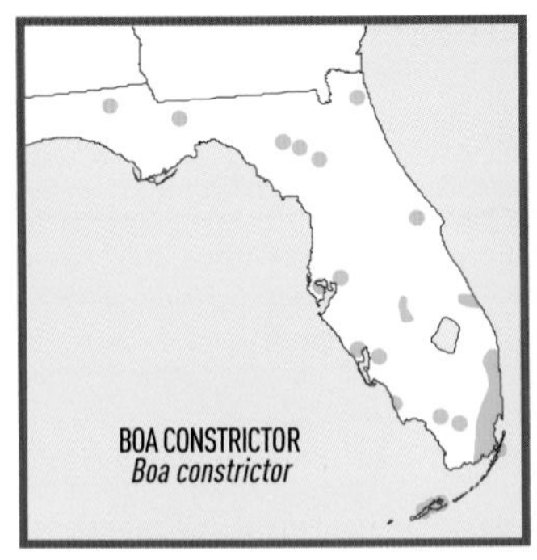

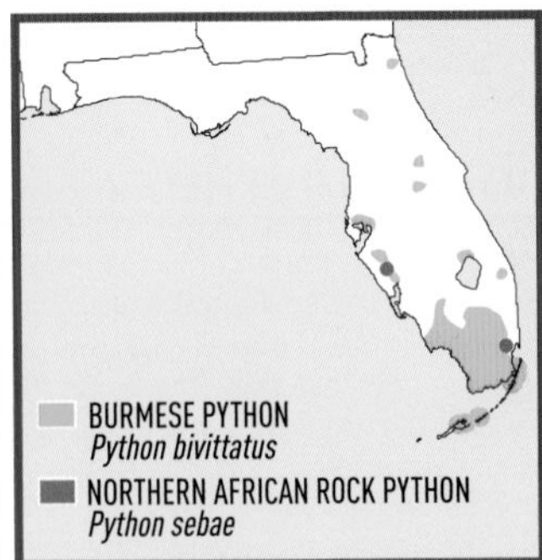

behind eyes, the last flaring posteriorly. Young 2 ft. (60 cm) at birth. **SIMILAR SPECIES:** See Burmese Python. **HABITAT:** Tropical hardwood hammocks, pine rockland, human-disturbed areas. **RANGE:** S. FL and some Keys; scattered records elsewhere in FL probably not established. Also introduced in PR. **NATURAL RANGE:** N. Mex. to n. Argentina.

BURMESE PYTHON *Python bivittatus* Non-native **FIG. 174**

10–13 ft. (3–5 m); largest FL snake 17 ft. 7½ in. (5.37 m) and 164½ lb. (74.6 kg); record 27 ft. (8.22 m) and 401 lb. (182 kg) for a captive snake. Irregular brown dorsal blotches separated by light areas extending to lower sides; top of triangular head with a dark brown "spearhead" outlined in buffy yellow; 2 light lines or patches under eyes border triangular subocular marks; labial scales in front of eyes often light. Center of belly unmarked, at least medially (lateral spots may be present). Hatchling 18–24 in. (45–60 cm). **SIMILAR SPECIES:** (1) Northern African Rock Python is very similar, but light areas separating dorsal blotches are not continuous with light lower sides, labial scales in front of eyes tend to be dark, and light lines extend from nostrils to lips. (2) Boa Constrictor has dorsal saddles that become more pronounced posteriorly and lacks a spearhead-shaped mark on top of head. **HABITAT:** Saline and freshwater glades in Everglades National Park, mangrove fringes, populated areas. **RANGE:** S. FL and some Keys; scattered records elsewhere in FL probably not established. **NATURAL RANGE:** S. Asia. **REMARK:** Often considered a subspecies of the Indian Python (*P. molurus*). **CONSERVATION:** Vulnerable (not applicable to introduced populations).

NORTHERN AFRICAN ROCK PYTHON *Python sebae* Non-native **FIG. 175**

10–13 ft. (3–5 m); record 23+ ft. (7+ m) and 143 lb. (65 kg), with many reports of larger individuals. Brown, olive, chestnut, and buffy yellow dorsal blotches form a broad, irregular middorsal stripe; light areas between blotches rarely continuous with light lower sides. Top of

FIG. 174. *The introduced Burmese Python* (Python bivittatus) *functions as a top predator in Florida, where it is known to consume a variety of vertebrates, including several endangered species.*

FIG. 175. *The Northern African Rock Python* (Python sebae) *recently has become established in Florida. This species and the Burmese Python* (P. bivittatus) *have bred in captivity, leading to unsubstantiated rumors of hybrid "super snakes" in Florida.*

triangular head with a dark brown "spearhead" outlined in buffy yellow; 2 light lines or patches border triangular subocular marks, a third light line often extends from nostrils to lips; labial scales in front of eyes usually dark. Belly with numerous black spots, producing a salt-and-pepper effect. Hatchling 18–24 in. (45–60 cm). **SIMILAR SPECIES:** See Burmese Python. **HABITAT:** Residential areas interspersed with undeveloped but disturbed habitats, agricultural areas, canals, lakes, seasonally flooded wetlands. **RANGE:** S. FL. **NATURAL RANGE:** Cen. and w. Africa.

HARMLESS EGG-LAYING SNAKES: Colubridae

About 120 genera and more than 800 species range across the Americas, Eurasia, and Africa. Snakes in the families Dipsadidae and Natricidae were until recently assigned to the family Colubridae as previously defined.

GLOSSY SNAKES and SCARLETSNAKES: *Arizona* and *Cemophora*

Both of these North American genera contain only one species.

GLOSSY SNAKE *Arizona elegans* **PL. 32, FIG. 157**

27–36 in. (68.8–90 cm); record 55⅞ in. (142 cm). Cream or buff, with fewer than 50 to more than 62 dark-edged brown blotches; lateral spots obscured in some large adults by brown or gray pigment, pale ground color sometimes restricted to small middorsal patches between large blotches. White or pale buff belly *unmarked*; pupils slightly elliptical. Two prefrontals (Fig. 157, p. 326); scales *smooth*; cloacal *undivided*. Hatchling 9½–11 in. (24–28 cm). **SIMILAR SPECIES:** (1) Ratsnakes have weakly keeled scales and a divided cloacal. (2) Prairie Kingsnake (a subspecies of the Yellow-bellied Kingsnake) has a belly marked with squarish blotches or with lowermost lateral spots encroaching on ends of ventrals. **HABITAT:** Sparse desert scrub, sagebrush flats, grasslands, sandhills, coastal scrub, chaparral slopes, sometimes oak-hickory woodland, generally open areas. **RANGE:** Disjunct; NE west to CA and south into Mex. **SUBSPECIES:** (1) **KANSAS GLOSSY SNAKE** (*A. e. elegans*); more than 50 body blotches, 29 or more scale rows, 211 or fewer ventrals in male, 220 or fewer in female. (2) **TEXAS GLOSSY SNAKE** (*A. e. arenicola*); 50 or fewer body blotches, 29 or more scale rows, 212 or more ventrals in male, 221 or more in female. (3) **PAINTED DESERT GLOSSY SNAKE** (*A. e. philipi*); 62 or more body blotches, 27 scale rows.

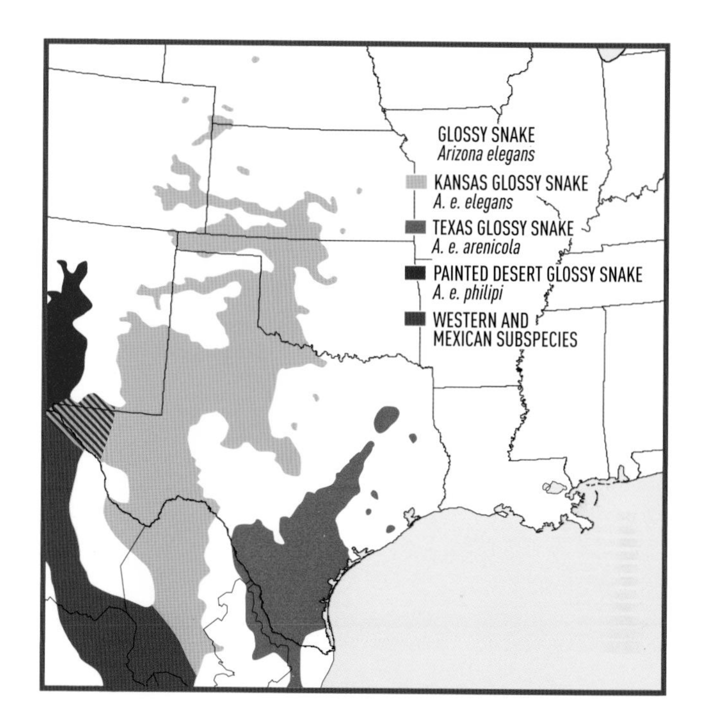

SCARLETSNAKE *Cemophora coccinea* **PLS. 32, 44, FIG. 159**

14–20 in. (36–51 cm); record 32½ in. (82.8 cm). Black-bordered red dorsal saddles separated by white, cream, or yellowish bands; belly *plain whitish (or yellowish)* (Fig. 159, p. 330); pointed snout *red*; pattern most distinct in young. Scales *smooth*; cloacal *undivided*. Hatchling 5–7 5/16 in. (12.5–18.6 cm). **SIMILAR SPECIES:** (1) Coralsnakes have a black-tipped snout and red and yellow rings are in contact. (2) Scarlet Kingsnake and Eastern and Western Milksnakes have a belly heavily invaded by black. **HABITAT:** Hardwood, mixed, or pine forests, adjacent open areas with soil suitable for burrowing. **RANGE:** Disjunct; NJ south through FL and west to e. OK and s. TX. **SUBSPECIES:** (1) **Florida Scarletsnake** (*C. c. coccinea*); 7 upper labials; first black body blotch separated from parietals by 1–4 scales. (2) **Northern Scarletsnake** (*C. c. copei*); usually 6 upper labials; first black body blotch touching parietals (at least along the Atlantic Coast). (3) **Texas Scarletsnake** (*C. c. lineri*); ventrals 184 or more (usually fewer in eastern subspecies); red blotches without black along lower edges.

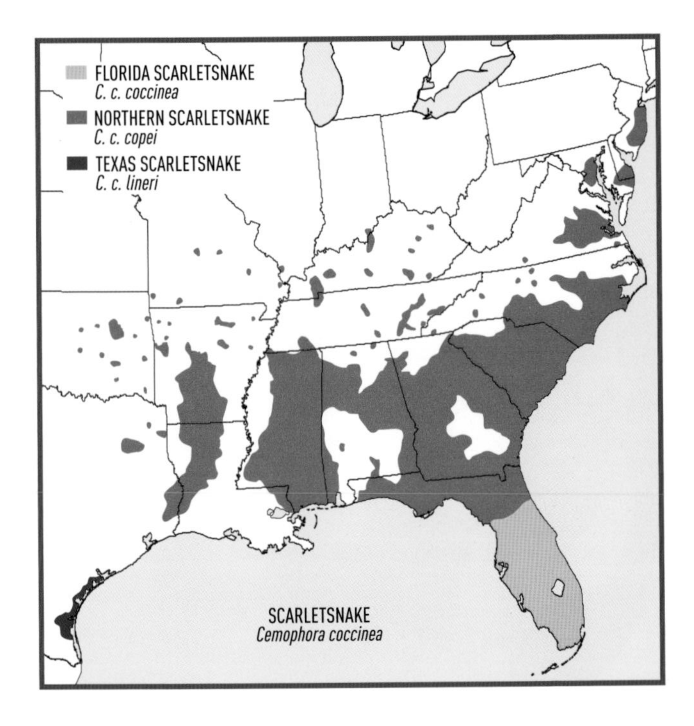

NORTH AMERICAN RACERS, COACHWHIPS, and WHIPSNAKES: *Coluber*

Twenty currently recognized species range from southern Canada through northern South America, southern Africa, and from the Horn of Africa to India. American species (except the North American Racer) until recently were placed in the genus *Masticophis*. Scales are *smooth*, the cloacal is *divided*, and *small lower preoculars are wedged between labials*.

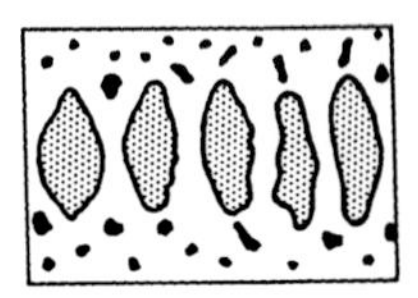

YOUNG RACERS
(*C. constrictor*)
of most subspecies.
Middorsal blotches

YOUNG MEXICAN RACER
(*C. c. oaxaca*)
Spots and crossbands

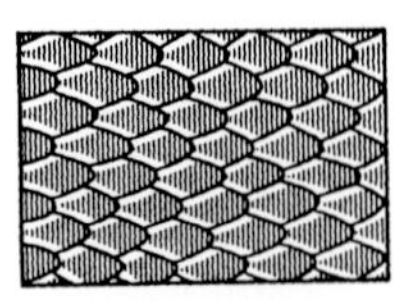

SCHOTT'S WHIPSNAKES
(*C. schotti*)
Scales of middorsal rows
with light edges

FIG. 176. *Dorsal patterns of racers and whipsnakes* (Coluber).

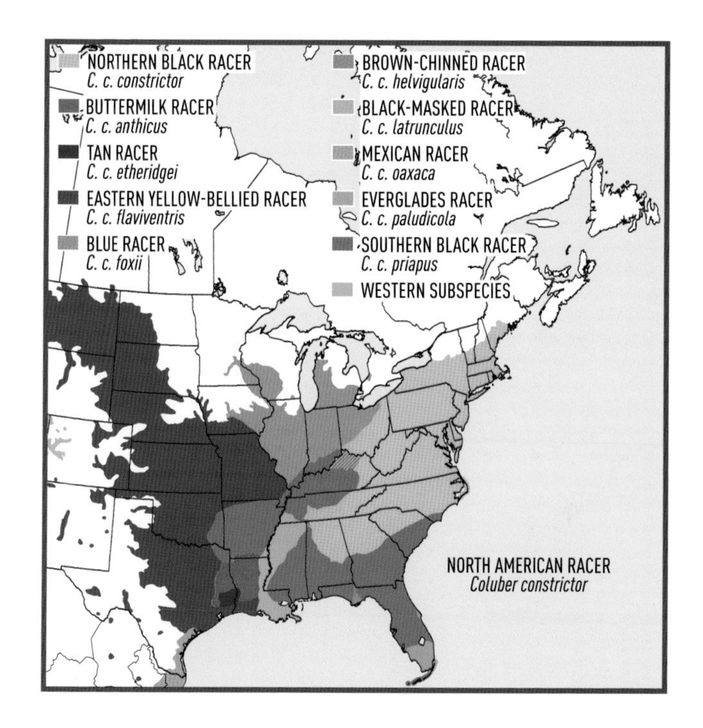

NORTH AMERICAN RACER *Coluber constrictor* **PL. 32, FIG. 176**

36–60 in. (90–152 cm); record 75 in. (190.5 cm). Uniformly black, brown, gray to blue or greenish or light tan above, sometimes speckled with lighter scales. Belly black to bluish gray to yellow or white; often some white on chin and throat. Gray or bluish gray juvenile strongly patterned with a middorsal row of dark gray, brown, or reddish brown blotches, small dark spots on flanks and venter (Fig. 176, facing page), or with scattered small dark spots joining anteriorly to form crossbands and unicolored posteriorly; tail virtually unpatterned; patterns increasingly indistinct as snakes grow, virtually disappearing at about 30 in. (76 cm). *Seventeen scale rows at midbody, 15 posteriorly.* **SIMILAR SPECIES:** (1) Coachwhip has 13 scale rows anterior to tail. (2) Ratsnakes and Speckled Racer have at least weak keels on some middorsal scales. (3) Greensnakes either have keeled scales or a maximum of 15 scale rows. (4) Kingsnakes have an undivided cloacal. **HABITAT:** Prairies, fields and grasslands, brushy areas and old fields, forest edges, open woodlands, environs of lakes and tamarack-sphagnum bogs. **RANGE:** S. ME through FL, west to the Pacific Coast, and south into Mex.; highly disjunct in the West. **SUBSPECIES:** (1) **Northern Black Racer** (*C. c. constrictor*); plain black above and below, usually some white on chin and throat, irises brown or dark amber. (2) **Buttermilk Racer** (*C. c. anthicus*); black, bluish, or olive with scattered white, yellow, buff, or pale blue scales, some slightly spotted, others heavily speckled. (3) **Tan Racer** (*C. c. etheridgei*); very similar to Buttermilk Racer; light tan with a variable number of pale spots. (4) **Eastern Yellow-bellied Racer** (*C. c. flaviventris*); plain brown, gray, olive, medium to pale dull green, or dull to dark blue; belly yellowish. (5) **Blue Racer** (*C. c. foxii*); blue to greenish or grayish blue, occasionally very dark (head darker, often darkest behind eyes), unspotted chin and throat white, belly bluish and paler than back. (6) **Brown-chinned Racer** (*C. c. helvigularis*); uniform black above and below except for plain or mottled light tan or brown chin and lips. (7) **Black-masked Racer** (*C. c. latrunculus*); slate gray, belly grayish blue, black masklike postocular stripes. (8) **Mexican Racer** (*C. c. oaxaca*); green or greenish gray, sides lighter, belly plain yellow to yellow-green; upper labials usually 8 (normally 7 in Eastern Yellow-bellied Racer); greenish juvenile with scattered small dark spots joining to form dark crossbands on neck and anterior body (Fig. 176, p. 368), body uniformly dark olive green toward tail. (9) **Everglades Racer** (*C. c. paludicola*); pale bluish, greenish, or brownish gray, belly whitish and usually with pale cloudy gray or blue markings; irises usually red; juvenile with a reddish cast, light chestnut, reddish, or pinkish dorsal spots, reddish or orange belly spots. (10) **Southern Black Racer** (*C. c. priapus*); similar to the Northern Black Racer but differs in structure of hemipenes; individuals from FL usually with considerable white on chin and throat, irises often bright red or orange; juveniles from FL often with reddish dorsal blotches and a belly that is reddish or pinkish posteriorly. **REMARK:** Probably a complex of species not corresponding to currently recognized subspecies.

COACHWHIP *Coluber flagellum* **PL. 32, FIG. 177**

42–60 in. (106.7–152 cm); record 102 in. (259 cm). Uniformly black or brown to light yellowish brown to very dark anteriorly and much lighter posteriorly or with dark narrow or broad crossbands; belly like back; scales on tail look like a *braided whip. Seventeen* scale *rows anteriorly, 13 near the vent*; cloacal *divided*. Juvenile with dark crosslines 1–2 scales wide and separated by width of 3 or more scales; head often darker than body. **SIMILAR SPECIES:** (1) North American Racer has 15 scale rows in front of vent. (2) Whipsnakes have 15 dorsal rows anteriorly. **HABITAT:** Grasslands, mesquite savannas, arid brushlands, open woodlands, swamps and valleys, especially in eastern portions of the range. **RANGE:** NC through FL, west to CA, and south into Mex. **SUBSPECIES:** (1) **Eastern Coachwhip** (*C. f. flagellum*); black or dark brown anteriorly to light brown posteriorly; extent of lighter color variable, only head and neck dark, half dark and half light, light or reddish only posteriorly, or light with narrow dark brown crossbands. (2) **Western Coachwhip** (*C. f. testaceus*); unicolored light yellow brown to dark brown or reddish (Fig. 177, facing page) or with narrow or broad (rare) dark crossbands. **REMARK:** Probably a complex of species not corresponding to currently recognized subspecies.

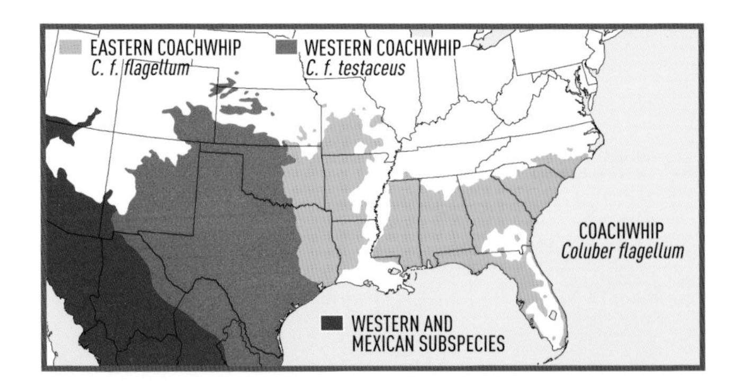

FIG. 177. *A reddish Western Coachwhip* (Coluber flagellum testaceus) *with its head raised in search of prey or potential enemies.*

SCHOTT'S WHIPSNAKE *Coluber schotti* **PL. 33, FIG. 176**

40–56 in. (102–142 cm); record 66 3/16 in. (168.1 cm). Bluish to greenish gray, 2 light lateral stripes on each side, one at edge of belly, the other on scale rows 3 and 4; stripes distinct or reduced to traces, especially on neck. Sides of neck reddish orange or throat dotted with orange. Belly yellow or whitish anteriorly, light bluish gray or olive posteriorly; underside of tail pink or red. Dorsal scales with paired light specks or light edges on 7–8 middorsal scale rows (Fig. 176, p. 368). Juvenile similar but more reddish. *Fifteen scale rows anteriorly, 13 near vent.* **SIMILAR SPECIES:** (1) Striped Whipsnake has light-bordered scales on head. (2) Coachwhip and North American Racer have 17 scale rows at least anteriorly. (3) Patch-nosed snakes have a dark-bordered middorsal stripe. **HABITAT:** Arid brushlands. **RANGE:** S. TX

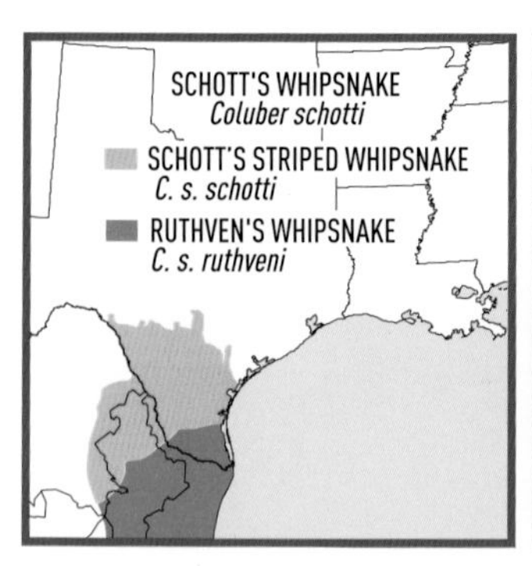

into Mex. **SUBSPECIES:** (1) **SCHOTT'S STRIPED WHIPSNAKE** (*C. s. schotti*); light longitudinal stripes strongly developed, sides of neck reddish orange, belly whitish anteriorly, stippled bluish gray posteriorly, underside of tail pink or salmon. (2) **RUTHVEN'S WHIPSNAKE** (*C. s. ruthveni*); pattern reduced to traces of narrow light stripes on neck or sides, throat dotted with dark orange, belly bright yellow anteriorly, light bluish gray or olive at midbody, pink posteriorly, underside of tail bright red.

STRIPED WHIPSNAKE *Coluber taeniatus* **PL. 33**

42–60 in. (106.7–152 cm); record 72 in. (182.9 cm). Black to very dark brown or reddish brown with distinct white to cream stripes often reduced to white patches strongest on neck, gradually less prominent posteriorly, sometimes producing a crossbanded effect; head scales usually with white outlines; belly white, cream, or yellowish; underside of tail *bright coral pink*. Juvenile usually reddish with a narrow light crossband behind head, longitudinal stripes on lower sides, most prominent on scale rows 3 and 4. *Fifteen scale rows anteriorly, 13 near vent.* **SIMILAR SPECIES:** See Schott's Whipsnake. **HABITAT:** Arid grasslands, thornscrub, semideserts, uplands. **RANGE:** Cen. and w. TX south into Mex. and northwest to WA. **SUBSPECIES:** (1) **DESERT STRIPED WHIPSNAKE** (*C. t. taeniatus*); 2 white lateral stripes strongly developed, upper stripe split by black line on center of fourth scale row. (2) **CENTRAL TEXAS WHIPSNAKE** (*C. t. girardi*); lateral stripes reduced to longitudinal white patches, strongest on neck, less prominent posteriorly, sometimes producing a crossbanded effect.

INDIGO SNAKES and NEOTROPICAL RACERS: *Drymarchon* and *Drymobius*

Five currently recognized species of *Drymarchon* range from the southeastern United States to Argentina, four currently recognized species of *Drymobius* from southern Texas to Peru and Venezuela.

EASTERN INDIGO SNAKE *Drymarchon couperi* **PL. 33, FIG. 158**

60–84 in. (152–213 cm); record 103½ in. (262.9 cm). Shiny bluish black above and below, chin and sides of head often reddish. Scales normally *smooth*, but some large males have faintly keeled scales on as many as 5 middorsal rows; cloacal *undivided*. Third from last upper labial wedge-shaped and cut off above by adjacent labials (Fig. 158, p. 328). *Fifteen* scale rows posteriorly. Juvenile like adult but more reddish on head and anterior belly. Hatchling 17–24 in. (43.2–61 cm). **SIMILAR SPECIES:** Other plain black snakes in range have keeled scales, a divided cloacal, or both. **HABITAT:** Sandhills, flatwoods, most hammocks, coastal scrub, dry glades, palmetto flats, prairie, brushy riparian and canal corridors, wet fields, even residential areas; frequently associated with Gopher Tortoise burrows. **RANGE:** S. GA through FL west to s. AL. Presumably introduced in s. MS. **REMARK:** Until recently considered a subspecies of *D. corais*, which occurs in S. America. **CONSERVATION:** (USFWS: Threatened).

CENTRAL AMERICAN INDIGO SNAKE *Drymarchon melanurus*

PL. 33, FIG. 158

60–78 in. (152–198 cm); record 100¼ in. (254.6 cm). Shiny bluish black above and below, anterior body often brownish, with indications of a pattern. Prominent *dark lines* extend down from eyes. Scales *smooth*; cloacal *undivided*. Third from last upper labial not cut off above by adjacent labials (Fig. 158, p. 328). Fourteen scale rows near vent. Hatchling 18–26 in. (45.7–66 cm). **SIMILAR SPECIES:** See Eastern Indigo Snake. **HABITAT:** Riparian corridors in thornbrush woodland and mesquite savanna, prairies, coastal sandhills, limestone deserts, tropical and subtropical forests in southern portions of the range. **RANGE:** S. TX to Cen. America. **SUBSPECIES: Texas Indigo Snake** (*D. m. erebennus*). **REMARKS:** Until recently considered a subspecies of *D. corais*, which occurs in S. America. Subspecies might not warrant formal recognition.

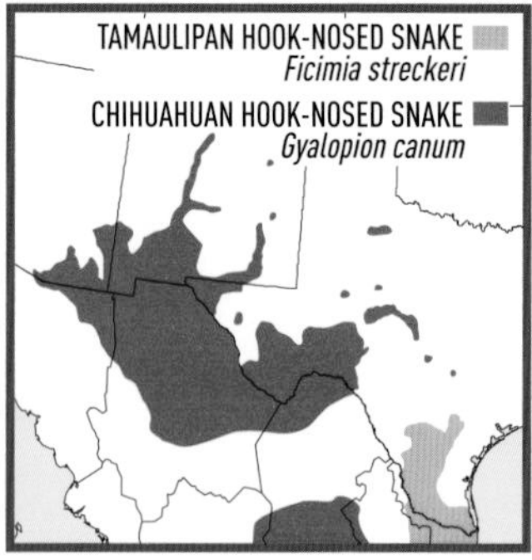

SPECKLED RACER *Drymobius margaritiferus* **PL. 33**

30–40 in. (76–102 cm); record 52¾ in. (133.9 cm). Scales with black outer edges, a central *yellow spot*, and a blue base, producing a greenish cast; black stripes behind eyes. Belly whitish or yellowish, subcaudal scales black-edged posteriorly (ventrals may be similarly marked). Juvenile similar, colors less vivid. Middorsal scale rows weakly *keeled*; outer rows smooth. Cloacal *divided*. Hatchling 6–10⅞ in. (15.2–27.7 cm). **SIMILAR SPECIES:** North American Racer and Kingsnakes have smooth scales. **HABITAT:** Near water, often in thickets of dense natural vegetation. **RANGE:** Extreme s. TX and w. Mex. south to n. S. America. **SUBSPECIES:** **NORTHERN SPECKLED RACER** (*D. m. margaritiferus*).

HOOK-NOSED SNAKES: *Ficimia* and *Gyalopion*

Seven currently recognized species of eastern hook-nosed snakes (*Ficimia*) range from southern Texas to Honduras, and two currently recognized species of western hook-nosed snakes (*Gyalopion*) range from Texas, New Mexico, and Arizona south into Mexico. Hook-nosed snakes in both genera have grooved teeth but no well-developed venom system; they are harmless to humans. The upturned snout serves as a "shovel" used to root in soil or debris while searching for prey.

TAMAULIPAN HOOK-NOSED SNAKE *Ficimia streckeri* **PL. 33**

7–11 in. (18–28 cm); record 19 in. (48.3 cm). Pale to medium brown or gray; top of head with relatively little pattern; 33–60 narrow brown or olive dorsal crossbands or blotches, often reduced to transverse rows of small dark spots. Rostral in broad contact with frontal. Scales *smooth*; cloacal *divided*. **SIMILAR SPECIES:** (1) Chihuahuan Hook-nosed Snake has fewer but more prominent dorsal markings, head with conspicuous crossbands, and rostrals extending posteriorly to prefrontals. (2) Hog-nosed snakes have keeled dorsal scales and a prominent keel behind the upturned snout. **HABITAT:** Thornbrush woodlands, especially near water or edges of agricultural fields, floodplains, lawns and gardens, tropical forests in Mex. **RANGE:** S. TX south into Mex.

CHIHUAHUAN HOOK-NOSED SNAKE *Gyalopion canum* **PL. 33**

7–11 in. (18–28 cm); record 15 1/16 in. (38.4 cm). Pale brown; dark-edged brown crossbands particularly prominent on top of head, 25–48 on body, 8–15 on tail. Rostral in contact with prefrontals. Scales *smooth*; cloacal *divided*. **SIMILAR SPECIES:** See Tamaulipan Hook-nosed Snake. **HABITAT:** Grassy foothills, lower desert communities, dry oak and pinyon-juniper woodlands at higher elevations. **RANGE:** Cen. TX to se. AZ and south into Mex.

KINGSNAKES and MILKSNAKES: *Lampropeltis*

These powerful constrictors range from southeastern Canada and Montana to Ecuador. Scales *smooth*, cloacal *undivided*. Recent studies of what were then called the Common Kingsnake and the Milksnake determined that both were species complexes composed of distinct lineages that do not correspond to traditionally recognized subspecies. Most are best diagnosed by range. Hatchling kingsnakes (except Scarlet and Short-tailed Kingsnakes) 6–12 in. (15.2–30.6 cm); hatchling Scarlet and Short-tailed Kingsnakes 4½–8 in. (11.4–20.3 cm); hatchling milksnakes 5–11 in. (12.5–28 cm).

GRAY-BANDED KINGSNAKE *Lampropeltis alterna* **PL. 34**

20–36 in. (51–90 cm); record 57¾ in. (147.1 cm). *Gray crossbands* alternate with narrow black crossbands or wide black crossbands with varying amounts of red or orange pigment; rarely almost solid black. Narrow white borders of black areas often obvious. Belly with black blotches that often fuse. Head noticeably wider than neck. **SIMILAR SPECIES:** (1) Milksnakes have black-bordered red rings separated by areas of white or yellow. (2) Texas Lyresnake has vertically elliptical pupils and a divided cloacal. **HABITAT:** Desert flats, rocky hillsides, canyons, escarpments, limestone ledges, roadcuts, mountain gaps. **RANGE:** W. TX and se. NM south into Mex.

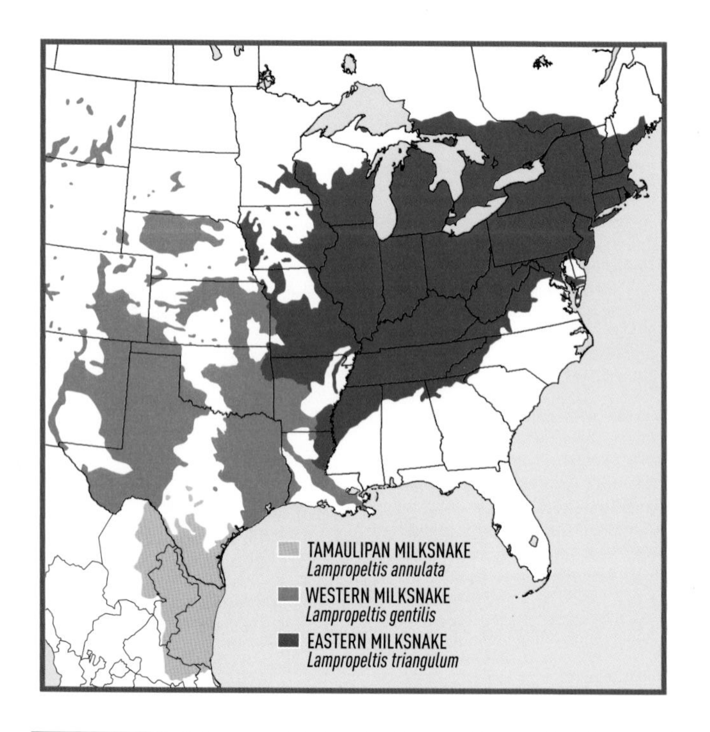

TAMAULIPAN MILKSNAKE *Lampropeltis annulata* **PL. 35, FIG. 159**

16–30 in. (41–76 cm). White, cream, or yellowish with red rings interrupted ventrally by black pigment (Fig. 159, p. 330). Head entirely or mostly black. Belly mostly black with extensions of light rings sometimes evident. Twenty-one or more midbody scale rows. **SIMILAR SPECIES:** See Western Milksnake. **HABITAT:** Coastal dunes, dry scrublands, river valleys, cultivated farmlands. **RANGE:** S. TX into n. Mex.

YELLOW-BELLIED KINGSNAKE *Lampropeltis calligaster* **PL. 34**

30–42 in. (76–106.7 cm); record 56¼ in. (143 cm). Brownish gray to tan or brown, usually with 60–75 dark-edged brown, reddish, or greenish spots or blotches on back and tail alternating with smaller dark markings on sides; sometimes dark with an obscure pattern of 4 dusky longitudinal stripes or even plain brown. Belly white or yellowish with brown checks, spots, or squarish blotches or clouded with brown. Juvenile strongly spotted. **SIMILAR SPECIES:** (1) Milksnakes have reddish blotches or rings surrounded by black and black markings on belly. (2) Ratsnakes have weakly keeled scales and a divided

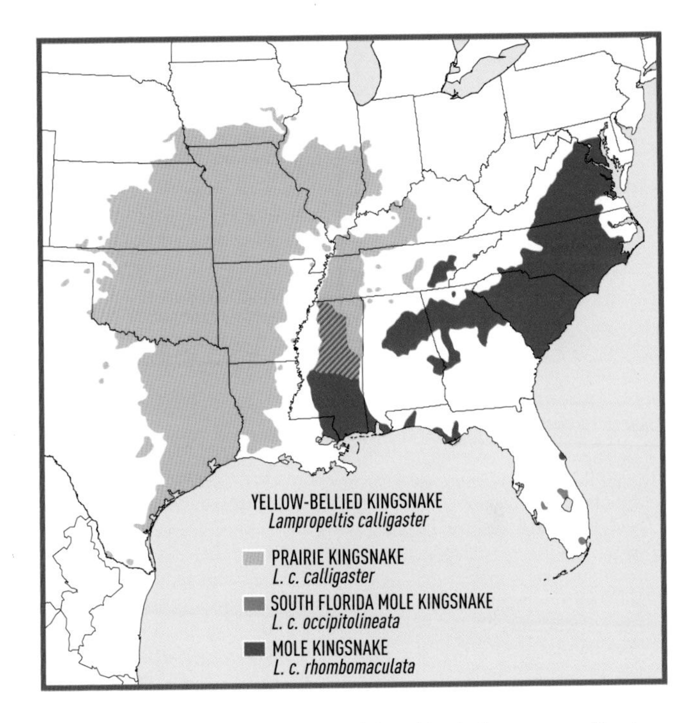

cloacal. (3) Glossy Snake has a plain white belly. **HABITAT:** Weedy fields, farmland, pastures, prairies, rocky hillsides, thickets, open woodlands, sandhills, pine flatwoods, landward side of barrier beaches, coastal salt-grass savannas, marsh borders, residential areas. **RANGE:** Disjunct; s. MD to FL and west to KS, OK, and TX. **SUBSPECIES:** (1) **PRAIRIE KINGSNAKE** (*L. c. calligaster*); brownish gray or tan; about 60 brown, reddish, or greenish, black-edged dorsal markings occasionally split dorsally; 2 alternating rows of smaller dark lateral markings sometimes fuse; older individuals often dark with an obscure pattern frequently including 4 dusky longitudinal stripes; belly yellowish with squarish brown blotches. (2) **SOUTH FLORIDA MOLE KINGSNAKE** (*L. c. occipitolineata*); light or dark brown, 75 or more dark-edged spots on back and tail and a network of dark lines on back of head; belly white or yellowish and checked, spotted, or clouded with brown. (3) **MOLE KINGSNAKE** (*L. c. rhombomaculata*); light to dark brown, sometimes with a greenish tinge, often yellowish on sides; usually with about 70 small, well-separated, dark-edged reddish brown spots on back and tail and smaller, fainter spots on sides; older individuals often plain brown or with 4 dusky longitudinal stripes; belly white or yellowish and checked, spotted, or clouded with brown.

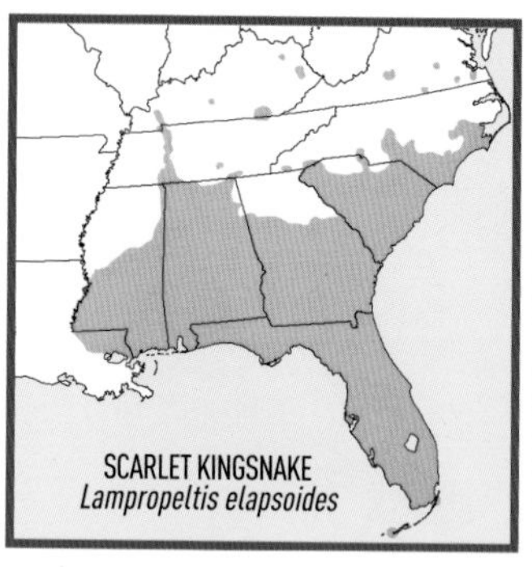

SCARLET KINGSNAKE *Lampropeltis elapsoides* **PL. 35, FIG. 159**

14–20 in. (36–51 cm); record 27 in. (68.6 cm); 29 1/8 in. (74 cm) in captivity. Red, black, and yellow rings around body usually *continue across belly* (Fig. 159, p. 330). Snout *red*, black across posterior edge of parietals; yellow and red rings separated by black. Snakes in s. FL often with middorsally enlarged black rings that cross red rings. Irises red. Nineteen scale rows at midbody. Juvenile often with white instead of yellow bands. **SIMILAR SPECIES:** (1) Milksnakes have 21 or more scale rows at midbody. (2) Scarletsnake has a plain white belly. (3) Harlequin Coralsnake has a black-tipped snout, and red and yellow bands are in contact. **HABITAT:** Woodlands, especially pines. **RANGE:** VA through FL and some Keys west to the Mississippi R. **REMARK:** Until recently considered a subspecies of what was considered a single species called the Milksnake.

SHORT-TAILED KINGSNAKE *Lampropeltis extenuata* **PL. 35**

14–20 in. (36–51 cm); record 25¾ in. (65.5 cm). Fifty to 80 small dark brown or black blotches separated along the dorsal midline by yellow, orange, or red areas. Belly blotched with brown or black. *Very short tail*, only 7–10 percent of total length. **HABITAT:** Chiefly dry pinewoods. **RANGE:** Cen. FL. **REMARK:** Until recently assigned to the genus *Stilosoma*.

WESTERN MILKSNAKE *Lampropeltis gentilis* **PL. 35, FIG. 159**

16–30 in. (41–76 cm); record 41½ in. (105.7 cm). White, cream, or yellowish with black-bordered red or orange rings with either *red, orange, or black extending onto venter*; black cutting off red rings on belly (Fig. 159, p. 330) (sometimes also middorsally, especially posteriorly). Head generally black, often with white or orange mottling on snout. Belly checked with black on white (Fig. 159, p. 330) or midventral area immaculate or with few scattered black marks. Twenty-one or more scale rows at midbody. **SIMILAR SPECIES:** Long-nosed Snake

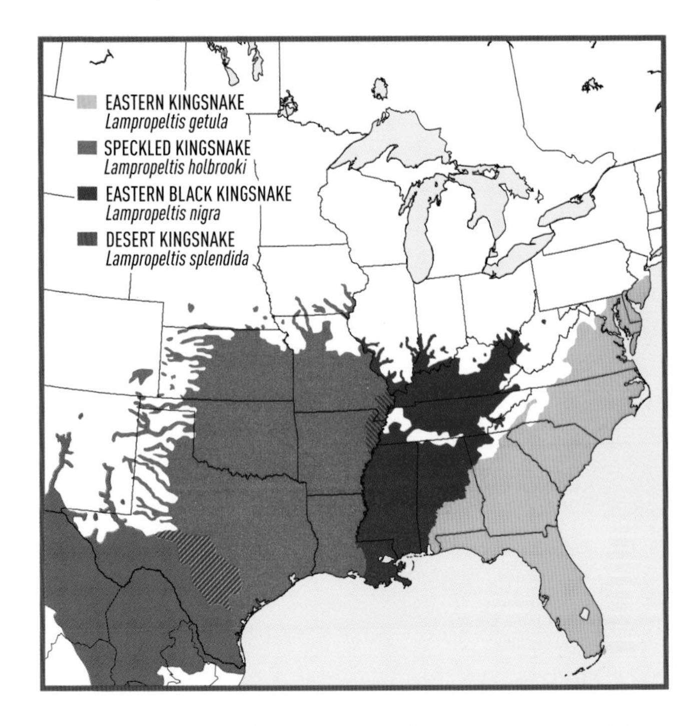

has broad black markings sprinkled with yellow spots, subcaudals mostly in a single row (in 2 rows in milksnakes). See also Eastern Milksnake. **HABITAT:** Open prairies, high plains, sand dunes, wooded stream valleys, rocky canyons, mountain slopes, wooded mountains. **RANGE:** Disjunct; MT and SD south to LA and west to AZ.

EASTERN KINGSNAKE *Lampropeltis getula* **PL. 34, P. 324**

36–48 in. (90–122 cm); record 82 in. (208.3 cm). "Chain" pattern of narrow white, yellow, or cream crossbands on dark ground color. Snakes north of FL black to dark brown with 17–36 light crossbands; snakes in peninsular FL light brown with stippling caused by light-centered scales and 22–54 wider bands; snakes in the Apalachicola lowlands may look like those in peninsular FL or lack crossbands or even show faint longitudinal stripes. Belly mostly black to dark brown, with alternating black to light brown patches forming a loose checkerboard, mostly light with scattered light brown spots (especially in FL), or ventral scales with dark posterior margins. Juvenile like adult. **SIMILAR SPECIES:** See Speckled Kingsnake. **HABITAT:** Coniferous forests, woodlands, mountain valleys, swamps, coastal marshes, river bottoms, farmland. **RANGE:** S. NJ through FL west to the Appalachians and s. AL.

FIG. 178. *The "salt-and-pepper" pattern of a Speckled Kingsnake* (Lampropeltis holbrooki) *is generated by small white or yellowish spots on the black or very dark brown dorsal scales.*

SPECKLED KINGSNAKE *Lampropeltis holbrooki* **PL. 34, FIG. 178**

36–48 in. (90–122 cm); record 72 in. (182.9 cm). Salt-and-pepper pattern results from white or yellowish spots in the center of each (or most) black or dark brown dorsal scales, but occasionally nearly all black, with light spots forming narrow whitish rows across back, or with larger light spots on scales producing a much lighter appearance (especially in cen. and w. TX). Head black with numerous light marks. Juvenile has distinct light dorsal crossbands with little or no spotting between them and spotted sides. **SIMILAR SPECIES:** Some Eastern and Eastern Black Kingsnakes can be distinguished only by using molecular methods or range. (1) Desert Kingsnake usually has few or no light marks on head and the demarcation between dark middorsal area and lighter sides is usually distinct. (2) North American Racer has a divided cloacal. (3) Ratsnakes and Speckled Racer have a divided cloacal and weakly keeled middorsal scale rows, and Speckled Racer has black stripes extending back from eyes. **HABITAT:** Upland wooded areas, stream valleys across open plains and prairies, coastal marshes, great river swamps of the lower Mississippi River Valley. **RANGE:** S. IA to the Gulf and west to se. CO and ne. NM.

EASTERN BLACK KINGSNAKE *Lampropeltis nigra* **PL. 34**

36–45 in. (90–114 cm); record 58¼ in. (148 cm). Nearly plain black (especially in northern portions of the range) to speckled (scale centers with yellow or white specks; especially in southern parts of the range) or with faint traces of a chainlike pattern. Top of head mostly black. Belly checkered black and white. Juvenile with chainlike pattern. **SIMILAR SPECIES:** See Speckled Kingsnake. **HABITAT:** Dry, rocky hills, open woods, stream valleys. **RANGE:** WV and IL south to the Gulf Coast; west of the Mississippi R. in s. LA and extreme e. AR and se. MO.

DESERT KINGSNAKE *Lampropeltis splendida* **PL. 34**

36–45 in. (90–114 cm); record 60 in. (152.4 cm). Profusely spotted laterally with white or yellowish dots, middorsal series of plain black or dark brown spots separated by rows of light dots. Head black or dark brown, first light pattern elements in form of a collar. *Belly chiefly black*. Juvenile with less dark pigment than adult, middorsal dark spots boldly outlined with yellow, rows of dark lateral spots. **SIMILAR SPECIES:** See Speckled Kingsnake. **HABITAT:** Deserts, semideserts, thornscrub, chaparral, desert grasslands, often near water. **RANGE:** Sw. TX to se. AZ and south into Mex.

EASTERN MILKSNAKE *Lampropeltis triangulum* **PL. 35, FIG. 159, P. 8**

24–36 in. (61–90 cm); record 52 in. (132.1 cm). Gray to tan or cream with large black-bordered brown, gray, or reddish brown middorsal blotches alternating with similar but smaller lateral blotches or with black-bordered red middorsal blotches extending well onto sides and lateral blotches small or virtually absent (at least on neck). Blotches do *not extend onto venter*. Head with a dark V- or Y-shaped mark that connects to the first body blotch or with the anterior portion of head entirely or partially black and posterior portion red and followed by a whitish or yellowish collar. Snakes in eastern portions of the range tend to be brownish, those in western portions reddish. Belly boldly checked (often irregularly) with black on white (Fig. 159, p. 330). Twenty-one or more scale rows at midbody. Juvenile with bright red blotches. **SIMILAR SPECIES:** (1) Western Milksnake can be distinguished only by using molecular methods or geographic distribution; identities of snakes from sw. MO, se. KS, ne. OK, nw. AR uncertain. (2) Scarlet Kingsnake has a red snout and 19 scale rows at midbody. (3) Yellow-bellied Kingsnake has brown blotches and brown or yellowish brown belly markings. (4) Copperheads have a virtually unmarked coppery-colored head, a single row of dorsal crossbands, and belly not like a checkerboard. (5) Coralsnakes have red and yellow bands in contact. (6) Watersnakes have keeled scales and a divided cloacal. (7) North American Racer has a divided cloacal. (8) Scarletsnake and Western Groundsnake have an unpatterned belly. (9) Red Cornsnake and Great Plains Ratsnake have a spearpoint-shaped mark between eyes, striped underside of tail, at least weakly keeled dorsal scales, and a divided cloacal. **HABITAT:** Fields, woodlands, rocky hillsides, river bottoms, open farmlands. **RANGE:** ME, s. Can., and MN south to n. AL and ne. LA.

FIG. 179. *The very slender, largely arboreal Rough Greensnake* (Opheodrys aestivus) *blends effectively with verdant vegetation. Its diet consists largely of insects and other small arthropods.*

GREENSNAKES: *Opheodrys*

ROUGH GREENSNAKE *Opheodrys aestivus* PL. 35, FIG. 179

22–32 in. (56–81 cm); record 45⅝ in. (116 cm). Bright green above; white, yellow, or pale greenish below. Scales *keeled*; cloacal *divided*. Juvenile grayish green. Hatchling 6¾–8¹⁵⁄₁₆ in. (17.2–22.8 cm). **SIMILAR SPECIES:** (1) Smooth Greensnake has smooth scales. (2) North American Racer sometimes is green but has smooth scales. **HABITAT:** Dense vegetation, often near or even overhanging water. **RANGE:** S. NJ through FL and the Keys, west to KS and south into Mex.

SMOOTH GREENSNAKE *Opheodrys vernalis* PL. 35

11⅞–20 in. (30.3–51 cm); record 31⅜ in. (79.7 cm). Bright green above; plain white or washed with pale yellow below, occasionally light olive brown (especially in se. TX) or tan (upper Midwest). Scales *smooth*; cloacal *divided*. Juvenile dark olive or bluish gray. Hatchling 3⁵⁄₁₆–6½ in. (8.4–16.5 cm). **SIMILAR SPECIES:** See Rough Greensnake. **HABITAT:** Grasslands, open areas. **RANGE:** Atlantic Provinces to VA and west to se. SK, UT, and NM; isolated populations in TX and Mex. (not mapped). Increasingly disjunct in southern and western portions of range. **REMARK:** Until recently placed in the genus *Liochlorophis*.

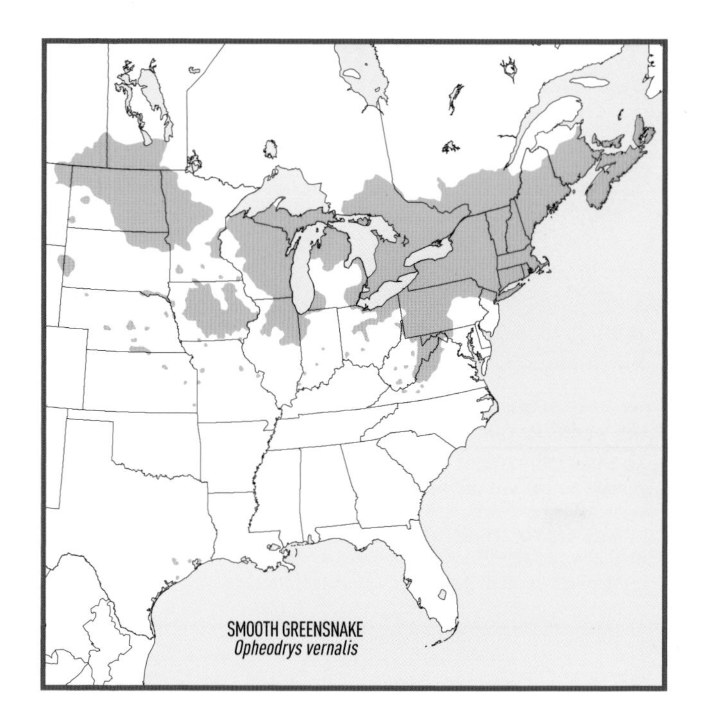

NORTH AMERICAN and DESERT RATSNAKES: *Pantherophis* and *Bogertophis*

Nine species in the genus *Pantherophis* and two species of *Bogertophis* range through much of eastern North America and south into Mexico. These snakes until recently were included in the genus *Elaphe*, a name now restricted to snakes in the Eastern Hemisphere. Ratsnakes (formerly *Elaphe obsoleta* or *Pantherophis obsoletus*) are a complex of species that do not correspond to previously recognized subspecies and are best distinguished on the basis of range.

These snakes are shaped in cross section like a loaf of bread, with the flat belly meeting the sides at an angle (Fig. 160, p. 332). Adults have several weakly *keeled* middorsal scales rows (keels slightly developed or lacking in young). The cloacal is *divided*. Most hatchlings are patterned with dark spots or blotches; some species and populations retain markings throughout life, whereas others become indistinct or lose the pattern altogether. Some bear four dark longitudinal stripes; similar stripes sometimes appear in individuals of species not normally striped. Hatchling foxsnakes and cornsnakes 7¹⁵⁄₁₆–15½ in. (20.2–39.4 cm); hatchling ratsnakes 10½–17 in. (26.6–43.2 cm).

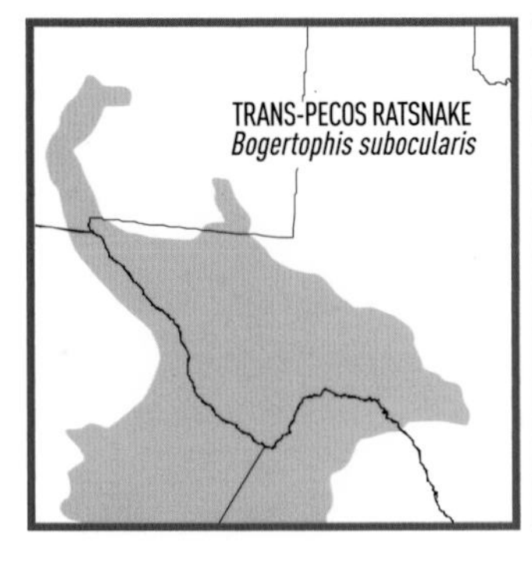

TRANS-PECOS RATSNAKE *Bogertophis subocularis* **PL. 35**

36–54 in. (90–137 cm); record 66 in. (167.6 cm). Tan, yellow, or olive yellow; no pattern on the olive or tan-colored head; typically 21–28 large black or brown H-shaped dorsal markings, most prominent anteriorly, and the sides of which form 2 dorsolateral stripes terminating on the neck. "Blond" variant in the lower Pecos R. drainage is lighter, lacks much or all of the pattern; snakes from extreme w. TX may have a steel gray ground color. Venter whitish, virtually unmarked except for indications of dusky stripes under tail. *Row of subocular scales between eyes and upper labials.* Juvenile paler than adult. **SIMILAR SPECIES:** Baird's Ratsnake has 4 vague dark stripes, juvenile has rounded or rectangular (not H-shaped) blotches. **HABITAT:** Dry, rocky terrain. **RANGE:** S. NM and w. TX south into Mex. **SUBSPECIES: NORTHERN TRANS-PECOS RATSNAKE** (*B. s. subocularis*).

EASTERN RATSNAKE *Pantherophis alleghaniensis* **PL. 36, FIGS. 161, 180**

42–72 in. (106.7–183 cm); record 101 in. (256.5 cm). Variable, ranging from plain black (especially in northern portions of the range), showing traces of spots when skin is distended, gray with distinct alternating middorsal and lateral brown or black blotches, to dusky gray, brown, or yellow with black or brown stripes (especially in southern portions of the range), orange with essentially no pattern (parts of s. FL), or combinations of blotches and stripes (some snakes from extreme s. FL and the Keys). Belly equally variable, uniformly dark or checkerboarded anteriorly, lightening posteriorly to diffused or clouded with gray or brown on white, yellowish, or even orange or orangish yellow (parts of s. FL). Head relatively long and narrow, top usually without any pattern, chin and throat plain white or cream, tongue black or red. Juvenile pale gray to pinkish buff or pinkish orange with gray or brown blotches and dark stripes extending back from eyes normally terminating at mouthline (Fig. 161, p. 334); blotches usually distinct, but not contrasting sharply with ground color in parts of s. FL. Some young snakes have an indication of dark

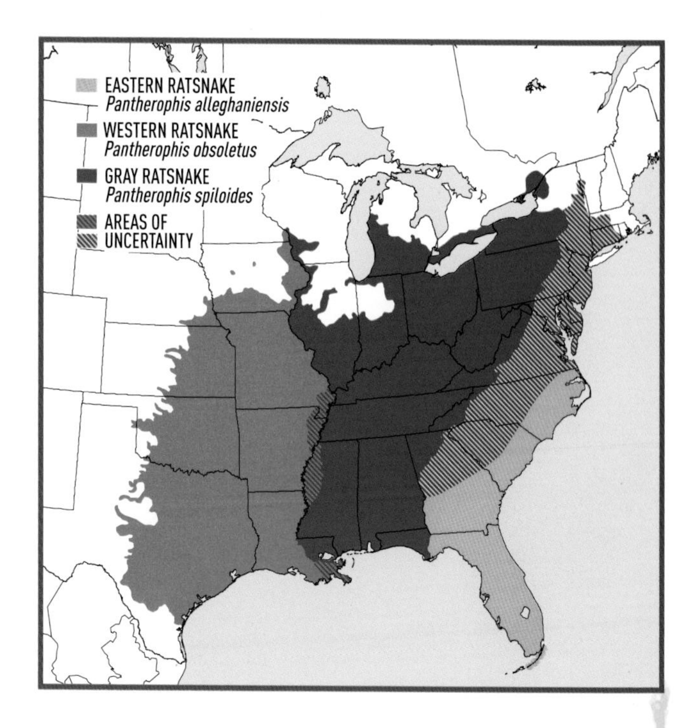

stripes beneath tail (usually much less prominent than in the Red Cornsnake) that typically fade as snakes grow (Fig. 180, p. 386). **SIMILAR SPECIES:** See Gray Ratsnake. **HABITAT:** Rocky, timbered hillsides, swamps, live-oak hammocks, cutover woods, fallow fields, farmlands, around barns and abandoned buildings, sawgrass, open prairies, trees or shrubs, along waterways in the Everglades. **RANGE:** NC through FL west to the Apalachicola R.; area of uncertainty with the Gray Ratsnake extends northward from GA into New England. **REMARKS:** Sometimes assigned to the genus *Scotophis*. Also called pilotsnake, chickensnake, or mountain blacksnake.

BAIRD'S RATSNAKE *Pantherophis bairdi* **PL. 36, FIGS. 161, 180**

33–54 in. (84–137 cm); record 62 in. (157.5 cm). Gray, grayish tan, or light brown, often with a yellowish or reddish orange cast, darker posteriorly; 4 vague brown *longitudinal stripes* (2 paramedian stripes darkest, lateral stripes often faint or broken); traces of dorsal and lateral spots, remnants of the juvenile pattern, often faintly discernible; occasional large individuals essentially patternless. Juvenile with 48 or more brown dorsal crossbands, additional crossbands on tail (Fig. 180, p. 386), alternating rows of smaller dark spots along each side, dark band across head anterior to eyes, dark postocular stripes stop

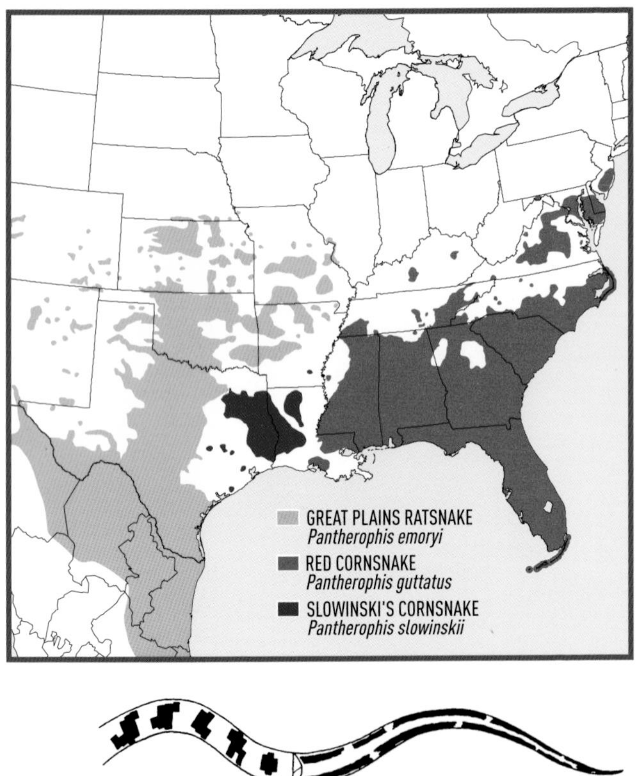

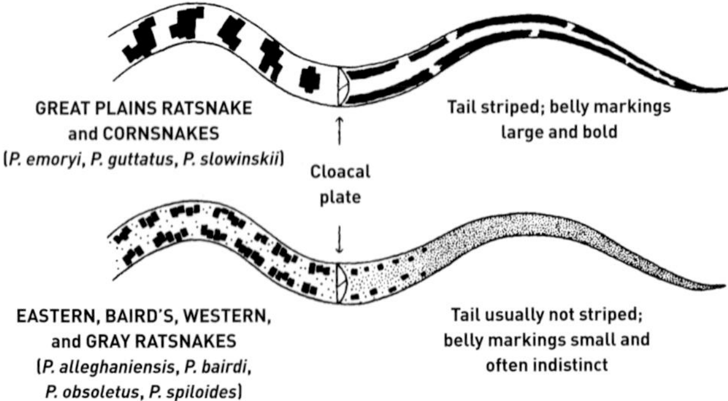

FIG. 180. *The venter and underside of the tail of young ratsnakes* (Pantherophis).

at mouthline (Fig. 161, p. 334). **SIMILAR SPECIES:** See Trans-Pecos and Western Ratsnakes. **HABITAT:** Desert scrub, rocky, wooded canyons, forested uplands. **RANGE:** Cen. and w. TX south into Mex.

GREAT PLAINS RATSNAKE *Pantherophis emoryi* **PL. 36, FIGS. 161, 180**

24–36 in. (61–90 cm); record 60¼ in. (153 cm). Usually light gray; 48–65 dark gray, brown, or olive brown blotches on body and tail often with narrow dark brown or black outlines (blotches on snakes in northern parts of the range more numerous and narrower, almost resembling transverse bands); first blotch on neck divided into 2 branches that extend forward and meet as a *spearpoint-shaped mark* between the eyes (Fig. 161, p. 334) that is sometimes incomplete or faint in large adults. Dorsum occasionally with 4 dusky longitudinal stripes. Extent of ventral pattern variable; belly checkered gray or black on whitish, underside of tail usually striped (Fig. 180, facing page). Juvenile more boldly patterned than adult, with dark-bordered stripes extending back from eyes usually continuing past the mouthline onto the neck (Fig. 161, p. 334). **SIMILAR SPECIES:** (1) Slowinski's Cornsnake tends to be brownish gray or tan, has fewer dark dorsal blotches and often a more distinctly patterned belly. (2) Ratsnakes lack a dark spearpoint between eyes and postocular dark stripes stop at mouthline. (3) Milksnakes and kingsnakes have an undivided cloacal, smooth scales, and lack stripes under tail. (4) Bullsnake, pinesnakes, and the Gophersnake have strongly keeled scales and an undivided cloacal. **HABITAT:** Canyons, rocky draws, hillsides, less frequently in open plains or prairies; often near water in arid regions. **RANGE:** Disjunct; sw. IL to e. UT south through TX into Mex. **REMARK:** Until recently considered a subspecies of the Red Cornsnake.

RED CORNSNAKE *Pantherophis guttatus* **PL. 36, FIGS. 161, 180, 181**

30–48 in. (76–122 cm); record 72 in. (182.9 cm). Orange and brown to gray (snakes in upland habitats tending to brown); 34–47 red or reddish brown blotches on body and tail outlined with black, first blotch on neck divided into 2 branches that extend forward and meet as a *spearpoint-shaped mark* between the eyes (Fig. 161, p. 334). Dorsum occasionally with 4 dusky longitudinal stripes. Belly checkered black on whitish, underside of tail usually striped (Fig. 180, facing page); belly entirely black in some individuals from northern portions of the

FIG. 181. *A Red Cornsnake* (Pantherophis guttatus) *in a defensive posture. Body-bridging (raising some coils) creates a larger image when facing a potential predator.*

range. Snakes from the FL Keys often with reduced black pigment dorsally and a less intense checkerboarded ventral pattern. Juvenile usually with rich reddish brown dorsal blotches, patches of orange between blotches along the middorsal line, dark-bordered stripes extending back from eyes usually continuing past the mouthline onto the neck (Fig. 161, p. 334). **SIMILAR SPECIES:** (1) Slowinski's Cornsnake and Great Plains Ratsnake tend to be gray, brownish gray, or tan, have a greater number of dark dorsal blotches, and lack brighter red and orange dorsal colors. (2) Other ratsnakes lack a dark spearpoint between the eyes, and postocular stripes stop at mouthline. (3) Milk- and kingsnakes have an undivided cloacal, smooth scales, and lack stripes under the tail. **HABITAT:** Pine and open hardwood forests, flatwoods, grasslands, open rocky areas, tropical hammocks. **RANGE:** Disjunct; s. NJ through the FL Keys and west to the Mississippi R.; disjunct populations in s.-cen. LA west of the Mississippi R. Widely introduced. **REMARKS:** Except in extreme s.-cen. LA, populations west of the Mississippi R. are now assigned to the Great Plains Ratsnake or Slowinski's Cornsnake. Tremendously popular pets, many pattern variants have been bred; such snakes should never be released into nature.

WESTERN RATSNAKE *Pantherophis obsoletus* **PL. 36, FIGS. 161, 180**

42–72 in. (106.7–183 cm); record 86 in. (218.4 cm). Variable; black with or without traces of a pattern (especially in northern portions of the range) to dark or medium gray to pale brown, yellowish, or gray with black, brown, or gray blotches. Skin between scales, even in very dark animals, may be red, orange, or yellow. Belly with dark medial blotches or dark flecking or peppering anteriorly, frequently lighter toward tail; often with light median lines under tail (Fig. 180, p. 386). Head relatively short and wide, top usually without any pattern, upper labials often with dark bars, blotched individuals with dark stripes extending back from eyes usually terminating at mouthline (Fig. 161, p. 334). Chin and throat plain white or cream. Juvenile like blotched adult. **SIMILAR SPECIES:** Baird's Ratsnake has 4 vague brown longitudinal stripes. See also Great Plains and Gray Ratsnakes. **HABITAT:** Rocky, timbered hillsides, flat farmlands, rocky canyons in western portions of the range, bayous, swamps, prairies, cutover woods, fallow fields, around barns and abandoned buildings. **RANGE:** Se. MN to s. LA west to NE, KS, OK, and TX; areas of uncertainty with Gray Ratsnake along the Mississippi R. **REMARKS:** Sometimes assigned to the genus *Scotophis*. Also called a pilotsnake.

WESTERN FOXSNAKE *Pantherophis ramspotti* **PL. 36, FIG. 161**

24–54 in. (61–137 cm); record 61 in. (155 cm). Yellowish to light brown; average of 37 chocolate brown to black spots and blotches (body only); brown to reddish brown head usually devoid of conspicuous markings (Fig. 161, p. 334). Belly yellowish, strongly checkered with black. Ventrals 216 or fewer. Juvenile paler than adult, blotches rich brown, narrowly edged with black or dark brown; bold head markings include a dark transverse line anterior to eyes and dark lines from eyes to angles of the jaw; lines on head fade with age. **SIMILAR SPECIES:** (1) Eastern Foxsnake has an average of 43 dorsal blotches and occurs east of the Mississippi R. (except in extreme e.-cen. MO). (2) Western Ratsnake has 221 or more ventrals. (3) Great Plains Ratsnake has a spearpoint-shaped mark between the eyes. (4) Milksnakes and kingsnakes have smooth scales and an undivided cloacal. (5) Bullsnake (a subspecies of Gophersnake) has a pointed snout, strongly keeled scales, and an undivided cloacal. **HABITAT:** Farmlands, prairies, stream valleys, open woodlands. **RANGE:** S. MN to n. MO and west to se. SD and e. NE; area of uncertainty with the Eastern Foxsnake along the Mississippi R. **REMARKS:** Sometimes assigned to the genus *Mintonius*. A larger foxsnake (70½ in; 179.1 cm) has been recorded, but whether this was a Western or Eastern Foxsnake is unknown.

SLOWINSKI'S CORNSNAKE *Pantherophis slowinskii* **FIGS. 161, 180**

24–52 in. (61–132 cm); record 59 in. (150 cm). Grayish brown or tan; 42–56 dark red, maroon, or brownish blotches on body and tail outlined with black or very dark brown, first blotch on the neck divided into 2 branches that extend forward and meet as a *spearpoint-shaped mark* between the eyes (Fig. 161, p. 334). Dorsal pattern occasionally obscure in very dark individuals. Belly checkered black on whitish or cream, frequently marked with smaller and more evenly paired rows of ventral blotches and some degree of peppering in intervening areas; underside of tail usually striped (Fig. 180, p. 386); belly entirely black in some individuals from northern portions of range. Juvenile with dark-bordered stripes extending back from eyes usually continuing past mouthline onto neck (Fig. 161, p. 334). **SIMILAR SPECIES:** See Great Plains Ratsnake. **HABITAT:** Pine-oak woodlands in TX; many snakes in n. LA in cultivated fields, pasturelands, barns, abandoned buildings. **RANGE:** LA and e. TX, presumably s.-cen. AR. **REMARK:** Until recently considered conspecific with the Red Cornsnake.

GRAY RATSNAKE *Pantherophis spiloides* **PL. 36, FIGS. 161, 180**

42–72 in. (106.7–183 cm); record 84¼ in. (214 cm). Variable; plain black with or without traces of a pattern (especially in northern portions of range) to dark or medium gray to pale brown (sometimes almost white) with distinct brown or gray blotches. Skin between scales, even in very dark animals, may be red, orange, or yellow. Belly with dark medial blotches or dark flecking or peppering; often with light median line under tail (Fig. 180, p. 386). Head relatively short and wide, top usually without any pattern, upper labials often with dark bars; blotched individuals with dark stripes extending back from eyes usually terminating at mouthline (Fig. 161, p. 334). Chin and throat plain white or cream. Juvenile like blotched adult. **SIMILAR SPECIES:** (1) Eastern Ratsnake is often very similar, impossible to distinguish in all cases, but tends to have more dorsal blotches (about 35 vs. 30 in Gray Ratsnake), somewhat longer, narrower head; they likely interbreed in northern portions of the area where ranges come into contact. (2) Western Ratsnake also is similar and often impossible to distinguish, but occurs largely west of the Mississippi R. (3) Red Cornsnake has a spearpoint-shaped mark between the eyes and postocular stripes continuing onto neck. (4) North American Racer and Coachwhip have smooth scales and a body round in cross section. (5) Watersnakes have strongly keeled scales. (6) Kingsnakes and milksnakes have smooth scales and an undivided cloacal. **HABITAT:** Rocky, timbered hillsides to flat farmlands, bayous and swamps, cutover woods, fallow fields, around barns and abandoned buildings. **RANGE:** Apalachicola R. in FL to the Mississippi R. north to sw. WI and se. ON; areas of uncertainty with the Eastern Ratsnake extend north from GA into New England and with the Western Ratsnake along the Mississippi R. **REMARKS:** Sometimes assigned to the genus *Scotophis*. Also called central ratsnake, pilotsnake, or oaksnake.

EASTERN FOXSNAKE *Pantherophis vulpinus* **FIG. 161**

24–54 in. (61–137 cm); record 67 1/8 in. (170.5 cm). Average of 43 chocolate brown to black spots and blotches (on body only) on yellowish to light brown ground color; head, usually devoid of conspicuous markings (Fig. 161, p. 334), brown to distinctly reddish. Belly yellowish and strongly checkered with black. Ventrals 216 or fewer. Scales *weakly keeled*; cloacal *divided*. Juvenile has a paler ground color than adult, blotches rich brown, narrowly edged with black or dark brown; bold head markings include a dark transverse line anterior to eyes and dark lines from eyes to angles of the jaw. Dark lines on head fade as snakes age. **SIMILAR SPECIES:** Gray Ratsnake has 221 or more ventrals. See also Western Foxsnake. **HABITAT:** Farmlands, prairies, stream valleys, woodlands, dune country in western portions of the range; snakes in eastern disjunct portions of the range occur primarily in extensive marshes near Lakes Erie and Huron. **RANGE:** Disjunct; ON to n.-cen. OH, Upper Peninsula of MI south to cen. IL and extreme e.-cen. MO; area of uncertainty with the Western Foxsnake along the Mississippi R. **REMARK:** See Western Foxsnake.

BULLSNAKES, PINESNAKES, and GOPHERSNAKES: *Pituophis*

Six currently recognized species range from southwestern Canada and southern New Jersey to Guatemala and the southern tip of Baja California. The three species in our area until recently were considered conspecific. All

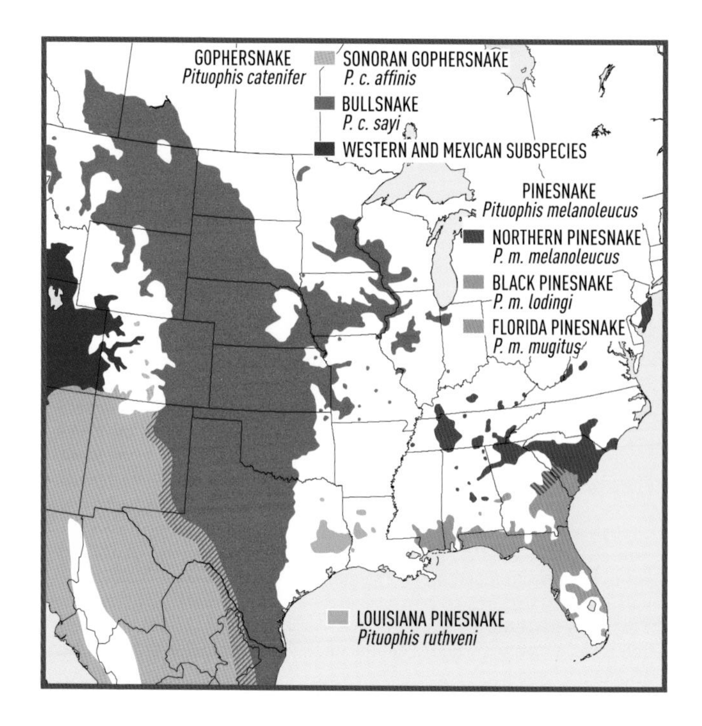

have a somewhat pointed snout, slightly upturned rostral, and *four prefrontal scales* (Fig. 162, p. 336; most other snakes have two). Scales are *strongly keeled*, and the cloacal is *undivided*. All of these species hiss loudly when disturbed. Hatchlings 12½–21½ in. (32 to 54.9 cm).

GOPHERSNAKE *Pituophis catenifer* **PL. 37**

37–72 in. (94.8–183 cm); record 105 in. (266.7 cm). Yellowish; 41 or more black, brown, or reddish brown dorsal blotches. Belly yellow with bold black spots, especially along edges; usually with *dark bands* and parallel yellow bands extending from eyes to angles of the jaw. **SIMILAR SPECIES:** (1) Ratsnakes have weakly keeled dorsal scales, a divided cloacal, and only 2 prefrontals. (2) Indigo snakes have only a few scale rows with weakly keeled scales and 2 prefrontals. (3) Coachwhip and North American Racer have smooth scales, a divided cloacal, and 2 prefrontals. (4) Glossy Snake has smooth scales and 2 prefrontals. (5) Kingsnakes have smooth scales and 2 prefrontals. **HABITAT:** Plains, prairies, desert, semiarid scrub. **RANGE:** N. IN to the Pacific Coast south into Mex. **SUBSPECIES:** (1) **Sonoran Gophersnake** (*P. c. affinis*); blotches brown or reddish brown anteriorly, almost black on and near tail. (2) **Bullsnake** (*P. c. sayi*); black, brown, or reddish brown dorsal blotches darkest and in strongest contrast with ground color near head and near and on tail; some individuals from eastern parts of the range have all dark blotches, those in arid regions are more pallid.

PINESNAKE *Pituophis melanoleucus* **PL. 37**

48–66 in. (122–168 cm); record 90 in. (228.6 cm). Variable; nearly uniformly black or dark brown above and below to very pale with dark blotches that vary from distinct to nearly absent. Juvenile with distinct blotches at least posteriorly, ground color often with a pink or orange tinge at least on ventrals. **SIMILAR SPECIES:** See Gophersnake. **HABITAT:** Dry sandy areas, dry mountain ridges, pine barrens, stands of pine or oaks, abandoned fields. **RANGE:** Disjunct; s. NJ to peninsular FL and west to e. TN and LA; presumably extirpated in extreme s. FL. **SUBSPECIES:** (1) **NORTHERN PINESNAKE** (*P. m. melanoleucus*); dull white, yellowish, or light gray with distinct dark blotches, black anteriorly, often brown near and on tail; juvenile paler, often with a pink or orange tinge. (2) **BLACK PINESNAKE** (*P. m. lodingi*); plain (or nearly plain) black or dark brown above and below, faint indications of blotches may be evident on or near tail, sometimes with few irregular white spots on throat or belly; snout and lips often dark russet brown; juvenile dark anteriorly, patterned posteriorly; venter tan to pink with some black blotches. (3) **FLORIDA PINESNAKE** (*P. m. mugitus*); tan or rusty brown with an indistinct pattern; dark blotches distinct only posteriorly and on tail; juvenile with distinct brown blotches; pattern intensity variable, but never as sharply defined as in the Northern Pinesnake.

LOUISIANA PINESNAKE *Pituophis ruthveni* **PL. 37**

48–56 in. (122–142 cm); record 70¼ in. (178.4 cm). Buff, but yellowish on and near tail. *Markings conspicuously different at opposite ends of body*, blotches toward head dark brown, obscuring ground color, often running together; those near and on tail brown or russet, clear-cut, well separated; 28–42 total body blotches in a middorsal row. Head buff, profusely spotted or splotched with dark brown. Belly boldly marked with black. **SIMILAR SPECIES:** See Gophersnake. **HABITAT:** Sandy, longleaf pinewoods. **RANGE:** W.-cen. LA and e. TX. Introduced in s. FL. **REMARK:** Previously considered a subspecies of the Pinesnake. **CONSERVATION:** Endangered.

LONG-NOSED SNAKES: *Rhinocheilus*

The only species in this genus occurs primarily west of our area.

LONG-NOSED SNAKE *Rhinocheilus lecontei* **PL. 37**

22–32 in. (56–81 cm); record 41 in. (104.1 cm). Strongly *speckled* red, black, and yellow, but snakes in arid areas more distinctly cross-

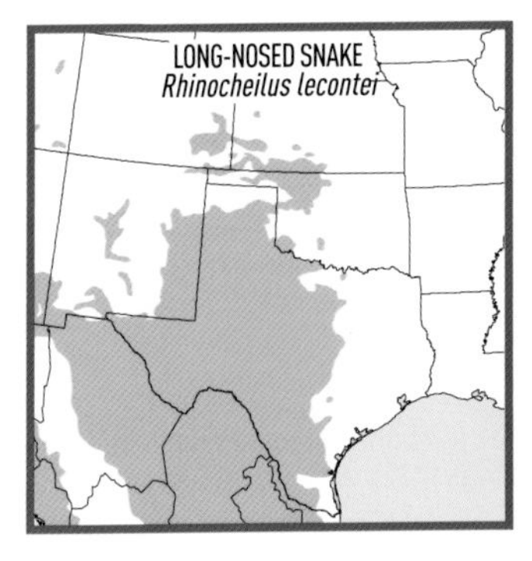

banded, with little or no red, little speckling in black fields. Head mostly black; snout red or pink and pointed, protruding, even upturned; lower jaw countersunk. Belly yellow or whitish, with few or no dark spots. *Scales under tail mostly in a single row.* Scales *smooth*; cloacal *undivided*. Juvenile generally paler (pink instead of red, whitish instead of yellow), speckling on sides partially developed or virtually absent. Hatchling 6½–10 in. (16.5–25.4 cm). **SIMILAR SPECIES:** (1) Scarletsnake has a plain belly, red dorsal saddles with black borders, and two rows of subcaudals. (2) Western Milksnake in range has considerable black on belly, no speckling on sides, and two rows of subcaudals. **HABITAT:** Deserts, dry prairies, arid river valleys, thornbrush, shrubland, sometimes oak-hackberry woodlands or even tropical habitats in southern portions of the range. **RANGE:** Sw. KS to CA and south into Mex.; isolated population in s. ID.

PATCH-NOSED SNAKES: *Salvadora*

Rostrals are large, curved upward, notched below, with free, slightly projecting flaps on each side. The *middorsal stripe is flanked by broad dark stripes*, scales are *smooth* (although some scales above the vent in adult males and large females are sometimes keeled), and the cloacal is *divided*.

BIG BEND PATCH-NOSED SNAKE *Salvadora deserticola* **PL. 37, FIG. 182**

24–32 in. (61–81 cm); record 45 in. (114.3 cm). Middorsal stripe brownish orange to tan; dorsolateral stripes black or dark brown; lateral ground color pale gray, with orange-brown encroaching on anterior corners of each scale; narrow dark lines on *fourth* rows of scales sometimes encroaching on third rows anteriorly and shifting entirely to third rows near tail; *2 or 3 small scales between posterior chin shields* (Fig. 182, p. 394). Belly peach. **SIMILAR SPECIES:** (1) Eastern Patch-nosed Snake has narrow dark lateral lines, if present, on third scale rows, posterior chin shields separated by width of only one scale. (2) Gartersnakes have keeled scales and an undivided cloacal.

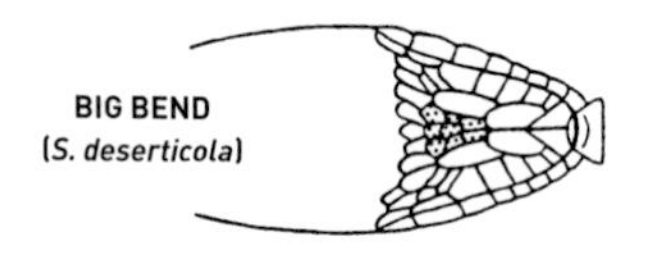

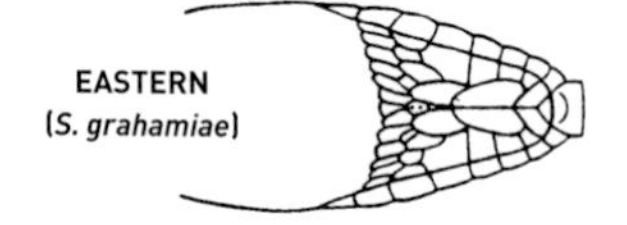

FIG. 182. *Chins of patch-nosed snakes* (Salvadora).

HABITAT: Desert flats and washes, into foothills and mesas surrounding higher mountains. **RANGE:** W. TX to se. AZ and south into Mex.

EASTERN PATCH-NOSED SNAKE *Salvadora grahamiae* **PL. 37, FIG. 182**

22–40 in. (56–102 cm); record 47 in. (119.4 cm). Pale gray, slightly olive, or brownish; 2 broad, dark olive brown to black dorsolateral stripes running length of body. Pale middorsal stripe matches ground color or varies from yellowish to pale orange. Narrow dark lines on *third* scale rows distinct, faint, or absent. Posterior chin shields *touch or are separated only by width of one scale* (Fig. 182, above). **SIMILAR SPECIES:** See Big Bend Patch-nosed Snake. **HABITAT:** Prairies, rocky terrain, isolated mountain areas, foothills, mesas, arid brushlands (especially farther south), cultivated areas. **RANGE:** Disjunct; n.-cen. TX to AZ and south into Mex. **SUBSPECIES:** (1) **MOUNTAIN PATCH-NOSED SNAKE** (*S. g. grahamiae*); pale middorsal stripe matching or only slightly brighter than coloration on sides, sometimes with faint narrow dark lines on third scale rows. (2) **TEXAS PATCH-NOSED SNAKE** (*S. g. lineata*); middorsal stripe yellowish to pale orange, distinct narrow dark lines on third scale rows (the second toward the tail).

NORTH AMERICAN GROUNDSNAKES: *Sonora*

Four currently recognized species range from the United States south into Mexico.

WESTERN GROUNDSNAKE *Sonora semiannulata* **PL. 38, FIG. 183**

8½–12 in. (21.5–30.6 cm); record 19 in. (48.3 cm). Extraordinarily variable; plain, crossbanded, black-collared, or longitudinally striped individuals often co-occur. Plain individuals brown, gray, or orange, with or without a dark head; crossbands black, dark brown, light brown, or orange, with or without a broad orange or reddish middorsal stripe. Head slightly wider than neck; loreal scales (Fig. 183, p. 396) usually present but sometimes fused with adjacent scales. Dorsal scales often slightly darker (longitudinally) along center. Belly white or yellowish; underside of tail sometimes crossbanded, often faintly. Dorsal scales *smooth*; 15 rows anteriorly, sometimes 14 or 13

posteriorly; cloacal *divided*. Hatchling 3⅛–4¾ in. (8–12.1 cm). **SIMILAR SPECIES:** (1) Flat-headed Snake and black-headed and crowned snakes lack crossbands or middorsal stripes, have a small head no wider than neck, and lack loreal scales. (2) Rough Earthsnake and brownsnakes have keeled scales. (3) Smooth Earthsnake has horizontally elongated loreals that touch eyes (no preoculars). **HABITAT:** Prairie and desert lowlands to dry upland forests, also dry river bottoms, sand hummocks, rocky hillsides. **RANGE:** Sw. MO south into Mex. and west to NV and CA.

BLACK-HEADED, CROWNED, and FLAT-HEADED SNAKES: *Tantilla*

Sixty-seven currently recognized species range from the southern United States to northern Argentina. Most have dark "caps"; all lack loreal scales; prefrontals may meet the second labials (Fig. 183, p. 396) or the tips of the postnasals may extend back to meet the preoculars, narrowly separating the second labials from the prefrontals (both conditions occasionally occur on opposite sides of the same individual). Scales are *smooth* and in 15 rows throughout the length of the body, and the cloacal is *divided*. Hatchling 3–3½ in. (7.5–9 cm). These snakes are vulnerable to desiccation so they usually occupy moist microhabitats.

MEXICAN BLACK-HEADED SNAKE *Tantilla atriceps* **PL. 38, FIG. 184**

5–8 in. (12.5–20.3 cm); record 9⅛ (23 cm). Tan to light brown, contrasting with black or dark brown cap. Posterior edge of cap usually *straight or slightly convex*, extending back only 1–2 scale lengths from parietals, not dipping below angles of the mouth, and often followed by a faint light line across neck. Belly pink or red. Seven upper labials; first lower labials usually meet beneath chin (as in Fig. 185, p. 398). **SIMILAR SPECIES:** See Smith's Black-headed Snake. **HABITAT:** Desert flats, grassland-thornbrush, wooded mountain canyons. **RANGE:** Disjunct in s. TX; main portion of the range is in Mex.

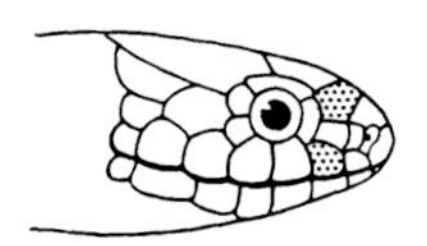

NORTHERN GROUNDSNAKES (*Sonora*)
A loreal scale separates second labial from prefrontal scale

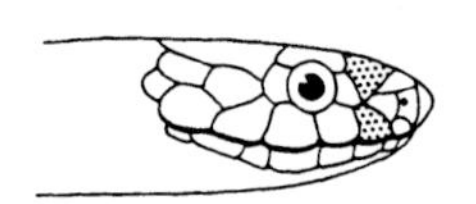

BLACK-HEADED SNAKES (*Tantilla*)
No loreal; second labial touches, or almost touches, prefrontal scale (see text)

FIG. 183. *Heads of the Western Groundsnake* (Sonora semiannulata) *and of black-headed, crowned, and flat-headed snakes* (Tantilla).

SOUTHEASTERN CROWNED SNAKE *Tantilla coronata* PL. 38, FIG. 184

8–10 in. (20–25.4 cm); record 13 in. (33 cm). Light or reddish brown, black cap followed by a light band across rear of head and a black collar 3–5 scales wide; dark pigment usually extends down from the cap to or almost to mouth both under the eyes and near back of head. Belly white or with a pinkish or yellowish tinge. **SIMILAR SPECIES:** See Florida Crowned Snake. **HABITAT:** Dry woodlands with logs or rotting stumps, sometimes mesic meadows, hardwood hammocks, wet margins of marshes, swamps, rivers. **RANGE:** Disjunct; VA to the FL Panhandle and west to KY and LA.

TRANS-PECOS BLACK-HEADED SNAKE *Tantilla cucullata* PL. 38, FIG. 184

8½–15 in. (21.5–38 cm); record 25⅝ in. (65.4 cm). Black hood covers dorsal and ventral surfaces of head or a black cap is followed by a light band or collar, sometimes interrupted middorsally by black, or with an anterior X-shaped mark and 3 black spots along the posterior margin and a black band 3½–4½ scales wide. Dorsum plain light or grayish brown. Often with whitish spots on upper labials posterior to eyes. Belly white. **SIMILAR SPECIES:** See Smith's Black-headed Snake.

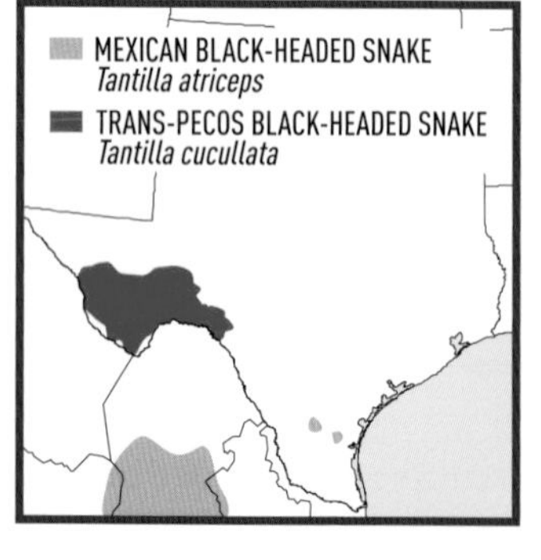

RANGE: W. TX. **REMARK:** Once considered 2 separate species or both considered subspecies of *T. rubra*, a name now restricted to snakes in s. Mex. and Guatemala.

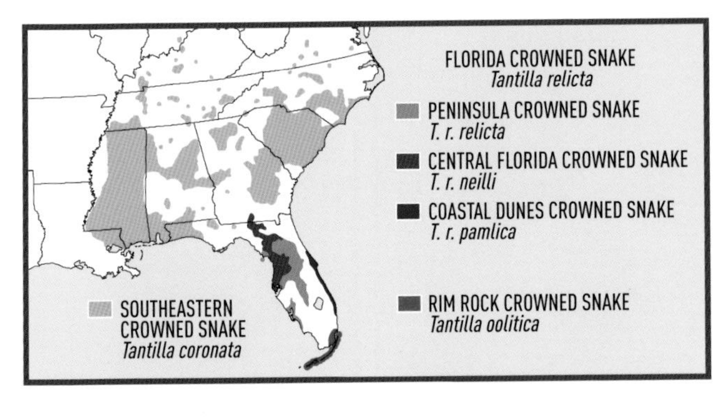

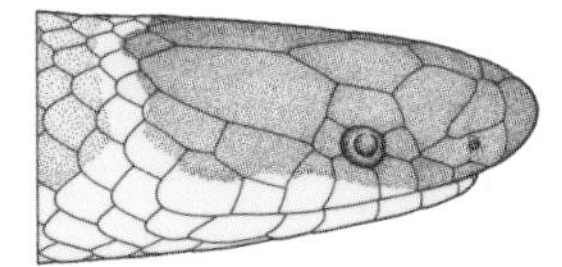

Mexican Black-headed (*T. atriceps*)

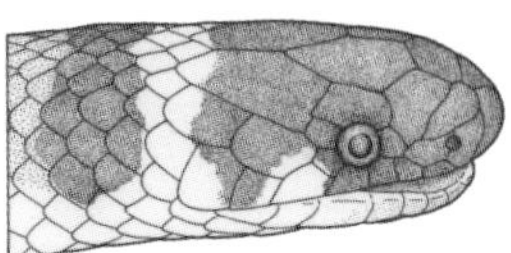

Southeastern Crowned (*T. coronata*)

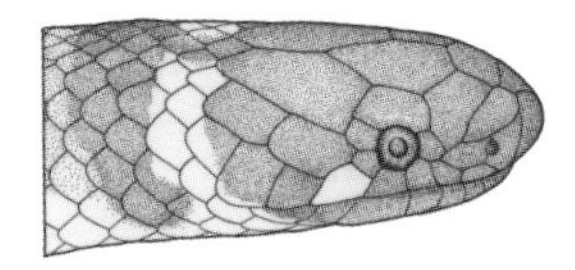

Trans-Pecos Black-headed (*T. cucullata*)

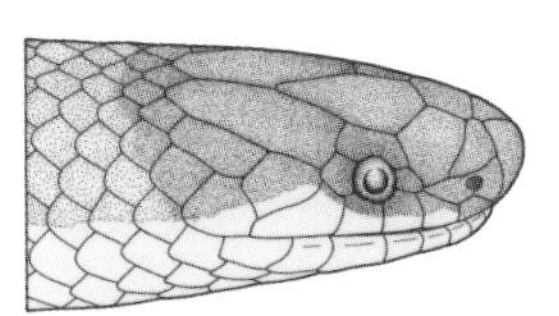

Flat-headed (*T. gracilis*)

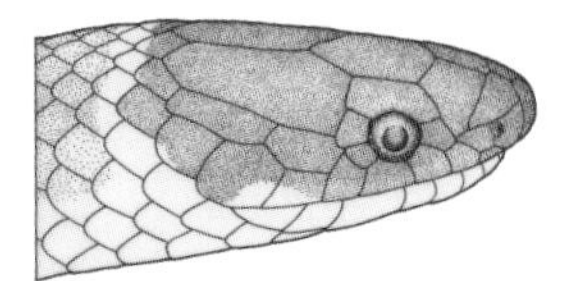

Smith's Black-headed (*T. hobartsmithi*)

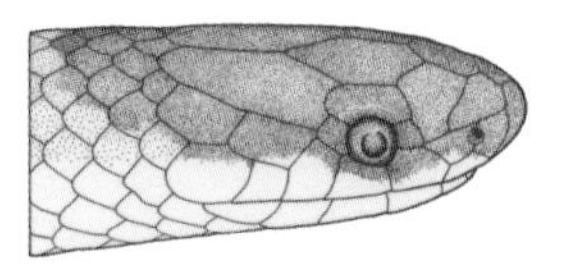

Plains Black-headed (*T. nigriceps*)

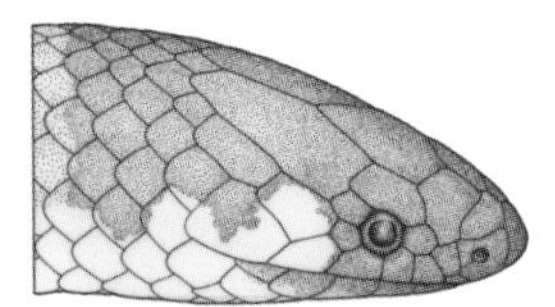

Rim Rock Crowned (*T. oolitica*)

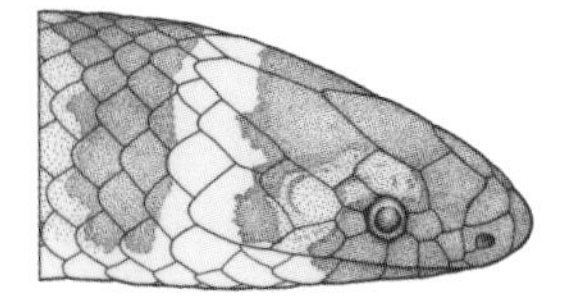

Florida Crowned (*T. relicta*)

FIG. 184. *Heads of black-headed, crowned, and flat-headed snakes* (Tantilla).

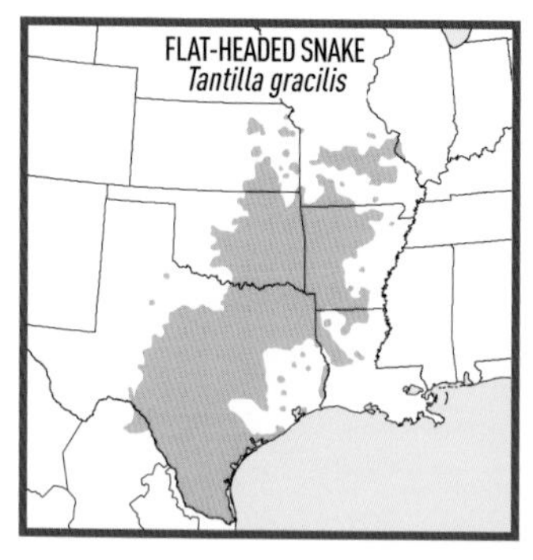

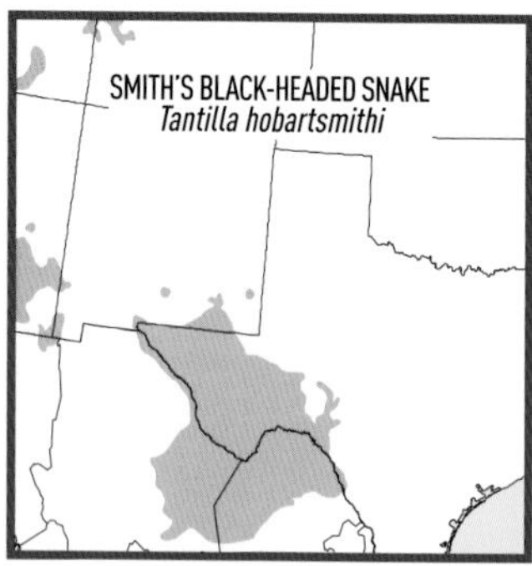

FLAT-HEADED SNAKE *Tantilla gracilis* **PL. 38, FIG. 184**

7–8 in. (18–20.3 cm); record 9⅞ in. (24.9 cm). Head usually only slightly darker than body (occasionally with a dark head); cap *concave* posteriorly. Dorsum plain golden or gray brown to reddish brown. Belly *salmon pink*. Six upper labial scales. Juvenile gray or brownish gray. **SIMILAR SPECIES:** (1) North American Brownsnake and Rough Earthsnake have keeled scales. (2) Smooth Earthsnake has weak keels on at least some middorsal scales. (3) Western Groundsnake has a cream or whitish belly and usually has loreal scales. (4) Threadsnakes have dorsal and ventral scales about the same size. See also Smith's Black-headed Snake. **HABITAT:** Rocky prairies, wooded hillsides, rocky forest edges, pine-oak uplands, oak-juniper brakes, pinewoods, moist deciduous forest, thorn woodlands, grass-brushlands. **RANGE:** S. IL and e. KS to s. TX and adjacent Mex.

SMITH'S BLACK-HEADED SNAKE *Tantilla hobartsmithi* **FIGS. 184, 185**

7–9 in. (18–23 cm); record 12 5/16 in. (31.3 cm). Tan to light brown, contrasting with a black or dark brown cap. Cap usually *straight or slightly convex* posteriorly, extending back only 1–3 scale lengths from parietals, not dipping below angles of the mouth, and followed by a subtle, narrow, light collar. Chin and throat light gray, belly and underside of tail reddish orange. Seven upper labials; first lower labials usually fail

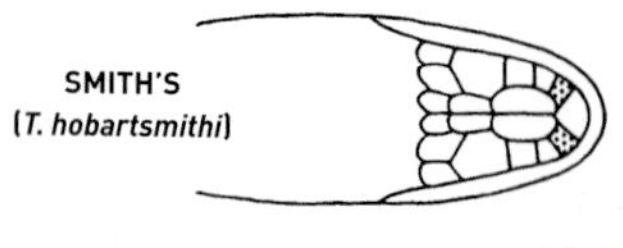

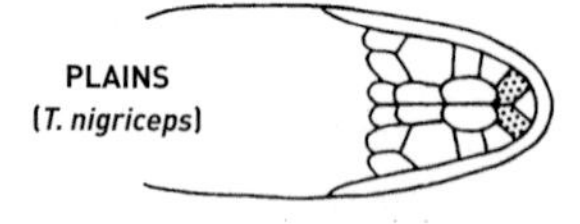

FIG. 185. *Scales on the chins of some black-headed snakes in the genus* Tantilla.

to meet beneath chin (Fig. 185, facing page). **SIMILAR SPECIES:** (1) Mexican Black-headed Snake very similar but hemipenes quite different; best distinguished by range. (2) Flat-headed Snake has 6 upper labials, black cap (when present) concave posteriorly. (3) Plains Black-headed Snake has first lower labials usually meeting beneath chin, head cap convex or pointed posteriorly. (4) Trans-Pecos Black-headed Snake has an entirely black head or a white collar, sometimes interrupted middorsally, followed by a broad black band, and a plain white belly. **HABITAT:** Open woodlands, mesquite-yucca grasslands, savannas, canyons, arroyos. **RANGE:** Disjunct; w. TX and adjacent Mex. west to CA, north to CO and UT.

PLAINS BLACK-HEADED SNAKE *Tantilla nigriceps* PL. 38, FIGS. 184, 185

7–10 in. (18–25.4 cm); record $16^{5}/_{8}$ in. (42.3 cm). Yellowish brown to brownish gray, black head cap *convex* or *pointed* posteriorly, extending back 2–5 scale lengths from parietals. Belly whitish with a broad midventral pink area. Seven upper labials; first lower labials usually meet beneath chin (Fig. 185, facing page). **SIMILAR SPECIES:** See Smith's Black-headed Snake. **HABITAT:** Plains, desert grasslands (including rangelands), shrublands, sandhills, rocky canyons, riparian zones along prairie streams, thornbrush woodlands. **RANGE:** Disjunct; sw. NE and e. WY south into Mex.

RIM ROCK CROWNED SNAKE *Tantilla oolitica* FIG. 184

7–9 in. (18–23 cm); record 11½ in. (29.2 cm). Black on head continuous from snout to neck, extending 3–8 scales behind parietals, except individuals from Key Largo, FL, sometimes have a broken light crossband separating the black cap from the black collar. **SIMILAR SPECIES:** See Florida Crowned Snake. **HABITAT:** Sandy soil in flatwoods, hammocks, vacant lots, pastures of rim rock area paralleling the Atlantic Coast. **RANGE:** S. FL and the Keys. **CONSERVATION:** Endangered.

FLORIDA CROWNED SNAKE *Tantilla relicta* PL. 38, FIG. 184

7–9 in. (18–23 cm); record 9½ in. (24.1 cm). Light or reddish brown; black on head and neck continuous, extending 3–8 scales behind parietals or with a light crossband separating a black cap and a black collar (crossband often interrupted by black at midline). Head mostly black, with whitish areas behind eyes, or with light areas on the snout, temporals, parietal, posterior labials. Head pointed or rounded, lower jaw sometimes partly countersunk into upper jaw. **SIMILAR SPECIES:** (1) Rim Rock Crowned Snake very similar but differs in structure of hemipenes and occurs only in extreme s. FL. (2) Southeastern Crowned Snake has mostly light labials behind eyes. (3) North American Brownsnake has a dark head followed by a light crossband and keeled scales. (4) Ring-necked Snake has a bright yellow, orange, or red belly boldly marked with black spots (at least in southern subspecies). **HABITAT:** Sandhills, sand pine scrub, coastal dunes, xeric and mesic hammocks. **RANGE:** Disjunct; peninsular FL to extreme s. GA. **SUBSPECIES:** (1) **Peninsula Crowned Snake** (*T. r. relicta*); usually with a light crossband separating the black head cap from a black collar (crossband often interrupted by black at midline), most of head black, including labials; head pointed and lower jaw partly countersunk into upper jaw. (2) **Central Florida Crowned Snake** (*T. r. neilli*); black on head and neck continuous, extending 3–8 scales behind parietals; labials all black or with whitish areas posterior to eyes; head narrowly rounded (not pointed), lower jaw not countersunk into upper jaw. (3) **Coastal Dunes Crowned Snake** (*T. r. pamlica*); light crossband and light areas usually present on the snout, temporals, parietals, posterior labials; black collar about 3 scales wide; head pointed and lower jaw countersunk into upper jaw.

LYRESNAKES: *Trimorphodon*

Seven currently recognized species range from the southern United States to Costa Rica. They are technically venomous but are not dangerous to humans (the venom is not effective on birds and mammals).

TEXAS LYRESNAKE *Trimorphodon vilkinsonii* PL. 38

18–30 in. (45.7–76 cm); record 41 in. (104.1 cm). Light brown or gray; 17–24 dark brown body saddles widest middorsally but narrow to width of only 1–2 scales on lower sides; similar but narrower markings on tail. Few dark smudges on head are remnants of a lyre-shaped pattern well developed in some related species. Belly white or gray to yellowish brown with 2 rows of dark blotches along sides. *Pupils of eyes vertically elliptical. Extra scale (lorilabial)* between loreal and upper labials. Scales *smooth*; cloacal *divided*. Juvenile similar to adult but more boldly patterned. Hatchling 8½ in.

(21.5 cm). **SIMILAR SPECIES:** (1) Gray-banded Kingsnake has round pupils and an undivided cloacal. (2) Eastern Black-tailed Rattlesnake has a somewhat similar pattern but a solid black tail (juvenile Eastern Black-tailed Rattlesnake or individuals without rattles are otherwise confusing). **HABITAT:** Dry, rocky terrain of mountains, canyons, hills, rock outcrops, fissured bluffs, arroyos, sometimes desert flats, especially in rock piles and slides on rocky slopes. **RANGE:** W. TX and s.-cen. NM south into Mex. **REMARK:** Until recently considered a subspecies of *T. biscutatus*, a name now restricted to snakes in s. Mex. and Cen. America.

REAR-FANGED SNAKES: Dipsadidae

Ninety genera and nearly 750 species range across the Americas. Enlarged, sometimes grooved teeth toward the back of the jaw are used to introduce venom, but these snakes generally are considered harmless to humans. Snakes in this family until recently were placed into the family Colubridae as then defined.

NORTH AMERICAN WORMSNAKES: *Carphophis*

COMMON WORMSNAKE *Carphophis amoenus* **PL. 38, FIG. 186**

7½–11 in. (19–28 cm); record 13 7/16 in. (34.2 cm). Brown above; *belly and adjacent 1–2 scale rows pink*; head pointed; tail ends in a sharp spine. Two prefrontals and 2 internasals sometimes fused (see Subspecies; Fig. 186, below). Scales *smooth*; cloacal *divided*. Hatchling 3¼–4 in. (8.3–10 cm), darker than adult. **SIMILAR SPECIES:** Other small brown snakes either have keeled scales or the belly color does not extend upward to involve one or more scale rows. **HABITAT:** Moist woodlands. **RANGE:** S. New England to SC and GA west to s. IL and LA; west of Mississippi R. in e. AR. **SUBSPECIES:** (1) **EASTERN WORMSNAKE** (*C. a. amoenus*); 2 prefrontals and 2 internasals. (2) **MIDWESTERN WORMSNAKE** (*C. a. helenae*); prefrontal scales fused with corresponding internasals.

EASTERN
(*C. a. amoenus*)
Prefrontals and internasals separate

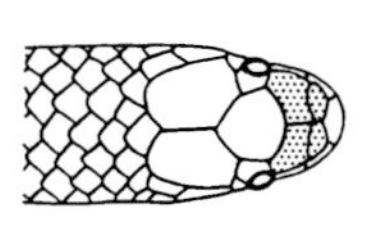

MIDWESTERN
(*C. a. helenae*)
Each prefrontal fused with the corresponding internasal

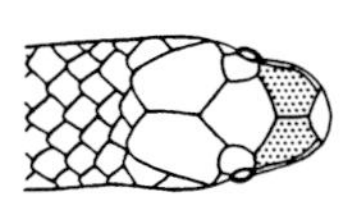

FIG. 186. *Head scales of subspecies of the Common Wormsnake* (Carphophis amoenus).

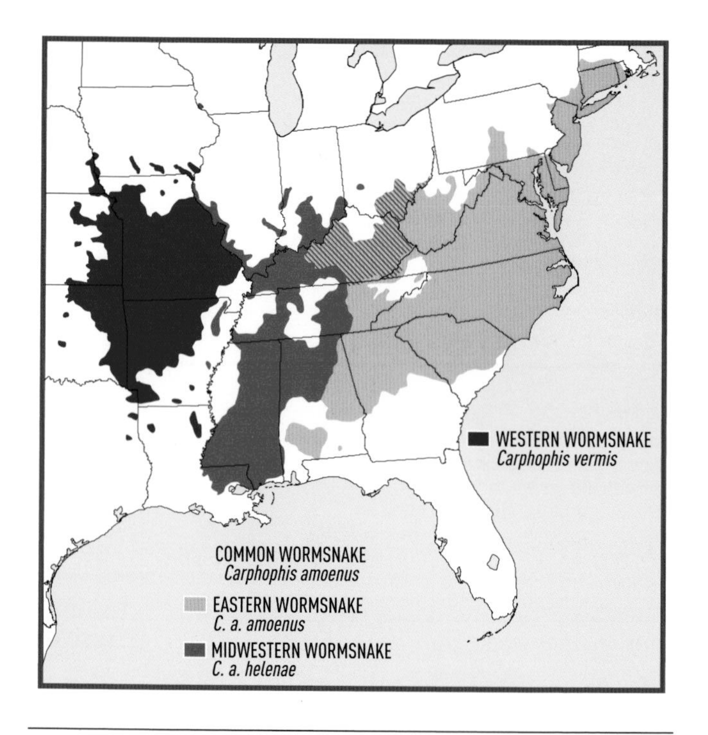

WESTERN WORMSNAKE *Carphophis vermis* **PL. 38, FIG. 187**

7½–11 in. (19–28 cm); record 15⅜ in. (39.1 cm). Purplish black above; pink of belly extends upward onto *third* scale rows. Head pointed. Scales *smooth*; cloacal *divided*. Hatchling 4–4¾ in. (10–12.1 cm). **SIMILAR SPECIES:** See Common Wormsnake. **HABITAT:** Woodlands, wooded riparian zones extending westward into prairies. **RANGE:** Disjunct; extreme sw. WI and se. NE south to n. LA and ne. TX. **REMARK:** Snakes in the disjunct area of ne. LA might be hybrids with the Common Wormsnake.

BLACK-STRIPED SNAKES: *Coniophanes*

Seventeen currently recognized species range from southern Texas to Peru.

REGAL BLACK-STRIPED SNAKE *Coniophanes imperialis* **PL. 38**

12–18 in. (30.6–45.7 cm); record 20 in. (50.8 cm). Broad black or dark brown stripes alternate with tan or brown stripes that brighten abruptly anteriorly. Thin whitish or yellowish lines from snout through top of eyes terminate near rear of head. Belly bright red or orange.

FIG. 187. *The bright pink belly color of the Western Wormsnake* (Carphophis vermis) *extends upward onto the third row of dorsal scales.*

Scales *smooth*; cloacal *divided*. **HABITAT:** Semiarid coastal plain, around buildings and in vacant lots in suburban areas. **RANGE:** Extreme s. TX to Cen. America. **SUBSPECIES: TAMAULIPAN BLACK-STRIPED SNAKE** (*C. i. imperialis*).

RING-NECKED SNAKES: *Diadophis*

The one species in this genus is almost certainly a complex with lineages not necessarily corresponding to currently recognized subspecies.

RING-NECKED SNAKE *Diadophis punctatus* **PL. 39, FIG. 188**

10–15 in. (25.4–38 cm); record 27 11/16 in. (70.6 cm) in our area. Black, bluish gray, or slate to light or dark brown or brownish gray, usually with at least a trace of a *light collar*. Belly yellow, often with dark spots (see Subspecies; Fig. 188, p. 405), turning red under vent and tail in some populations. Dorsal scales *smooth*, in 15 or 17 rows; cloacal *divided*. Young often darker than adult. Hatchling 3½–5½ in. (9–14 cm). **SIMILAR SPECIES:** North American brownsnakes and the Red-bellied Snake have keeled scales. **HABITAT:** Rocky, wooded hillsides; cutover areas with stones, logs, bark slabs; moist areas near swamps and springs; damp hillsides; flat, poorly drained pinewoods; open prairies in western portions of the range, but limited to moister forested uplands or vicinity of water in arid regions. **RANGE:** NS to se. SD south to the FL Keys, west to the Pacific Coast and south into Mex.

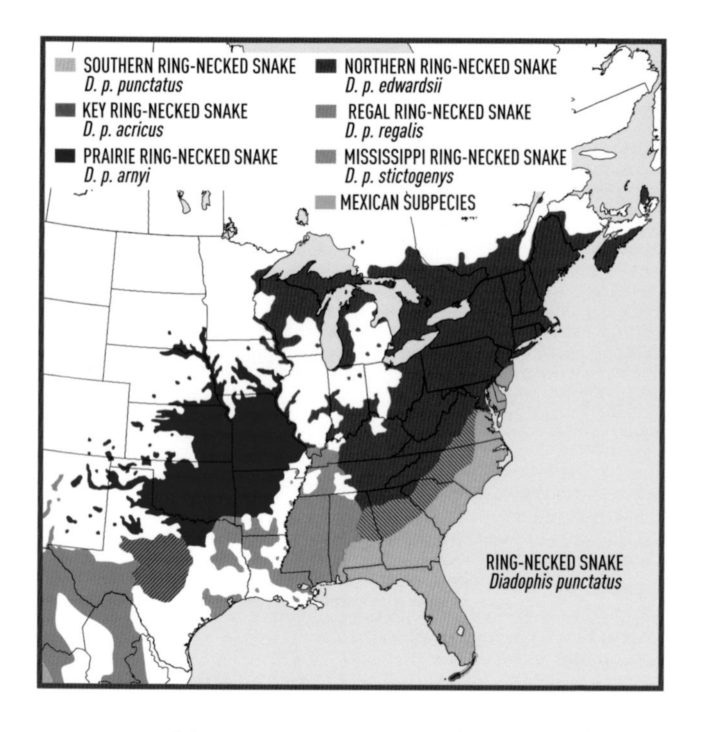

SUBSPECIES: (1) **Southern Ring-necked Snake** (*D. p. punctatus*); light brown to nearly black; belly yellow, changing to red posteriorly and under tail in peninsular FL; neck ring normally interrupted dorsally by dark pigment; spotted belly with large half-moonlike marks in a central row; small black spots on chin and lower lip; 15 scale rows. (2) **Key Ring-necked Snake** (*D. p. acricus*); slate gray anteriorly, black posteriorly, head pale grayish brown, virtually no neck ring, chin and labials faintly spotted, 15 scale rows. (3) **Prairie Ring-necked Snake** (*D. p. arnyi*); dark head color extends around or across angles of the jaw, slightly forward on lower jaw; neck ring sometimes interrupted; yellow venter turns red posteriorly; belly spots numerous, highly irregular; usually 17 scale rows anteriorly. (4) **Northern Ring-necked Snake** (*D. p. edwardsii*); bluish black, bluish gray, slate, or brownish, golden collar complete; belly uniformly yellow, occasionally with a row or partial row of small black midventral dots; 15 scale rows. (5) **Regal Ring-necked Snake** (*D. p. regalis*); among largest Ring-necked Snakes, to 33⅝ in. (85.7 cm) west of our area; plain greenish, brownish, or pale slate gray (pale bluish gray in juvenile); neck ring usually absent, sometimes partially indicated or complete in Trans-Pecos (TX) populations; belly yellow, yellow extending onto first or first and second scale rows, irregularly spotted with black, turning to bright orange-red near and under tail; usually 17 scale rows anteriorly and

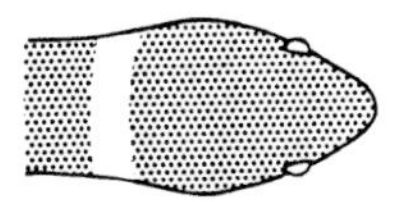

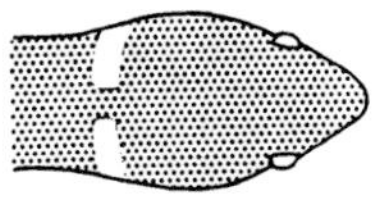

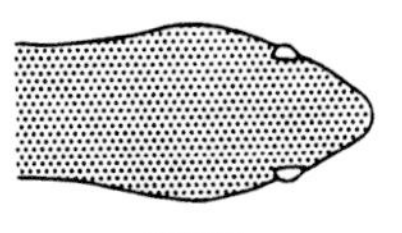

FIG. 188. *Neck and ventral patterns of subspecies of the Ring-necked Snake* (Diadophis punctatus).

midbody. (6) **Mississippi Ring-necked Snake** (*D. p. stictogenys*); neck ring narrow, often interrupted; belly spots irregular, usually grouped along midline in attached or separate pairs.

MUDSNAKES and RAINBOW SNAKES: *Farancia*

RED-BELLIED MUDSNAKE *Farancia abacura* **PL. 39, P. i**

40–54 in. (102–137 cm); record 81½ in. (207 cm). Shiny, iridescent black; belly red or pink, extending onto lower sides to form red bars or triangles, black dorsal color often extending across belly. Scales *smooth*, except keeled above vent. Cloacal usually *divided*. Young has a sharp tailtip, blunt in adult. Hatchling 6¼–10⅝ in. (15.9–27 cm). **SIMILAR SPECIES:** Rainbow Snake is striped above and has 2 rows of black marks on the belly. **HABITAT:** Swamps, wet lowlands. **RANGE:** Se. VA through FL west to e. TX and north in the Mississippi River Valley to s. IL and IN. **SUBSPECIES:** (1) **Eastern Mudsnake** (*F. a. abacura*); belly color extends upward on lower sides to form 53 or more triangular red bars on body (not including tail); upward red extensions in some juveniles may cross neck (rarely entire back). (2) **Western Mudsnake** (*F. a. reinwardtii*); belly color extends upward on lower sides to form 52 or fewer red bars with rounded tops (not including tail).

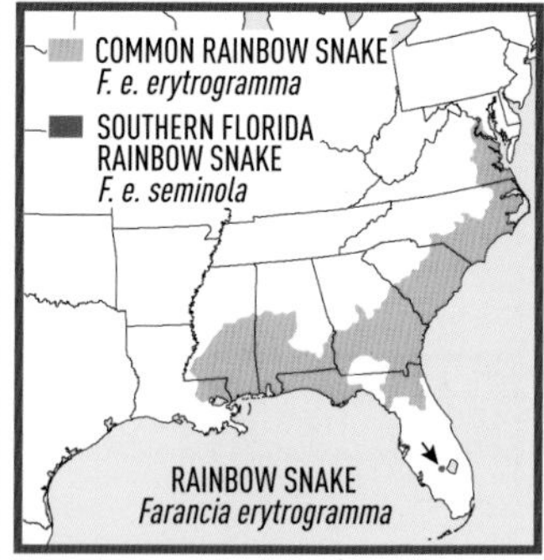

RAINBOW SNAKE *Farancia erytrogramma* **PL. 39**

27–48 in. (68.8–122 cm); record 68¼ in. (173.3 cm). Iridescent red and black stripes; belly red or pink with 2 rows of black spots and often an extra row of smaller median spots. Scales *smooth*, sometimes with keeled scales above vent. Cloacal usually *divided*. Hatchling 7¾–11 in. (19.7–28 cm). **SIMILAR SPECIES:** See Red-bellied Mudsnake. **HABITAT:** Usually in or near water, including larger rivers, especially streams in cypress swamps; in northern parts of the range, snakes sometimes are plowed up in sandy fields near water. **RANGE:** S. MD to s.-cen. FL and e. LA; isolated population in s. FL. **SUBSPECIES:** (1) **COMMON RAINBOW SNAKE** (*F. e. erytrogramma*); as above. (2) **SOUTHERN FLORIDA RAINBOW SNAKE** (*F. e. seminola*); black pigment predominates on the venter and extends upward onto lower rows of dorsal scales.

NORTH AMERICAN HOG-NOSED SNAKES: *Heterodon*

Strictly North American snakes with a *keeled upturned rostral*, *keeled dorsals*, and a *divided cloacal*. When disturbed, snakes (especially the Eastern Hog-nosed Snake) often flatten the head and neck, hiss loudly, and inflate the body. If the threat fails, they might roll over, open the mouth, and feign death. Consequently they are called hissing adders, blow vipers, spreading adders, hissing sand snakes, spread-head moccasins, and puff adders. Hatchlings 5–10 in. (12.5–25.4 cm) in the East, slightly larger in the West.

DUSTY HOG-NOSED SNAKE *Heterodon gloydi* **FIG. 189**

15–25 in. (38–63.5 cm); record "almost 3 ft." (91.4 cm). Brown to brownish gray, dark brown to black middorsal blotches; blotches from head to above vent *fewer than 32 in male, fewer than 37 in female*. Venter and underside of tail *black*, but with white or yellow blotches (Fig. 189, p. 408). *Nine or more* small azygous scales behind rostral. **SIMILAR SPECIES:** (1) Mexican Hog-nosed Snake has 2–6 azygous

scales. (2) Plains Hog-nosed Snake has more than 35 dark dorsal blotches in male, more than 40 in female. (3) Eastern Hog-nosed Snake has underside of tail usually lighter than belly. (4) Hook-nosed snakes have smooth scales and a depression instead of a raised keel behind the upturned snout. **HABITAT:** Dry prairies, semideserts, especially with sandy soils. **RANGE:** Disjunct; extreme w. IL and s. KS south through e. TX. **REMARKS:** Until recently, Dusty, Plains, and Mexican Hog-nosed Snakes were considered conspecific. A record "Western" Hog-nosed Snake (39½ in.; 100.6 cm) has been recorded, but whether it was a Dusty, Plains, or Mexican Hog-nosed Snake is unknown.

MEXICAN HOG-NOSED SNAKE *Heterodon kennerlyi* **FIG. 189**

15–25 in. (38–63.5 cm); record 29 15/16 in. (76 cm). Like Dusty Hog-nosed Snake, except with *2–6* small azygous scales behind rostral. **SIMILAR SPECIES:** See Dusty Hog-nosed Snake. **HABITAT:** Dry prairies, semideserts, especially with sandy soils. **RANGE:** S. TX to se. AZ south into Mex. **REMARK:** See Dusty Hog-nosed Snake.

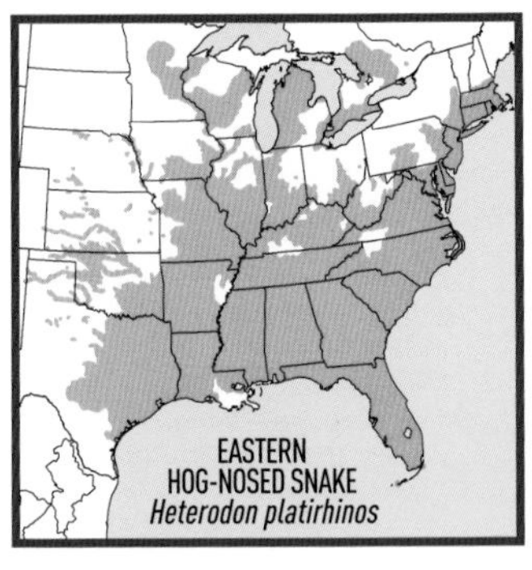

PLAINS HOG-NOSED SNAKE *Heterodon nasicus* **PL. 39, FIG. 189**

15–25 in. (38–63.5 cm); record 36⅛ in. (91.8 cm). Tan, gray, or grayish brown, contrasting sharply with dark brown to black middorsal blotches; blotches from head to above vent *more than 35 in male, more than 40 in female*. Venter and underside of tail *black*, but with white or yellow blotches (Fig. 189, below). *Nine or more* small azygous scales behind rostral. **SIMILAR SPECIES:** See Dusty Hog-nosed Snake. **HABITAT:** Prairies, especially areas with sandy soils. **RANGE:** Disjunct; IL to se. AB south to NM and TX; isolated populations indicative of a once much wider distribution. **REMARK:** See Dusty Hog-nosed Snake.

EASTERN HOG-NOSED SNAKE *Heterodon platirhinos* **PL. 39, FIG. 189**

20–33 in. (51–84 cm); record 49¹⁵⁄₁₆ in. (126.8 cm). Yellow, brown, gray, olive to orange or red, usually spotted but sometimes uniformly black, brown, or nearly plain gray. Belly yellow, light gray, or pinkish,

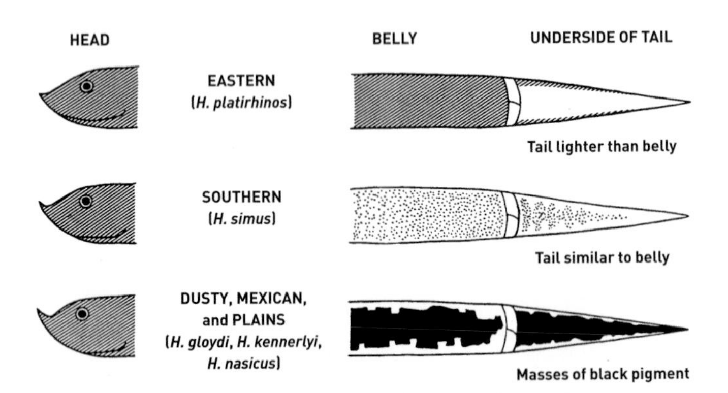

FIG. 189. *Head shapes and ventral patterns of hog-nosed snakes* (Heterodon).

mottled with gray or greenish (rarely black); underside of tail *lighter than belly* (Fig. 189, facing page). **SIMILAR SPECIES:** (1) Southern Hog-nosed Snake has belly not conspicuously darker than underside of tail. (2) Dusty, Mexican, and Plains Hog-nosed Snakes have a black belly with white or yellow patches. **HABITAT:** Open woodlands, fields, river valleys, especially areas with sandy soil. **RANGE:** S. NH to MN south to the FL Keys and TX.

SOUTHERN HOG-NOSED SNAKE *Heterodon simus* **PL. 39, FIG. 189**

14–20 in. (36–51 cm); record 24 in. (61 cm). Snout sharply upturned (Fig. 189, facing page). Brown to grayish brown, with irregular black or dark brown spots. Underside of tail *not* conspicuously lighter than belly (Fig. 189, facing page). Hatchling 6¼–6¾ in. (16–17.3 cm). **SIMILAR SPECIES:** See Eastern Hog-nosed Snake. **HABITAT:** Sandy woods, fields, groves, dry river floodplains, hardwood hammocks, especially areas with sandy soils. **RANGE:** Se. NC to s.-cen. FL west to s. MS. Introduced in s. FL. **CONSERVATION:** Vulnerable.

NIGHTSNAKES and CAT-EYED SNAKES: *Hypsiglena* and *Leptodeira*

Seven currently recognized species of nightsnakes range from the southern United States to Costa Rica, 11 currently recognized species of cat-eyed snakes from extreme southern Texas to northern Argentina and Paraguay. All have *vertically elliptical pupils*, *smooth* scales, and a *divided* cloacal.

CHIHUAHUAN NIGHTSNAKE *Hypsiglena jani* **PL. 39**

14–16 in. (36–41 cm); record 24 3/16 in. (61.5 cm). Light brown or grayish; brown or dark gray middorsal spots alternate with smaller lateral spots on body; bold elongated blotches on nape and sides of neck, latter extending backward and downward from eyes. Belly immaculate white or yellowish. Hatchling 5–7 in. (12.5–18 cm). **HABITAT:** Arid or semiarid regions, especially rocky areas. **RANGE:** S. KS to e. AZ and south into Mex. **SUBSPECIES: TEXAS NIGHTSNAKE** (*H. j. texana*). **REMARK:** Until recently considered a subspecies of *H. torquata*, a taxon now restricted to Mex.

CAT-EYED SNAKE *Leptodeira septentrionalis* **PL. 39**

18–24 in. (45.7–61 cm); record 38¾ in. (98.5 cm). Cream and yellow to reddish tan, contrasting sharply with broad dark brown or black crossbands (saddles) that extend almost completely across the back. Head *broad*, much wider than neck. Juvenile more boldly patterned than adult, ground color orange-tan. Hatchling 8½–9½ in. (21.5–24 cm). **HABITAT:** Arid thornscrub, often near streams or other bodies of water. **RANGE:** S. TX to n. S. America. **REMARK:** Rear fangs generally pose no threat, but best handled with caution.

LITTERSNAKES: *Rhadinaea*

Twenty currently recognized species range from the southern United States to Panama.

PINE WOODS LITTERSNAKE *Rhadinaea flavilata* **PL. 39**

10–13 in. (25–33 cm); record 15⅞ in. (40.3 cm). Rich golden to light reddish brown, paler on lower sides; head darker; *dark lines through eyes.* Usually 7 upper labials; labials virtually plain white or yellowish in most snakes from FL, usually speckled with dark pigment in western and especially northern parts of the range. Often with a suggestion of narrow dark middorsal and 2 lateral stripes on scale rows 2 and 3 (third rows only toward tail). Belly plain white, pale yellow, or yellow-green. Scales *smooth*; cloacal *divided*. Hatchling 5½–6½ in. (14–16 cm). **HABITAT:** Damp woodlands, chiefly pine flatwoods, occasionally hardwood hammocks, coastal islands. **RANGE:** Disjunct; NC to e. LA, including most of FL. **REMARK:** Also called the Yellow-lipped Snake.

HARMLESS LIVE-BEARING SNAKES: Natricidae

More than 30 genera and 220 species range across eastern North America, Europe, the Indian subcontinent, southeastern Asia, and Africa. Snakes in this family until recently were placed into the family Colubridae as then defined.

KIRTLAND'S SNAKES: *Clonophis*

The single species in this genus occurs only in our area.

KIRTLAND'S SNAKE *Clonophis kirtlandii* **PL. 40**

14–18 in. (36–45.7 cm); record 26 in. (66.2 cm). Belly reddish with prominent rows of *round black spots* down each side; dorsum reddish

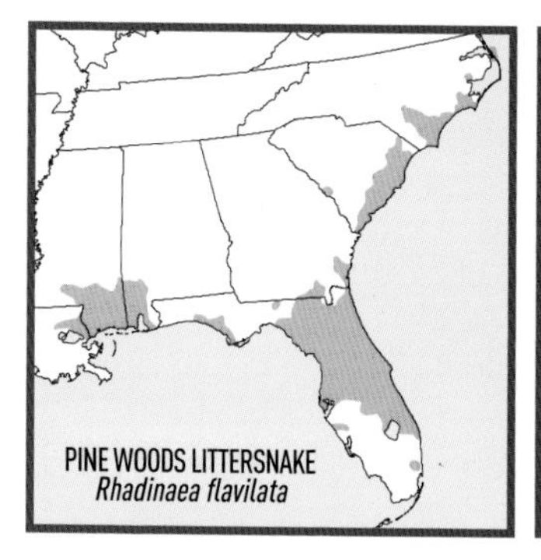

brown, 4 rows of dark spots not always conspicuous; spots of 2 central rows occasionally small and indistinct, especially posteriorly, ground color may appear as a light middorsal stripe. Scales *keeled*; cloacal *divided*. Young dark and virtually unicolored above; belly deep red; 4½–6½ in. (11.5–16.5 cm) at birth. **SIMILAR SPECIES:** Red-bellied Snake lacks conspicuous black spots on back or belly. **HABITAT:** In or near streams, wet meadows, open swamp-forests, sparsely wooded slopes, adjacent meadows, parks (most urban populations extirpated); often associated with crayfish burrows. **RANGE:** Disjunct; w. PA to ne. MO and south into KY and extreme ne. TN; many populations extirpated.

EARTHSNAKES: *Haldea* and *Virginia*

Earthsnakes occur only in the eastern United States. Loreals are *horizontal* (Fig. 165, p. 350), and the cloacal is usually *divided*.

ROUGH EARTHSNAKE *Haldea striatula* **PL. 44**

7–10 in. (18–25.4 cm); record 12¾ in. (32.4 cm). Cone-shaped head with a pointed snout; light gray or brown to reddish brown. *Five upper labials; internasals fused.* Scales *keeled*; cloacal rarely undivided. Young darker and grayer than adult, often with a pale gray to white band across back of head; 3⅛–4¾ in. (8–12.1 cm) at birth. **SIMILAR SPECIES:** (1) Smooth Earthsnake has 6 upper labials and smooth or weakly keeled scales. (2) North American brownsnakes have no loreal scales. (3) North American wormsnakes, Ring-necked Snake, and Pine Woods Littersnake have smooth scales. **HABITAT:** Woodlands to open prairie. **RANGE:** VA to n. FL west to TX and north to se. KS and s. MO; apparently absent from much of the Mississippi R. floodplain. **REMARK:** Until recently assigned to the genus *Virginia*.

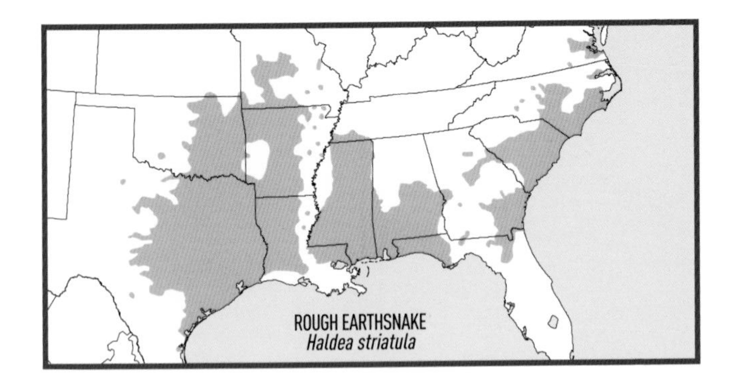

SMOOTH EARTHSNAKE *Virginia valeriae* **PL. 44**

7–10 in. (18–25.4 cm); record 15⅜ in. (39.3 cm). Gray to reddish brown; slightly darker from eyes to nostrils; a faint light middorsal stripe sometimes indicated. Many scales bear very faint light lines that look like keels. Belly plain white or yellowish. *Six upper labials.* Scales *smooth or weakly keeled*, 15 or 17 rows. Young 3⅛–4½ in. (8–11.5 cm) at birth. **SIMILAR SPECIES:** Western Groundsnake has smooth scales in 15 rows. See also Rough Earthsnake. **HABITAT:** Old fields, open woodlands. **RANGE:** Disjunct; NJ to n. FL west to IA and TX; isolated population in s. FL. **SUBSPECIES:** (1) **Eastern Smooth Earthsnake** (*V. v. valeriae*); gray or light brownish gray; 15 scale rows, scales mostly smooth, faint keels near tail; often with tiny black dorsal spots scattered or in 4 rows. (2) **Western Smooth Earthsnake** (*V. v. elegans*); reddish to grayish brown, belly whitish with a pale greenish yellow tint in adult; scales weakly keeled, 17 scale rows. (3) **Mountain Earthsnake** (*V. v. pulchra*); reddish brown to dark gray; scales weakly keeled, in 15 rows anteriorly, 17 at midbody and posteriorly. **REMARK:** Mountain Earthsnake might be a distinct species.

SWAMPSNAKES: *Liodytes*

Three currently recognized species, all from eastern North America, until recently were assigned to other genera. The cloacal is *divided*.

STRIPED SWAMPSNAKE *Liodytes alleni* **PL. 41**

13–20 in. (33–51 cm); record 27¾ in. (70.5 cm). Shiny brown; broad yellowish stripes along lower sides, 3 relatively inconspicuous dark dorsal stripes. Belly yellowish, usually unmarked; belly and lateral stripes occasionally orange or orange-brown. Dark midventral markings, when present, vary from few scattered smudges to a defined row of spots. *One internasal scale*, nasals *meet* middorsally on snout. Scales *smooth* except keeled at vent and on top of tail. Young 6¼–7 in.

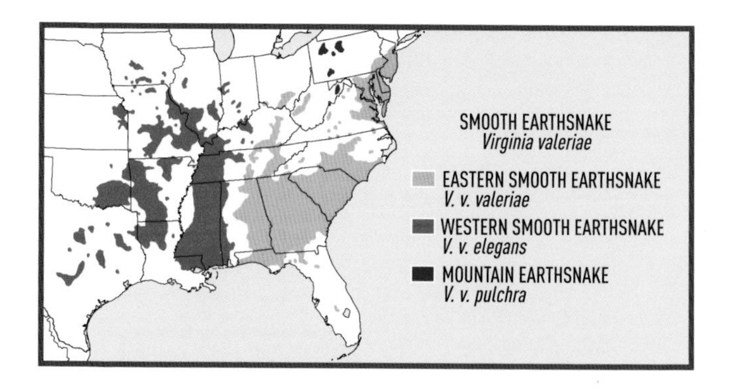

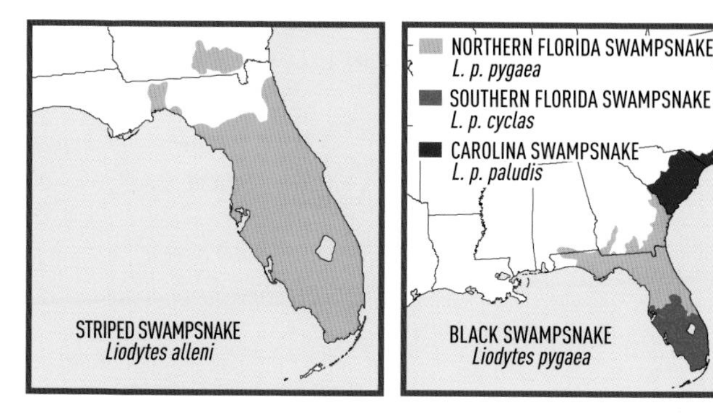

(15.9–18 cm) at birth. **SIMILAR SPECIES:** Saltmarsh Watersnake and gartersnakes have keeled scales. **HABITAT:** Sloughs, marshes, bay heads, sphagnum bogs. **RANGE:** Disjunct; s. GA through peninsular FL. **REMARK:** Until recently assigned to the genus *Regina*.

BLACK SWAMPSNAKE *Liodytes pygaea* **PL. 42**

10–15 in. (25.4–38 cm); record 21$\frac{7}{8}$ in. (55.5 cm). Shiny black, belly red with black marks. Scales *smooth*, but scales on 3–5 lowermost rows bear light longitudinal lines that look like keels. Young 4$\frac{1}{4}$–5$\frac{15}{16}$ in. (10.7–15.1) at birth. **SIMILAR SPECIES:** Red-bellied Snake has keeled scales. **HABITAT:** Swamps, ponds, marshes, grassy wet prairies, sluggish streams, ditches, canals, lakes; sometimes in salt marshes or brackish water; often associated with introduced water hyacinths. **RANGE:** Coastal NC through FL. **SUBSPECIES:** (1) **Northern Florida Swampsnake** (*L. p. pygaea*); belly plain red or with pair of black bars on the base of each ventral scale; ventrals 118–124. (2) **Southern Florida Swampsnake** (*L. p. cyclas*); short triangular black marks at front edge of ventral scales; ventrals 117 or fewer. (3) **Carolina Swampsnake** (*L. p.

paludis); pair of black bars on each ventral scale; ventrals 127 or more. **REMARK:** Until recently assigned to the genus *Seminatrix*.

GLOSSY SWAMPSNAKE *Liodytes rigida* **PL. 41**

14–24 in. (36–61 cm); record 32¹¹⁄₁₆ in. (83 cm). Shiny, more or less plain brown or olive brown; dark stripes sometimes faintly evident on back or (more strongly) on lower sides. *Two rows of black spots* on belly distinct even in large individuals, central portions often clouded with dark pigment. Scales *keeled*. Young 6½–9 in. (16.5–22.7 cm) at birth. **SIMILAR SPECIES:** Gartersnakes, Striped Swampsnake, Queensnake, Lined Snake, and Saltmarsh Watersnake have prominent light lateral stripes. **HABITAT:** Slow lowland waters, swamps, marshes, sphagnum bogs, ponds, lakes, bayous, rice fields, canals, drainage ditches, mucky areas along streams, floodplains; grassy or wooded uplands near wetlands. **RANGE:** Disjunct; e. VA through n.-cen. FL and west to e. TX. **SUBSPECIES:** (1) **EASTERN GLOSSY SWAMPSNAKE** (*L. r. rigida*); narrow dusky stripes follow edges of scales on sides of throat; subcaudals usually 54 or fewer in female, 62 or fewer in male. (2) **DELTA SWAMPSNAKE** (*L. r. deltae*); one preocular scale on at least one side of head (2 on both sides in other subspecies); number of subcaudals subtracted from number of ventrals usually 81 or more in female, 73 or more in male (fewer in other subspecies). (3) **GULF SWAMPSNAKE** (*L. r. sinicola*); no pattern on sides of throat; subcaudals usually 55 or more in female, 63 or more in male. **REMARK:** Until recently assigned to the genus *Regina*.

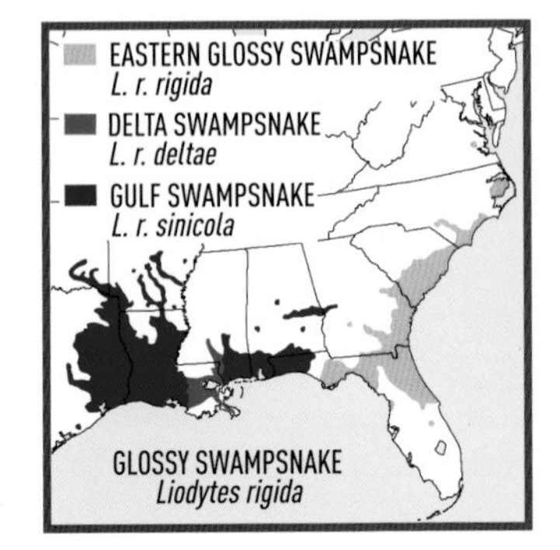

NORTH AMERICAN WATERSNAKES: *Nerodia*

Endemic to the eastern and central United States and Mexico; one species also occurs in Cuba. All have *keeled* scales and bear live young. The cloacal is almost always *divided*. Watersnakes sometimes are confused with venomous cottonmouths (Fig. 166, p. 352)—always be careful around aquatic snakes in the range of cottonmouths.

SALTMARSH WATERSNAKE *Nerodia clarkii* **PL. 40**

15–30 in. (38–76 cm); record 36¾ in. (93.3 cm). Extremely variable. Snakes along the Gulf Coast have 2 dark brown and 2 tan or yellowish stripes on sides, but stripes on snakes from peninsular FL usually

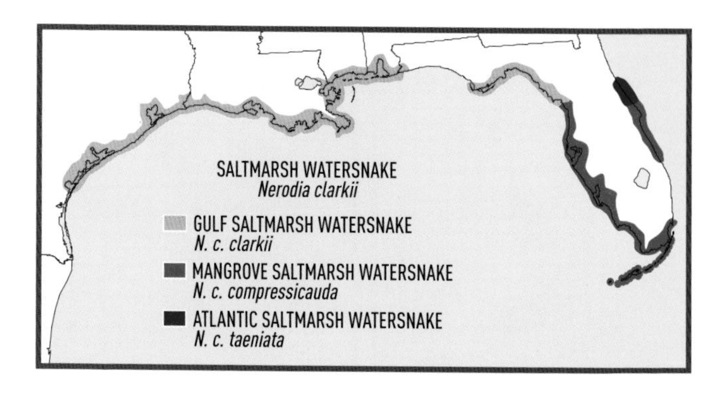

evident only on neck; also, some snakes have dark spots or crossbands on light ground color, others are uniformly black, red, orange, or straw colored. Ventral pattern, when present, a central row of white or yellow spots on darker ground color, sometimes with extra rows of smaller light spots on each side. Dorsal scale rows 21–23. Young 7¾–10½ in. (19.7–26.6 cm) at birth. **SIMILAR SPECIES:** (1) Plain-bellied Watersnake occasionally similar to unicolored Saltmarsh Watersnake, but brown or gray above and rarely encountered in brackish water. (2) Queensnake, Graham's Crayfish Snake, and swampsnakes have 19 scale rows. (3) Gartersnakes have an *undivided* cloacal and a *single* light lateral stripe. See also cottonmouths. **HABITAT:** Coastal salt meadows, swamps, marshes; the only watersnake routinely in salt- or brackish water; rarely enters fresh water (some ponds in the FL Keys). **RANGE:** FL coast to s. TX; northern coast of Cuba (not mapped). **SUBSPECIES:** (1) **Gulf Saltmarsh Watersnake** (*N. c. clarkii*); 2 dark brown and *2* tan or yellowish lateral stripes; venter brown or reddish brown, with a *central row of large white or yellow spots*, sometimes with extra rows of smaller light spots on each side. (2) **Mangrove Saltmarsh Watersnake** (*N. c. compressicauda*); greenish, with dark spots or crossbands, stripes may appear on neck; sometimes almost plain black, occasionally straw colored, red, or orange-red. (3) **Atlantic Saltmarsh Watersnake** (*N. c. taeniata*); light, striped anteriorly, but remainder of body with dark blotches; belly with a median row of broad light yellowish spots. **REMARK:** Sometimes with aberrant patterns, possibly due to interbreeding with freshwater species washed into coastal areas after hurricanes. **CONSERVATION:** Atlantic Saltmarsh Watersnake (USFWS: Threatened).

MISSISSIPPI GREEN WATERSNAKE *Nerodia cyclopion* **PL. 40, FIG. 163**

30–45 in. (76–114 cm); record 51 in. (129.5 cm). Greenish or brownish, usually with at least a suggestion of a dark pattern. Belly gray or brown, with *light* spots or half-moons. *Rows of scales between eyes and upper labials* (Fig. 163, p. 342). Young 9–11 in. (23–28 cm) at birth. **SIMILAR SPECIES:** See Florida Green Watersnake. **HABITAT:** Quiet wa-

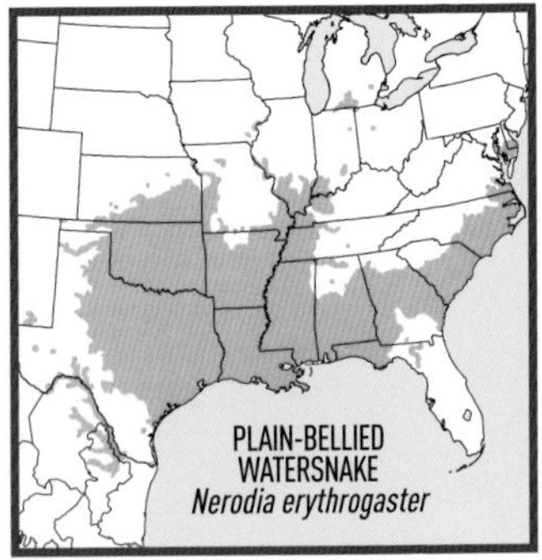

ters, edges of lakes and ponds, swamps, rice fields, marshes, bayous, other waterways; occasionally in brackish water. **RANGE:** Extreme s. IL to the Gulf.

PLAIN-BELLIED WATERSNAKE *Nerodia erythrogaster* **PLS. 40, 41**

30–48 in. (76–122 cm); record 64³/₈ in. (163.6 cm). Unicolored, distinctly patterned, or with traces of a pattern as dark-bordered light transverse bands. Brown (pale reddish to rich chocolate brown) or gray to greenish gray, especially on lower sides, but snakes (particularly from western portions of the range) often retain a blotched dorsal juvenile pattern. *Belly and underside of tail plain red or orange to yellow*; dorsal color sometimes extends onto edges or posterior borders of ventrals. Juvenile boldly patterned, ground color usually lighter than in adult, sometimes pinkish; lateral and larger middorsal blotches *alternate forward to head* (or nearly so). Cloacal rarely undivided. Young 8–12 in. (20.3–30.6 cm) at birth. **SIMILAR SPECIES:** (1) Uniformly brown watersnakes of several species usually have a strongly patterned belly. (2) Common Watersnake has lateral blotches joining dorsal blotches, forming crossbands on neck and anterior body, and the Southern Watersnake has crossbands on entire body; both have a distinct ventral pattern. (3) Florida Green Watersnake has scales between eyes and upper labials. (4) Saltmarsh Watersnake is restricted to brackish water, occasionally unicolored with an unmarked belly, but dorsal and ventral color is essentially the same in such snakes. (5) Red-bellied Snake and Black Swampsnake often have a plain dark dorsal surface and red venter, but adults are about the size of a patterned young Plain-bellied Watersnake. **HABITAT:** Usually larger, more permanent bodies of water, river bottoms, swamps, marshes, edges of ponds and lakes; western populations often near semipermanent water, ditches, cattle tanks, intermittent streams; will follow rivers into arid country. **RANGE:** Delmarva Peninsula to n. FL west to w. OK, NM, TX, and adjacent Mex., north along the Mississippi R. to se. IA; isolated populations in the upper Midwest. **REMARK:** Previously recognized subspecies are no longer considered valid.

SOUTHERN WATERSNAKE *Nerodia fasciata* **PL. 40, FIGS. 163, 190**

22–42 in. (56–106.7 cm); record 62½ in. (158.8 cm). Gray, tan, yellow, or reddish; red, brown, or black *crossbands* (often black-bordered) vary in number, color, and intensity; *dark stripes from eyes to angles of the jaws* (Fig. 163, p. 342); venter with vermiculate markings (extreme se. GA and FL) or squarish spots along sides of belly. Often darkens with age, markings become obscure, resulting in virtually all-black snakes (especially in southern parts of the range), but usually with patches of red or other light colors on lower sides. Juvenile with prominent crossbands contrasting strongly with pale ground color; 7–10½ in. (18–26.6 cm) at birth. **SIMILAR SPECIES:** (1) Common Watersnake usually has dark crossbands

FIG. 190. *The Broad-banded Watersnake* (Nerodia fasciata confluens), *a subspecies of the Southern Watersnake, occupies virtually all types of freshwater habitats, but populations near the northern extent of the range along the Mississippi River Valley often are densest in swamplike situations.*

only anteriorly, belly markings typically include dark or reddish half-moons. (2) Mississippi and Florida Green Watersnakes have rows of scales between eyes and upper labials. See also cottonmouths. **HABITAT:** Virtually all types of freshwater habitats, including streams, rivers, ponds, lakes, swamps, marshes, cypress bays; to very edge of salt or brackish water along the Gulf. **RANGE:** NC south through FL east to TX and north along the Mississippi R. to extreme s. IL. **SUBSPECIES:** (1) **BANDED WATERSNAKE** (*N. f. fasciata*); numerous red to brown or black crossbands, squarish spots along sides of belly. (2) **BROAD-BANDED WATERSNAKE** (*N. f. confluens*); 11–17 black to rich red-brown crossbands separated by yellow and irregular in shape and arrangement, frequently running together. (3) **FLORIDA WATERSNAKE** (*N. f. pictiventris*); secondary lateral dark spots between crossbands; venter with vermiculate red or black markings. **REMARK:** Interbreeds occasionally with the Common Watersnake.

FLORIDA GREEN WATERSNAKE *Nerodia floridana* **PL. 41, FIG. 163**

30–55 in. (76–140 cm); record 74 in. (188 cm). Greenish or brownish (sometimes reddish in s. FL) without distinctive markings. Belly plain whitish or cream, except near vent and under tail, where marked with *light* spots or half-moons. *Rows of scales between eyes and upper labials* (Fig. 163, p. 342). Young brownish or greenish olive with about 50 black or dark brown lateral bars and similar, less conspicuous dorsal markings; belly yellow, but with black markings near and under tail; 8¾–10 in. (22.4–25.4 cm) at birth. **SIMILAR SPECIES:** Mississippi Green Watersnake has a dark belly with light markings. No other watersnakes in range have rows of scales between eyes and upper labials, and most have distinct patterns on back, belly, or both. See also cottonmouths. **HABITAT:** Everglades and the Okefenokee region, other swamps, marshes, quiet bodies of water; sometimes in brackish water. **RANGE:** Disjunct; s. SC and adjacent GA, most of FL, adjacent s. GA, and extreme se. AL.

BRAZOS RIVER WATERSNAKE *Nerodia harteri* **PL. 41**

20–30 in. (51–76 cm); record 35½ in. (90.2 cm). Brownish gray or slightly greenish; *4 rows of dark brown dorsal spots* alternate to produce a checkerboard appearance. Belly pink or orange, *2 rows of dark dots* along sides. Usually *2 rows* of small scales between posterior

chin shields. Young 6¾–10 in. (17.2–25.4 cm) at birth. **SIMILAR SPECIES:** Concho Watersnake is reddish, has less prominent dorsal spots, dots on belly inconspicuous or absent, and a single row of scales between posterior chin shields. **HABITAT:** Fast-flowing, rocky streams, banks of impoundments. **RANGE:** Brazos R. system in cen. TX.

CONCHO WATERSNAKE *Nerodia paucimaculata* **NOT ILLUS.**

20–30 in. (51–76 cm); record 25⅝ in. (65.2 cm). Reddish brown; *4 rows of dark brown dorsal spots*, often poorly defined, alternate to produce a checkerboard appearance. Belly pink or orange, immaculate or slightly marked. *One row* of small scales between posterior chin shields. Young 6¾–10 in. (17.2–25.4 cm) at birth. **SIMILAR SPECIES:** See Brazos River Watersnake. **HABITAT:** Fast-flowing, rocky streams, banks of impoundments. **RANGE:** Concho R. system in cen. TX.

DIAMOND-BACKED WATERSNAKE *Nerodia rhombifer* **PL. 41, FIG. 191**

30–48 in. (76–122 cm); record 69 in. (175.3 cm). Light brown or dirty yellow; dark brown *chainlike markings* surround vaguely diamond-shaped light areas (Fig. 191, p. 420). Belly yellow, marked with black or dark brown spots or half-moons. Adult male with numerous raised protuberances (papillae) under chin. Young strongly patterned, belly often brightly tinged with orange; 8¼–13⅛ in. (21–33.2 cm) at birth. **SIMILAR SPECIES:** Brown Watersnake has dark dorsal blotches alternating with lateral blotches but no chainlike pattern. See also cottonmouths. **HABITAT:** Large lakes, rivers, ditches, cattle tanks; extends along rivers far into arid areas in western parts of the range. **RANGE:** Sw. IN and se. IA south to the Gulf and into Mex.; largely absent from the Central Highlands in MO, AR, and e. OK. **SUBSPECIES: NORTHERN DIAMOND-BACKED WATERSNAKE** (*N. r. rhombifer*). **REMARK:** Populations east and west of the Mississippi R. might represent distinct species.

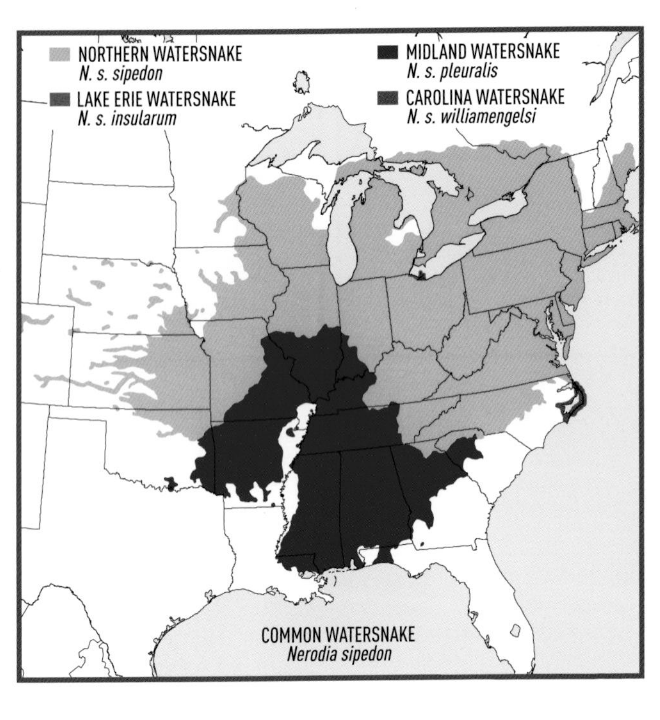

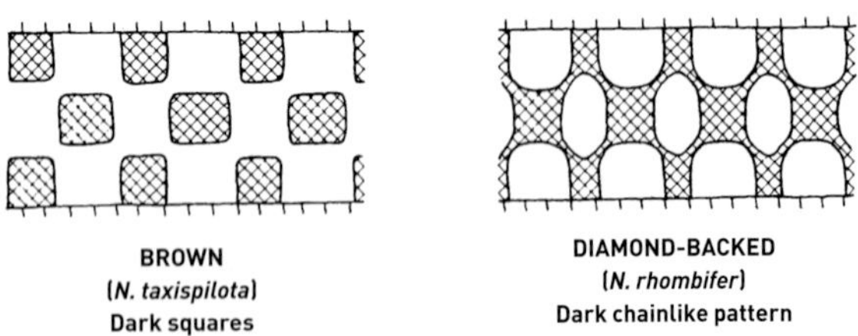

FIG. 191. *Diagrammatic dorsal patterns of two species of watersnakes in the genus* Nerodia.

COMMON WATERSNAKE *Nerodia sipedon* **PL. 41, P. xiv**

22–42 in. (56–106.7 cm); record 59 in. (150 cm). Pale gray to dark brown; reddish brown to black dorsal markings are usually *bands anteriorly but alternating dorsal and lateral blotches posteriorly* (rarely completely crossbanded); large adults often dark with an obscure pattern, appearing almost plain black or dark brown. Venter usually with half-moons; other distinct markings in a regular pattern, scattered randomly, represented by dusky areas, or entirely absent; sometimes almost uniformly stippled with gray except for yellow, orange, or pinkish midventral stripe. Juvenile pale gray or light brown with a very dark pattern; 7½–10¾ in. (19–27.3 cm) at birth. **SIMILAR SPECIES:** See Southern and Plain-bellied Watersnakes. **HABITAT:** Any body of water, including swift-flowing streams in uplands. **RANGE:** ME to MN and e. CO south to LA and the FL Panhandle. **SUBSPECIES:** (1) **Northern Watersnake** (*N. s. sipedon*); usually 30 or more dark dorsal markings (bands and blotches) from neck to vent, separated by no more than one scale row; light spaces between lateral dark markings usually fewer than 2½ scale rows wide; reddish- or brownish-centered half-moon and other distinct markings on belly usually present and extending past the 50th ventral scute; dark pattern on posterior venter continuous to tailtip; first 3–15 tail markings (counting back from the vent) usually composed of alternating dorsal blotches and lateral bars. (2) **Lake Erie Watersnake** (*N. s. insularum*); pale gray (often greenish or brownish), dorsal pattern reduced or absent (p. xiv); belly white or yellowish, sometimes with a central pink or orange tinge. (3) **Midland Watersnake** (*N. s. pleuralis*); usually 30 or fewer dark dorsal markings (bands and blotches) from neck to vent, separated by more than one scale row; light spaces between dark lateral markings usually greater than 2½ scales rows wide; tail markings usually consist of complete rings; occasionally crossbands continue throughout length of body; belly markings tend to be paired and rarely broken. (4) **Carolina watersnake** (*N. s. williamengelsi*); very dark, often almost plain black above; half-moons on belly black, except anterior to the 50th ventral scale where some may have brown or reddish centers; light spaces between dark lateral markings usually no more than 1½ scales wide at midbody, 1½ scales wide on neck. **REMARK:** Interbreeds occasionally with the Southern Watersnake.

BROWN WATERSNAKE *Nerodia taxispilota* **PL. 41, FIG. 191**

30–60 in. (76–152 cm); record 69½ in. (176.6 cm). Brown; large, squarish, dark brown middorsal blotches alternate with similar lateral series of blotches (Fig. 191, facing page); often very dark, blotches only slightly darker than ground color. Belly yellow to brown, boldly marked with dark brown or black spots and half-moons. Head wider than neck, heart- or diamond-shaped when viewed from above. Young 7½–14 in. (19–36 cm) at birth. **SIMILAR SPECIES:** Mississippi and Florida Green Watersnakes have rows of scales between eyes and upper labials. See also Diamond-backed Watersnake

and cottonmouths. **HABITAT:** Quiet waters of swamps and rivers; occasionally enters brackish marshes. **RANGE:** Se. VA through FL to sw. AL.

CRAYFISH SNAKES: *Regina*

Eastern North American snakes with relatively small heads (compared with watersnakes). They have *19* scale rows, scales are *keeled*, and the cloacal is *divided*.

GRAHAM'S CRAYFISH SNAKE *Regina grahamii* **PL. 41, FIG. 207, P. 494**

18–28 in. (45.7–71 cm); record 47 in. (119.4 cm). Broad yellow stripes on *scale rows 1, 2, and 3*, narrow black stripes where lowermost rows of scales meet ventrals (latter often zigzagged or irregular); sometimes with a dark-bordered, pale middorsal stripe. Belly yellowish, either plain or marked with dark dots or a dull dark central area. Dark variant in IA with a brown dorsum and obscure pattern, belly deep olive buff, chin and throat yellow. Young 7½–10 in. (19–25.4 cm) at birth. **SIMILAR SPECIES:** See Queensnake. **HABITAT:** Ponds, streams, sloughs, bayous, swamps. **RANGE:** Disjunct; IL and IA south to the Gulf.

QUEENSNAKE *Regina septemvittata* **PL. 41**

15–24 in. (38–61 cm); record 36¼ in. (92.2 cm). Brown; 3 very narrow dark dorsal stripes (difficult to see except after shedding), yellow lateral stripes on *second and upper half of first scale rows*. Belly yellow, with *4 brown stripes*, outer 2 larger and on edges of ventrals (plus lower half of first scale row), most prominent toward neck (tend to run together posteriorly, especially in adult). Snakes in southern parts of the range nearly unicolored, with indications of a pattern only in neck region. Juvenile has clearly defined ventral stripes usually extending to tail; 6 13/16–10½ in. (17.3–26.6 cm) at birth. **SIMILAR SPECIES:** (1) Graham's Crayfish Snake has broad yellow lateral stripes on scale rows 1, 2, and 3, belly either plain or with a single dark area or row of central spots. (2) Glossy Swampsnake and Lined Snake have double rows of black ventral spots. (3) Gartersnakes have an undivided cloacal and usually have a middorsal light stripe. (4) Gulf Saltmarsh Watersnake (a subspecies of the Saltmarsh Watersnake) has a brown or black belly with 1–3 rows of light central spots. **HABITAT:** Any aquatic habitat, especially streams. **RANGE:** W. NY to se. WI south to the Gulf; disjunct populations west of the Mississippi R. in AR and sw. MO, latter likely extirpated (not mapped). **REMARK:** Some lineages might warrant recognition as distinct species.

NORTH AMERICAN BROWNSNAKES: *Storeria*

Five currently recognized species range from Canada to Honduras. Scales are *keeled*, the cloacal is *divided*, and *loreals are absent* (Fig. 192, below). Young 2¾–4⁷⁄₁₆ in. (7–11.3 cm) at birth.

DEKAY'S BROWNSNAKE *Storeria dekayi* **PL. 42, FIG. 193**

9–13 in. (23–33 cm); record 20¾ in. (52.7 cm). Light yellowish brown or gray to dark brown or deep reddish brown; about *4 middorsal scale rows* almost always lighter than sides, *2 parallel rows of blackish dorsal spots*, sometimes linked across back by narrow dark lines, small dark lateral spots often inconspicuous; dark *downward streaks or spots* usually present on sides of head (Fig. 193, p. 424). Belly pale yellowish, brownish, or pinkish, unmarked except for one or more small black dots along sides of each ventral. *Scale rows 17.* Young with a conspicuous yellowish collar, often darker than adult, spotted pattern scarcely evident. **SIMILAR SPECIES:** (1) Red-bellied Snake has 15 scale rows, belly usually red. (2) Florida Brownsnake has 15 scale rows. (3) Earthsnakes have long, horizontal loreals. (4) Western Groundsnake, wormsnakes, and Ring-necked Snake have smooth scales. (5) Gartersnakes have an undivided cloacal and usually have light lateral stripes. **HABITAT:** Bogs, swamps, freshwater marshes, moist woodlands, parks, cemeteries, beneath trash in empty lots, even in urban centers. **RANGE:** S. ME to the FL Panhandle west to the Great Plains and south into Mex. **REMARK:** Previously recognized subspecies probably represent clinal variation.

NORTH AMERICAN BROWNSNAKES (*Storeria*)

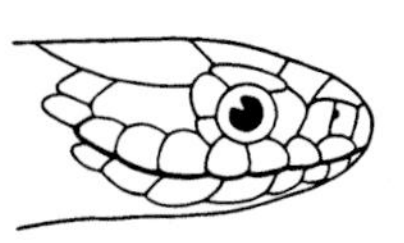

No loreal; postnasal scale touches preocular

FIG. 192. *The head of a North American brownsnake* (Storeria).

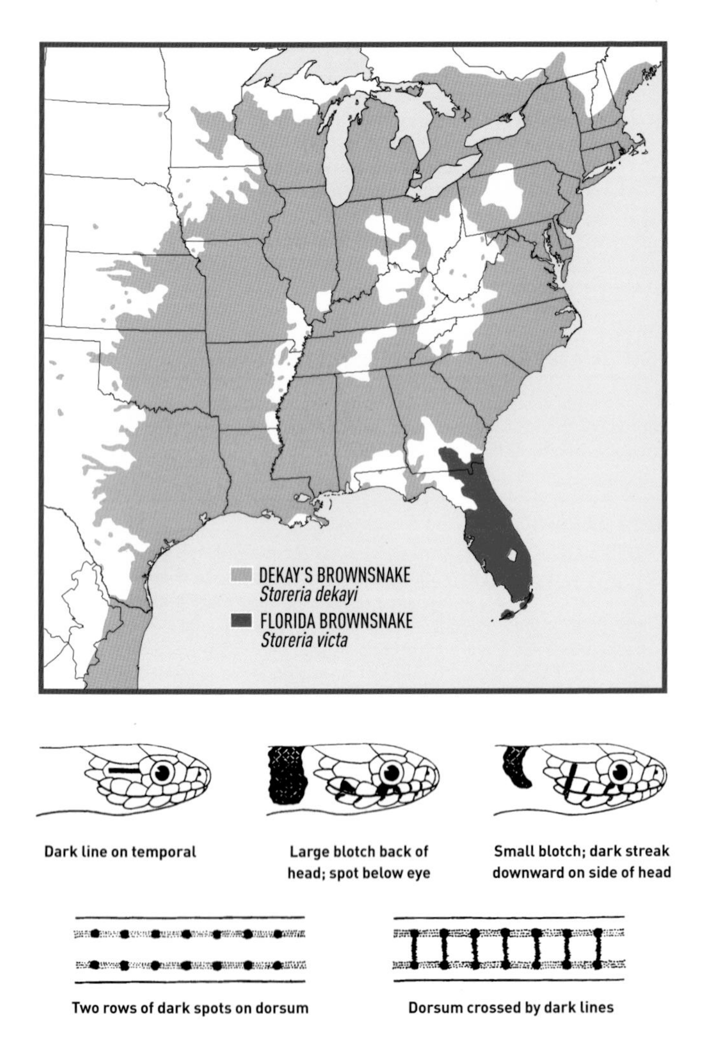

FIG. 193. *Variation in head and dorsal patterns of Dekay's Brownsnake* (Storeria dekayi).

RED-BELLIED SNAKE *Storeria occipitomaculata* **PL. 42, FIG. 194**

8–10 in. (20.3–25.4 cm); record 16⅝ in. (42.2 cm). Brown (sometimes gray, occasionally black); indications of 4 narrow dark stripes, broad, fairly light middorsal stripe, or both; *3 pale nape spots* usually present

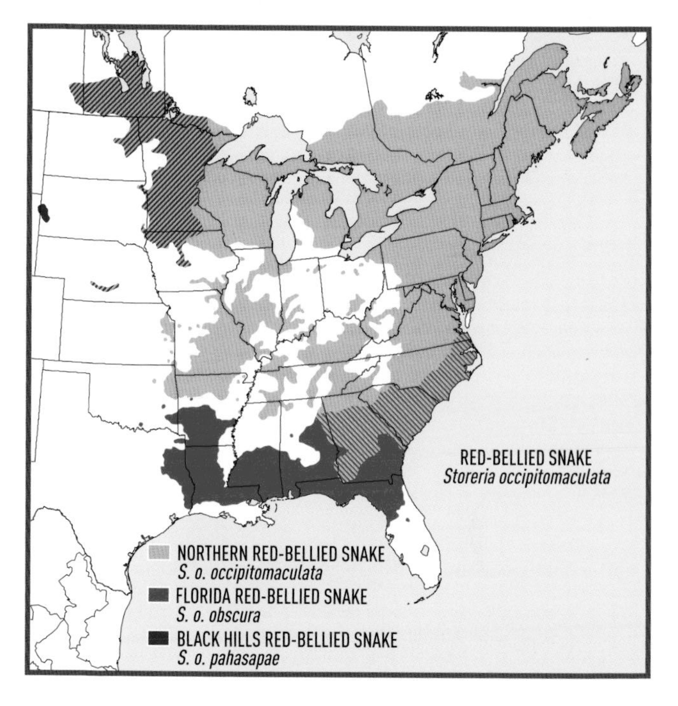

FIG. 194. *The aptly named Red-bellied Snake* (Storeria occipitomaculata) *is very secretive and is rarely encountered in the open. This small snake feeds almost entirely on slugs.*

(occasionally fused to form a collar, especially in FL). Belly *plain red* (occasionally orange to pale yellow, very rarely blue-black). Two preoculars. *Scale rows 15*. Young similar to adult but darker. **SIMILAR SPECIES:** (1) Kirtland's Snake has a double row of black spots on belly. (2) Black Swampsnake has smooth scales. See also Dekay's Brownsnake. **HABITAT:** Open woods, sphagnum bogs. **RANGE:** Disjunct; NS to cen. FL west to se. SK and e. TX; isolated areas in the Black Hills of SD and WY, cen. NE, and OK. **SUBSPECIES:** (1) **NORTHERN RED-**

BELLIED SNAKE (*S. o. occipitomaculata*); nape spots usually well defined, moderate amount of black on sides and back of head, light marks on fifth upper labials bordered below by black. (2) **FLORIDA RED-BELLIED SNAKE** (*S. o. obscura*); nape spots fused to form a collar, top and sides of head black, light spots on fifth upper labials extending down to edge of mouth. (3) **BLACK HILLS RED-BELLIED SNAKE** (*S. o. pahasapae*); light nape spots small or lacking, no light spots on fifth upper labials.

FLORIDA BROWNSNAKE *Storeria victa* **PL. 42**

9–13 in. (23–33 cm); record 19 in. (48.3 cm). Similar to Dekay's Brownsnake but markings reduced; *broad light band* across back of head, heavy *dark pigment on labials* below eyes, usually a *double row of small black spots* on sides of belly. *Scale rows 15.* Young similar to adult but darker, with a prominent light band across back of head. **SIMILAR SPECIES:** Pine Woods Littersnake and crowned snakes have smooth scales. See also Dekay's Brownsnake. **HABITAT:** Bogs, marshes, swamps, ponds, sloughs, upland hammocks, pineland prairies. **RANGE:** Se. GA through the FL Keys. **REMARK:** Until recently considered a subspecies of Dekay's Brownsnake.

GARTERSNAKES and RIBBONSNAKES: *Thamnophis*

Thirty-four currently recognized species range from southern Canada to Costa Rica. Scales are strongly *keeled* and the cloacal is almost always *undivided.* Most have yellowish (occasionally orange) longitudinal stripes. For positions of lateral stripes, count upward from the ventral scales (Fig. 164, p. 348). These snakes are often near water, especially in the arid or semiarid West where watersnakes are absent; they are less aquatic in the East.

SHORT-HEADED GARTERSNAKE *Thamnophis brachystoma* **PL. 42**

14–18 in. (36–45.7 cm); record 22¾ in. (57.8 cm). Head *short, no wider* than neck; lateral stripes on rows 2 and 3 (on neck), occasionally lower part of row 4, often bordered by narrow dark lines. Dark spots between stripes lacking or faintly indicated. *Scale rows 17.* Young 4 15/16–6¼ in. (12.4–15.9 cm) at birth. **SIMILAR SPECIES:** Butler's Gartersnake has a larger head and 19 scale rows. See also Common Gartersnake. **HABITAT:** Meadows, old fields. **RANGE:** Sw. NY, nw. PA, and extreme e. OH. Introduced in Pittsburgh, elsewhere in PA, and s.-cen. NY.

BUTLER'S GARTERSNAKE *Thamnophis butleri* **PL. 42**

15–20 in. (38–51 cm); record 29 in. (73.7 cm). Head *small,* slightly wider than neck; lateral stripes sometimes orange, on *row 3 and adjacent halves of rows 2 and 4* (on neck). Double row of black spots between stripes often indistinct. *Scale rows 19.* Young 5–7 in. (12.5–18 cm) at birth. **SIMILAR SPECIES:** See Short-headed and Common Gartersnakes. **HABITAT:** Open, prairielike areas. **RANGE:** Disjunct; OH and

s. ON to se. WI and IN. **REMARK:** Interbreeds with the Plains Gartersnake in se. WI.

BLACK-NECKED GARTERSNAKE *Thamnophis cyrtopsis* **PL. 42**

16–28 in. (41–71 cm); record 43 in. (109.2 cm). Head gray (usually bluish gray); *large black blotches or rows of large dark spots on sides of neck*; lateral stripes on scale rows 2 and 3, yellowish anteriorly, whitish or pale tan posteriorly. Snakes from the Big Bend area of TX mostly black or very dark brown between light stripes, frequently obscuring large black spots; those from extreme w. TX brown with smudgy black spots. Belly unmarked, whitish or slightly brownish or greenish. Usually 19 scale rows at midbody. Young brighter than adult; 8–10½ in. (20.2–26.5 cm) at birth. **SIMILAR SPECIES:** Checkered Gartersnake usually has 21 scale rows at midbody, lateral stripes only on scale row 3 near head. See also Common Gartersnake. **HABITAT:** Desert flats, rocky hillsides, forested mountains; often associated with limestone ledges, usually near water. **RANGE:** Edwards Plateau of s.-cen. TX to s. CO and se. UT south into Mex. **SUBSPECIES:** (1) **WESTERN BLACK-NECKED GARTERSNAKE** (*T. c. cyrtopsis*); large black blotches on sides of neck usually separated by a middorsal stripe, which is orange anteriorly but fades to dull yellow posteriorly. (2) **EASTERN BLACK-NECKED GARTERSNAKE** (*T. c. ocellatus*); rows of large dark spots on sides of neck invade light lateral stripes on rows 2 and 3, causing them to be wavy.

TERRESTRIAL GARTERSNAKE *Thamnophis elegans* **PL. 42**

18–30 in. (45.7–76 cm); record 38 3/16 in. (97 cm). Brown, brownish green, greenish buff, or gray; belly often heavily marked with black posteriorly. Light middorsal stripe fairly well defined, lateral light stripes on scale rows 2 and 3; dark lateral markings *rounded*; usually 8 upper labials, 21 scale rows at midbody. Young 7–8 in. (18–20.3 cm) at birth. **SIMILAR SPECIES:** (1) Plains Gartersnake has lateral stripes on scale rows 3 and 4. (2) Common Gartersnake in range has red skin between dorsal scales. (3) Checkered Gartersnake has lateral stripes confined to scale row 3 near head. **HABITAT:** Usually near water. **RANGE:** W. SD and extreme w. OK west to CA and the Pacific Northwest. **SUBSPECIES: WANDERING GARTERSNAKE** (*T. e. vagrans*). **REMARK:** Evolutionary lineages do not correspond to currently recognized subspecies.

CHECKERED GARTERSNAKE *Thamnophis marcianus* **PL. 42**

18–24 in. (45.7–61 cm); record 42 13/16 in. (108.8 cm). Yellowish curves or triangles on sides of head behind mouth followed by *large dark blotches*; middorsal stripe usually with jagged edges; black squarish spots often invade light stripes, producing a *checkerboard pattern*; lateral stripes on row 3 near head, on rows 2 and 3 posteriorly. *Scale rows 21*. Young 6¼–9¼ in. (15.9–23.5 cm) at birth. **SIMILAR SPECIES:** Plains Gartersnake may have a suggestion of yellowish curves behind the head, but lateral stripes involve row 4 and the middorsal stripe usually is straight-edged. **HABITAT:** Grasslands, deserts, rarely far from water. **RANGE:** S. KS to se. CA and south into Mex. **SUBSPECIES: MARCY'S CHECKERED GARTERSNAKE** (*T. m. marcianus*).

WESTERN RIBBONSNAKE *Thamnophis proximus* **PL. 42**

20–30 in. (51–76 cm); record 49⅞ in. (126.8 cm). *Slender, long-tailed* (usually slightly less than one-third total length). Three stripes generally distinct; lateral stripes on rows 3 and 4. *Parietal spots large and in contact*. Belly unmarked. Upper labials usually 8. Young 9–11 13/16 in. (23–29.8 cm) at birth. **SIMILAR SPECIES:** (1) Eastern Ribbonsnake with parietal spots, if present, faint and not in contact. (2) Gartersnakes with lateral stripes on rows 3 and 4 have tails generally less than one-fourth total length. (3) Graham's Crayfish Snake and swampsnakes have a divided cloacal and usually have bold ventral markings. **HABITAT:** Streams, ditches, lakes, ponds, cattle tanks, rarely far from water. **RANGE:** WI and IN to LA west to e. NM and south into Mex. **SUBSPECIES:** (1) **ORANGE-STRIPED RIBBONSNAKE** (*T. p. proximus*); dorsum black, narrow orange middorsal stripe. (2) **ARID LAND RIBBONSNAKE** (*T. p. diabolicus*); dorsum olive gray to olive brown, middorsal stripe orange, ventrolateral stripes brown. (3) **GULF COAST RIBBONSNAKE** (*T. p. orarius*); dorsum olive brown, broad gold middorsal stripe. (4) **RED-STRIPED RIBBONSNAKE** (*T. p. rubrilineatus*); dorsum olive brown to olive gray, red middorsal stripe sometimes orange on neck (rarely throughout), nar-

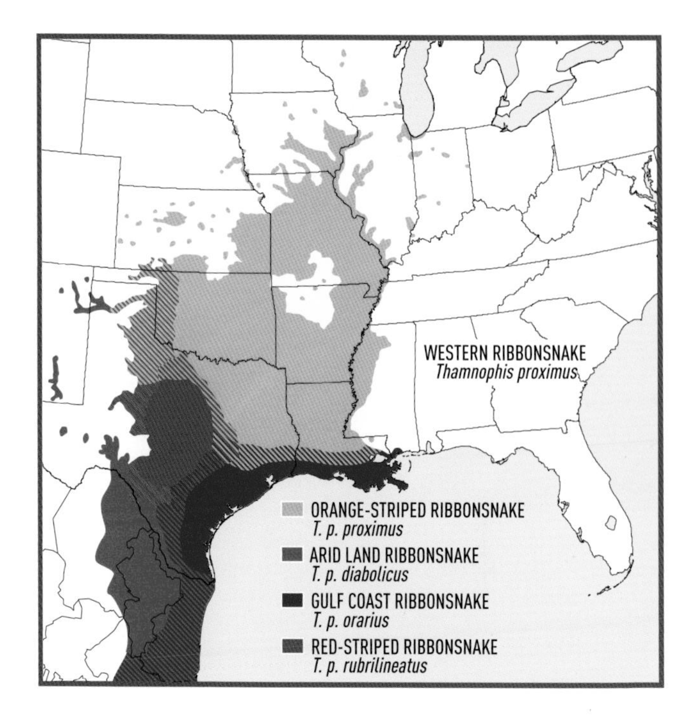

row dark ventrolateral stripes sometimes present. **REMARK:** Interbreeds in some areas with the Eastern Ribbonsnake.

PLAINS GARTERSNAKE *Thamnophis radix* **PL. 43, FIG. 195**

15–28 in. (38–71 cm); record 43 in. (109.3 cm). Brown, greenish (rarely reddish), occasionally very dark with obscured markings. Middorsal stripe bright yellow or orange; pale lateral stripes sometimes greenish or bluish, on scales rows 3 and 4; double alternating rows of black spots between stripes, row of black spots *below* lateral stripes. Black bars on lips; rows of dark, often poorly defined spots along sides of belly. Maximum 21 scale rows. Young 5¹⁵⁄₁₆–7½ in. (15.1–19 cm) at birth. **SIMILAR SPECIES:** (1) Most gartersnakes in range have 19 scale rows and lateral stripes involving scale

FIG. 195. *As its name implies, the Plains Gartersnake* (Thamnophis radix) *ranges far into open prairies, but is typically encountered near bodies of water. The diagnostic characters of this species (bright orange middorsal stripe, pale lateral stripes on scale rows 3 and 4, double alternating rows of black spots between the stripes, a row of black spots below the lateral stripes, and black bars on the lips) are clearly evident in this individual.*

row 2. (2) Checkered Gartersnake has light curved bands behind the mouth followed by broad dark blotches. (3) Ribbonsnakes are slender, long-tailed (at least one-fourth total length). **HABITAT:** Open prairies, often near ponds, sloughs, river valleys. **RANGE:** Nw. IN to s. AB south to n. TX and ne. NM; disjunct population in OH. **REMARK:** Interbreeds with Butler's Gartersnake in se. WI.

EASTERN RIBBONSNAKE *Thamnophis sauritus* **PL. 43**

18–26 in. (45.7–66 cm); record 40¹⁵/₁₆ in. (104 cm). *Slender, long-tailed* (usually slightly less than one-third total length). Lateral stripes on rows 3 and 4. *Parietal spots, if present, faint and not in contact.* Lips unpatterned; normally 7–8 upper labials. Belly plain yellowish or greenish. Young 7¼–9 in. (18.4–23 cm) at birth. **SIMILAR SPECIES:** See Western Ribbonsnake. **HABITAT:** Seldom far from streams, ponds, marshes, bogs, swamps. **RANGE:** WI to NS south through the FL Keys, west to the Mississippi R. and north to IN; increasingly patchy, especially in the Northeast. **SUBSPECIES:** (1) **Common Ribbonsnake** (*T. s. sauritus*); 3 bright stripes contrast with dark ground color; stripes normally yellow, but middorsal stripe sometimes with an orange or greenish tinge; double rows of black spots between stripes often ob-

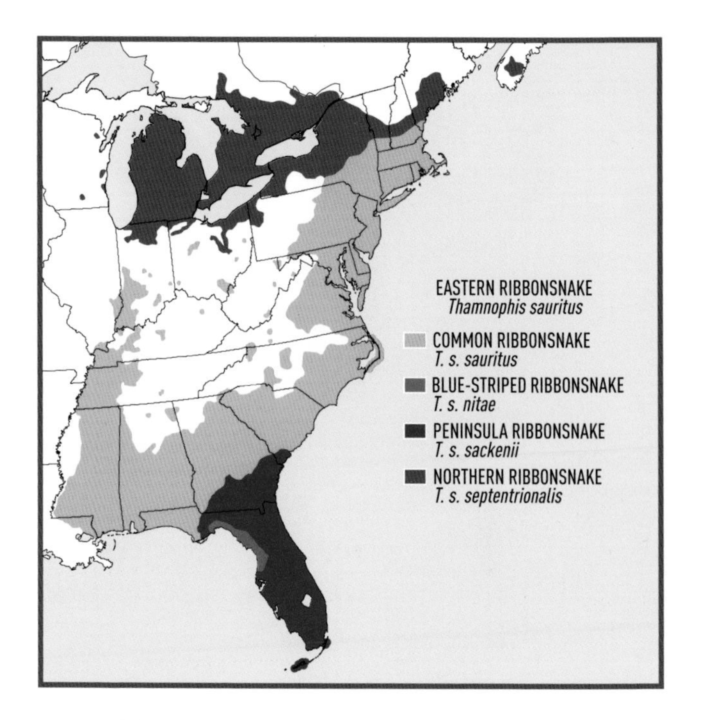

scure. (2) **Blue-striped Ribbonsnake** (*T. s. nitae*); black or dark brown with narrow sky blue or bluish white lateral stripes; obscure middorsal stripe sometimes present; 8 upper labials. (3) **Peninsula Ribbonsnake** (*T. s. sackenii*); tan or brown, pattern often obscure, narrow light lateral stripes, middorsal stripe less distinct, occasionally absent or represented only by a short line on neck; often 8 upper labials; snakes from the Lower FL Keys with only 7 upper labials, distinct yellow or orange middorsal stripe bordered by narrow black stripes. (4) **Northern Ribbonsnake** (*T. s. septentrionalis*); smaller and darker, dorsum black or dark brown, yellow middorsal stripe often partly obscured by brown pigment; tail usually less than one-third total length. **REMARK:** Interbreeds in some areas with the Western Ribbonsnake.

COMMON GARTERSNAKE *Thamnophis sirtalis* **PL. 43**

18–26 in. (45.7–66 cm); record 54 in. (137.2 cm). Black, dark brown, green, or olive, either stripes or spots predominate. Usually 3 yellowish stripes (sometimes brownish, greenish, bluish) and double rows of alternating black spots between stripes, spots sometimes invading stripes; occasionally stripeless or uniformly black; lateral stripes usually *confined to rows 2 and 3*. Belly greenish or yellowish, with 2 rows of

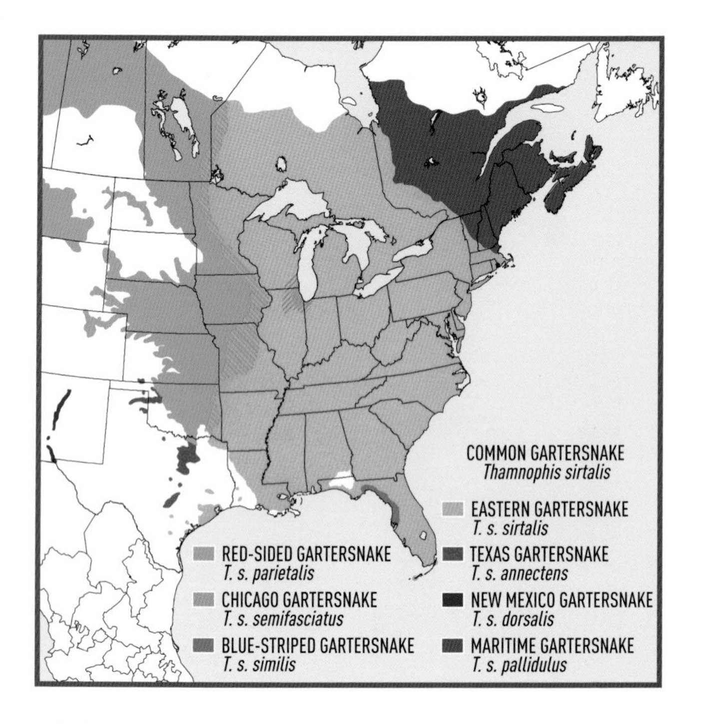

indistinct black spots partially hidden under overlapping portions of ventrals. Red or orange skin between scales especially in western parts of the range. *Scale rows 19*. Young 5–9 in. (12.5–23 cm) at birth. **SIMILAR SPECIES:** (1) Most other gartersnakes in range have lateral stripes involving row 4, at least on the neck. (2) Short-headed Gartersnake has a small head and 17 scale rows. (3) Black-necked Gartersnake has large black blotches or rows of large black spots on sides of the neck. (4) Checkered Gartersnake has light curved bands behind the mouth followed by broad dark blotches. (5) Lined Snake has a double row of black half-moons on belly. (6) Graham's Crayfish Snake, Queensnake, Glossy Swampsnake, and Saltmarsh Watersnake have a divided cloacal and most have a strongly patterned belly. (7) North American brownsnakes lack light lateral stripes and loreal scales. **HABITAT:** Meadows, marshes, woodlands, pine flatwoods, hillsides, prairie swales, along streams and drainage ditches, sometimes in city lots, parks, residential areas; in western portions of the range, snakes follow intermittent watercourses far into arid areas. **RANGE:** ON through FL and some Keys west to the Pacific Coast; isolated populations in NM and adjacent w. TX. **SUBSPECIES:** (1) **Eastern Gartersnake** (*T. s. sirtalis*); usually without red skin between scales. (2) **Texas Gartersnake** (*T. s. annectens*); usually has a broad orange middorsal stripe; lateral stripes on row 3 plus adjacent parts of rows 2 and 4 on anterior third of body. (3) **New Mexico Gartersnake** (*T. s. dorsalis*); red between scales subdued or reduced, stripes prominent, general appearance often greenish; belly bluish or brownish, unmarked or with dark spots along edges; young occasionally rusty orange. (4) **Maritime Gartersnake** (*T. s. pallidulus*); gray, tan, or yellow middorsal stripe absent or present only anteriorly and poorly developed; whitish, gray, or tan lateral stripes more distinct, but often merging into ground color below; dorsolateral ground color cinnamon brown, yellowish olive, or olive gray with alternating rows of black or dark brown spots forming a checkered pattern; belly whitish anteriorly, dusky gray posteriorly. (5) **Red-sided Gartersnake** (*T. s. parietalis*); olive, brown, or black, red or orange bars vary in size and intensity, dark pigment sometimes invades much of belly; middorsal and lateral stripes yellow, orange-yellow, bluish, or greenish. (6) **Chicago Gartersnake** (*T. s. semifasciatus*); lateral stripes regularly interrupted anteriorly by vertical black crossbars formed by fusion of black spots above and below light stripes; one or more narrow black lines may cross nape. (7) **Blue-striped Gartersnake** (*T. s. similis*); black or dark brown with sky blue or bluish white lateral stripes, middorsal stripe often obscure. **REMARK:** Probably a complex of species with lineages not corresponding to currently recognized subspecies.

LINED SNAKES: *Tropidoclonion*

The only species in this genus occurs largely in our area.

LINED SNAKE *Tropidoclonion lineatum* **PL. 44**

8¾–15 in. (22.4–38 cm); record 21½ in. (54.4 cm). Gray or grayish tan, light middorsal stripe whitish, yellow, orange, or light gray; lateral stripes on rows 2 and 3. Belly with a *double row of bold black half-moons*. Scales *keeled*; cloacal *undivided*. Young 3¾–4¾ in. (9.5–12.1 cm) at birth. **SIMILAR SPECIES:** (1) Some gartersnakes have dark spots on the belly, but never as large, dark, or clearly defined as in the Lined Snake. (2) Glossy Swampsnake has a divided cloacal and similar ventral markings, but a dorsal pattern, if present, of dark stripes. (3) Graham's Crayfish Snake has a divided cloacal. **HABITAT:** Open prairies, sparsely timbered areas, city lots, parks, cemeteries, abandoned trash dumps. **RANGE:** Disjunct; s. WI to se. SD south to NM and TX.

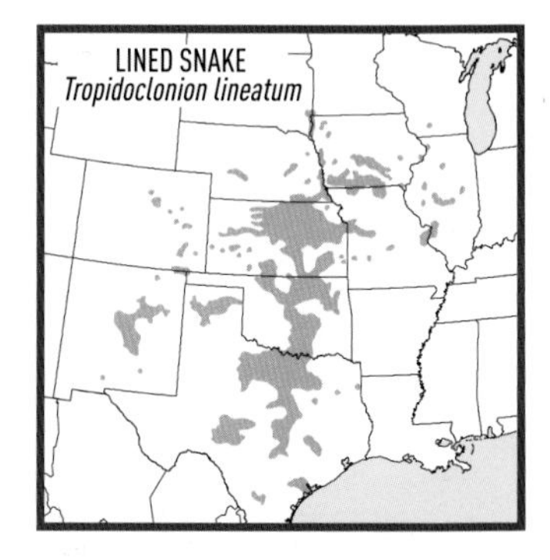

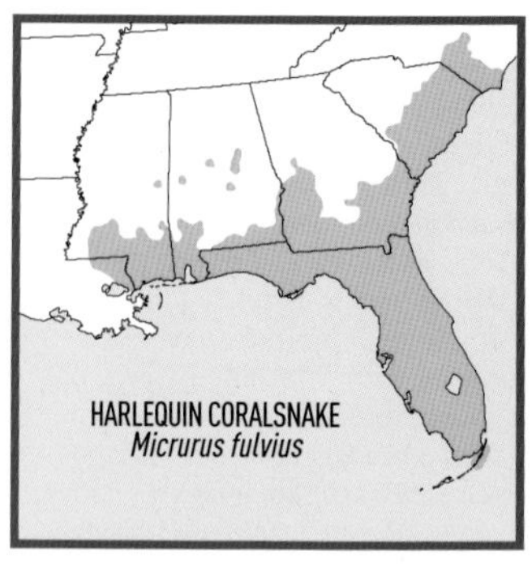

CORALSNAKES: Elapidae

AMERICAN CORALSNAKES: *Micrurus*

Most of the more than 350 species in the family are in the Eastern Hemisphere. ALL ARE DANGEROUSLY VENOMOUS. American coralsnakes range from the southern United States to central Argentina. Coralsnakes in the United States have rings completely encircling the body, and the *red and yellow rings are in contact* (Fig. 159, p. 330). Hatchlings 6½–9⁷⁄₁₆ in. (16.5–23.9 cm).

HARLEQUIN CORALSNAKE *Micrurus fulvius* **PL. 44**

20–30 in. (51–76 cm); record 47½ in. (120.7 cm). Tip of snout black, followed by a yellow band across the head. Black ring on neck *does not touch* parietals. Red rings dotted or spotted with black, dark markings often concentrated into pairs of large black spots in each red ring (lacking in some snakes from s. FL). **SIMILAR SPECIES:** Scarletsnake and Scarlet Kingsnake have a predominately red snout and red and yellow rings separated by black. **HABITAT:** Pinewoods, open, dry, or sandy areas to moister pond and lake borders, hammocks. **RANGE:** Se. NC south through FL and the Upper Keys and west to MS and e. LA. **REMARK:** Also called the Eastern Coralsnake.

TEXAS CORALSNAKE *Micrurus tener* **PL. 44, FIG. 196**

20–30 in. (51–76 cm); record 47¾ in. (121.7 cm). Tip of snout black, followed by a yellow band across the head. Black ring on neck *extends onto* posterior tip of parietals. Red rings dotted or spotted with black. **SIMILAR SPECIES:** (1) Scarletsnake and Western Milksnake have red on the snout and red and yellow rings are separated by black. (2) Tamaulipan Milksnake has a black head, but red and yellow rings are separated by black. (3) Western Groundsnake lacks a black snout and

FIG. 196. *A hatchling Texas Coralsnake* (Micrurus tener) *with an egg from the same clutch. Coralsnakes are the only venomous snakes in our area that lay eggs. Note that the red and yellow rings are in contact. In the South, children are taught that "red and yellow kill a fellow."*

broad black rings, and the belly is unmarked. (4) Long-nosed Snake has a red snout and only one row of subcaudals, is prominently speckled, and rings are incomplete and do not extend onto the belly. **HABITAT:** Moist lowland areas, often along streams or canals, cedar brakes, rocky canyons, rocky hillsides. **RANGE:** Sw. AR and LA through TX south into Mex. **SUBSPECIES: TEXAS GULF-COAST CORALSNAKE** (*M. t. tener*). **REMARK:** Until recently considered a subspecies of the Harlequin Coralsnake.

VIPERS: Viperidae and PITVIPERS: Crotalinae

More than 220 species of pitvipers range from southern Canada to southern Argentina and from extreme southeastern Europe through southern and central Asia to Malaysia. ALL ARE DANGEROUSLY VENOMOUS. *Heads are distinctly wider than the neck, deep facial pits are present between the eyes and nostrils, scales under the tail are in a single row,* at least anteriorly, and the pupils are *vertically elliptical* (Fig. 166, p. 352).

AMERICAN MOCCASINS: *Agkistrodon*

Eight currently recognized species occur in North and Central America. Scales are *weakly keeled*, and the cloacal is *undivided*. Young have a yellow or greenish tailtip.

EASTERN COPPERHEAD *Agkistrodon contortrix* **PL. 45, FIG. 197**

BROAD-BANDED COPPERHEAD *Agkistrodon laticinctus* **PL. 45**

24–36 in. (61–90 cm); record 53 in. (134.6 cm). Pinkish, tan, grayish tan, pale reddish brown; top of head coppery red; distinct reddish brown or brown crossbands often with dark edges. Crossbands of the Eastern Copperhead often *hourglass*-shaped (when viewed from above, wide on sides and narrow middorsally), sometimes broken or offset middorsally, especially in southern populations; often edged in white in the Midwest; small dark spots sometimes present between crossbands. Crossbands uniformly broad in the Broad-banded Copperhead, with pale areas beneath each band in populations from w. TX and Mex. Belly whitish, very pale tan, or grayish tan with dark rounded spots along sides or with gray, brown, or black markings producing a dusky, marbled, clouded, or distinct pattern, the last especially in populations from w. TX. Juvenile usually paler than adult; *narrow* dark lines through eyes divide the dark head from paler lips. Young 7–10$^{13}/_{16}$ in. (18–27.5 cm) at birth. **SIMILAR SPECIES:** (1) Eastern Milksnake has markings widest middorsally, belly usually a black-and-white checkerboard, scales smooth. (2) Yellow-bellied Kingsnake has middorsal blotches alternating with smaller lateral blotches, scales smooth. (3) Hog-nosed snakes have an upturned snout. (4) Watersnakes have strongly keeled scales. (5) Ratsnakes with patterns have large dark middorsal blotches alternating with smaller lateral

FIG. 197. *The forked tongue of an Eastern Copperhead* (Agkistrodon contortrix) *gathers chemical information and transfers it to a sensory organ in the roof of the mouth. The facial pit in front of the eye is heat-sensitive, helping the snake find and strike prey that is either warmer or cooler than its surroundings.*

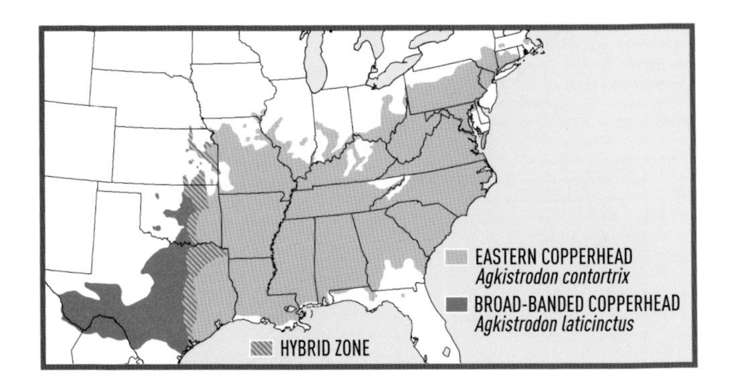

blotches. See also cottonmouths. **HABITAT:** Hilly deciduous forests, usually near rock outcrops, floodplains, edges of swamps in the South, mesic situations near water in the arid West. **RANGE:** Eastern Copperhead: MA to n. FL and west to e. KS, OK, and TX; Broad-banded Copperhead: cen. OK through w. TX and into Mex. Broad hybrid zone from w. MO and e. KS south through TX. **REMARK:** Until recently considered a single, variable species.

NORTHERN COTTONMOUTH *Agkistrodon piscivorus* **PL. 45**

FLORIDA COTTONMOUTH *Agkistrodon conanti* **PL. 45, FIG. 198**

30–48 in. (76–122 cm); record 74½ in. (189.2 cm). Olive, brown, or black above, generally lighter below (often dark brown or black above and below in western portions of the range). Crossbands obvious to absent, often with dark borders and lighter centers. Head markings distinct in the Florida Cottonmouth to essentially absent in the Northern Cottonmouth; dark cheek stripes, when present, bordered above and below by narrow light lines, a pair of dark stripes at front of lower jaw, a pair of dark vertical lines at tip of snout. Pattern on head or body most evident in juvenile; old adults often dark and unpatterned. Juvenile with light-centered dark brown to reddish brown crossbands; *broad dark bands* through eyes; 8–13 in. (20.3–33 cm) at birth. **SIMILAR SPECIES:** Several nonvenomous North American watersnakes are sometimes difficult to distinguish from cottonmouths. The latter tend to swim with most of the body at the surface (watersnakes usually swim with only the head and a small portion of the body above the surface); cottonmouths often stand their ground or crawl slowly away (watersnakes usually flee quickly or drop into water); cottonmouths vibrate the tail when excited (watersnakes do not); aroused cottonmouths often throw the head back and gape, revealing the whitish interior of the mouth (hence the name "cottonmouth"). (1) Copperheads are lighter, more reddish, retain a banded pattern throughout life, often have hourglass-shaped crossbands and narrow dark lines through the eyes. (2) North American watersnakes have strongly keeled scales, a divided cloacal, double row of scales under tail, and no facial pits.

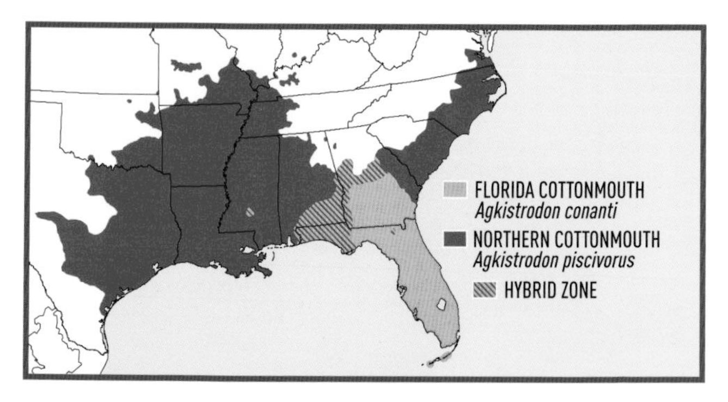

FIG. 198. *A young Florida Cottonmouth* (Agkistrodon conanti) *in a defensive posture. When alarmed, cottonmouths arrange themselves in a tight coil and open the mouth to reveal the cottony white lining. Despite their alarming behavior, cottonmouths are not overly aggressive and generally will flee if given the chance.*

HABITAT: Swamps, sloughs, bayous, rice fields, ponds, streams, marshes, river bottoms, lowland floodplains, tidal stream courses, dune and beach areas, clear upland brooks, drainage ditches in some cities, brackish waters, sometimes salt marshes; may forage in upland areas far from water. **RANGE:** Northern Cottonmouth: se. VA west to TX and north along the Mississippi R. to s. IL and IN; Florida Cottonmouth: peninsular FL. Broad hybrid zone across s. GA west to MS. **REMARK:** Until recently considered a single, variable species.

RATTLESNAKES: *Crotalus* and *Sistrurus*

Rattlesnakes occur only in the Western Hemisphere. Thirty-six currently recognized species in the genus *Crotalus* and three species of *Sistrurus* range from southern Canada to northern Argentina and Uruguay. New segments are added to rattles each time a snake sheds (Fig. 200, facing page). Rattlesnakes in the genus *Crotalus* have heads covered largely with small scales, whereas those in the genus *Sistrurus* have nine large scales on top of the head (Fig. 167, p. 354). Scales are *keeled*, the cloacal is *undivided*.

FIG. 199. *An Eastern Diamond-backed Rattlesnake* (Crotalus adamanteus) *in an iconic defensive posture with elevated head and body ready to strike while the rattle sounds an alarm.*

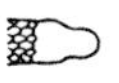

Button

Young rattle, tapering in size, button still retained

Old rattle, terminal joints broken off

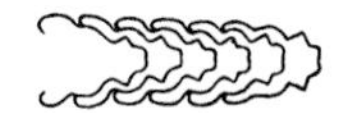

Cross section showing interlinking arrangement

FIG. 200. *Rattles, thought to have evolved to deter large grazing mammals from stepping on snakes, consist of loosely interlocking horny segments that strike against one another to produce a buzzing sound when the tail is vibrated.*

EASTERN DIAMOND-BACKED RATTLESNAKE

Crotalus adamanteus **PL. 46, FIG. 199**

33–72 in. (84–183 cm); record 99 in. (251.5 cm). Olive, brown, or almost black; dark brown or black diamond-shaped dorsal marks strongly outlined by rows of cream-colored or yellowish scales. Two prominent light lines on face, vertical light lines on snout. Young 12–15 1/16 in. (30.6–38.1 cm) at birth. **SIMILAR SPECIES:** Timber Rattlesnake has dark, narrow, often V-shaped dorsal crossbands. **HABITAT:** Flatwoods, hammocks, coastal scrub, mixed pine-hardwood successional woodlands, abandoned farms and fields. **RANGE:** Se. NC through FL and the Keys to extreme e. LA.

WESTERN DIAMOND-BACKED RATTLESNAKE *Crotalus atrox*

PL. 47, FIG. 168

30–72 in. (76–183 cm); record 92 in. (233.7 cm). Brown or gray (sometimes with reddish or yellowish tones); diamonds, often darker along margins, not clearly defined; head and body may have a dusty appearance. Light stripes behind eyes meet mouthline in front of angles of the jaw (Fig. 168, p. 356). Tail with *black and white (or light gray) rings*; proximal rattle segment uniformly colored. Juvenile has more sharply defined diamonds; 9½–14 in. (24–36 cm) at birth. **SIMILAR SPECIES:** (1) Mohave Rattlesnake has narrower black tail rings, light lines behind

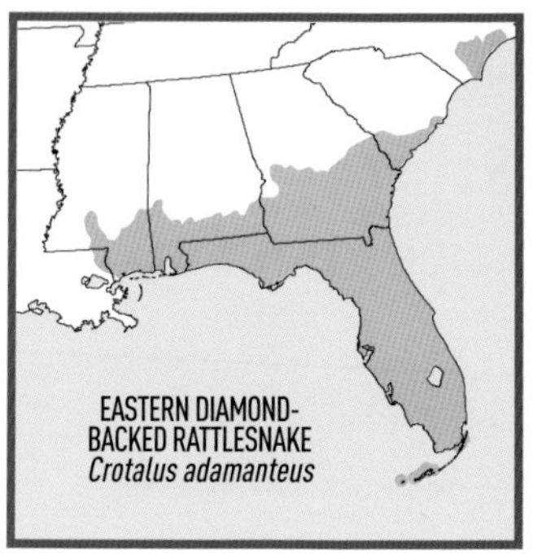

eyes meeting the mouthline, diamonds largely unicolored, outlined by white scales, lower half of proximal rattle segment paler than upper half, enlarged scales on snout and between supraoculars. (2) Prairie Rattlesnake has light-bordered dorsal blotches not diamond-shaped, lacks black and white tail rings, and light lines behind eyes pass above corners of the mouth. **HABITAT:** Arid and semiarid regions from plains to mountains, sandy flats to rocky uplands, including rocky cliffs, canyons, open desert, grasslands, shrublands, woodlands, open pine forests, river bottoms, coastal islands. **RANGE:** W.-cen. AR to s. CA south into Mex. **REMARKS:** Responsible for more serious snakebites and fatalities than any other North American snake. Sometimes called a coontailed rattler.

TIMBER RATTLESNAKE *Crotalus horridus* **PL. 46, FIG. 168**

36–60 in. (90–152 cm); record 74½ in. (189.2 cm). "Yellow" snakes yellow or brown with black or dark brown, usually narrow crossbands; crossbands V-shaped or broken anteriorly to form rows of dark dorsal and lateral spots. "Black" snakes with heavy stippling of black or very dark brown; completely black individuals not unusual. Yellow and black variants most common in the northeastern uplands. Snakes west of the Mississippi R. north of the Ozarks gray, yellow, tan, or brown with black or dark brown crossbands, with a broad rusty middorsal stripe. Snakes in southern portions of the range pinkish buff, pale gray, or beige with black crossbands, a broad reddish middorsal stripe splitting crossbands anteriorly. Both western and southern snakes have broad dark stripes behind eyes (Fig. 168, p. 356). Tail black. Juvenile has distinct crossbands; 7¾–16 in. (19.7–41 cm) at birth. **SIMILAR SPECIES:** (1) Eastern and Western Diamond-backed Rattlesnakes have diamond-shaped dorsal markings. (2) Eastern Massasauga and Pygmy Rattlesnake have 9 large scales on top of head. **HABITAT:** Timbered areas near rocky outcroppings, ledges, rockslides in the Northeast; rocky bluffs, bluff prairies, open oak-hickory forests in the Midwest; river bottoms, floodplains, swampy areas, cane thickets in the South; western populations follow wooded

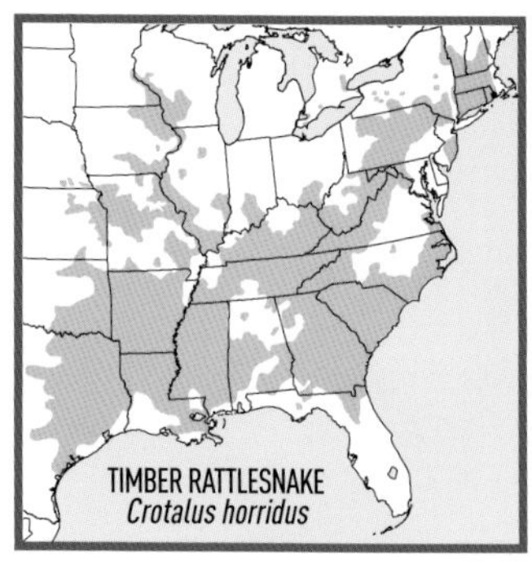

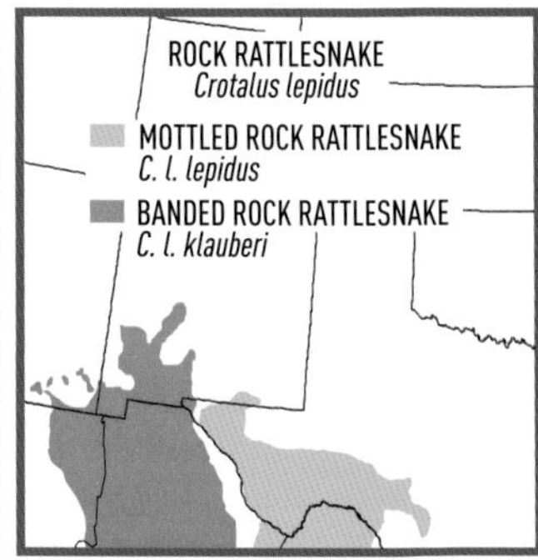

stream valleys into prairies. **RANGE:** New England and NY south to n. FL west to se. MN and cen. TX. Many populations, especially in Northeast, extirpated. **REMARKS:** Might be a species complex. Also called a velvet-tailed or canebrake rattler.

ROCK RATTLESNAKE *Crotalus lepidus* **PL. 47**

15–24 in. (38–61 cm); record 32⅝ in. (82.9 cm). Gray, bluish gray, greenish gray, tan, brown, or pinkish; narrow, widely spaced crossbars often less distinct anteriorly. Crossbars light-bordered and distinct or obscured by profuse stippling. Dark stripes from eyes to angles of the jaw distinct to absent. Young 6½–8½ in. (16.5–21.5 cm) at birth. **SIMILAR SPECIES:** Eastern Black-tailed Rattlesnake has a black tail and wider, closely spaced, light-centered crossbands vaguely diamond-shaped anteriorly. **HABITAT:** Mountainous areas, talus slopes, gorges, rimrock, limestone outcrops, rocky streambeds, mesquite grasslands, rocky desert flats and canyons, open tropical forests in Mex. **RANGE:** S.-cen. and w. TX to se. AZ south into Mex. **SUBSPECIES:** (1) **Mottled Rock Rattlesnake** (*C. l. lepidus*); profuse dark stippling may form pseudo-crossbands between dark crossbars; snakes from the Big Bend and Davis Mtns. usually pink or buff, with 18–19 dark crossbars; those from the Edwards Plateau-Stockton area usually gray, with 22–23 dark crossbars inconspicuous anteriorly but progressively more prominent toward the tail; dark stripes from eyes to angles of the jaw distinct. (2) **Banded Rock Rattlesnake** (*C. l. klauberi*); widely spaced black or brown crossbars contrast strongly with greenish to bluish green color of male or gray to bluish gray color of female; little dark spotting between crossbars; dark stripes from eyes to angles of the jaw vague to absent.

EASTERN BLACK-TAILED RATTLESNAKE *Crotalus ornatus* **PL. 47**

30–42 in. (76–106.7 cm); record 52 in. (132.1 cm). Cream, yellow, grayish, olive, greenish, or dark rust; black or very dark brown irregular dorsal crossbands edged with whitish scales *enclosing one or 2*

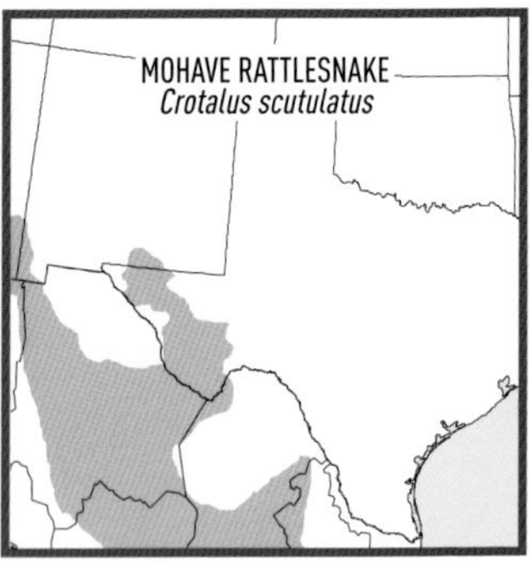

patches of lighter colored scales; snakes on old lava flows and in upland areas often dark. Crossbands anteriorly to roughly midbody vaguely diamond-shaped. Scales *unicolored* (not partly dark and partly light). *Tail solid black*. Juvenile with dark crossbands visible on tail; 6½–11¾ in. (16.5–29.7 cm) at birth. **SIMILAR SPECIES:** See Rock Rattlesnake and Texas Lyresnake. **HABITAT:** Rock piles and slides, wooded canyons, vicinity of cliffs. **RANGE:** NM and w. TX south into Mex.; disjunct populations in cen. TX. **REMARK:** Until recently thought to be conspecific with the Western Black-tailed Rattlesnake (*C. molossus*), which occurs west of our area.

MOHAVE RATTLESNAKE *Crotalus scutulatus* **PL. 47, FIG. 168**

24–36 in. (61–90 cm); record 48⅝ in. (123.6 cm); a frequently cited record of 54 in. (137.3 cm) cannot be verified. Gray, greenish, or brownish; whitish scales surround largely unicolored diamond-shaped dorsal markings; sometimes quite pale, often relatively dark. Light lines behind eyes pass *above and beyond* corners of the mouth (Fig. 168, p. 356). *Tail ringed* with black and white (or light gray), *black rings much narrower than white rings*. Lower half of proximal rattle segment *paler* than upper half. *Enlarged scales* on snout and between supraoculars. Young 10–11 in. (25.4–28 cm) at birth. **SIMILAR SPECIES:** See Western Diamond-backed Rattlesnake. **HABITAT:** Uplands, lower mountain slopes, including barren deserts, grasslands, open woodlands, scrublands, usually not in rocky terrain or densely vegetated areas. **RANGE:** Extreme w. TX to s. NV and CA south into Mex. **SUBSPECIES: NORTHERN MOHAVE RATTLESNAKE** (*C. s. scutulatus*).

PRAIRIE RATTLESNAKE *Crotalus viridis* **PL. 47, FIG. 168**

35–45 in. (89–114 cm); record 59⅝ in. (151.5 cm). Greenish gray, olive green, or greenish brown (sometimes light brown or yellowish); white-bordered dark brown blotches broad anteriorly, narrower and joining with lateral markings to form crossbands near tail. Two oblique light lines on head, those behind eyes passing *above* corners of the mouth

(Fig. 168, p. 356). Young 8½–11 in. (21.5–28 cm) at birth. **SIMILAR SPECIES:** See Western Diamond-backed Rattlesnake. **HABITAT:** Prairies, arid basins, wooded mountains. **RANGE:** Extreme w. IA to MT south through NM and w. TX into Mex. **REMARK:** Until recently considered conspecific with several species ranging to and along the Pacific Coast.

EASTERN MASSASAUGA *Sistrurus catenatus* **PL. 46, FIG. 201**

18½–30 in. (47.2–76 cm); record 39½ in. (100.3 cm). Gray or brownish gray (occasional adults, especially in the Northeast, entirely black above and below); row of large black to light brown middorsal blotches and 3 lateral rows of smaller dark spots. Belly *black* with irregular light marks (Fig. 201, p. 444). Juvenile paler than adult; 7–9 15/16 in. (18–25.3 cm) at birth. **SIMILAR SPECIES:** (1) Western Massasauga has a light belly with dark marks. (2) Pygmy Rattlesnake has a slender tail, tiny rattle, and usually a reddish middorsal stripe. **HABITAT:** Wetlands (bogs, fens, swamps, marshes, peatlands, wet meadows, floodplains), open savannas, prairies, old fields, dry woodlands; in wetlands most of the year, but adjacent drier uplands in summer. **RANGE:** Disjunct; w. NY and s. ON southwest to w. IL and e. IA (extirpated populations in e. MO presumably belong to this species).

PYGMY RATTLESNAKE *Sistrurus miliarius* **PL. 46**

15–21 in. (38–54 cm); record in captivity 32 13/16 in. (83.2 cm). Brown to light gray or grayish brown (sometimes reddish); middorsal and 1–3 lateral rows of dark spots, often with a *reddish brown middorsal stripe*; dark stippling may obscure markings, especially on the head of snakes in southern parts of the range. *Tail very slender*; *rattle tiny*. Belly whitish to cream, moderately flecked with brown or gray or heavily marked with black or dark brown, sometimes uniformly brownish black, at least posteriorly. Young 3–7½ in. (7.8–19 cm) at birth. **SIMILAR SPECIES:** See Eastern Massasauga. **HABITAT:** Moist or wet lowlands, wet prairies, savannas, pastures, palmetto-pine flatwoods, swamps, borders of cypress ponds, near lakes and marshes,

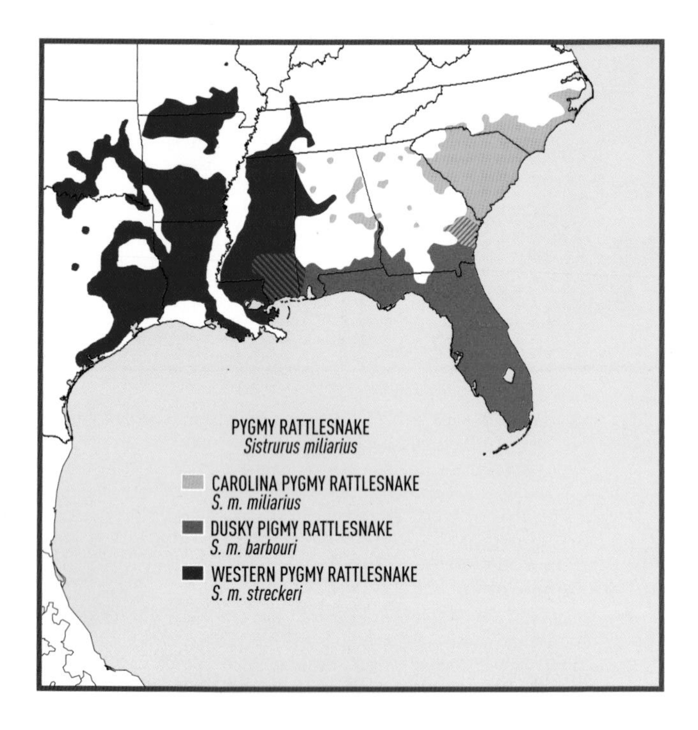

along rice-field canals and roadside ditches, hardwood-dominated floodplains, sandhills, mixed pine-hardwood forest, hilly second-growth forests, scrub pinewoods. **RANGE:** E. NC through FL west to e. OK and TX. **SUBSPECIES:** (1) **Carolina Pygmy Rattlesnake** (*S. m. miliarius*); brown or light gray (frequently reddish in e. NC), 1–2 lateral rows of distinct dark spots, usually with a reddish middorsal stripe; venter cream, moderately flecked with brown or gray. (2) **Dusky Pygmy Rattlesnake** (*S. m. barbouri*); dark stippling may largely obscure markings, especially on head; reddish brown middorsal stripe frequently prominent, sometimes subdued or only on neck; usually 3 lateral rows of dark spots; venter whitish, heavily blotched or flecked with black or dark brown, sometimes uniformly brownish black, at least posteriorly. (3) **Western Pygmy Rattlesnake** (*S. m. streckeri*); pale grayish brown; middorsal dark spots often irregular in shape, frequently forming short transverse bars; 1–2 conspicuous lateral rows of dark spots; the reddish middorsal stripe sometimes absent. **REMARK:** Often called a ground rattler.

WESTERN MASSASAUGA *Sistrurus tergeminus* **PL. 46, FIGS. 201, 202**

18½–30 in. (47.2–76 cm); record 34¾ in. (88.3 cm). Light gray or tan to grayish tan; middorsal row of dark to light brown blotches and 3 lat-

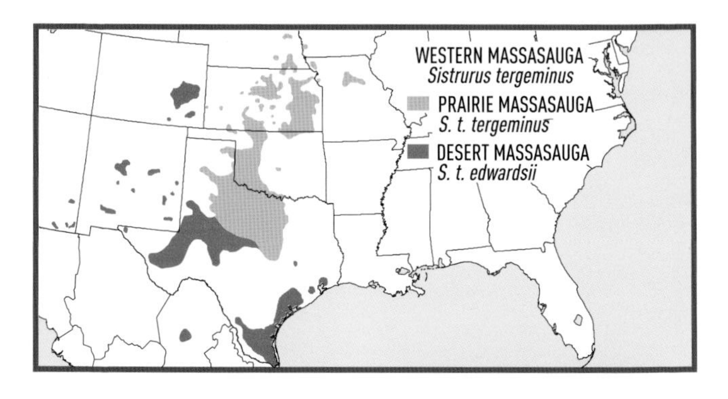

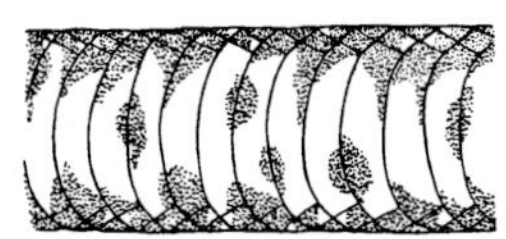

WESTERN (*S. tergeminus*)
Light in coloration

EASTERN (*S. catenatus*)
Chiefly black

FIG. 201. *Bellies of massasaugas* (Sistrurus)*.*

eral rows of smaller dark spots. Belly light, at least anteriorly, with dark marks varying in number and size (Fig. 201, above). Juvenile paler than adult; 6–8½ in. (15–21.6 cm) at birth. **SIMILAR SPECIES:** See Eastern Massasauga. **HABITAT:** Old fields, dry woodlands, grassy wetlands, rocky hillsides, mesquite and scrub plains, thornbrush, oak-grasslands, dry prairies, desert grasslands, coastal sand dunes, often near water. **RANGE:** Disjunct; N.-cen. MO and sw. IA southwest through cen. TX to the Gulf and west to extreme se. AZ. **SUBSPECIES:** (1) **Prairie Massasauga** (*S. t. tergeminus*); light gray or tan gray contrasting strongly with dark brown blotches; belly light, at least anteriorly, with dark marks varying in number and size; juvenile often pinkish ventrally and toward tailtip. (2) **Desert Massasauga** (*S. t. edwardsii*); light gray or tan gray with brown blotches; belly nearly white, virtually (or completely) unmarked. **REMARK:** Until recently considered a subspecies of the Eastern Massasauga.

FIG. 202. *The Western Massasauga* (Sistrurus tergeminus) *primarily inhabits open areas, often in dry habitats but frequently near water. However, populations at the eastern edge of the range in Missouri often are associated with grassy wetlands, where viable populations are concentrated in national and state wildlife refuges managed largely for waterfowl.*

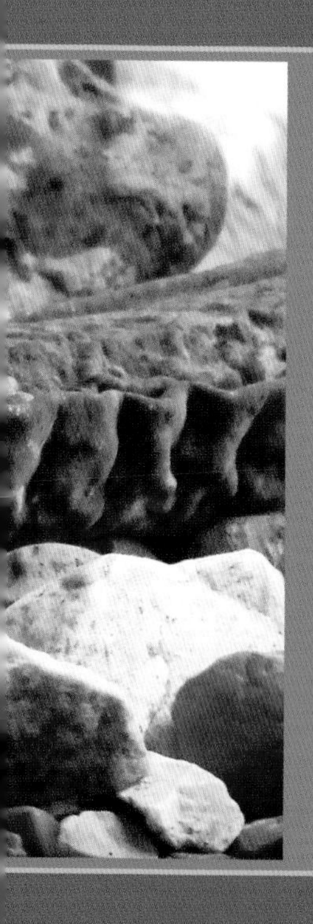

ACKNOWLEDGMENTS
GLOSSARY
REFERENCES
PHOTO CREDITS
PHOTO CAPTIONS
INDEX

We collectively thank everyone who contributed information used in the previous editions of this guide. Their help was and remains invaluable. Most were acknowledged by name in the first three editions, and we apologize again for anyone whose name was omitted.

Since the publication of the third edition of this guide in 1991 and the expanded version in 1998, we have received many comments with new distribution records, maximum sizes, and a wealth of other data. We are deeply grateful, but somewhat embarrassed to admit that space limitations preclude us from listing all of the names of those who wrote or called us.

We once again acknowledge Tom R. Johnson, whose contributions to the color plates in the third edition greatly enhanced that effort. We did not attempt to amend the plates for the current volume, instead relying on photographs to illustrate most of the species not represented there. Many of these are newly described taxa or species that recently have established breeding populations in the area covered by this guide. Photo credits are listed on p. 458, and we thank each person who graciously shared their images.

Errol D. Hooper Jr. provided several line drawings and illustrations new to this guide. In particular, his figures of map turtles (Figs. 92 and 93) and of Flat-headed, crowned, and black-headed snakes (Fig. 184) dramatically enhance the likelihood of correctly identifying those species.

Many herpetologists generously gave of their time to review species accounts, provide reference materials, and alert us to taxonomic revisions. We are particularly grateful to AJ Gutman and Joseph C. Mitchell, who critically read early drafts of the entire manuscript. Joe signed on for a second round, and he, Richard E. Daniel, and Travis W. Taggart read later revisions. Their collective comments greatly improved the final product. Others who contributed include Aaron M.

Bauer, Brian I. Crother, C. Kenneth Dodd Jr., Darrel R. Frost, H. Carl Gerhardt Jr., Matthew E. Gifford, S. Blair Hedges, Robert W. Henderson, John B. Iverson, Kenneth L. Krysko, Peter V. Lindeman, Jeffrey E. Lovich, John R. MacGregor, Walter E. Meshaka Jr., Leroy Nuñez, Sara Ruane, Laurie J. Vitt, and George R. Zug.

Travis W. Taggart prepared the maps for this book—a task considerably more difficult and time-consuming than we had anticipated—and Sarah Taggart prodded him to get them done. We consulted innumerable published and online references and solicited help from many people, who willingly shared their knowledge, experience, and, in some instances, unpublished data. Many people also invested countless hours helping us refine the maps. For their contributions, we are grateful to Aaron M. Bauer, Russell A. Blaine, Jeff Boundy, Jeffrey T. Briggler, Donald J. Brown, Gary S. Casper, Richard E. Daniel, C. Kenneth Dodd Jr., Arthur C. Echternacht, Brian Edmond, Kevin M. Enge, Vincent R. Farallo, Daniel D. Fogell, Paul Frese, Darrel R. Frost, J. Whitfield Gibbons, Matthew E. Gifford, Scott D. Gillingwater, Brad M. Glorioso, David M. Green, Craig Guyer, James H. Harding, Toby J. Hibbits, Troy D. Hibbits, Kelly J. Irwin, John B. Iverson, Robert L. Jones, Edmund D. Keiser, Bruce A. Kingsbury, Kenneth L. Krysko, Jennifer Y. Lamb, Michael J. Lanoo, Peter V. Lindeman, Jeffrey E. Lovich, John R. MacGregor, Debora L. Mann, Tom M. Mann, Walter E. Meshaka Jr., David A. Mifsud, Brian T. Miller, Joseph C. Mitchell, Randall Mooi, John J. Moriarty, Matthew L. Niemiller, Thomas K. Pauley, Christopher A. Phillips, John G. Phillips, Kory G. Roberts, Curtis J. Schmidt, A. Floyd Scott, Christopher A. Sheil, Donald B. Shepard, Greg Sievert, Geoffrey R. Smith, Michael A. Steffen, Stanley E. Trauth, Renn Tumlison, R. Wayne Van Devender, Terry L. Vandeventer, Terry J. VanDeWalle, Laurie J. Vitt, Timothy Warfel, Gregory J. Watkins-Colwell, Erick R. Wild, and Robert A. Young.

We also thank Lisa A. White, Director of Guidebooks at Houghton Mifflin Harcourt. She and the staff at HMHCo were invariably pleasant and immensely patient when dealing with our myriad queries during the preparation of this guide. Beth Burleigh Fuller, Senior Production Editor, supervised the phenomenally complex production process. Elizabeth C. Pierson was our copy editor and her attention to detail exceeded all expectations. Brian Moore, Associate Art Director, and Eugenie Delaney laid out the book and were incredibly tolerant of the many last-minute changes. Laney Everson and Jennifer Puk, editorial associates, helped in many ways. Taryn Roeder, publicist, and Katrina Kruse, marketing manager, have worked diligently to make this project a successful endeavor.

Finally, we extend credit to the hundreds of herpetologists whose published and online works we consulted, often repeatedly. We are deeply grateful to them and the legacy of their collected works.

Robert Powell
Roger Conant
Joseph T. Collins

On a more personal level, I wish to acknowledge my coauthors, to whom this guide is dedicated and who are not here to see its completion. Like many young herpetologists, I was inspired to become a herpetologist by Roger's early editions of this guide, and I'll never forget how gracious he was to an aspiring neophyte when we first met many years ago. I mourn the loss of his often lyrical prose and keen insights, nearly all of which fell victim to the huge increase in the number of species and the need to keep this book small enough to take into the field.

I first met Joe when he was collection manager in the herpetology department at the University of Kansas Museum of Natural History (now the KU Biodiversity Institute & Natural History Museum). Even then he went out of his way to help a young colleague just starting an academic career. We became good friends over the years, although we frequently agreed to disagree about taxonomic changes and their implications. As for Roger, I was sad to see how many of the natural history and behavioral observations he contributed to the third edition were lost to accommodate the growing number of species in the region covered by this guide.

Suzanne L. Collins not only provided several exquisite photographs but was instrumental in passing to me the authorship of this edition. I thank her for her trust. Darrel R. Frost also contributed to my taking on this task and graciously helped in many ways, not the least of which was to serve as a taxonomically literate resource when I was faced with recent taxonomic changes.

I thank the late Dean E. Metter, whose herpetology class at the University of Missouri first exposed me to the wonders of amphibians and reptiles, and Joseph P. Ward, whose advice caused me to enroll in that course. Too many friends and students to mention individually have accompanied me into the field over the years, but I must single out John S. Parmerlee Jr., with whom I've suffered through many scorching hot days and cold, rainy nights in search of salamanders, frogs, turtles, lizards, and snakes.

I also appreciate the support provided by Avila University in the form of release time and scheduling to accommodate this and my many other herpetological endeavors. Administrators, colleagues, and many students have been exceptionally tolerant of my interests and supportive of my efforts to pursue them. Particularly relevant to this book, I thank Liz Bradish, office manager of the School of Science and Health, who graciously copied, scanned, mailed, and did whatever else was needed to help me complete the task.

Finally, I thank my wife, Beverly, whose undying forbearance of my many trips (often without her) and the occasional loose snake in the house must assure her a little piece of paradise to call her own.

ROBERT POWELL

FIG. 203. *This Florida Softshell* (Apalone ferox) *was photographed on Big Pine Key, where the population might have been introduced from peninsular Florida.*

For names of scales and other anatomical nomenclature, see Figs. 1, 2, 88, and 169. For definitions of other words and terms used in herpetology, see Harvey B. Lillywhite's *Dictionary of Herpetology* (2008; Krieger Publishing Co., Malabar, FL).

Anterior. In front of.

Atlantic Provinces. The four Canadian provinces of New Brunswick, Prince Edward Island, Nova Scotia (collectively known as the Maritime Provinces), and Newfoundland and Labrador.

Azygous. Odd or irregular, not paired (in reference to scales).

Barbels. Small, fleshy downward projections of skin on the chin and/or throat of some turtles.

Boss. A raised rounded area on the head between the eyes or on or near the snout in some toads and spadefoots.

Canthus rostralis. The ridge from the eye to the tip of the snout that separates the top of the snout from the side.

Cirri. Downward projections of the upper lip below the nostrils in males of certain lungless salamanders (Plethodontidae). The nasolabial groove extends downward to near the tip of each cirrus.

Cline. A gradual gradient across the range (e.g., clinal variation in size, large in northern populations and smaller toward the south), as opposed to abrupt changes over small distances that might indicate boundaries between species.

Cloaca. The common chamber into which the urinary, digestive, and reproductive canals discharge their contents, and which opens to the exterior through the vent.

Conspecific. Belonging to the same species.

Costal grooves. Vertical grooves on the flanks of salamanders. The spaces between grooves are called costal folds.

Cranial crests. The raised ridges on the head of most toads; these may be interorbital (between the eyes) or postorbital (behind the eyes).

Cusp. Toothlike projections on the jaws of some turtles.

Dorsal. Pertaining to the back of an animal.

Dorsolateral. Pertaining to the area between the back and the sides of the body.

Dorsum. Back of an animal.

Endemic. Indigenous to a certain area or region where the taxon evolved.

Excrescence. Rough, wartlike growths on the hindlimbs of some male newts during the breeding season.

Extant. Still existing.

Extinction. The permanent loss of all individuals of a species.

Extirpation. The loss of one or more populations, although other populations might be extant.

Family. A group of genera related through a single common ancestor.

Femoral pores. Small openings on the underside of the thighs in some lizards; these secrete a waxlike material.

Fossorial. Adapted for burrowing.

Genus. A group of species related through a single common ancestor (plural: genera).

Granules. Very small scales.

Gravid. Pertaining to a female bearing eggs (ordinarily in the oviducts); the term for a female bearing young is *pregnant*.

Growth rings. Concentric subcircular areas on the scutes of some turtles. Each ring represents a season's growth. Growth rings, if present, are most evident in young turtles; they usually cannot be counted in adults.

Gular. Pertaining to the throat; also the most anterior scutes (or scute) on a turtle's plastron.

Hemipenes. The paired copulatory organs of male squamates (singular: hemipenis).

Keel. A ridge down the back (or along the plastron) of some turtles. Also, a longitudinal ridge on the dorsal scales of some snakes.

Labial. Pertaining to the lip; often used in reference to scales along the lips of reptiles.

Lateral. Pertaining to the side.

Melanism. Abundance of black pigment, sometimes resulting in all-black or nearly all-black animals; the opposite of albinism (lack of pigment).

Mental gland. A gland, often large, situated on the chin of some salamanders; secretions are used to sexually stimulate females of the same species.
Mesic. Moist.
Middorsal. Pertaining to the center of the back.
Midventral. Pertaining to the center of the belly.
Nasolabial groove. A groove in lungless salamanders (Plethodontidae) extending downward from the nostril and sometimes following cirri that extend beyond the mouthline.
Neotenic. Pertaining to some salamanders that remain aquatic throughout life and retain larval features such as external gills while becoming sexually mature (noun: neoteny).
Ocelli. Round, eyelike spots (singular: ocellus).
Papillae. Fingerlike projections; also small raised protuberances under the chin of some snakes.
Paravertebral. On either side of the midline of the back.
Parotoid glands. Paired glands on the shoulder, neck, or behind the eyes in most toads; enlarged and prominent in many species.
Parthenogenesis. Development of an egg without fertilization by a male; applicable to all-female species of lizards (and one species of snake).
Posterior. Behind.
Postocular. Behind the eye.
Preocular. In front of the eye.
Reticulate: Netlike.
Rugose. Wrinkled or warty.
Scute. Any enlarged scale; sometimes called a shield or plate.
Serrate. Sawlike, such as the edges of the shells of some turtles.
Spatulate. Flat and rounded (shaped like the blade of a kitchen spatula).
Subcaudals. Scales beneath the tail; sometimes shortened to "caudals."
Subocular. Beneath the eye.
Supraocular. Above the eye.
Supraorbital semicircle. A row of small scales separating the supraoculars from the median head scutes in certain lizards.
Suture. A seam; the boundary between scales or scutes.
Taxon. The technical term for a unit or category used in classification (e.g., species, genus, family) (plural: taxa).
Taxonomy. The science of classifying and naming living things.

Tibia. Pertaining to the portion of the leg from the heel to the knee (usually of toads and frogs); derived from the name of the larger bone in that part of the leg.

Troglobite. A cave-dweller restricted to subterranean habitats (adjective: troglobitic).

Tubercles. Small knoblike projections.

Tuberculate. With raised projections.

Unisexual. All-female populations that reproduce by parthenogenesis.

Vent. Cloacal aperture.

Venter. The belly of an animal.

Ventral. Pertaining to the undersurfaces of an animal (e.g., the belly or lower surface of the tail).

Vermiculate: Wormlike.

Vocal sac. An inflatable sac on the throat or at the sides of the neck in male toads and frogs; single in most species, paired in others. Used as a resonating chamber when calling.

Xeric. Arid or dry.

FIG. 204. *The Plains Leopard Frog* (Lithobates blairi) *generally is associated with various bodies of water in open habitats. The spot on top of the snout, light spots on the eardrums, and dorsolateral ridges interrupted anterior to the groin and inset medially differentiate this species from the other leopard frogs.*

We have not listed every relevant reference used in compiling this guide, but instead provide a short list of key works that readers might find useful. These and various regional, state, and provincial guides were instrumental in writing the species accounts and the introductions to the various sections. Although not cited individually, many authors will see references to their work, for which we are deeply indebted.

Altig, R., and R. W. McDiarmid. 2015. *Handbook of Larval Amphibians of the United States and Canada*. Ithaca, NY: Comstock Publishing.

Altig, R., R. W. McDiarmid, K. A. Nichols, and P. C. Ustach. 1998. *Tadpoles of the United States and Canada: A Tutorial and Key*. Laurel, MD: Patuxent Wildlife Research Center (www.pwrc.usgs.gov/tadpole/).

Crother, B. I. (ed.). 2012. Scientific and standard English and French names of amphibians and reptiles of North America north of Mexico, with comments regarding confidence in our understanding. 7th ed. *SSAR Herpetological Circular* 39: 1–92 (Updated regularly at http://ssarherps.org/publications/north-american-checklist/north-american-checklist-of-scientific-and-common-names/).

Dodd, C. K. Jr. 2002. *North American Box Turtles: A Natural History*. Norman: University of Oklahoma Press.

———. 2013. *Frogs of the United States and Canada*. 2 vols. Baltimore, MD: Johns Hopkins University Press.

Elliot, L., C. Gerhardt, and C. Davidson. 2009. *The Frogs and Toads of North America. A Comprehensive Guide to Their Identification, Behavior, and Calls*. Boston, MA: Houghton Mifflin Harcourt.

Ernst, C. H., and E. M. Ernst. 2003. *Snakes of the United States and Canada*. Washington, DC: Smithsonian Institution Press.

———. 2011. *Venomous Reptiles of the United States, Canada, and Northern Mexico*. 2 vols. Baltimore, MD: Johns Hopkins University Press.

Ernst, C. H., and J. E. Lovich. 2009. *Turtles of the United States and Canada*. 2d ed. Baltimore, MD: Johns Hopkins University Press.

Gibbons, J. W., and M. E. Dorcas. 2004. *North American Watersnakes: A Natural History*. Norman: University of Oklahoma Press.

Gregoire, D. R. 2005. *Tadpoles of the Southeastern United States Coastal Plain*. U.S. Geological Survey Report. Gainesville, FL: Florida Integrated Science Center (http://fl.biology.usgs.gov/armi/Guide_to_Tadpoles/SEARMITadpoleGuide.pdf).

Krysko, K. L., J. P. Burgess, M. R. Rochford, C. R. Gillette, D. Cueva, K. M. Enge, L. A. Somma, J. L. Stabile, D. C. Smith, J. A. Wasilewski, G. N. Kieckhefer III, M. C. Granatosky, and S. V. Nielsen. 2011. Verified nonindigenous amphibians and reptiles in Florida from 1863 through 2010: Outlining the invasion process and identifying invasion pathways and stages. *Zootaxa* 3028: 1–64.

Lillywhite, H. B. 2008. *Dictionary of Herpetology.* Malabar, FL: Krieger Publishing.

Lindeman, P. V. 2013. *The Map Turtle and Sawback Atlas: Ecology, Evolution, Distribution, and Conservation*. Norman: University of Oklahoma Press.

Meshaka, W. E. Jr. 2011. A runaway train in the making: The exotic amphibians, reptiles, turtles, and crocodilians of Florida. Monograph 1. *Herpetological Conservation and Biology* 6: 1–101 (http://herpconbio.org/Volume_6/Monographs/Meshaka_2011.pdf).

Petranka, J. W. 1998. *Salamanders of the United States and Canada*. Washington, DC: Smithsonian Institution Press.

Pough, F. H., R. M. Andrews, M. L. Crump, A. H. Savitzky, K. D. Wells, and M. C. Brandley. 2015. *Herpetology*. 4th ed. Sunderland, MA: Sinauer Associates.

Powell, R., J. T. Collins, and E. D. Hooper Jr. 2012. *Key to the Herpetofauna of the Continental United States and Canada*. 2d ed. Lawrence: University Press of Kansas.

Rossman, D. A., N. B. Ford, and R. A. Seigel. 1996. *The Garter Snakes: Evolution and Ecology*. Norman: University of Oklahoma Press.

Stebbins, R. C. 2003. *A Field Guide to Western Reptiles and Amphibians*. 3d ed. Boston, MA: Houghton Mifflin.

Vitt, L. J., and J. P. Caldwell. 2013. *Herpetology. An Introductory Biology of Amphibians and Reptiles.* 4th ed. San Diego, CA: Academic Press.

See also species accounts in the *Catalogue of American Amphibians and Reptiles* published by the Society for the Study of Amphibians and Reptiles (zenscientist.com/index.php/pdflibrary1/Open-Access-Journals/caar/) and information provided on the website of the Center for North American Herpetology (www.cnah.org/).

Photographs are listed in the order in which they appear in the book.

RICHARD D. BARTLETT: Fig. 116. Bibron's Sand Gecko (*Chondrodactylus bibronii*), Fig. 121. Asian Flat-tailed House Gecko (*Hemidactylus platyurus*), Fig. 122. Mourning Gecko (*Lepidodactylus lugubris*), Fig. 130. Veiled Chameleon (*Chamaeleo calyptratus*), Fig. 144. Common Sun Skink (*Eutropis multifasciata*), Fig. 151. Giant Ameiva (Ameiva ameiva), Fig. 152. Laredo Striped Whiptail (*Aspidoscelis laredoensis* complex), Fig. 154. Argentine Giant Tegu (*Salvator merianae*), Fig. 155. Nile Monitor (*Varanus niloticus*).

JEFFREY T. BRIGGLER: Fig. 8. Ringed Salamander (*Ambystoma annulatum*), Fig. 59. Blanchard's Cricket Frog (*Acris blanchardi*), Fig. 100. Three-toed Box Turtle (*Terrapene triunguis*), Fig. 132. Eastern Collared Lizard (*Crotaphytus collaris*), Fig. 187. Western Wormsnake (*Carphophis vermis*), Fig. 190. Broad-banded Watersnake (*Nerodia fasciata confluens*), Fig. 194. Red-bellied Snake (*Storeria occipitomaculata*), Fig. 202. Western Massasauga (*Sistrurus tergeminus*), p. 445. Common Five-lined Skink (*Plestiodon fasciatus*), Fig. 207. Graham's Crayfish Snake (*Regina grahamii*).

JOSEPH BURGESS: Fig. 135. St. Vincent Bush Anole (*Anolis trinitatis*), Fig. 136. Gray's Spiny-tailed Iguana (*Ctenosaura similis*).

SUZANNE L. COLLINS (CENTER FOR NORTH AMERICAN HERPETOLOGY): Fig. 14. Apalachicola Dusky Salamander (*Desmognathus apalachicolae*), Fig. 20. Black Mountain Salamander (*Desmognathus welteri*), Fig. 22. Junaluska Salamander (*Eurycea junaluska*), Fig. 34. Tellico Salamander (*Plethodon aureolus*), Fig. 39. Western Slimy Salamander (*Plethodon albagula*), Fig. 42. Cumberland Plateau Salamander (*Plethodon kentucki*), Fig. 45. Southern Appalachian Salamander (*Plethodon teyahalee*), Fig. 64. Illinois Chorus Frog (*Pseudacris illinoiensis*), Fig. 65. New Jersey Chorus Frog (*Pseudacris kalmi*), Fig. 79. Plains Spadefoot (*Spea bombifrons*), Fig. 94. Alabama Red-bellied Cooter (*Pseudemys alabamensis*), Fig. 95. Rio Grande Cooter (*Pseudemys gorzugi*), Fig. 97. Red-eared Slider (*Trachemys scripta elegans*), Fig. 139. Prairie Lizard (*Sceloporus consobrinus*), Fig. 143. Western Green Lacerta (*Lacerta bilineata*), Fig. 149. Broad-headed Skink (*Plestiodon laticeps*), Fig. 153. Giant Whiptail (*Aspidoscelis motaguae*), Fig. 179. Rough Greensnake (*Opheodrys aestivus*), Fig. 196. Texas Coralsnake (*Micrurus tener*), p. 452. Spectacled Caiman (*Caiman crocodilus*).

ADAM CRANE: p. 446. Eastern Hellbender (*Cryptobranchus alleganiensis*).

C. KENNETH DODD JR.: p. 172. Suwannee Cooter (*Pseudemys suwanniensis*).

KEVIN M. ENGE: Fig. 128. Oriental Garden Lizard (*Calotes versicolor* complex), Fig. 146. Ocellated Skink (*Chalcides ocellatus*), Fig. 173. Boa Constrictor (*Boa constrictor*).

VINCENT R. FARALLO: Fig. 41. Peaks of Otter Salamander (*Plethodon hubrichti*).

DANTÉ FENOLIO (WWW.ANOTHECA.COM): p. ix. Loggerhead Musk Turtle (*Sternotherus minor*), p. 14. Georgia Blind Salamander (*Eurycea* wallacei), Fig. 18. Imitator Salamander (*Desmognathus imitator*), Fig. 24. Georgia Blind Salamander (*Eurycea wallacei*), Fig. 25. Cascade Caverns Salamander (*Eurycea latitans* complex), Fig. 26. Texas Blind Salamander (*Eurycea rathbuni*), Fig. 27. Barton Springs Salamander (*Eurycea sosorum*), Fig. 28. Comal Blind Salamander (*Eurycea tridentifera*), Fig. 29. Valdina Farms Salamander (*Eurycea troglodytes* complex), Fig. 31. West Virginia Spring Salamander (*Gyrinophilus subterraneus*), Fig. 56. Houston Toad (*Anaxyrus houstonensis*).

FIG. 205. *The Gopher Tortoise* (Gopherus polyphemus) *digs extensive burrows that are used by many other animals. This species is listed as Vulnerable on the IUCN Red List of Threatened Species. The principle threat is habitat destruction attributable to rampant development. Many tortoises are buried alive in their burrows under new buildings and parking lots.*

H. CARL GERHARDT JR. (NATUREVIEWMEDIA.COM): p. viii. Carpenter Frog (*Lithobates virgatipes*), p. 8. Eastern Milksnake (*Lampropeltis triangulum*), p. 104. Pine Barrens Treefrog (Hyla andersonii), Fig. 68. Crawfish Frog (*Lithobates areolatus*), Fig. 204. Plains Leopard Frog (*Lithobates blairi*).

CHRISTOPHER GILLETTE: p. i. Red-bellied Mudsnake (*Farancia abacura*), p. 166. American Alligator (*Alligator mississippiensis*), Fig. 126. Moorish Gecko (*Tarentola mauritanica*), Fig. 127. Peters's Rock Agama (*Agama picticauda*), p. 324. Eastern Kingsnake (*Lampropeltis getula*), Fig. 170. Javanese Filesnake (*Acrochordus javanicus*), Fig. 198. Florida Cottonmouth (*Agkistrodon conanti*), Fig. 199. Eastern Diamond-backed Rattlesnake (*Crotalus adamanteus*), p. 458. Diamond-backed Terrapin (*Malaclemys terrapin*), Fig. 205. Gopher Tortoise (*Gopherus polyphemus*).

RICHARD E. GLOR: Fig. 133. Hispaniolan Green Anole (*Anolis chlorocyanus*).

EVAN H. CAMPBELL GRANT (USGS): Fig. 43. Shenandoah Salamander (*Plethodon shenandoah*).

DAVID HECKARD: Fig. 120. Mediterranean Gecko (*Hemidactylus turcicus*) and Sri Lankan House Gecko (*H. parvimaculatus*).

JOHN B. JENSEN: p. x. Barbour's Map Turtle (*Graptemys barbouri*).

MATTHEW JEPPSON: Fig. 138. Texas Horned Lizard (*Phrynosoma cornutum*), Fig. 178. Speckled Kingsnake (*Lampropeltis holbrooki*), Fig. 195. Plains Gartersnake (*Thamnophis radix*).

TOM R. JOHNSON: Fig. 105. Spiny Softshell (*Apalone spinifera*), Fig. 206. Mink Frog (*Lithobates septentrionalis*).

KENNETH L. KRYSKO: p. ii–iii. American Crocodile (*Crocodylus acutus*), Fig. 117. Golden Gecko (*Gekko badenii*), Fig. 123. Madagascan Day Gecko (*Phelsuma grandis*), Fig. 124. Gold Dust Day Gecko (*Phelsuma laticauda*), Fig. 125. Ringed Wall Gecko (*Tarentola annularis*), Fig. 129. Common Butterfly Lizard (*Leiolepis belliana*), Fig. 131. Oustalet's Chameleon (*Furcifer oustaleti*), Fig. 145. African Five-lined Skink (*Trachylepis quinquetaeniata*), Fig. 171. Brahminy Blindsnake (*Indotyphlops braminus*), Fig. 174. Burmese Python (*Python bivittatus*), p. 448. Common House Gecko (*Hemidactylus frenatus*).

TRIP LAMB: Fig. 46. Patch-nosed Salamander (*Urspelerpes brucei*).

CLAUDIA MACKENZIE-KRYSKO: Fig. 118. Tropical House Gecko (*Hemidactylus mabouia*).

MATTHEW L. NIEMILLER: Fig. 30. Berry Cave Salamander (*Gyrinophilus gulolineatus*).

LORI OBERHOFER (NATIONAL PARK SERVICE): Fig. 175. Northern African Rock Python (*Python sebae*).

ROBERT POWELL: p. vi. Northern Curly-tailed Lizard (*Leiocephalus carinatus*), p. xiv. Lake Erie Watersnake (*Nerodia sipedon insularum*), p. 5. Prairie Ring-necked Snake (*Diadophis punctatus arnyi*), p. 234. Green Anole (*Anolis carolinensis*), Fig. 203. Florida Softshell (*Apalone ferox*), p. 456. Green Sea Turtle (*Chelonia mydas*), p. 462. Painted Turtle (*Chrysemys picta*), p. 465. American Alligator (*Alligator mississippiensis*).

RICHARD A. SAJDAK: Fig. 73. Green Frog (*Lithobates clamitans*), Fig. 166. Northern Cottonmouth (*Agkistrodon piscivorus*).

DONALD S. SIAS: Fig. 55. Great Plains Toad (*Anaxyrus cognatus*), Fig. 181. Red Cornsnake (*Pantherophis guttatus*), Fig. 197. Eastern Copperhead (*Agkistrodon contortrix*).

MICHAEL A. STEFFEN: Fig. 23. Ouachita Streambed Salamander (*Eurycea subfluvicola*).

ROBERT WAYNE VAN DEVENDER: p. xi. Blue Ridge Gray-cheeked Salamander (*Plethodon amplus*), p. 1. Georgetown Salamander (*Eurycea naufragia*), Fig. 9. Reticulated Flatwoods Salamander (*Ambystoma bishopi*), Fig. 17. Dwarf Black-bellied Salamander (*Desmognathus folkertsi*), Fig. 19. Flat-headed Salamander (*Desmognathus planiceps*), Fig. 21. Southern Two-lined Salamander (*Eurycea cirrigera*), Fig. 32. South Mountain Graycheeked Salamander (*Plethodon meridianus*), Fig. 35. Cheoh Bald Salamander (*Plethodon cheoah*), Fig. 36. Eastern Red-backed Salamander (*Plethodon cinereus*), Fig. 40. Shenandoah Mountain Salamander (*Plethodon virginia*), Fig. 44. Big Levels Salamander (*Plethodon sherando*), Fig. 49. Red River Mudpuppy (*Necturus louisianensis*), Fig. 50. Red-spotted Newt (*Notophthalmus viridescens viridescens*), Fig. 51. Eastern Lesser Siren (*Siren intermedia intermedia*), Fig. 156. Florida Wormlizard (*Rhineura floridana*), Fig. 177. Western Coachwhip (*Coluber flagellum testaceus*).

FIG. 206. *The Mink Frog* (Lithobates septentrionalis) *has a northerly distribution, chiefly in eastern and midwestern Canada, although it occurs in some northern states. The common name is attributable to skin secretions said to have an odor like that of a mink.*

PHOTO CAPTIONS

P. i: The Red-bellied Mudsnake (*Farancia abacura*), like this juvenile, is docile and rarely bites, but it might press the pointed tailtip against your hand in an attempt to escape. This behavior led to the marvelous myth of the "hoop snake" with its poisonous tail. According to legend, the snake grabs its tail in its mouth to form a hoop and rolls downhill toward an unlucky victim. At the last second, the snake releases its tail, straightens out like a spear, and flies tail-first at its target.

Pp. ii–iii: The American Crocodile (*Crocodylus acutus*) reaches the northernmost end of its range in coastal areas of southern Florida. Despite an extensive distribution that extends to northern South America, the American Crocodile is listed as Vulnerable on the IUCN Red List of Threatened Species. Principal threats are habitat loss and illegal hunting for skins and meat.

P. vi: The Northern Curly-tailed Lizard (*Leiocephalus carinatus*) is established in peninsular Florida and on many Keys. It often is associated with rocky areas or artificial equivalents such as sidewalks, parking lots, and stonewalls, frequently digging burrows under pavement or large rocks.

P. viii: The Carpenter Frog (*Lithobates virgatipes*) sometimes is called the Sphagnum Frog because of its close association with sphagnum bogs and swamps and sphagnum borders of lakes and ponds.

P. ix: Like other musk turtles, the Loggerhead Musk Turtle (*Sternotherus minor*) is strongly aquatic.

P. x: Barbour's Map Turtle (*Graptemys barbouri*) is threatened by habitat degradation, overharvesting, and predation. Channel modification, dredging, barge traffic, and pollution threaten riverine habitats; overgrowth of sandy nesting sites causes nests to become increasingly clumped; and removal of dead trees and snags eliminates basking sites.

P. xi: The Blue Ridge Gray-cheeked Salamander (*Plethodon amplus*) and the other gray-cheeked salamanders until recently were considered color variants of the Red-cheeked Salamander (*P. jordani*).

P. xiv: A "mating ball" of Lake Erie Watersnakes (*Nerodia sipedon insularum*) usually comprises one female and a variable number of males. This subspecies of the Common Watersnake was originally described by Roger Conant and William M. Clay in 1937. Restricted to islands in Lake Erie, it until recently was listed as Threatened by the U.S. Fish and Wildlife Service and remains protected by the Canadian Species at Risk Act and Ohio state regulations. Interestingly, some of the species' recovery can be attributed to the introduction of the non-native round goby (*Neogobius melanostomus*), a fish that now accounts for about 90 percent of the snake's diet.

P. 1: The Georgetown Salamander (*Eurycea naufraugia*), like a number of related neotenic brook salamanders in central Texas, has a restricted range. Found only in the San Gabriel River drainage in and near Georgetown, Williamson County, Texas, the species is listed as Endangered on the IUCN Red List of Threatened Species. Populations in the city of Georgetown are largely extirpated. Threats include residential development and quarrying with a concomitant negative effect on groundwater quality.

P. 5: When threatened, several species of small snakes like these Prairie Ring-necked Snakes (*Diadophis punctatus arnyi*) roll over and expose the bright venter while coiling the tail and hiding the head beneath the body. The red color on the underside of the tail presumably diverts a predator's attention away from the snake's head. These individuals have recovered after being handled and quickly sought shelter.

P. 8: Milksnakes were found to be a complex of species that did not correspond to traditionally recognized subspecies. This is an Eastern Milksnake (*Lampropeltis triangulum*) from west of the Mississippi River. Snakes in these western populations are much more brightly colored than their eastern counterparts.

P. 14: The Georgia Blind Salamander (*Eurycea wallacei*) is found only in cave pools, streams, and submerged passages of the Upper Floridian Aquifer in southwestern Georgia and adjacent northwestern Florida.

P. 104: The Pine Barrens Treefrog (*Hyla andersonii*) is restricted to the pine barrens of New Jersey and comparable habitats in the Southeast. Rarely seen, these frogs are best found by homing in on calling males during the breeding season.

P. 166: The American Alligator (*Alligator mississippiensis*), here photographed underwater, was once on the brink of extinction but is now thriving. The species is becoming reestablished in many portions of its historic range, a success story largely attributable to state and federal protection, efforts to preserve habitats, and a reduced demand for alligator products.

P. 172: Chiefly aquatic turtles, such as these Suwannee Cooters (*Pseudemys suwanniensis*), often have considerable algal growth on their shells. This frequently obscures patterns and renders identification more difficult.

P. 234: The Green Anole (*Anolis carolinensis*) is the only anole native to our area, but at least nine West Indian species are established in Florida. This male has extended its dewlap, which is used to communicate during courtship and territorial defense.

P. 324: The Eastern Kingsnake *(Lampropeltis getula)* is a powerful constrictor that preys on a variety of vertebrates, including other snakes. The blue-gray eye indicates that the clear protective scale covering the eye is loosening and that this snake will soon shed its skin (along with the ocular scale).

P. 446: A flattened head and body allow the Eastern Hellbender (*Cryptobranchus alleganiensis*) to seek shelter under rocks. Lateral skinfolds facilitate respiration by increasing the surface area for gas exchange through the skin.

P. 448: The Common House Gecko (*Hemidactylus frenatus*) is widely introduced in tropical and subtropical habitats around the world. In some areas, this gecko appears to be displacing native or other introduced species of house geckos.

P. 452: The Spectacled Caiman (*Caiman crocodilus*) is firmly established in southern Florida. These introduced populations can be traced to released "pets."

P. 456: The *Green* in the name of the Green Sea Turtle (*Chelonia mydas*) refers to the greenish fat used to make turtle soup.

P. 458: The Diamond-backed Terrapin (*Malaclemys terrapin*) rarely strays far from salt or brackish water.

P. 462: The Painted Turtle (*Chrysemys picta*), along with sliders and cooters, often is called a basking turtle because of its propensity to sun itself. This raises the body temperature, which facilitates digestion.

P. 465: The broadly rounded snout of this American Alligator (*Alligator mississippiensis*) is easily distinguished from the long tapering snout of the American Crocodile (*Crocodylus acutus*).

P. 493: This juvenile Common Five-lined Skink (*Plestiodon fasciatus*) has distinct light lines and a bright blue tail. The brightly colored tail appears to divert predators from the head. The tail will readily regenerate if lost, although it will lack the bright color. The pattern fades with age as adults presumably become more adept at avoiding danger. Females often retain some semblance of stripes, but many males are uniformly brown.

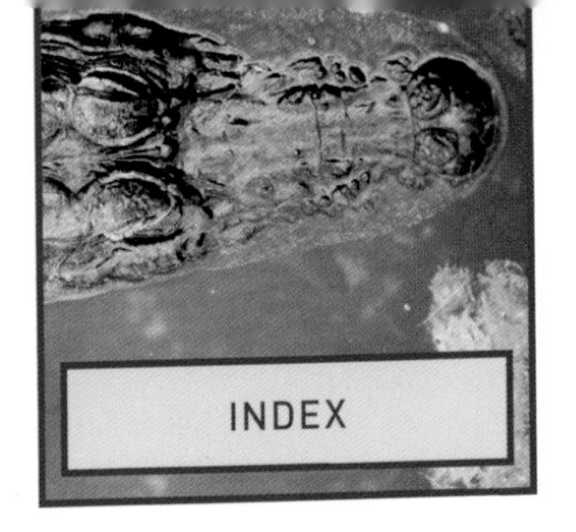

We have organized this index to avoid repeating words such as *salamander*, *frog*, *toad*, *turtle*, *lizard*, and *snake*. In looking for common names, work backward. If you wish to find an entry for a species or subspecies—Blanchard's Cricket Frog, for example—look first under *Frog* and then for *Blanchard's Cricket*. For groups (e.g., families or genera)—cricket frogs, for example—look first under *Frogs* and then for *cricket*. In looking for scientific names, look first for the name of the genus and then for the name of the species or subspecies.

Page numbers of species accounts or maps are in lightfaced type, whereas those of illustrations (plates, photographs, line drawings) are in boldfaced type. Note that references to species in accounts of other taxa (i.e., similar species) are not listed.

Robert Powell is professor of biology at Avila University in Kansas City, Missouri, and a research associate at the University of Kansas Natural History Museum in Lawrence. He has written hundreds of scientific articles and notes (many with student coauthors) and is coauthor or coeditor of seven books, including *Key to the Herpetofauna of the Continental United States and Canada* (1998, second edition 2012; with Joseph T. Collins and Errol D. Hooper Jr.).

The late Roger Conant retired in 1973 as director of the Philadelphia Zoo and as its curator of reptiles, a post he had held for nearly 40 years. Subsequently, he served as an adjunct professor of biology at the University of New Mexico in Albuquerque. The Boy Scouts of America's "Reptile and Amphibian Study" merit badge pamphlet is dedicated to the memory of Dr. Conant, who wrote the first edition of that book and many others, including a monograph on copperheads, cottonmouths, and their relatives.

The late Joseph T. Collins was the herpetologist with the Kansas Biological Survey and herpetologist emeritus at the University of Kansas Natural History Museum, where he worked for 30 years. He was the founder and director of the Center for North American Herpetology and the author of many articles and books, most recently *Amphibians, Reptiles, and Turtles in Kansas* (2010; with Suzanne L. Collins and Travis W. Taggart).

The late Isabelle Hunt Conant was official photographer at the Philadelphia Zoo and her photographs appeared in scores of publications. She worked closely with her husband during many field trips in the United States and Mexico.

Tom R. Johnson was state herpetologist for the Missouri Department of Conservation until 2000. He is an accomplished photographer and illustrator and author of *The Amphibians and Reptiles of Missouri* (1987, revised edition 2000).

Errol D. Hooper Jr. is a scientific illustrator whose drawings have been published in many articles and books, including *Key to the Herpetofauna of the Continental United States and Canada* (1998, second edition 2012) and *Amphibians, Reptiles, and Turtles in Kansas* (2010).

Travis W. Taggart is Adjunct Curator of Herpetology at Fort Hays State University's Sternberg Museum of Natural History, president and executive director of the Center for North American Herpetology, and author of many articles and books, most recently *Amphibians, Reptiles, and Turtles in Kansas* (2010; with Joseph T. Collins and Suzanne L. Collins).

FIG. 207. *Graham's Crayfish Snake* (Regina grahamii) *feeds on a variety of aquatic or semiaquatic prey, but specializes on crayfish, exhibiting a strong preference for those that have recently shed their protective exoskeletons.*

PETERSON FIELD GUIDES®

Roger Tory Peterson's innovative format uses accurate, detailed drawings to pinpoint key field marks for quick recognition of species and easy comparison of confusing look-alikes.

BIRDS

Birds of North America

Birds of Eastern and Central North America

Western Birds

Eastern Birds

Feeder Birds of Eastern North America

Hawks of North America

Hummingbirds of North America

Warblers

Eastern Birds' Nests

PLANTS AND ECOLOGY

Eastern and Central Edible Wild Plants

Eastern and Central Medicinal Plants and Herbs

Western Medicinal Plants and Herbs

Eastern Forests

Eastern Trees

Western Trees

Eastern Trees and Shrubs

Ferns of Northeastern and Central North America

Mushrooms

North American Prairie

Venomous Animals and Poisonous Plants

Wildflowers of Northeastern and North-Central North America

MAMMALS

Animal Tracks

Mammals

Finding Mammals

INSECTS

Insects

Eastern Butterflies

Moths of Northeastern North America

REPTILES AND AMPHIBIANS

Eastern Reptiles and Amphibians

Western Reptiles and Amphibians

FISHES

Freshwater Fishes

SPACE

Stars and Planets

GEOLOGY

Rocks and Minerals

PETERSON FIRST GUIDES®

The first books the beginning naturalist needs, whether young or old. Simplified versions of the full-size guides, they make it easy to get started in the field, and feature the most commonly seen natural life.

Astronomy

Birds

Butterflies and Moths

Caterpillars

Clouds and Weather

Fishes

Insects

Mammals

Reptiles and Amphibians

Rocks and Minerals

Seashores

Shells

Trees

Urban Wildlife

Wildflowers

PETERSON FIELD GUIDES FOR YOUNG NATURALISTS

This series is designed with young readers ages eight to twelve in mind, featuring the original artwork of the celebrated naturalist Roger Tory Peterson.

Backyard Birds

Birds of Prey

Songbirds

Butterflies

Caterpillars

PETERSON FIELD GUIDES® COLORING BOOKS®

Fun for kids ages eight to twelve, these color-your-own field guides include color stickers and are suitable for use with pencils or paint.

Birds

Butterflies

Dinosaurs

Reptiles and Amphibians

Wildflowers

Seashores

Shells

Mammals

PETERSON REFERENCE GUIDES®

Reference Guides provide in-depth information on groups of birds and topics beyond identification.

Behavior of North American Mammals

Birding by Impression

Molt in North American Birds

Owls of North America and the Caribbean

Seawatching: Eastern Waterbirds in Flight

PETERSON AUDIO GUIDES

Birding by Ear: Eastern/Central

Bird Songs: Eastern/Central

PETERSON FIELD GUIDE / *BIRD WATCHER'S DIGEST* BACKYARD BIRD GUIDES

Identifying and Feeding Birds

Hummingbirds and Butterflies

Bird Homes and Habitats

The Young Birder's Guide to Birds of North America

The New Birder's Guide to Birds of North America

DIGITAL

App available for Apple and Android.

Peterson Birds of North America

E-books

Birds of Arizona

Birds of California

Birds of Florida

Birds of Massachusetts

Birds of Minnesota

Birds of New Jersey

Birds of New York

Birds of Ohio

Birds of Pennsylvania

Birds of Texas